NASA's Scientist-Astronauts

David J. Shayler and Colin Burgess

NASA's Scientist-Astronauts

Published in association with
Praxis Publishing
Chichester, UK

Mr David J. Shayler, FBIS
Astronautical Historian
Astro Info Service
Halesowen
West Midlands
UK
www.astroinfoservice.co.uk

Mr Colin Burgess, BIS
Spaceflight Historian
Bonnet Bay
New South Wales
Australia

SPRINGER–PRAXIS BOOKS IN SPACE EXPLORATION
SUBJECT *ADVISORY EDITOR*: John Mason B.Sc., M.Sc., Ph.D.

ISBN 10: 0-387-21897-1 Springer Berlin Heidelberg New York

Springer is part of Springer-Science + Business Media (*springeronline.com*)

Library of Congress Control Number: 2006930295

Apart from any fair dealing for the purposes of research or private study, or criticism or review, as permitted under the Copyright, Designs and Patents Act 1988, this publication may only be reproduced, stored or transmitted, in any form or by any means, with the prior permission in writing of the publishers, or in the case of reprographic reproduction in accordance with the terms of licences issued by the Copyright Licensing Agency. Enquiries concerning reproduction outside those terms should be sent to the publishers.

© Copyright, 2007 Praxis Publishing Ltd

The use of general descriptive names, registered names, trademarks, etc. in this publication does not imply, even in the absence of a specific statement, that such names are exempt from the relevant protective laws and regulations and therefore free for general use.

Cover design: Jim Wilkie
Project Copy Editor: Mike Shayler
Typesetting: Originator Publishing Services Ltd, Gt Yarmouth, Norfolk, UK

Printed in Germany on acid-free paper

Contents

Foreword	xv
Authors' Preface	xvii
Acknowledgements	xxiii
List of Illustrations	xxvii
List of Tables	xxxiii
List of Abbreviations	xxxv
Other Works	xli
Prologue	xliii

1 The Wrong Stuff	1
Organising the effort	1
A manned satellite project	2
Orbital piloted spaceship of the Soviet Union	3
Security over science?	3
Who should or could fly?	7
Requirements for astronaut selection – the USAF approach	7
Requirements for astronaut selection – the NASA approach	8
The first cosmonauts	12
Pilot-astronauts not scientist-astronauts	13
Science and manned space flight	13
NASA's long-term planning 1959–64	14
In a packed programme	17
Science and manned orbital space flight 1961–76	19
Salyut, Skylab and Spacelab – orbital research labs for scientists?	22
References	23

Contents

2	**Scientists as Astronauts**	25
	An essential part of future exploration	26
	Under careful study	27
	Taking immediate steps	29
	Reasonably strong case for immediate selection	30
	Selecting the selection board	32
	A change in selection criteria	34
	A new breed of astronaut	35
	Going through the process	36
	NASA's astronaut selection process	37
	Scientists as cosmonauts	39
	Voskhod – the first opportunities	40
	Academy of Sciences Cosmonaut Group	42
	Lack of assignments	43
	Demise of the scientist-cosmonaut group	43
	Waiting for the call	44
	Military scientists	44
	Physician cosmonauts	45
	Other selections	46
	Science not a priority	46
	Changes in selection	48
	A good career move?	48
	References	48
3	**The Scientific Six**	51
	A gamble for glory	51
	A propaganda machine	52
	Testing the candidates	52
	Garriott's diary	54
	The chosen few	57
	Owen K. Garriott	57
	In the footsteps of pioneers	58
	An interesting proposition	61
	Edward G. Gibson	62
	An inauspicious start	63
	Changing careers	65
	Joseph P. Kerwin	66
	Just like Copernicus	67
	Flight surgeon school	70
	F. Curtis Michel	71
	A career in science	72
	Rice University	73
	Harrison H. Schmitt	76
	Hereditary interest in geology	76
	Looking at the Moon	79

		Contents	vii

	Duane E. Graveline	80
	Early influences	80
	Flight surgeon	81
	A time of devastation	84
	Other roads to travel	86
	The "almost" scientist-astronauts (1965)	88
	References	91
4	**School for Scientists**	**93**
	Flight training	93
	Screaming Purvis	97
	Technical assignments and the AAP Office	99
	Work begins in earnest	100
	General training	102
	General training plan – 1966	102
	General training overview	102
	Science and technology summary courses	103
	Operational briefings	105
	Spacecraft systems training	106
	Wilderness and survival training	109
	Control task training	111
	Launch vehicle abort training	114
	Aircraft flight programme	115
	A hectic diary	115
	References	116
5	**The Excess Eleven**	**117**
	A second selection	117
	The screening process	118
	The Group Six selection	121
	Joseph P. Allen IV	122
	A distinguished heritage	122
	Deciding on a future	124
	Philip K. Chapman	126
	Growing up in Australia	126
	International Geophysical Year	128
	Anthony W. England	130
	A family on the move	131
	A real turning point	133
	Karl G. Henize	135
	Just like Daniel Boone	135
	The skies and a thesis	137
	Donald L. Holmquest	139
	A strong educational discipline	139
	Applying to NASA	142

William B. Lenoir	142
A natural-born engineer	143
Research for Apollo	145
John A. Llewellyn	146
Early influences	148
Working in Ottawa	149
F. Story Musgrave	150
A childhood filled with despair	150
Settling into the Marine Corps	152
Brian T. O'Leary	154
The influence of the heavens	154
Overcoming the obstacles	157
Robert A.R. Parker	157
Astronomy beckons	159
Reasons against selection	160
William E. Thornton	161
A fascination with anything aeronautical	162
Introducing electronics into medicine	164
The other "almost" scientist-astronauts (1967)	165
References	168

6 "Flying Is Just Not My Cup of Tea" — 171

Knuckling down to the task	173
Back to school	174
Flight training	177
Strapping on the jets	181
Eleven becomes ten, then nine	183
Looking to the future	189
Jobs on the line	189
Losing the Moon	194
Putting things in perspective	196
References	197

7 A Geologist on the Moon — 199

Supporting Apollo	199
Vacuum testing Apollo	201
Chamber testing the Block I CSM	202
Chamber testing the Block II CSM	202
Qualifying the Lunar Receiving Laboratory	204
An experiment package for the Moon	205
After Apollo?	206
Apollo or Skylab	206
Supporting the landings	207
Mission scientist for the Moon	210
A stroll or a ride?	212

Lost missions and a crew change ... 212
    An uncertain future ... 213
    Juggling the rockets ... 214
    A difficult decision is made ... 215
Selecting the last landing site ... 215
    A place called Taurus-Littrow ... 216
    A crew is formed ... 217
Setting off for the final time ... 219
    The Moon looms larger ... 221
    A "go" for landing ... 223
A geologist walks on the Moon ... 224
    Preparing for the task ... 224
    The proudest moment ... 225
    Finding orange soil ... 226
Last steps on the lunar surface ... 230
    Heading home ... 230
    Deep-space EVA ... 233
    Journey's end ... 235
What the future may hold ... 238
The end of the beginning ... 240
References ... 241

8 Laboratories in the Sky ... 243
Michel resigns ... 243
    A dissatisfied customer ... 244
    Turning to Apollo Applications ... 245
    Possibilities fade ... 245
    Looking back ... 247
Skylab – A space station for America ... 248
Applying Apollo to other goals ... 248
    Mercury–Gemini–Apollo–the Moon ... 249
    Applying skills to AAP ... 250
Supporting AAP ... 250
Science pilots for Skylab ... 253
    Skylab assignments ... 256
Supporting Skylab ... 258
    Skylab support roles ... 258
    Dr. Bill and SMEAT ... 261
Science pilot training ... 264
    Reviewing the Skylab training programme ... 265
Skylab – Human experience ... 266
    The first manned mission (Skylab 2 – 25 May–22 Jun 1973) ... 267
    The second manned mission (Skylab 3 – 28 Jul–25 Sep 1973) ... 270
    The third manned mission (Skylab 4 – 16 Nov 1973–18 Feb 1974) ... 272

x Contents

        Skylab Rescue – a fifth mission?. 276
        Skylab B 279
    References 281

9 Shuttling into Space 283
    Space Shuttle – A Reliable Access to Space? 283
        "An entirely new type of space transportation system" 284
        Reorganising the scientist-astronaut office. 286
    Simulating Spacelab. 288
        Shuttle's laboratory 288
        Ground and airborne simulations 291
    Airborne Science/Spacelab Experiment System Simulation (ASSESS). . 294
        Learjet simulation programme 1972–4 294
        Learjet 4 simulation mission 294
    Origins of ASSESS 295
        Scientist-astronauts' role on Space Shuttle missions 296
        ASSESS-I 298
        ASSESS-II 299
        Defining the role of mission specialist 299
        ASSESS-II crew assignments 303
        Training for ASSESS-II 305
        ASSESS-II in flight 307
    Spacelab medical simulations. 309
        Spacelab Medical Development Test I 310
        Spacelab Medical Development Test II 315
        Spacelab Medical Development Test III. 316
        SMD-III – an overview 321
    The value of participation. 327
        Mission specialists for the Shuttle. 328
        Other early Spacelab assignments 329
        Selecting the first Spacelab crew 329
    References 329

10 The Long Wait 333
    Supporting the Shuttle 333
        Thirty-five new guys. 334
        "America's greatest flying machine" 336
    STS-5: we deliver 340
        Assigning the first mission specialists. 340
        The challenge and the responsibility 342
        The first operational Shuttle mission 342
        Upgrading the Columbia. 344
        A laid-back approach to launch 345
        Welcome to space 346
        We deliver! 348

No EVA this time	349
Flying for work, not comfort	352
Experiments and hardware	353
STS-6: the challenge of EVA	356
Musgrave's STS-6 training load	358
Challenger flies	358
Story's story	361
Medicine takes precedence over Earth science	363
STS-8: Dr. Bill flies	366
A workaholic astronaut	366
Dr. Bill's orbital clinic	368
First Shuttle night launch and night landing	370
Thornton's "chamber of horrors"	371
Reality of space flight	374
A long wait and a short wait	375
STS-9/Spacelab 1	375
Occupying the Spacelab module	376
More doctors than pilots	377
A busy schedule	380
Problems and progress	387
Monkeying around with the media	391
A fire on landing	393
STS 51-A: we deliver and pick up – twice	396
Deployment and retrieval	397
Flight-specific EVA training	398
Satellites for sale – the fourteenth Shuttle mission	400
"Mighty Joe" returns to space	401
A bone-rattling lift-off	401
A butter cookie for good luck	403
Flying free	404
Having your hands full	405
Fun in space	407
STS 51-B: Spacelab 3 and those monkeys	407
The second Spacelab mission	408
Thornton's return	409
Monkeys and men	410
Problem after problem	412
Running around the world	413
Back on the ground	413
STS 51-F: Spacelab 2 and three scientist-astronauts	414
False starts but a fine mission	416
A long preparation	416
Spain or Earth orbit?	419
Karl flying high	421
Taking the last chance to fly	422

	Another trip into space?	427
	References	428
11	**Ending of Eras**	431
	Moving on – life after space flight	431
	Joe Kerwin – Skylab–Shuttle–Space Station	432
	Astronaut Office – circa spring 1984	433
	After Spacelab 1	433
	Lenoir departs – and comes back	434
	Joe Allen and the ISF	435
	CB points of contact for Flight Data File – November 1985	438
	After Challenger	439
	Tony England – Losing Sunlab and back to teaching	439
	Karl Henize – new mountains to climb	440
	Owen Garriott – EOM and SPEDO	442
	Return-to-flight and a return to space	444
	Bill Thornton	444
	An astronomer for Astro	448
	Forty days from Halley's Comet	448
	Temporary duty in Washington	449
	Astro-1 flies – eventually	449
	Parker's role on Astro-1	451
	Back to Washington	453
	Six missions and thirty years	456
	Education and mission support	456
	Military Musgrave	458
	Servicing Hubble	462
	Back in the pool	464
	Improving Musgrave's ratio	466
	The last flight	468
	"You can't fly anymore!"	472
	All good things come to an end	474
	References	475
12	**Science Officers on ISS**	477
	Building a dream	478
	From imagination to reality	479
	Science on ISS	482
	ISS science officer	482
	Science officer – a job description	482
	NASA's first ISS science officer	484
	Saturday morning science – ISS Science Officer Two	486
	A reduced role – ISS science officers 2003–5	486
	Is the "science officer" really a science officer?	487
	Future roles?	489

Are science officers today's scientist-astronauts?. 489
        Memories from orbit 494
References 495

Appendix 1 – Chronology of the NASA Scientist-Astronaut Programme . . 497

Appendix 2 – Scientist-Astronaut Careers and Experience 505

Appendix 3 – Spaceflight Records and EVA Experience 507

Appendix 4 – Profiles of the Seventeen 511

Appendix 5 – Where Are They Now? 519

Bibliography 527

Index 533

Foreword

Were we really "fish out of water"? Did we ever "fit in" with a group of highly skilled and dedicated, hard-driving, egocentric test pilots? Yes and yes, I think.

There were only two groups of designated "Scientist-Astronauts" ever selected by NASA – the fourth astronaut group of six and the sixth group of eleven. Young men back then, we were solicited from the science and research communities, our credentials carefully evaluated by the National Academy of Sciences (NAS), our numbers further winnowed by NASA examiners and finally sent in small groups for very thorough and extensive physical examinations at the US Air Force School for Aerospace Medicine in San Antonio, Texas. Of those seventeen finally selected, six left the programme before achieving a single flight into space. Reasons ranged from a very public divorce (a lesson not lost on the incumbent astronauts), to an inability to reach the requisite piloting skills, to decisions that time lost in accomplishing the space tasks would prove detrimental to, or irretrievably retard, later careers in science.

Of the eleven remaining, one walked on the Moon, three flew long-duration, world-record-setting missions on Skylab, and the rest – all from the second group – eventually flew on from one to six Shuttle missions.

At the time of the first scientist-astronaut selection in 1965, there were only twenty-eight pilot-astronauts in the Astronaut Office and two of them were not on flight status. We were already four years into the decade in which President Kennedy declared we would reach the Moon. We had only four and one-half years to go and were just starting the Gemini programme. In mid-1965 NASA had still not carried out a manned rendezvous in space – an intricate technological accomplishment absolutely crucial to the success of the Apollo programme. Is it any wonder that those pilots, test pilots and programme managers wanted help from the most experienced flying personnel that could be found?

Over the course of the next five years, I believe the remaining scientist-astronauts came to be fully accepted by the corps of pilot-trained astronauts. Indeed, our special skills in geology, medicine and other sciences were recognised as bringing an extra dimension to the capabilities of the Astronaut Office. As fully trained pilots, we all flew T-38 aircraft on training missions and for quick movement between working locations around the country, exactly the same as any pilot-astronaut. Some of us had graduated as the top pilot in their flight squadrons of sixty or seventy much younger Air Force students. In our semi-annual flight examinations, our performance continued to measure quite well against the other highly qualified pilots in the Astronaut Office.

In my personal experience, I can understand the reluctance of many to accept a bunch of new "science types" into what was clearly a pilot-focused activity. Yet never once did I experience any discrimination because of a prior academic and research background. In fact, it was almost the opposite, as if every chance was provided to allow me to demonstrate the capability to work in their arena, and when that was provided, confidence was extended. In the end, I believe that our personalities were not all that different from the test pilots, only in our first calling.

Unfortunately, the term "Scientist-Astronaut" was never used again in selections beyond Group 6 in 1967. Undoubtedly, many of those selected in later years as a mission specialist (MS) would qualify for this description. I believe this was principally a political decision, made at a higher level, to avoid a distinction between those selected with a research or scientific background and those other competent individuals coming from a military technical background. I believe some of the vitality and perceived capability of the Astronaut Office has been lost as a consequence.

But what follows will be more a personal story of the individual scientists and researchers, the many obstacles they encountered, and how they morphed into essential elements of the total space flight programme, as it reached to the Moon and then to long-duration orbital flights on Skylab, the Space Shuttle, and Shuttle–Spacelab.

Owen K. Garriott, Ph.D.
NASA Scientist-Astronaut
Skylab 3 STS-9/Spacelab 1

Authors' Preface

This book is the result of an email discussion between the two authors about five years ago, exploring the possibility of a cooperative project. Both authors had recently completed solo projects and discovered a common interest in both the Skylab missions and the two groups of scientist-astronauts that were chosen in the 1960s for the Apollo Applications Program. The role of the non-pilot astronaut within NASA had changed during the transition from Apollo to the Shuttle and the designation "Mission Specialist" rather than "Scientist-Astronaut" was more than just a change of title.

Space Shuttle flights had shown that a broad knowledge base was advantageous for flying multiple missions, and that pure "science" research activities were the primary role of the payload specialist, the "part-time" astronaut who was selected for a specific mission or investigation, rather than as a career astronaut. During the 1980s and 1990s, the role of the mission specialist became one of team leader on dedicated scientific research missions, evolving into what became known as the payload commander. After the experiences of American astronauts on Mir, and with the introduction of the International Space Station (ISS) featuring further long duration residencies by American astronauts, the need to focus science research objectives and operations became clear.

In 2002, with ISS capable of supporting a long duration resident crew without a docked Shuttle, and with a broad range of scientific experiments and support hardware on the Destiny module, ISS-5 NASA flight engineer Peggy Whitson was designated the first NASA ISS science officer. This role was not the same as the payload commander, mission specialist, or scientist-astronaut before it, but like them, built upon the experiences of these earlier roles to become a focal point for US science on the station, while still balancing operational and habitation issues. The loss of Columbia in early 2003 suspended the expansion of science operations until 2006 at the earliest, but even with a two-person caretaker resident crew, US astronauts took on the dual role of either ISS commander or flight engineer, along with that of NASA

xviii **Authors' Preface**

science officer. From 2006, it is hoped that the role will be changed again (depending upon the next Shuttle flight), once a full, three-person resident crew is restored to the station. Only then will the lost dreams of the scientist-astronauts selected four decades before be fully realised, as preparations for a return to the Moon and out to Mars are developed and, hopefully, scientist-astronaut participation on the lunar and Martian surfaces becomes a reality.

Dave Shayler

In 1970, I obtained a new space book for Christmas, entitled *America's Astronauts and their Spacecraft*. It revealed brief biographies of most of the astronauts selected between 1959 and 1969, including the active scientist-astronauts. While the backgrounds of most of the astronauts featured a military flying career, these more academic astronauts interested me and revealed an expansion of the programme that, at that time, added to my information on the huge space stations, manned bases on the Moon and expeditions to Mars that were anticipated over the next couple of decades. To me, a young teenager, high on the excitement of the first Moon landings the year before, it became clear that on a space station, a moonbase or a long flight to Mars, the skills of a scientist would be more useful than those of a pilot. The work these men would do would secure even further expansion of our human space flight goals. The careers of these scientist-astronauts would be worth following over the next decade, as much as the pilot-astronauts assigned to Apollo.

In further research, I found that some of the scientist-astronauts had left before completing pilot school, and with the talk of cutting back Apollo in the wake of Apollo 13 to less than the ten landings planned, I hoped that some of these scientist-astronauts would get the chance to fly – anywhere. Several names interested me from those early days. Coming from the United Kingdom, with no chance of a manned space programme of our own, having an astronaut (and geologist) named Tony England suggested we might at least claim the feat of placing "England" on the Moon! Sadly, the only actual "British astronaut", Tony Llewellyn (born in Wales), had to withdraw from the programme due to his difficulties with piloting jets. Still, Bill Thornton was married to an English woman, which interested me, and two other scientist-astronauts had interesting backgrounds and potential careers; Joe Allen and Karl Henize. Two years later, I watched Jack Schmitt explore Taurus-Littrow during Apollo 17, lamenting the end of the programme, but becoming fascinated at the same time with the potential of Skylab and the assignment of scientist-astronauts to it.

What was unimaginable in 1970–1973 was that just a decade or so later, I would meet several of these scientist-astronauts many times, and become good friends with not only the men, but in many cases their families as well. Astronauts Allen, Thornton, Henize, Garriott and Gibson in particular have provided insights into the workings of the space programme, life in space, and how to adapt to life after space, becoming ambassadors for space exploration. Their enthusiasm and openness, despite their long years of frustration while waiting to fly and then only flying a few missions, was a story that needed to be told. These were not the famous "fighter-jocks" of the "Right Stuff" era of the Mercury programme; these were a different breed of space explorers, whose sense of adventure was geared towards science and understanding, as well as exploration and discovery.

In 1988, I had the opportunity to visit Johnson Space Center in Houston, Texas, for the first time. It was just prior to the Return-to-Flight mission of STS-26 and the resumption of Shuttle operations. The "NASA family" was buzzing; interviews were arranged and personal tours organised at a time when the lack of flight operations afforded access that, twenty years later, is now impossible. It was a great way to really understand how NASA worked. Official interviews with Story Musgrave and Bill Thornton at JSC were supplemented by off-site interviews with Thornton, Joe Allen and Karl Henize, and social evenings with the Thorntons and Henizes.

Having written a book on Challenger, in which Thornton, Henize, and Musgrave had flown, and with an interest in the EVAs by Joe Allen, I had the information and understanding to be taken seriously. During that same 1988 visit, I met Jerry Carr for the first time and began to gain an insight into Skylab that has grown over the years to include discussions with many of the men assigned to that programme, including the scientist-astronauts.

From the seventeen selected, only one scientist-astronaut flew on Apollo and walked on the Moon. Others could have. Just three flew on Skylab. Others could have; to the second Skylab that is now in the Air and Space Museum in Washington. After years of waiting, eight flew on the Shuttle and participated in some of the most fascinating missions of the early Shuttle era. Though one scientist-astronaut's career covered the whole Apollo to Shuttle/Mir era, none of them made it to another space station. ISS was a mission too far, but its NASA science officers owe their chance in orbit to the pioneering work completed by the unique team of seventeen scientists. They challenged not only space, but also the military pilot hierarchy of the NASA Astronaut Office, and began to change the traditional perception of the American astronaut. Their role in space history deserves to be recalled and applauded.

Colin Burgess

In October 1993, I flew from Australia to attend the ninth Planetary Congress of the Association of Space Explorers in Vienna, Austria. Although my interest in the human space flight programme had been ignited more than thirty years earlier with the drama-filled orbital flight of Mercury astronaut John Glenn, any opportunities to meet and talk with these space explorers, flown or not, had been very few and far between in our land Down Under.

I was attending the Congress with fellow space flight enthusiasts Bert Vis and Simon Vaughan, as accredited historians, and for the next week, we were able to meet and talk with a veritable cornucopia of astronauts and cosmonauts. While many friendships and contacts were forged during that week-long forum, there was also one sobering piece of news relayed to those attending. Scientist-astronaut Dr. Karl Henize, selected in 1967, but who had to wait eighteen years for his first space mission aboard STS 51-F/Spacelab 2, had perished while attempting to scale the north face of Mount Everest as part of an organised expedition. The news was truly devastating. At the time of his only mission, aged 58, he had become the oldest person to fly into space, and everyone knew the mountain known as Everest was simply another challenge he undertook, respectful of its dangers, in a life filled with prodigious and prestigious accomplishments.

When David Shayler and I finally shook hands on this book in London after email discussions on the project, one of our first resolutions was to dedicate the book to Karl Henize.

It would prove to be a marvellous amalgamation of interests; while David has an encyclopaedic knowledge of space flight missions and technology, my interest lies more in the people who flew those missions, and their motivations to do so. Research for this book not only brought me in touch with all sixteen surviving scientist-astronauts, but ultimately resulted in some wonderful and enduring friendships. As my first contribution to this effort, I set out to write factual pre-NASA biographies of all seventeen men from Group 4 and Group 6, of precisely four double-spaced typed pages each. These were forwarded to them (and the family of Karl Henize), not only to ask questions, but to ensure complete accuracy, so that future historians will come to regard this book as the ultimate authority on these two groups of men and their many fine achievements.

Just as my British co-author took immeasurable pride in having Tony Llewellyn announced as a member of the Group 6 scientist-astronauts in 1967, so, too, was Australia celebrating the selection of Dr. Philip Chapman. Like Llewellyn, Chapman would not get a chance to fly, although he did serve as a mission scientist and support crew member for Apollo 14 and Apollo 16. When he resigned in 1972, it was less to do with his abilities than with being involved in a faltering programme with a lack of foreseeable flight opportunities, many years before the envisaged Space Shuttle became operational. Since his resignation, Chapman has been an outspoken critic of many of NASA's past and present policies and directions, as well as the incumbent office politics and a well-documented, prevailing disdain for space science that became transparently obvious in the attitudes of many pilot-astronauts with whom they shared their astronaut training.

One of those swept up in contemporary office politics was Dr. Duane Graveline, who had been selected in the first scientist-astronaut group in 1965. Former NASA physician, Dr. Fred Kelly, once described Graveline as "a man I considered head and shoulders above all the other scientist-astronaut selectees," yet he ignominiously fell victim to NASA policies of that era, when his wife sued for divorce following his selection. A media-sensitive NASA dragged him back to Houston from the beginning of his flight training and showed him the door. Although several pilot-astronauts' marriages were on the rocks, these men were America's new Cold War heroes, and they were portrayed by the popular press as impeccable, All-American, church-going family men, with lives untainted by controversy. Everyone knew that any hint of divorce would see an end to their astronaut career. It's the way things were at the time, and Graveline became the first victim of this career-ending policy. He was a shattered man, with a shattered career. Recently, he told me that NASA still insisted on taking his official portrait just hours after bluntly informing him that he was out of the programme. The photograph shows him glumly holding a model of the Lunar

Excursion Module in front of the American flag. "I think what I was feeling at that very moment is written all over my face," was his reflection.

Ultimately, there were tribulations and many triumphs for all seventeen scientist-astronauts, and we have allowed them to express their innermost feelings in all areas. On behalf of David and myself, I thank them all, and the family of Karl Henize, for allowing us access to them in their busy schedules, and for their cooperation, advice and even friendships of lasting tenure. We sincerely hope we have done their lives and accomplishments justice in this book.

Acknowledgements

For their assistance in conducting research, supplying photographs and checking facts for us, we would like to thank Walt Sipes, Hart Sastrowardoyo, Bruce Rogalska, Peter Smith, Dr. John B. Charles, Lawrence McGlynn, Anne Lenehan, Michael Cassutt, and Francis and Erin French.

For information on the life of Karl Henize, many thanks to his family; Caroline and Vance Henize, and Roddy Seekins. For information on Bob Parker, many thanks to Sonia Parker at the Reuben H. Fleet Science Center, San Diego

For supplying anecdotal material, Dr. Alex Dessler from the University of Arizona, Dr. Loren Acton from the Montana State University, and Professor Bob Bless from the University of Wisconsin. Further assistance from the University of Wisconsin was given by staff members and researchers Leonard Black, Steve Masar and Kerri Canepa.

We would also like to thank Jody Russell from NASA's Media Resource Center, Johnson Space Center, Houston, Sally Little, NASA Marshall Space Flight Center and to Professor C. William Birky for details of his family and career background.

The following astronauts were instrumental in providing an insight not only into their flights, but also into the workings of NASA during their time there: Joe Allen, Ed Gibson, Karl Henize and Bill Thornton. In addition, we wish to thank all the members of the two groups for their time during various interviews, correspondence and assistance in the compilation of this book. Much of this assistance has been listed as references and sources. We must also thank the various candidates from the two groups for supplying additional information about their quest to become a scientist-astronaut.

The cooperation of the staff at the Public Affairs Office and the History Office at NASA JSC in Houston over many years has been of considerable help in accessing archival data on the scientist-astronauts. This has been continued by the staff of the University of Houston at Clear Lake, the custodians of the NASA JSC History

Collection. Staff members of Rice University, Houston, have given extensive support in researching the Curt Michel Collection held there.

Thanks go to Jerry Carr for access to his personal archive that included details of the Group 4/5 academic and survival training programme during 1966–1967.

The two authors' personal archives, collected over forty years, have been indispensable in conducting research, but there was also a network of colleagues and friends willing to share snippets of information and suggestions for further research to add that little bit of extra detail to the facts. This select group includes Rex Hall, Mike Cassutt, and Bert Vis, each of whom has helped indirectly in the compilation of this volume over many years.

We must of course thank Owen Garriott for the superb foreword, for supplying the authors with a copy of his diary notes on his selection as well as other vital information and quotable material, and for clarifying a few points in the draft phase.

Thanks also to Mike Shayler our copy editor, whose professional work over an extended time period (and time zones) along with skilful photo preparation is appreciated. Once again the support and understanding of Clive Horwood, Chairman of Praxis in difficult times for us all is recognised as is his continued support and belief in his authors. Thanks to the staff of Springer-Verlag in both London and New York for post-production support; to Neil Shuttlewood and staff at Originator, for their typesetting skills; to Jim Wilkie for his continued skills in preparing the cover for the project, and the printers for the final result.

Both authors wish to thank the support and encouragement of our immediate families in the compilation of this book, especially Bel Edge and Pat Burgess.

This has been a very personal project for both authors, who have a long interest in the lives and careers of the NASA astronauts (whether or not they ever made it into orbit) and the families who support them. All of their stories need to be told, as it was a team effort that got them though the selection and training, into space, out to the Moon and safely home. The seventeen scientist-astronauts selected in 1965 and 1967 were just as important to the overall NASA effort to reach the Moon and laying the foundations for what followed as the other groups selected between 1959 and 1969. Thanks to their efforts, sacrifices and determination this important story can now be told.

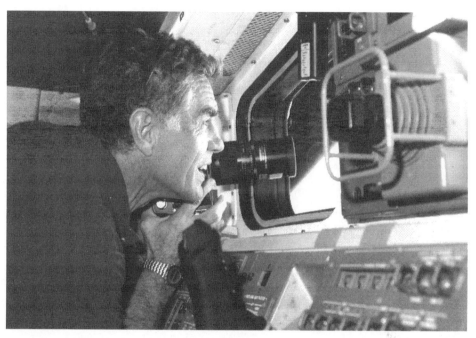

Karl Henize, with camera, at the aft flight deck windows of Space Shuttle Challenger during the STS 51-F Spacelab 2 mission in 1985. An astronomer by profession, the delight on his face reflects his feelings at finally making it to orbit after waiting eighteen years to fly in space.

To Dr. Karl G. Henize (1926–1993)
Mission Specialist STS 51-F (Spacelab 2) 1985

This book is also dedicated to the other sixteen men who were chosen as NASA's scientist-astronauts. We further dedicate this work to the families who supported them in their quest for space and to the candidates who almost became fellow scientist-astronauts.

List of Illustrations

Dedication: Karl Henize at the window of Challenger during STS 51-F xxv

1 The Wrong Stuff

The Soviet single-seat Vostok spacecraft	4
The American single-seat Mercury spacecraft	6
The first seven American astronauts – all pilots	10
Sergey Korolyov and Yuri Gagarin	15
US President John F. Kennedy	16
Wernher von Braun	18

2 Scientists as Astronauts

Eugene Shoemaker with the first group of scientist-astronauts	28
Allen and Lenoir in the T-38 jet	31
Five flight surgeons attend training in 1964	33

3 The Scientific Six

The six chosen for the 1965 scientist-astronaut group	57
A press conference for the new recruits	58
Owen Garriott	59
Ed Gibson	63
Joe Kerwin	67
The formal group portrait following the removal of Graveline	69
Curt Michel	71
Harrison Schmitt	77
Duane Graveline	84
Graveline addresses a forum in 2004	87

4 School for Scientists

Gibson and colleagues on a geological field trip	104
Michel studying geological samples	105
Michel and colleagues during jungle survival training	110
Relaxing during a break in jungle training	111
Schmitt and colleagues preparing to build a shelter	112
Schmitt enjoys a light-hearted moment	113

5 The Excess Eleven

The official group portrait of the 1967 selection	120
The second group of scientist-astronauts at a less formal gathering	121
Joe Allen	123
Phil Chapman	127
Tony England	132
Karl Henize	136
Don Holmquest	140
Bill Lenoir	144
Tony Llewellyn	147
Story Musgrave	151
Brian O'Leary	155
Bob Parker	158
Bill Thornton	162

6 "Flying is Just not My Cup of Tea"

Seven members of the 1967 group examine a model workshop	174
The same seven standing before a huge Saturn F-1 engine	175
Chapman enters a centrifuge	176
Chapman examines a Lunar Module mock-up	178
Chapman inside the Lunar Module simulator	179
Chapman stands in the centrifuge gondola at the MSC	179
O'Leary and Allen are strapped into the centrifuge gondola	180
Chapman takes a coffee break during pilot training	183
A thumbs-up from Parker in the T-38	184
Henize during desert survival training	186
Chapman and Allen undertake jungle survival training	187
Lenoir and Musgrave prepare to spend a night in their shelter	188
Four of the Group 6 astronauts practicing in rafts	189
Chapman wearing an Apollo pressure suit	192

7 A Geologist on the Moon

Stowage tests in the Lunar Excursion Module simulator 1966	200
A crowded console during the 1970 Apollo 13 crisis	208
Kerwin and Mattingly monitor communications with Apollo 13	209
The Apollo 15 hammer and feather experiment	211
Schmitt and Cernan review traverse maps	217
Schmitt and Cernan riding the Rover test vehicle	218
Schmitt briefs members of the press on the Apollo 17 landing site	219
Apollo 17 lifts off	220
The Apollo 17 CSM photographed from the Lunar Module	222
Station 4 – the location of the orange soil	227
Schmitt collects a soil sample at Station 5	228
A panorama of Station 7	229
Schmitt stands next to the American flag	231
The lunar surface through the LM window	232
The trans-Earth EVA to retrieve film canisters	234
Cernan and Schmitt – exhausted but exhilarated	235
Splashdown at the end of Apollo 17	236
The Apollo 17 crew address a joint session of the US Congress	238
Harrison Schmitt today	239

8 Laboratories in the Sky

The SMEAT crew	262
The Skylab SMEAT crew about to start their 56-day test	263
The Skylab 2 crew of Weitz, Conrad and Kerwin	267
Conrad and Kerwin on EVA during Skylab 2	268
The Skylab 2 crew present mementoes to US and Soviet leaders	269
The Skylab 3 crew of Garriott, Lousma and Bean	270
The Skylab 3 crew in the 1 G Skylab trainer	271
Deke Slayton talks to Bean prior to launch	272
Garriott enjoys a meal aboard Skylab	273
Garriott on EVA during the second manned mission	274
The Skylab 4 crew of Gibson, Carr and Pogue arrive at KSC in their T-38s	275
Gibson at the Apollo Telescope Mount console	276
A selection of Skylab on-board images	277

9 Shuttling into Space

The interior of the 1974 SMS module	289
Musgrave feeding the plants	289
A 1974 Scuba diving training exercise with three female participants	293
The crew for the second Spacelab medical simulation	308
The mock-up of the aft flight deck of the Shuttle	309
The SMD mock-up configured for the second test	310
A selection of experiments conducted by Story Musgrave	311
SMD hardware for the 1977 test	314

xxx List of Illustrations

Monitoring the progress of the simulation 317
The emblem of SMD-III 318
Dr. Bill with his monkeys 320
The crew for the SMD-III test 321
The rats are prepared for SMD-III 322
Thornton inside the Spacelab mock-up for the third test 324
Thornton at one of the work stations 325
The crew celebrates the end of the simulation 327

10 The Long Wait

Gibson in 2003 335
Kerwin and Anna Fisher at the Neutral Buoyancy Simulator in 1980 337
Kerwin and Fisher evaluating a Hubble Servicing-type mission at the NBS in 1980 .. 338
The "Ace Moving Company" in starburst formation during STS-5 340
Allen and Lenoir practice their EVA underwater 341
Lenoir and Allen traverse the mock-up Shuttle payload bay 343
Lenoir and Allen ready to enter Columbia 344
Lenoir in the White Room prior to launch 345
Allen in his seat on the mid-deck of Columbia 347
The first two mission specialists in space 348
Lenoir and Allen conducting biomedical tests 350
Allen participates in experiments 351
Lenoir participates in a vision test 353
Joe Allen and the sheer joy of space flight 354
Demonstrating the physics of liquids in space 355
Overmyer gets a trim from Lenoir 356
The STS-6 crew receiving flight instructions for their T-38 manoeuvres 357
Musgrave dons the Hard Upper Torso of the Shuttle EVA suit 359
Karol Bobko assists Musgrave to prepare for an underwater EVA simulation .. 360
The STS-6 crew participate in a simulated launch 362
Musgrave on EVA during STS-6 363
A close-up of Musgrave showing the EVA equipment 364
Musgrave monitoring experiments during STS-6 365
Thornton checking biomedical experiment data 367
Thornton measures the leg volume of Dick Truly 370
The STS-8 crew displays an official US postal cover 372
The STS-8 crew participate in a post-flight news broadcast 374
A press conference for the crew of STS-9 376
Garriott prepares for a simulation in the KC-135 aircraft 378
Garriott demonstrates a ham-radio receiver 379
Garriott at work in the aft flight deck mock-up at JSC 381
Garriott and Merbold at work in the first Spacelab during STS-9 383
Parker and Garriott at work in Spacelab 1 388
Garriott operates amateur radio equipment 390
Playing cards on the "roof" of Spacelab 1 391
"See no evil, hear no evil, speak no evil" 396
Allen about to enter Discovery for his second space flight 398

The crew of STS 51-A pose for their in-flight portrait	399
Allen and Dale Gardner each flew the MMU	402
Allen holds the Palapa satellite while tethered to the end of the robot arm	404
Two satellites "For Sale"	406
The STS 51-A crew marks their achievements during the mission	408
The STS 51-B crew pose for photographs	409
Launch day breakfast for the STS 51-B crew	412
Thornton, Taylor Wang and Don Lind in Spacelab 3	414
The crew for STS 51-F takes a break in training	415
The science crew for Spacelab 2 visits the University of Birmingham in England	417
Inspecting the hardware at the University of Birmingham	418
Challenger's flight deck controls featuring the Abort-to-Orbit control	420
Spacelab 2 science payload deployed from the payload bay	421
A humorous photo of the STS 51-F crew	423
Musgrave draws blood from Tony England during STS 51-F	424
Musgrave photographs a plant growth experiment	425
Henize and England sampling carbonated beverages in orbit	426
England takes the opportunity to look out of the window	428

11 Ending of Eras

Joe Allen in 2002	435
Allen receives an award	437
Henize displays a model of the Spacelab 2 payload	441
Thornton with UK astronaut Helen Sharman during a trip to the West Midlands, England	445
Thornton is the proud recipient of the 2003 North Carolina Award for Science	447
Parker briefs the payload specialists during training for STS 61-E	450
Parker and Durrance during simulations for STS-35	451
The STS-35 crew during emergency egress training at KSC	452
Parker using fire-fighting equipment during emergency egress training	453
Parker operating the Instrument Pointing System during STS-35	454
Bob Parker in 2002	455
Pre-flight breakfast for the STS-33 crew	457
The crew of STS-44	458
Musgrave conducts medical experiments on Mario Runco during STS-44	460
Musgrave works with Henricks on the STS-44 medical programme	461
Musgrave looking out of the window	462
Musgrave in the launch and entry suit	463
Musgrave in the White Room of Pad 39B	465
Musgrave and Jeff Hoffman on the first of the STS-61 EVAs	467
Hoffman reflected in Musgrave's visor on EVA	468
Musgrave holds on to one of Hubble's handrails during STS-61 EVA	469
The crew of STS-80 includes Musgrave on his sixth and final flight	470
Musgrave working on the aft flight deck during STS-80	472
Story Musgrave in 2004	473
Musgrave's biography was released in 2004	475

12 Science Officers on ISS

The US Destiny laboratory module under construction in 1997	480
Destiny shown attached to ISS in February 2001	481
A close-up of Destiny taken during STS-100	483
Peggy Whitson, the first NASA Science Officer	485
ISS-6 Flight Engineer Don Pettit, the second NASA Science Officer	487

List of Tables

1	Selection boards for the NASA Astronaut Class of 1965 (Group 4)	34
2	Group 6 Academic Training Programme October 1967–February 1968	171
3	Story Musgrave's planned vs. actual training hours (STS-6)	358
4	Bill Thornton's planned vs. actual training hours (STS-8)	369
5	Experiments carried out on STS-9/Spacelab 1	385
6	ISS Main Element launches	480
7	ISS Resident Crews (2000–2006)	481

List of Abbreviations

AAAS	American Association for the Advancement of Science
AAP	Apollo Applications Program
ABF	Air Bearing Floor
ACR	Active Cavity Radiometer
AEPI	Atmospheric Emission Photometric Imaging
AES	Apollo Extension System
AF	Air Force
AFB	Air Force Base
AFROTC	Air Force Reserve Officer Training Corps
AKM	Apogee Kick Motor
ALAE	Atmospheric Lyman-Alpha Emission
ALSEP	Apollo Lunar Surface Experiment Package
AN	Akademiya Nauk – the Soviet Academy of Sciences
ANARE	Australian National Antarctic Research Expeditions
AOSO	Advanced Orbiting Solar Observatory
APU	Auxiliary Power Unit
ASSESS	Airborne Science/Spacelab Experiment System Simulation
ASTP	Apollo-Soyuz Test Project
ATM	Apollo Telescope Mount
ATO	Abort-To-Orbit
BIG	Biological Isolation Garment
BIS	British Interplanetary Society
CAG	Carrier Air Group
CalTech	California Institute of Technology
CAP	Civil Air Patrol
CB	This is an internal JSC mailing code for the Astronaut Office. It stands for Directorate level C department B

CDMS	Command Data Management System
CEIT	Crew Equipment Interface Test
CFES	Continuous Flow Electrophoresis System
CHASE	Coronal Helium Abundance Spacelab Experiment
CM	Command and Service Module
CMP	Command Module Pilot
CSC	Civil Service Commission
CSM	Command and Service Module
CTSc	Candidate of Technical Sciences
CVT	Concept Verification Test
DCPS	Dynamic Crew Procedures Simulator
DFVLR	Deutsche Forschunesund Versuchsanstalt für Luft und Raumfahrt – the Federal German Aerospace Research Establishment
DoD	Department of Defense
DPS	Data Processing Systems/software
DSc	Doctorate of Science
DSO	Detailed Supplementary Objectives
EASEP	Early Apollo Surface Experiment Package
EDO	Extended Duration Orbiter
EEG	Electro-EncephaloGraph
WFPCII	Wide Field Planetary Camera II (2)
EKGs	Electro-cardioGrams (EKGs)
EMU	Shuttle Extravehicular Mobility Unit
EO	Experiment Operator
EOM	Earth Observation Mission
EPS	Electrical Power System/subsystem
EREP	Earth Resources Experiment Package
ESA	European Space Agency
ESRO	European Space Research Organisation
EST	Eastern Standard Time
ESTEC	European Space TEchnology Centre
ETS	Educational Testing Service
EV	Extra Vehicular
EVA	Extra Vehicular Activity
FAUST	Far Ultraviolet Space Telescope
FCC	Federal Communications Commission
FD	Flight Day
FDA	Federal Drug Agency
FDF	Flight Data File
FE	Flight Engineer
FGB	Functional Cargo Bloc
FOD	Flight Operations Directorate
FTD	Foreign Technology Division
FY	Financial Year

G&N	Guidance & Navigation
Gidromettsentr	USSR Weather Service
GIRD	*Gruppa Isutcheniya Reaktivnovo Dvisheniya* – Group for Investigation of Reactive Motion
GMVK	Gosudarstvennaya Mezh Vedomstvennaya Komissiya – the Russian State Interdepartmental Comission
GPC	General Purpose Computer
GSFC	Goddard Space Flight Center
HFMS	High Fidelity Mechanical Simulator
HOSC	Huntsville Operations Support Center
HST	Hubble Space Telescope
ICBM	Inter-Continental Ballistic Missile
IFEP	In-Flight Experiments Panel
IKI	Institute Kosmichelskikh Issledovany – Space Research Institute
IMBP	Institute of Medical and Biological Problems
IMSS	In-flight Medical Support System
IP	Instructor-Pilot
IPS	Instrument Pointing System
ISF	Industrial Space Facility
ISO	Imaging Spectrometric Observatory
ISS	International Space Station
IUS	Inertial Upper Stage
IVA	Intra Vehicular Activity
IWCS	Improved Waste Collection System
IWG	Investigators Working Group
IZMIRAN	Institut Zemovno Magnetizma Ionosfery i rasprostraneniya Radiovoln Akademii Nauk – Institute for Terrestrial Magnetism, Ionosphere and Radio Wave Propagation of the AN
JIS	Joint Integrated Simulations
JPL	Jet Propulsion Laboratory
JSC	Johnson Space Center
JSWG	Joint Spacelab Working Group
KSC	Kennedy Space Center
LEM	Lunar Excursion Module
LES	Launch Escape System
LEVA	Lunar ExtraVehicular Activity
LM	Lunar Module (for Apollo) or Long Module (for Spacelab)
LMP	Lunar Module Pilot
LPI	Lunar and Planetary Institute
LRL	Lunar Receiving Laboratory
LSD	Life Science Directorate
LSPF	Life Sciences Payload Facility
LSPMS	Life Science Payload Mission Simulation

MA-6	Mercury-Atlas-6 (John Glenn's flight)
MAI	Moscow Aviation Institute
MCC	Mission Control Center
MCC-H	Mission Control Center – Houston
MDA	Multiple Docking Adapter
MDF	Manipulator Development Facility
MECO	Main Engine Cut-Off
MIT	Massachusetts Institute of Technology
MMP	Mission Module Pilot
MMPI	Minnesota Multiphasic Personality Inventory
MMU	Manned Manoeuvring Unit
MOCR	Mission Operations Control Room
MOL	US Air Force Manned Orbiting Laboratory
MOLAB	MObile LAB
MPG	Mission Planning Group
MS	Mission Specialist
MSC	Manned Spacecraft Center
MSE	Manned Spaceflight Engineer
MSEP	Mercury Scientific Experiment Panel
MSFC	Marshall Space Flight Center
MSG	Mission Steering Group
MTVT	Manned Thermal Vacuum Tests
NAA	North American Aviation
NACA	National Advisory Committee for Aeronautics
NAS	National Academy of Sciences
NASA	National Aeronautics and Space Administration
NBC	JSC Neutral Buoyancy Simulator
NBS	Neutral Buoyancy Simulator
NII	Nauchno Issledovatelsky Institut – the Russian Scientific Research Institute
NIIAS	Scientific Research Institute for Automatic Systems
NKVD	People's Commissariat for Internal Affairs
NOAA	National Oceanographic and Atmospheric Administration
NPO	Nauchno Priozvodstvennoye Obedineniye – the Russian Scientific Production Association
NRC	National Research Council
NROTC	Naval Reserve Officer Training Corps
NSBRI	National Space Biomedical Research Institute
NSF	National Science Foundation
NSSTC	National Space Science & Technology Center
OFT	Orbital Flight Test
OKB-1	*Opytnoe Konstructorskoe Byuro* – Experimental Design Bureau
OMSF	Office of Manned Space Flight
ORFEUS-SPAS	ORFEUS – Orbiting Retrievable Far-and-Extreme

	Ultraviolet Spectrometer; SPAS – Shuttle Pallet Applications Satellite
OSS	Office of Space Science
OWS	Orbital WorkShop (the Skylab station)
PABF	Precision Air Bearing Floor
PAM	Payload Assist Module
PAO	Public Affairs Office or Officer
PC	Payload Commander
PDSTS	Physical Diagnosis Self-Teaching System
PFTA	Payload Flight Test Article
PI	Principal Investigator
PICPAB	Phenomena Induced by Charged PArticle Beams
PKM	Perigee Kick Motor
PLSS	Portable Life Support System
POCC	Payload Operations Control Center
POISE	Panel On In-flight Scientific Experiments
PS	Payload Specialist
PSAC	President's Science Advisory Committee
QRP	A ham expression for very low power (no exact translation)
RAAF	Royal Australian Air Force
RAF	Royal Air Force
RAHF	Research Animal Holding Facility
RAN	Russian Academy of Sciences
RCS	Reaction Control System
RKK Energiya	Raketno Komicheskaya Korporatsiya – the Russian Rocket-Space Corporation
RLE	MIT's Research Laboratory of Electronics
RMS	Remote Manipulator System
ROTC	Reserve Officer Training Corps
S-IVB	Saturn-4B – the third stage of a Saturn V and the second stage of a Saturn IB
SA	Scientist-Astronaut
SAFER	Simplified Aid For EVA Rescue
SAIC	Science Applications International Corporation
SAIL	Shuttle Avionics Integration Laboratory
SAL	Skylab Air Lock
SAM	Surface-to-Air Missile
SAREX	Space Amateur Radio EXperiment
SAS	Space Adaptation Syndrome
SASTPC	Scientific Association for the Study of Time in Physics in Cosmology
SEPAC	Space Experiments with Particle ACcelerators
SES	Space Environment Simulator
SFC	Space Flight Center
SIGINT	SIGnal INTelligence satellite

SII	Space Industries Incorporated
SIR	Synthetic Imaging Radar
SLIC	Student Life and Interest Committee
SM	Short Module
SMD	Spacelab Mission Development
SMEAT	Skylab Medical Experiment Attitude Test
SMS	Shuttle Mission Simulator
SO	ISS Science Officer
SolCon	Solar Constant
SolSpec	Solar Spectrum
SOS	Space Operations Simulator
SP	Spacelab Pallet
SPAC	Space Program Advisory Council
SPAN	SPacecraft ANalysis
SPEDO	Solar Powered Extended Duration Orbiter
SPICE	Spacelab Payload Integration and Coordination in Europe
SPS	Service Propulsion System (Apollo)
SRL	Systems Research Laboratories
SRB	Solid Rocket Booster
SSME	Space Shuttle Main Engine
STS	Space Transportation System (Shuttle)
TDRS	Tracking and Data Relay Satellite
TRW	Formerly Thompson Ramo Wooldridge Inc.
TsPK	Cosmonaut training centre
TsVNIAG	Tsentralny Voyenny Nauchno Issledovatelsky Aviatsionny Gospital – Central Military Scientific Research Aviation Hospital
UHV	Ultra High Vacuum
UNC	University of North Carolina
US PS	United States Payload Specialist
USAF	United States Air Force
USF	University of South Florida
USGS	US Geological Survey
USN	United States Navy
UTMB	University of Texas Medical Branch
UW	University of Wisconsin
VAB	Vehicle Assembly Building
VASIMIR	VAriable Specific Impulse Magnetoplasma Rocket
VVS	Voenno-Vozdushnye Sily
VWFC	Very Wide Field Camera
WCS	Waste Collection System
WETF	Weightless Environment Training Facility
WMS	Waste Management System
WTC	Wyatt Technology Corporation
XMAS	Extended Mission Apollo System

Other works

Other space exploration books by Colin Burgess:

Space: The New Frontier (1987), ISBN 0-86896-361-5
Oceans to Orbit: The Story of Dr. Paul Scully-Power (1995), ISBN 0-949853-53-4
Australia's Astronauts: Three Men and a Spaceflight Dream (1999), ISBN 0-7318-0831-2
Teacher in Space: Christa McAuliffe and the Challenger Legacy (2000), ISBN 0-8032-6182-9

With Kate Doolan and Bert Vis:

Fallen Astronauts: Heroes Who Died Reaching for the Moon (2003), ISBN 0-8032-1332-8

Other space exploration books by David J. Shayler:

Challenger Fact File (1987), ISBN 0-86101-272-0
Apollo 11 Moonlanding (1989), ISBN 0-7110-1844-8
Exploring Space (1994), ISBN 0-600-58199-3
All About Space (1999), ISBN 0-7497-4005-X

With Harry Siepmann:

NASA Space Shuttle (1987), ISBN 0-7110-1681

Other books by David J. Shayler in this series:

Disasters and Accidents in Manned Spaceflight (2000), ISBN 1-85233-225-5
Skylab: America's Space Station (2001), ISBN 1-85233-407-X
Gemini: Steps to the Moon (2001), ISBN 1-85233-405-3
Apollo: The Lost and Forgotten Missions (2002), ISBN 1-85233-575-0
Walking in Space (2004), ISBN 1-85233-710-9

With Rex Hall:

The Rocket Men (2001), ISBN 1-85233-391-X
Soyuz: A Universal Spacecraft (2003), ISBN 1-85233-657-9

With Rex Hall and Bert Vis:

Russia's Cosmonauts (2005), ISBN 0-38721-894-7

With Ian Moule:

Women in Space: Following Valentina (2005), ISBN 1-85233-744-3

With Andy Salmon and Mike Shayler:

Marswalk: First Steps on a New Planet (2005), ISBN 1-85233-792-3

Prologue

On 29 July 1958, Dwight David Eisenhower, the thirty-fourth president of the United States, signed the National Aeronautics and Space Act of 1958, officially establishing America's civilian space agency. The National Aeronautics and Space Administration, which rapidly became known by the more familiar acronym NASA, would commence operations on 1 October that year. Primarily, it comprised around 8,000 employees recruited from a government research agency known as the National Advisory Committee for Aeronautics (NACA), based in Langley, Virginia.

There had been a lot of questions asked of the Eisenhower administration after the humiliating shock of the first Sputnik satellite launch, and then Sputnik II, which had carried a small dog into orbit. The president tried to be reassuring, but he fell short. A nation's pride in its accomplishments and abilities had been badly bruised. Privately, Eisenhower knew that America was already creating an impressive missile booster programme, a bristling arsenal based on captured wartime rockets and using German scientists. He also knew that the first generation of military spy satellites was under advanced development. However this was top-secret, classified stuff, and he couldn't reveal either programme to the public. One thing he could do to assuage public perception was to set up a civilian space agency – open and accountable, assigned to carrying out sophisticated manned and unmanned flight programmes, and steadfastly dedicated to the task of taking the new high ground of space.

The heads of President Eisenhower's Science Advisory Committee had argued that science should be assured of a place in setting up the new space agency. In its submitted report, *Introduction to Outer Space*, the committee aggressively argued that resolving scientific and physiological questions should be a priority, that the value of launching satellites and sending rockets into space had to be measured in scientific terms, and that the United States should not arbitrarily engage in some sort of long-term race into space with the Soviet Union.

James Killian was Eisenhower's chief scientific adviser and head of the President's Science Advisory Committee (PSAC), and he strongly endorsed this position with the president. His views were shared by many of his colleagues, but overwhelmingly and ultimately it was not the popular choice. Quite simply, America was not prepared to adopt a subservient or secondary position in space flight technology and was intent on beating the smug Russians. A rational, objective schedule and a role for science played little or no part in many influential opinions.

Shortly after the creation of NASA, a search was initiated for suitable candidates interested in becoming the space agency's first group of astronauts. Several career skills were listed for evaluation early on in the process, including those that might have scientific or medical applications. But ultimately, it was decided to restrict the selection to military test pilots. The seven men eventually selected as NASA's Project Mercury astronauts unexpectedly became adored national heroes, even before they had left the Washington press conference that had announced their names to the world.

The characters and carefully polished images of those Original Seven astronauts set a superlative for the patriotic, hero test pilot, who would risk his life for God, America, the military and his family in order to conquer and explore space for the free world. These so-called "All-American Boys" would become the nation's much-lauded Cold War warriors in an age of uncertainty, international tension and a highly unpopular conflict gathering ugly momentum in Southeast Asia. Years later, an appropriate new sobriquet would be thrust upon them by popular author Tom Wolfe; thereafter they would become known as members of an exclusive brotherhood with an almost indefinable quality, known as "The Right Stuff."

Two years after the creation of NASA, the Democrats boldly threw themselves into the 1960 presidential election, their hopes riding on the shoulders of a popular young senator from Massachusetts, John Fitzgerald Kennedy. With his brother Bobby's support, Kennedy took up the daunting challenge of defeating Republican candidate Richard Nixon, employing innovative political tactics and shamelessly exploiting a receptive media. Kennedy's running mate Lyndon Baines Johnson, the Democratic majority leader in the senate, made competitiveness in space a big issue, and Nixon found he could not match Johnson's enthusiasm, or Kennedy's stirring rhetoric. Like Eisenhower, Nixon was unable to reveal what was really happening with America's military-based rocket and satellite programmes.

The election was extremely close, but Kennedy and Johnson prevailed. Now they had to make good on their promises to haul America squarely into the unknown frontier of a rapidly escalating space race.

As good as the Mercury Seven's flying skills and Right Stuff qualities might have been, these men were essentially trained, top-notch pilots, generally with sound mechanical or engineering backgrounds and skills. They were not scientists. Although several experiments were carried out during Project Mercury – mostly physiological – science was never at the forefront of mission planning, nor in the astronauts' minds. Their primary objectives were to survive each given assignment and to report on the engineering aspects of their flight into space. They would tolerate being plastered with sticky sensors, and having their body wastes retrieved for post-flight scrutiny under the microscopes of doctors, but they did not regard themselves as potential polyps or lunar lab rats. These were test pilots, aviators, fighter jocks – noble warriors for the working day. Science was the thing that got them where they had to go, and they appreciated this, but their job was to check out the mighty beasts of carriage, and to make them less dangerous for the next guy in the flight rotation.

On 12 April 1961, NASA stood proud and ready to send an American astronaut into space for the first time. The historic launch of US Navy Commander Alan B. Shepard, Jr., was a mere three weeks away. But this was the day that the United States narrowly lost the undeclared race, when the Soviet Union strapped a young Air Force lieutenant named Yuri Gagarin into a spherical Vostok spacecraft and launched him on a single, attention-grabbing orbit of the Earth.

Not only were the citizens of the United States shaken out of their complacency, but an influential media was unimpressed by the fact that the nation had been caught off-guard. Journalists and commentators began hammering home the point that this was not only a challenge to America's leadership by an old enemy, but also to the nation's perceived supremacy in space. Congressional advocates began campaigning for a competitive national effort to outdo the Soviets in space, as did many highly placed government officials.

In a bid to see what could be done, President Kennedy sought the advice of his vice president and also serving chairman of the National Space Council, Lyndon Johnson. Johnson was instructed to conduct a survey of the national space programme and to make a determination about creating a dynamic new project that would elevate the United States into a superior and possibly unassailable role in manned space flight. The vice president began consulting with the NASA hierarchy, key members of Congress, and Defense Department officials. This resulted in the promulgation of a lengthy report that finally reached the desk of the president. He read the report, and liked what he saw. On 25 May 1961, less than three weeks after America had finally launched Alan Shepard into space on a fifteen-minute sub-orbital mission, an inspired President Kennedy made a dramatic national commitment before a special assembly of Congress. The goal, he stated, would be nothing less than placing a man on the Moon and returning him to the Earth before the end of that decade. With this presidential declaration came the impetus to expand the space programme well beyond the short, one-man Mercury missions.

Although the objective was the Moon, there was no immediate call for a protracted programme of exploration, or even scientific research. Many interpreted the Kennedy plan as a simple mandate – to get American astronauts to the Moon and bring them back safely. No only that, but to do it before the Soviet Union. Science and exploration were incidental to the task. They would come later – perhaps.

1

The Wrong Stuff

"Here is how we shall go to the Moon. The pioneer expedition, fifty scientists and technicians, will take off from the space station's orbit in three clumsy-looking but highly efficient rocket ships." Dr. Wernher von Braun.[1]

During the first half of the twentieth century, in the fifty years preceding the dawn of the space age, the dreams of science fiction were often portrayed as the dawn of science fact. The human exploration of space was perceived to be within reach and with it, expeditions to the Moon and planets. Countless designs and proposals for human-tended space bases were suggested as the most efficient way to support an integrated exploration of space over the long term. These permanent facilities were identified as stations in space (or, more commonly, space stations) from which specialist crews could conduct Earth and astronomical observations, and which would serve as weather forecasting or communication satellites, or as a base for construction or refuelling. To achieve this, these "specialist crews" would incorporate the skills of doctors, meteorologists, engineers, chemists, biologists, physicists, astronomers, and geologists, rather than crew members with purely piloting skills. It was understood, however, that the first human explorers of space would be drawn from the military piloting community.

ORGANISING THE EFFORT

The true dawn of the space age occurred on 4 October 1957 with the launch of the world's first artificial satellite, Sputnik 1, by the Soviet Union. A month later, Sputnik 2 carried the first living creature into orbit – the dog Laika. It would not be long before humans followed that same hazardous pathway into the heavens and in both the Soviet Union and the United States of America, selection criteria were being debated in order to select the first groups of candidates to undergo space flight training.

A manned satellite project

In March 1956, the United States Air Force (USAF) initiated Project 7969, entitled "Manned Ballistic Rocket Research System", a staged programme to develop a small recoverable capsule from orbit that would lead to a larger, manned capsule recovery. In addition to the USAF, other proposals for manned space missions came from the US Army and US Navy, the National Advisory Committee for Aeronautics (NACA) and a number of aerospace companies. Existing ballistic missiles, converted to carry a human payload with adequate safety and escape methods, were proposed as the safest and quickest way to place an American in space, rather than the more complicated and longer-range proposals of manned rocket research aircraft (such as the X-15 or X-20) or lifting body technology then under development.

During the period 1956–8, USAF efforts in manned orbital space flight were in study projects rather than a designated or approved programme, although they did generate submissions from several aerospace companies concerning the design of the spacecraft, its subsystems and infrastructure that had direct implications for the eventual pioneering manned American spacecraft. In comparable studies, NACA had been working on the problems of orbital space flight and recovery since early 1952. This work was focused into one civilian space agency as a result of Senate Resolution 25, dated 6 February 1958, for a "Special Committee on Space and Astronautics to form legislation on a national program for space exploration." The National Aeronautics and Space Act of 1958 was signed by President Eisenhower on 29 July and led to the activation, on 1 October, of the National Aeronautics and Space Administration (NASA). They would be responsible for the management and development of non-military space activities, including the manned satellite project. Presentations on this project were made to Dr. Keith T. Glennan, the first NASA administrator, seven days into the activities of the new space agency, and approval of the project was driven by Glennan's desire to "get on with it."[2]

On the other side of the world, a different approach was being followed to place a citizen of the Soviet Union into space. Early designs for the human exploration of space were proposed in the pioneering theoretical studies of the "Father of Cosmonautics", Konstantin Tsiolkovsky, in the early years of the twentieth century. In the 1930s, Sergey Korolyov became involved in early developments of Soviet rocketry with GIRD (*Gruppa Isutcheniya Reaktivnovo Dvisheniya* – the Group for Investigation of Reactive Motion) as well as Scientific Research Institute NII-3, and in 1946 was appointed Chief Designer of bureau OKB-1 (*Opytnoe Konstructorskoe Byuro* – Experimental Design Bureau). Years later, he was the driving force behind what evolved into the Soviet space programme in the 1950s and 1960s, responsible for the development of the programme and the many space firsts the Soviets achieved between 1957 and 1966. Studies into launching a Soviet citizen into space, though sub-orbital, were conducted during 1945–8 under the guidance of Mikhail Tikhonravov, using a series of manned vertical rocket flights to explore the upper atmosphere. The following year, Korolyov worked with medical specialists in the Air Force Institute of Aviation Medicine to launch dogs into space, a programme he considered to be a stepping stone for placing humans in orbit. Alongside the biomedical studies on both canines and Air Force test subjects came developments in Soviet rocketry and the creation of a suitable launch site. The development of the launch site (and the R-7 ICBM as a potential manned launch vehicle) shortened the programme of vertical flights and brought manned space flights forward to earlier than the 1964–6 estimate.[3]

Orbital piloted spaceship of the Soviet Union

Initial work at OKB-1 focused on vertical flights of a single person on a short "suborbital" trip using the scientific version of the military R-2 ballistic missile. Under the technical leadership of engineer Nikolay P. Belov, similar work was being carried out by Tikhonravov at another scientific research institute, NII-4. This work continued from 1955 for some years, until the advent of the larger R-7 indicated that higher and longer flights into orbit were possible. Running simultaneously to these studies were efforts to place the first satellite into orbit, which resulted in the launch of Sputnik 1. This satellite, however, was not the original choice for the pioneering mission into space. "Object D" was the fifth (in the Russian alphabet) to be assigned for launch on the R-7, after four nuclear warhead variants. This was an unguided artificial satellite, with a mass of roughly 900–1,350 kg and a scientific payload of 200–300 kg, planned for launch by 1957, within the eighteen months of the International Geophysical Year. Identified as the "simplest satellite" in early planning, it was far more complicated than envisaged, resulting in delays and the fear that the Soviets might lose the race to place an object in orbit to the Americans. Korolyov decided to delay Object D until a simpler design had been launched to achieve the goal of becoming the first artificial Earth satellite. Simple Satellite 1 (PS-1) became Sputnik 1 and a month later, PS-2 was adapted to include a canister for Laika and a suitable life support system. The original payload intended as the first satellite was finally launched as Sputnik 3 on 15 May 1958.

At the same time, studies and designs for a vehicle that would support a human in space were being pursued in the Planning Section of OKB-1, based on designs for an unmanned reconnaissance satellite codenamed Object OD-2. By 22 May 1959 this was renamed "Vostok", now a programme of several variants including Object 1K (Vostok 1), the unmanned research and development experimental version; Object 2K (Vostok 2), which would be the unmanned reconnaissance satellite; Object 3K (Vostok 3), the design intended for manned space flights; and Object 4K (Vostok 4), a design concept for an unmanned spacecraft with high resolution photographic capability. By April 1960, the unmanned reconnaissance satellite was renamed Zenit, while the manned variant retained the name Vostok. By this time, the selection process for the first group of cosmonauts was under way.

Security over science?

On 17 December 1958, the US manned satellite programme was renamed Project Mercury. The director of the Space Task Group, Dr. Robert Gilruth, had preferred the name "Project Astronaut", which emphasised the inclusion of a man in the satellite, but this was not chosen due to concerns that the personality of "the astronaut" would be emphasised over the mission he would carry out.[4]

When considering who would be most suitable to become "the astronauts", it is worth recalling the political climate of the late 1950s. President Eisenhower had strongly urged that a civilian American satellite be part of the International Geophysical Year, under the auspices of a civilian space agency, leaving the Department of Defense to focus on the development of long range ballistic missiles. He adhered to a strict division between military projects and civilian "scientific" projects, but by 1958 and the creation of NASA, both the USAF and US Army had (short-lived) plans for their own man-in-space programmes. It seems contradictory that a president who fostered the idea of separating civilian and military space goals to different agencies would also decide that the first American astronauts for the civilian Project Mercury should come from a military test pilot background.

The idea of launching an Earth-orbiting satellite was not regarded by the Eisenhower administration as a politically significant event. Even after the launch of Sputnik 1 and its psychological impact on the American public, Eisenhower stated, "I see nothing at this moment ... that is significant in that development as far as security is concerned."[5] His later comment that, "The Russians have only put one small ball in the air," was explained in his memoirs as part of a desire to relieve the wave of near hysteria that the launch of the Sputnik had caused in the US. By reacting

A Vostok spacecraft similar to those that carried the first six Soviet cosmonauts into orbit. At front is the spherical descent module that housed the cosmonaut, and from which they would be automatically ejected prior to landing.

with alarm, Eisenhower would have undermined confidence in the United States at home and abroad over their failure to pursue a missile and space capability much sooner.

In the two weeks following the launch of Sputnik, scientists held more meetings with Eisenhower than they had done in the previous nine months. In a move to allow scientists more access to national science policy making, the President's Science Advisory Committee (PSAC) was established in November 1957. Though restricted in its activities and influence, seventeen members of the PSAC were selected from the elite of "hard" scientists in the fields of physics, chemistry, engineering and mathematics. They were chosen from members of the National Academy of Science, the Massachusetts Institute of Technology (MIT) Radiation Laboratory and the Los Alamos Scientific Laboratory. This new committee focused on national security matters from the point of view of science and technology and, following the launch of Sputnik, it was tasked by the president with determining the national goals of a future space programme.

As a result, the group promoted science-driven space goals over the research-orientated space agency. At the time, a number of leading scientists were apprehensive that intensive political involvement (or interference) in the programme would drive the national space goals towards "programmes [that were] larger than professionally warranted." Members of the group also indicated their doubts about man in space as a useful scientific instrument and when asked to prioritise the national goals into a timetable of "early", "later" and "still later" time frames, manned orbital flight was the final item on the list in the "still later" category. The group also indicated that it was not wise to focus on space science at the expense of other scientific endeavours and that a balance between space, science and technology was a far more sustainable approach than an all-out diversion of funds to a space programme driven by politics.

After he left office as Eisenhower's first Scientific Advisor, James P. Killian (the president of MIT) stated that "many thoughtful citizens were convinced that the really exciting space discoveries could be accomplished better by instruments than by man." Recognising the Soviets' repeated use of technology and politics for propaganda, he acknowledged that the Soviet space spectaculars were an intended measurement of the index of national strength, but in the long run, it would be a balance of science and technology (not military might) that would sustain a rewarding programme. Killian argued that although America should not be content with being second best, the national space goals should not be politically driven by the Soviets (or indeed, by the US government).

Another to express serious concerns about the national security implications of the Soviets orbiting Sputnik was Senate Majority Leader, and Chairman of the Prepared Subcommittee of the Senate Armed Services Committee, Lyndon B. Johnson. In the period 1957–8, Johnson often identified himself with "space issues", focusing on ensuring America's supremacy in space in both the civilian and military aspects of exploration. This was not only for the benefit of the nation, but also for his own possible nomination in the 1960 Presidential campaign. Conservatism in space policy decisions was evident to the end of Eisenhower's administration, and the signing of the National Aeronautics and Space Act of 1958 that created NASA was, in

Engineers inspect and test a "boilerplate" Mercury spacecraft. The flight (and much modified) version of this vehicle would one day carry six NASA astronauts into space.

part, a comparison between the different views of the president and his would-be successor, Lyndon Johnson. This conservatism was not without its critics, but Eisenhower's "space for peace" plan won out as, in the 1958 Act, there was no clear military justification for placing a man in orbit. However, air force and army planners (and to some extent the US Navy) were formulating plans for manned ballistic vehicles, large space-based outposts and even man-tended bases on the Moon. All of these were immensely costly and, under a budget-conscious administration, never ventured much further than paper studies, while NASA's somewhat primitive manned programme in low Earth orbit received the go-ahead it needed.

Eisenhower had originally intended his final budget message in 1960 to include the statement that there would be no further manned space flight activities beyond Project Mercury which, with the limited objective of sustaining a man in space for at least twenty-four hours, would fulfil all demands for a manned space programme. But at the eleventh hour, NASA administrator T. Keith Glennan and Deputy Administrator Hugh Dryden successfully amended the obvious termination of future manned programmes to a more open statement, indicating that any further manned programmes after Mercury would be based on "further tests and experimentation ... to establish scientific reasons for extending manned space flight."

WHO SHOULD OR COULD FLY?

As the development of both the spacecraft and the launch vehicles was being pursued in the Soviet Union and United States, there were discussions and arguments over which type of person would both cope with and survive space flight. Since there had never been anything to equal this type of endeavour before, the closest approximation was to be found in aeronautical research. As many considered exploring space to be something of an extension to flying, the air force biomedical departments in both countries began to investigate the expected stresses and strains of putting a human on a rocket, blasting him into space, sustaining him while there and, most importantly, bringing him back to Earth alive and in good health.

Requirements for astronaut selection – the USAF approach

"The analysis of staffing requirements ... assumes that the space crewman will be an operator of the spacecraft and will be linked with its control system as a crucial part of a man-machine complex. If he were only a passive occupant of an automatically guided ballistic missile, perhaps as a collector of scientific data, the problem would be somewhat different. However, the difference would be primarily one of emphasis, with the same general framework of reference. In either case it must be assumed that the occupant has been placed on board to permit necessary and useful functions. It seems unlikely that men will be sent aloft in ballistic missiles merely as subjects for physiological and psychological experiments in which animals might be used just as well." So wrote Dr. Saul B. Sells and Major Charles A. Berry (who later became head of medical operations at the NASA Manned Spacecraft Center in Houston) in the 1959 book *Man in Space* (by Kenneth Gants, published by Duell, Sloan and Pearce). The book was a collection of contemporary papers detailing the USAF programme for developing a profile for a spacecraft crew. They quoted[6] detailed proposals for the selection and training of "space pilots" in aptitude and skill requirements; biological, medical and physical requirements; and specific tolerances of anticipated physiological stresses.

Refinements by Sells and Berry a couple of years later supported this view and added that the selection of candidates with "appropriate background and experience may substantially reduce the training and indoctrination program." Thinking along the lines of a pilot flying the vehicle, in the mode of the X-15 or X-20 rather than a ballistic flight where an experimenter performed or gathered science data, the candidates, in the view of Sells and Berry, had to be able to perform a range of functions with alertness, speed and accuracy. Such functions included piloting a high performance ultrasonic aircraft through the atmosphere during boost, controlling it in orbit and re-entry glide, and guiding it to a landing. They would also be responsible for obtaining and interpreting information concerning vehicle operations, cabin-environment conditioning, personnel function and external conditions. Further, the candidates would need to be capable of making rapid and accurate compensations and decisions, anticipating difficulties by advance planning and action, and checking, testing, observing and reporting data concerning the spacecraft, its personnel and the environment.

The authors reasoned that such functions would require personnel with special skills, particularly proficiency and experience as a pilot of high-speed, high-performance jet and rocket aircraft. This would be combined with detailed engineering understanding of the operation and maintenance of power plants and the control and environmental conditioning equipment of the spacecraft, as well as medical and physiological training in human performance and function (with a particular emphasis on survival and efficiency in the spacecraft). Candidates would also require detailed understanding of the operation and maintenance of all communication and (interestingly) all scientific observational equipment, together with a specific understanding of the navigational and astronomical aspects of the mission plan.

The authors concluded: "It would be uneconomical to consider any personnel for (the space) program that are not already highly experienced in high-performance jet or rocket aircraft. Test pilots frequently have, in addition, engineering and scientific training and interest. A requirement for proficiency in high-performance flight and in the engineering, physiological, communication, scientific and navigational skills needed would narrow the selection problem greatly." The article also pointed out that altitude flying necessitates aptitude and performance tests in both physiological training and cabin protection systems, including dealing with life support systems, in-flight emergencies and escape scenarios. Furthermore, assessments of flying proficiency, as well as flying records, would be readily available for military test pilots.

This paper focused on the potential for USAF pilots flying X-20-type vehicles into space in the 1960s. Though such a programme was envisaged and a team of test pilots selected, no such flights took place, but these evaluations did have an influence on the selection of the first NASA astronaut candidates between 1959 and 1966.

Requirements for astronaut selection – the NASA approach

In November 1958, a month after the civilian space agency NASA had been formed, planning to select the first candidates for the Mercury programme began. The first decision to be made concerned the type of candidate that would be most suitable. Experience gained in the manned rocket aircraft research programme, high-altitude balloon and parachute descent projects, and a variety of research by the USAF into aerospace medicine, had defined a number of parameters with regard to the stresses and strains of high-speed flight and deceleration. At the Space Task Group headquarters in the Langley Research Center, Virginia, meetings between the space agency, the military and industry resulted in a selection process that would propose 150 men, from which thirty-six would be chosen for physiological and physical testing. From these, a group of twelve would participate in a nine-month training and evaluation programme, at the end of which six (in the event, seven) would be selected to make flights aboard the Mercury spacecraft.[7]

The Mercury spacecraft was capable of completing a limited and pre-programmed mission on a totally automated basis, but due to the unproven reliability of automated systems over prolonged operations, it also included a totally manual capability. This would allow the astronaut to take over manual control if required and also give him the capability of participating in all aspects of the flight to expand its test and operational envelope. This meant an "active piloting capability" over a totally automated flight programme, and helped the astronauts argue for pilot control of the spacecraft, as they would if they were flying an aircraft. Despite jibes from the test piloting community at Edwards AFB in California (where the rocket research programme was being directed) and media cartoons and tongue-in-cheek comments about them following monkeys into space, the astronauts would provide active input into the development of the spacecraft and its mission during their training and preparations for flight, thus rebuffing the sarcastic theory that they would be just "spam in a can".

With a piloting capability for the vehicle, it was logical that any candidate should have already demonstrated similar capabilities in their career. In fact, the Mercury mission plans identified several tasks that the "astronaut" would have to be capable of performing, including sequence monitoring, systems management, attitude control, and research observations. In addition to these, they would have to be physically fit enough to endure various stressful conditions. As a result, it was decided early in the process that only males between the ages of 25 and 35 (later raised to 39 because too few applicants could meet the other requirements) should be considered, due to the expected high physical stresses that would be encountered during the flight. Equally restricting was the design of the spacecraft, which was limited by the capabilities of the Atlas and Redstone ballistic missile launch vehicles and meant that the applicants could be no taller than 180 cm (5 feet 11 inches).

Further, the candidates considered for astronaut selection in the Mercury programme had to possess a good knowledge of engineering and of operational procedures found typically in aircraft or missiles, and a high degree of intelligence and psychomotor skills that were similar to those required to fly high-performance aircraft. They had to demonstrate an above-average tolerance to stress and an ability to make decisions and to work with others, showing both emotional maturity and a strong motivation towards team objectives. A clean medical record was essential, along with a tolerance for physiological stress. Although a high educational degree was not required (few pilots had attained one), the very nature of the programme in investigating new fields of science, engineering and technology meant that the candidates had to demonstrate sound general scientific knowledge and research skills.

Using US government employment procedures, the selection was based on the technical requirements of the Mercury programme and the applicant's personal qualifications and experiences, so it was not possible to identify ethnic origin or gender. However, in the 1950s, there were very few black or female jet pilots, let alone test pilots.

An aeromedical team, comprising leading air force medical officials and representatives from the Space Task Group, NASA Headquarters and the Special Committee on Life Sciences, established a list of duties that the astronaut might be expected to perform during the mission. Principally, he had to be able to survive and return (always useful!) and to clearly demonstrate man's capability to be launched into space by rocket. He would need to withstand the crushing acceleration forces

10 The Wrong Stuff

associated with lift-off, fly in orbit and perform simple tasks under the potentially disorientating conditions of weightlessness. He would also function as the back-up to the automated systems, offering further redundancy for mission safety and success and, drawing upon his previous talents, operate as an engineering observer, "as well as a true test pilot to improve flight systems and its components." A significant inclusion was that the astronaut must be able "to serve as the scientific observer ... to go beyond what instruments and satellites can observe and report."

The next task was to identify from where such experience could be obtained. The categories considered by the Space Task Group were aircraft pilots, balloonists, submariners, deep sea divers, scuba divers, mountain climbers, Arctic and Antarctic explorers, flight surgeons and scientists (including physicists, astronomers and meteorologists).[8] Candidates with flying experience were considered, as were those with three years' work in appropriate sciences, holders of a PhD or medical degree, or those who could demonstrate three years' experience as an aircraft, balloon or submarine commander, pilot, navigator or communication officer, plus engineers or those with comparable technical positions. All this demonstrated the desire to

America's Mercury astronauts. From left, in their standard "CCGGSSS" alphabetical order: Navy Lt. M. Scott Carpenter, USAF Capt. L. Gordon Cooper, Jr., Marine Lt. Col. John H. Glenn, Jr., USAF Capt. Virgil I. (Gus) Grissom, Navy Lt. Cdr. Walter M. Schirra, Jr., Navy Lt. Cdr. Alan B. Shepard, Jr., USAF Capt. Donald (Deke) Slayton.

offer the chance of flying into space to the broadest, yet most qualified range of individuals available. However, time constraints imposed by the programme meant that a protracted selection process was impractical, and the selection board reasoned that test pilots would already meet most of the criteria, particularly graduates of the USAF or Navy test pilot schools.

Their reasoning was sound: the test pilot group would be already familiar with stress levels, forces of acceleration and deceleration, reduction of pressure, vertigo, and other flying phenomena on a daily basis. They would be familiar with partial or full pressure suits, and have training for (or experience in) emergency situations, parachute training, and in some cases, combat situations. Being military pilots, their education and security clearance would already be appropriate, while retaining their active flight status required mandatory regular physical examinations and sustained good health. In December 1958, in a meeting with President Eisenhower, the idea of restricting the selection of the nation's first astronauts to military test pilots was suggested, which went against the president's push for a "civilian space agency". However, Eisenhower agreed. Years later, Dr. Robert Gilruth, Director of the Space Task Group and subsequently first Director of the Manned (Johnson) Spacecraft Center, admitted that the decision to recruit military test pilots was one of the best that Eisenhower ever made. Test pilots already screened for security (in a programme where certain national and defence "secrets" would be part of their daily working environment) favoured neither a purely military nor solely civilian space programme, but a blend of both.

The final selection process for the first group of NASA astronauts is beyond the scope and direction of this book, but once the decision was made to restrict selection to military pilot astronauts, any possibility of considering candidates with a principally scientific background was quietly abandoned. The section of the first seven Mercury astronauts in April 1959, the subsequent media coverage, and their spellbinding exploits, quickly gave rise to the sublime myth of "the astronaut programme." Extending the test piloting "Right Stuff" beyond the atmosphere, the professional elite of the flying profession – the test pilots – also supported the pioneering notion of the early manned flights into space, testing the prototype vehicle so that later spacecraft could fully explore the new environment.

With the development of the Gemini programme to gain operational experience of techniques in preparation for the Apollo programme, the criteria of group selections in 1962, 1963 and 1966 were influenced by flying credentials rather than scientific capabilities, underlining the "testing" nature of the programme. This persisted for the initial missions leading up to and achieving the first landing on the Moon. Even in the parallel military programme, piloting skills were paramount in the selection of crew members for the rocket research planes, such as the X-20 and the USAF Manned Orbiting Laboratory (MOL) programme. When that programme was cancelled in 1969 without any manned flights taking place, the MOL astronauts who transferred to NASA primarily had a flying career background, not one involving research or experimentation. The exception was Don Peterson, who had gained a Master's degree in nuclear engineering before working towards a PhD. In the USAF, he had served as a nuclear systems analyst as well as being a pilot.

12 The Wrong Stuff

It would require a change of policy in both NASA and the government to allow the selection of more scientifically-trained astronaut candidates, rather than applying the standard military piloting criterion, which was also somewhat governed by the environment of the "Cold War".

The first cosmonauts

While the USAF was conducting research into high-speed and extreme-altitude flight in the 1950s, creating baseline data for future manned flights into space, a comparable programme of biomedical research supporting strategic and military goals (but with potential for space exploration) was taking place in the Soviet Union. For the Soviets, the choice of exactly who would fly the first pioneering mission in space was not dissimilar to that faced by their American counterparts, coming down to aviators, submariners, polar explorers, parachutists or mountaineers. Eventually, and for essentially the same reasons as the Americans (including military security clearance), the selection of the first Soviet cosmonauts would fall to jet pilots. One additional qualification specified was a proficiency in parachuting, a requirement for the Vostok programme since the cosmonauts would eject from their spacecraft as it descended under parachute to Earth at the end of the mission.

Selection criteria were also similar to those of the American Mercury astronauts, although the basic requirements were lower for the Soviets. Candidates had to be qualified to jet pilot third class (basic operational flying criteria in the Soviet Union) and aged under 30, with height (170 cm) and weight (70 kg) defined by the limitations of the spacecraft and launch vehicle. Their flying experience was much less than that required of the American candidates, and was set between 200 and 1,000 hours flying time, not necessarily all in jets (as opposed to 1,500 hours in high-performance jets for the Americans). One significant factor in these parameters was that Soviet rocketry at that time focused more on automated systems than those of the Americans. When asked about the number of candidates required, Sergey Korolyov suggested that three times that of the Americans would be required, indicating a group of around twenty to twenty-four. When the first medical screening produced insufficient numbers of qualified candidates, the age and height restrictions were relaxed even more to include a number of older, more experienced pilots with greater flying skills (though still not test piloting skills), as well as engineering experience. The first group of Soviet cosmonauts – all pilots – was selected on 20 February 1960, almost a year after the American Mercury astronauts.[9]

Again, technical parameters, engineering test objectives and the unknown factors of human space flight were paramount in defining who would be the "first cosmonauts" and, as with the Americans, these set the criteria for future Air Force pilot selections for many years (although some Air Force groups did include a number of candidates with other specialities as well as flying experience, as indeed did the later NASA pilot selections of the 1960s).

Pilot-astronauts not scientist-astronauts

The reasoning behind not selecting scientists for the first trips into space was straightforward. It was clear that the number of unknowns were far greater on these pioneering missions, and while there might be a whole universe waiting for dedicated scientific research, it first had to be proven that man could indeed leave Earth, survive in space, and make it back in one piece. It had been demonstrated many times while testing new aircraft that one tiny error or a minute failure in the hardware could result in a catastrophic incident and even the death of the occupant. Several times while rocket research was still in its infancy in the 1950s, a successful ignition or launch of an unmanned vehicle was suddenly followed by a massive explosion, or the complete break-up of the booster and payload. Test pilots accepted the risks and trained hard to overcome any such failures. Astronaut Virgil (Gus) Grissom is often quoted as saying that space was a risky business, but that the exploration was worth the risk, even if this meant the death of some of the explorers. His words became even more poignant following his death in a launch pad fire during training in January 1967.

With the early programmes focused on engineering tests, development and evaluation, any science on manned spacecraft would have to take a back seat to safety and mission success. Although some science would eventually be conducted on these early missions, the inclusion of a science-trained crew member would have to wait until enough operational experience had been built up, space flight had been proven to be survivable, and larger and more versatile vehicles with more complex objectives became available.

SCIENCE AND MANNED SPACE FLIGHT

In late October 1958, an American Special Committee on Space Technology made several recommendations to the newly-created NASA regarding a civilian space programme.[10] These included the observation: "The major objectives of a civil space research program are scientific research in the physical and life sciences, advancement of space flight technology, development of manned space flight capability and exploitation of space flight for human benefits. Inherent in the achievement of these objectives is the development and unification of new scientific concepts of unforeseeable broad impact." In reviewing the prospects for manned space flight, the document added: "Instruments for the collection and transmission of data on the space environment have been designed and put into orbit about the Earth. However, man has the capability of correlating disparate events and unexpected observations, a capacity for overall evaluation of situations, and the background knowledge and experience to apply judgement that cannot be provided by instruments; and in many other ways the intellectual functions of man are a necessary complement to the observing and recording functions of complicated instrument systems. Furthermore, man is capable of voice communications for sending detailed descriptions and receiving information whereby the concerted judgment of others may be brought to bear on unforeseen problems that may arise during flight."

The 1958 document also included this comment: "Although it is believed that a manned satellite is not necessary for the collection of environmental data in the vicinity of the Earth, exploration of the solar system in a sophisticated way will require a human crew."

NASA's long-term planning 1959–64

A year later, on 16 December 1959, NASA issued its first ten-year plan, which included the long-term goal of manned exploration of the Moon and nearby planets. It also predicated the first launch of an astronaut in sub-orbital flight (Mercury-Redstone) in 1960, attainment of manned orbital flight (Mercury-Atlas) between 1961and 1962 and the first launches in a programme leading to manned circumlunar flight and a permanent near-Earth space station (1965–7). Then, after 1970, manned flight(s) to the surface of the Moon.[11]

By December 1960, NASA was submitting its request for funding in the 1962 fiscal year, and revealed for the first time the Apollo programme that would follow Mercury to attain at least circumlunar flight, if not develop the technologies and experience to land men on the Moon. The report also indicated the use of a Saturn C-2 launch vehicle (capable of launching 18,144 kg into low Earth orbit), utilising this capability to orbit a small laboratory that would stay aloft "for two weeks or more." Though smaller, instrumented craft could acquire more scientific information using lesser launch vehicles, the larger laboratory "might be of value as a life science laboratory to acquire physiological and psychological data on humans, to study life support mechanisms, to perform biological studies and to carry out engineering tests under gravity free conditions."

For increasingly complex Apollo missions deeper into space, the family of Saturn launch vehicles would be superseded by the larger Nova rockets with a nuclear rocket stage, to support manned lunar surface exploration and, it was hoped, initial manned planetary expeditions in the 1970s.

The argument about the degree to which man adds an element of variety and quality to scientific observations and attainments over fully automated systems was also presented. Finally, the report stated: "Though man in space could not be justified purely on scientific grounds, most of the motivation and drive for the exploration of space comes from the dream of man getting into space himself."

The following month, an *ad hoc* Committee on Space reported to the president elect (John F. Kennedy) on scientific and technical issues relating to the nation's space programme.[12] This report reflected widespread scepticism within the scientific community over the value, and at that time the feasibility, of human space flight. It stated that, in planning space activities, scientific objectives must be assigned a prominent place and that the greatest wisdom and foresight should be applied to "the selection of the scientific missions and the scientists assigned to carry them out." This was not necessarily advocating scientist-astronauts, but rather principal investigators. However, it could be interpreted to suggest that the best people to perform science in space were not necessarily pilot-astronauts, but career scientists. The report went on to state, "NASA has not fulfilled all of the ... requirements (in the scientific exploration of

Chief designer Sergey Korolyov (left) talks with a beaming Yuri Gagarin, the first person to fly into space.

space) satisfactorily. The main obstacle here has been the lack of a strong scientific personality in the top echelons of its organisation." In its assessment of the pending entry of man into space (via Project Mercury), the report predicted that, "Some day it may be possible for men in space to accomplish important scientific or technical tasks. For the time being, however, it appears that space exploration must rely on unmanned vehicles. Therefore, a crash program aimed a placing a man into orbit at the earliest possible time cannot be justified solely on scientific or technical grounds. Indeed, it may hinder the development of our scientific and technical program."

This argument continued in the wake of the 12 April 1961 launch of Yuri Gagarin, and in President Kennedy's proclamation to send an American to the Moon "within a

16 The Wrong Stuff

On 25 May 1961, US President John F. Kennedy delivered his Special Address to Congress on Urgent National Needs, committing the nation to a manned lunar landing by the end of the decade. Vice President Lyndon B. Johnson (left background) looks on.

decade", science was not mentioned. In a memo for the vice president (Lyndon B. Johnson) dated 20 April 1961, Kennedy asked where America stood in space and whether a laboratory in space or a manned lunar flight could be achieved before the Soviets.

In his 28 April reply, Johnson, referring to CIA-provided intelligence estimates, stated that the Soviet Union had demonstrated the launch capability for placing a space laboratory in orbit and for placing payloads on the Moon. Based on their success at launching Gagarin ahead of the first American astronauts, their capacity for sending men to the Moon was plausible, but with extra effort, Americans could be the first to touch down on the lunar surface. The following day, Wernher von Braun sent a memo to Johnson, stating that America had an excellent chance of beating the Soviet Union to the Moon. However, he reasoned that the probability of the Soviets launching a space laboratory ahead of the Americans was strong, until the planned Saturn C-1 launch vehicle was debuted in 1964, at which point several astronauts could be launched into orbit simultaneously in an enlarged capsule "that could serve as a small 'laboratory' in space."[13]

Less than a month later, the president made his historic, seminal speech before Congress, committing America to the Moon landing programme. The arms race that evolved into a space race had now clearly turned into a Moon race, and science on the Moon played no part in this massive undertaking. It was simply a race to get American astronauts there and back ahead of the Soviet Union, and the science would only follow once this primary goal had been achieved. As there was still so much to develop and understand about space flight beyond Project Mercury, the ensuing Gemini and early Apollo programmes leaned more towards test programmes than scientific explorations and this was reflected in the engineer, test and jet pilot candidates of the 1962 and 1963 NASA selections.

As plans for the lunar programme developed, ideas for future objectives in the civilian space programme began to emerge.[14] Some of the major capabilities already existing or under development during the compilation of the Future Programs Task Group report (1964) included manned flight in Earth orbit for between one and two weeks (Gemini/Apollo Block I), manoeuvring and rendezvous (Gemini/Apollo Block II), and lunar orbit, landing and return (Apollo Block II). Additionally, NASA was looking at future applications involving Apollo-type hardware in Earth orbit for between one and two months; operations with crew in equilateral, polar and synchronised orbits; rendezvous inspection, repair and rescue; lunar mapping; and extended stays on the surface of the Moon of between three and fourteen days. All these fell under what became the Apollo Applications Program (known variously as Apollo A, Apollo X, Apollo AES, and AAP).[15] In pursuit of these goals, a range of manned Earth-orbital experiments was proposed. This covered such fields as biosciences, physical sciences and astronomy/astrophysics. There were also Earth-orientated applications, including atmospheric science and technology, and communications. Support for space operations included advanced technology and subsystems, operational techniques and subsystems, and biomedical and behavioural investigations.

The long-term development that the agency was planning under manned space exploration programmes included conceptual take-off and landing of space vehicles (such as a reusable Space Shuttle) and flexible Earth-orbital operations. Other concepts included a large permanent space laboratory, roving lunar vehicles and the creation of research bases on the Moon, and manned planetary exploration.

In a packed programme ...

In early 1963, the Gemini programme was manifested for ten or eleven manned flights (it evolved into ten), while early Apollo mission planning predicted four manned Earth-orbital missions launched by the Saturn 1 beginning in 1965; two to four manned Apollo Earth-orbital missions with the Saturn 1B from 1966; and at least six manned Earth-orbital and lunar-orbital flights with the Saturn V commencing in 1967. These would all lead to the first landing missions (two to six) between 1968 and 1969 and were designed to achieve the primary goal of landing Americans on the Moon before 1970. None of these missions were dedicated to scientific exploration of

Wernher von Braun, whose Saturn family of rockets took America to the Moon and launched its first space station, but which – unlike earlier plans – only carried four scientist-astronauts into space.

the Moon, which fell under the mandate of the Apollo Applications Program, or AAP.

Early plans for AAP in 1965 predicted three Saturn 1B "wet workshop" (an expended fuelled stage) missions, three Saturn V "dry workshop" (an unfuelled stage) launches, and four independent flights of the Apollo Telescope Mount (ATM), with at least two or three Saturn 1B-launched missions transporting three astronauts to each workshop. This equated to around eighteen to twenty manned flights, which would be added to the proposed ten to eighteen manned Apollo missions assigned to the lunar landing programme, and suggested extended Apollo lunar landing missions after the initial goal had been achieved. This would be a launch manifest of some fifty manned missions (and 140 flight seats) between 1965 and 1975 under the Gemini, Apollo and AAP programmes.

By 1964, some of the Apollo Earth-orbital missions had been cancelled, due to a change from unmanned stage-by-stage testing to "all-up testing", eliminating hardware that would not be used in the manned lunar missions. Unflown Mercury astronaut and interim Director of Flight Crew Operations Donald (Deke) Slayton, in planning crews for the missions leading to the Moon, foresaw the need for ten crews for Gemini and eight for Apollo *prior* to the first landing attempt. As Slayton explained in his 1994 biography, "My mission was to create a pool of guys who had the necessary experience in rendezvous and docking, EVA and long duration, before I had to select which three would attempt the first landing."[16] In the summer of 1964, there were twenty-six astronauts available for assignment (twenty-eight if the medically grounded Alan Shepard and Slayton were counted, as they hoped to be restored to flight status in time for Apollo). This was more than enough for Gemini and the early Apollo missions leading up to the first landing, but with increasing murmurs of discontent coming from within the scientific community, the pressure for selecting scientists as astronauts was growing.

"I didn't have anything against scientists, or doctors," Slayton wrote, "but I wasn't quite sure what I was supposed to do with them on flight crews." In his mind, it took two or three astronauts to get the Apollo spacecraft to where it was planned to be, maintain it in space and make sure it returned home safely. "There was no room or requirement for what would basically be a 'passenger', in other words, a scientist flying to operate experiments and not fly the vehicle. If something goes wrong with the spacecraft, you need to come home quickly, and every member of the crew would be required to 'fly' the spacecraft." Slayton's point would be very clearly demonstrated during the life-threatening Apollo 13 incident in 1970.

With the gradual increase in "post-Apollo" operations under Apollo Applications, however, the need for additional astronauts *after* the first landing had been achieved was clear. Therefore, the crewing requirements for the proposed AAP programme had to be planned. Allowing time for recruitment, selection and training and a period of technical support roles prior to their first flights, astronauts selected in 1965–7 would probably not get to fly before 1968–9 at the earliest. In fact for some it turned out to be 1970, with a number waiting between sixteen and nineteen years to fly their first space mission. Several of the earlier astronauts would be retiring, but with more flights it was clear that pilot-astronauts would still be required, as well as the first scientist-astronauts. However, as they would be flying on Apollo-type spacecraft for some years to come, any candidate without flying experience would have to undergo the USAF jet pilot training course before any astronaut assignment could be considered.

SCIENCE AND MANNED ORBITAL SPACE FLIGHT 1961–76

With the development of manned space flight came the opportunity to assign small, simple experiments to the missions, as secondary objectives to the main purpose of the mission. Flying experiments on the early spacecraft was limited by available volume, the lifting capability of the launch vehicle, restrictions on electrical power, data recording systems, manoeuvring capabilities, and the limited time available to the crew to activate or operate the experiment and to record the data they collected. Any experiment, successful or not, could not compromise the mission or the integrity of the spacecraft. Nor could it endanger the lives or health of any crew member.

Vostok & Voskhod: The primary objective of the Soviet Union's first manned space flight programmes was to develop the infrastructure and experience of orbiting cosmonauts in spacecraft of increasing capabilities and capacity, sustaining them for up to three weeks and returning them safely to Earth. As with the American pioneering space programme, there were limitations to the Soviet spacecraft that precluded extensive scientific investigations, other than very basic biomedical studies on the crew, visual observations of the Earth and Moon, and astronomical studies. Radiation measurements were taken and biological specimens were carried to support the data gathered on the biomedical parameters of each cosmonaut. When Dr. Boris Yegorov flew on the first manned Voskhod mission, it allowed a professionally trained physician to perform limited research on himself and his two colleagues, albeit on a flight that lasted barely twenty-four hours.[17]

Mercury: As Project Mercury evolved, a growing number of people within both NASA and the American scientific community requested that scientific experiments be flown, as part of the overall national space science programme. In addition to biomedical studies on the human crew member in space, a range of Earth observations, photographic, radiation detection, and technology studies were devised for the short-duration Mercury orbital missions. The earlier sub-orbital missions, lasting just over fifteen minutes, were far too concerned with engineering tests and qualification profiles (and astronaut safety) to warrant adding science, apart from basic medical and visual observations.

Seventeen experiments (excluding biomedical studies) were assigned to the four orbital missions, with several flying on more than one mission. As the flights increased in duration, so did the number of scientific tasks assigned to them.[18] Experiments for MA-6 (Glenn) were photographic in nature, but in April 1962, the Mercury Scientific Experiment Panel (MSEP) was formed. It included representatives from different branches of the Mercury programme and the NASA space science programme, whose aim was to develop suitable experiments, nominate them for flight and prioritise which would be flown on MA-7 (initially assigned to Slayton, then Carpenter) and MA-8 (Schirra). In October 1962, the MSEP was replaced by the In-Flight Experiments Panel (IFEP), which continued the role of the MSEP for MA-9 (Cooper) and into early experiment planning for Project Gemini and Project Apollo.

Gemini: With the extended duration capability of the Gemini spacecraft, its enlarged volume, manoeuvring capability and crew of two, an expanded experiment programme was included in the Gemini programme from 1963. Fifty-four experiments were flown under the categories of medical, engineering, Department of Defense, and scientific studies. Those assigned to the first five manned flights (where the primary objectives were manned certification, EVA, rendezvous and docking, and extending the duration) were Category B experiments, or secondary objectives. Should they potentially impede a launch, they could be easily removed from the mission without serious consequences. It was important in planning and assigning these experiments that serious consideration be given to the amount of time that a crew would have direct contact with the hardware or be able to retrieve the result data. One of these experiments (assigned to the final Gemini missions GT-10, -11 and -12) was the S13 UV Astronomical Camera, whose principal investigator was Dr. Karl Henize of Dearborn and Northwestern Universities. He was subsequently selected as one of the second group of scientist-astronauts in 1967, and had briefed the three Gemini crews on his experiment during their preparations for flight.[19]

Apollo: The primary objective of landing a man on the Moon remained the driving force behind Apollo through to 1969, but in addition to qualifying the hardware, it was recognised that some missions could also include scientific experiments that were not directly related to the lunar exploration programme. With the original Block I missions of Apollo 1 and Apollo 2 each planned for up to fourteen days, several on-board experiments were assigned in addition to the primary goal of checking the spacecraft and its systems for the first time with astronauts aboard. Apollo 2 offered the opportunity to fly repeat experiments, or to re-fly an experiment that had failed during Apollo 1. Nine medical, two scientific and one technology experiment were assigned to Apollo 1, but after a review of the science package, four of the medical and the single technology experiment were found to be unsuitable and were dropped (with an alternative medical investigation added instead). Apollo 2 was manifested for fourteen experiments; eight medical and six scientific.

The details of these experiments and the saga of the Apollo 1 and 2 science programme and missions have already been covered in this series,[20] but it is worth recalling the CB (Astronaut Office – the pilots) concerns about flying too many science experiments, in addition to the extensive engineering tests planned for both missions, on what were essentially the maiden flights of these vehicles. The ever-increasing experiment load on Apollo 2 was of greatest concern. It has been stated that Grissom wanted to delete anything non-engineering from Apollo 1 and lump it instead onto Apollo 2, but that flight's commander (Schirra) was, if anything, famously more anti-science than Grissom, even though he did complete a range of science experiments on his Mercury mission. Apollo 2 was cancelled in November 1966 and after the Apollo 1 pad fire of January 1967, Block I missions were terminated, allowing the programme to proceed to Block II (Apollo lunar mission capability). Apollo 7, to which Schirra's crew had been reassigned after the launch pad tragedy, was a successful 1968 flight which qualified the Command and Service Module in Earth orbit, but carried few "experiments" other than engineering tasks. Its five experiments included two on Earth terrain and weather photography, and three medical experiments that required no in-flight crew activity.

Photography continued to be a major objective on all Apollo missions and this included further Earth observation photographic experiments on Apollo 9. During Apollo 8 and 10–13 and the challenge to achieve the first lunar landings, the major objective was to get to the Moon in proven spacecraft, complete the mission and return safely. Little thought was therefore given to science outside of lunar operations, photography and surface activities. But with Apollo 14–17, the opportunity arose to fly several experiments to investigate the phenomena of microgravity during translunar and trans-Earth coasts. These experiments would serve as precursors for more extensive studies on the Skylab space station, where some of the first scientist-astronauts (already in training) would conduct extensive studies in a space laboratory converted from leftover Apollo hardware.

Soyuz: This programme has been the mainstay of Soviet/Russian manned space flight since the mid-1960s.[21] For the initial manned flights, the techniques of rendezvous, docking and crew transfer were developed alongside the qualification of the spacecraft. Part of this development was also connected to the manned lunar programme (Zond/N1/N3), which should have competed with America's Apollo programme in the race to the Moon but which included no manned flight activities prior to its cancellation in 1974. Since 1971, Soyuz (meaning "union") has been used as a ferry craft carrying crew and cargo payloads to and from a series of space stations. Between 1969 and 1976, it supported a small series of solo flights carrying a range of experiments and research, including:

Soyuz 6	1969 Oct	First experiments in space welding
Soyuz 9	1970 Jun	Medical studies during an extended duration flight of 18 days
Soyuz 13	1973 Dec	Astrophysical and biological research
Soyuz 16	1974 Dec	Apollo-Soyuz Test Project (ASTP) dress rehearsal mission
Soyuz 19	1975 Jul	ASTP docking mission with American Apollo
Soyuz 22	1976 Sep	Earth resources observations.

Though these flights and experiments were directly related to the ensuing Soviet manned space station programme known as Salyut ("salute"), there were no Soviet scientists among the crews. A small corps of scientists from the Academy of Sciences, not unlike the selection of NASA's scientist-astronauts, was considered for inclusion on space station missions for a short period. Unfortunately, none would ever make it into space.

Salyut, Skylab and Spacelab – orbital research labs for scientists?

After losing the Moon race, the Soviet Union turned its attention to creating a series of manned space stations, starting with Salyut and its military version called Almaz (1971–91), their successor Mir (1986–2001) and from 2000, parts of the International Space Station (ISS). The Americans used leftover Apollo hardware to support the Skylab space station (1973–9) and its three visiting crews, each of which featured a scientist-astronaut. Though other Skylab stations and even larger orbiting space laboratories were planned, nothing came to fruition until the first elements of the ISS were transported into orbit and assembled in the 1990s. Meanwhile the Space Shuttle has only provided support for short-term space research missions, including the Spacelab series. A number of NASA astronauts also flew to Mir for long duration missions (1995–8) and since 2000, American astronauts and Russian cosmonauts have flown to the ISS as long-duration resident crew members. However very few of these spacefarers can be classified as true "scientists".

The story of professional scientists flying in space in either the American or Russian programmes has been one of trial and frustration, success and achievement. More than fifty years on, the early dreams of teams of scientists working on research bases in space, the Moon or at Mars, as envisaged by von Braun in 1952, have yet to be fully realised.

REFERENCES

1. "Man on the Moon: The Journey," *Colliers' Magazine*, 18 October 1952, pp. 52–59.
2. *Project Mercury: A Chronology*, NASA SP-4001, 1963, pp. 1–27.
3. *Challenge to Apollo*, Asif Siddiqi, NASA SP-2000-4408, 2000, pp. 119–195.
4. *Origins of NASA Names*, NASA SP-4402, Helen T. Wells, Susan H. Whiteley, and Carrie E. Karegeannes, pp. 106–109, NASA 1976.
5. *The Real Stuff: A History of NASA's Astronaut Recruitment Program*, Joseph P. Atkinson Jr. and Jay M. Shafritz, Prager Publishers 1985, pp. 18–28.
6. Beyer D.H. and Sells S.B.: Selection and training of personnel for space flight, *Journal of Aviation Medicine*, **28**: 1–6, 1957.
7. NASA Astronauts, by David J. Shayler, in *Who's Who in Space*, International Space Station Edition, by Michael Cassutt, Macmillan 1999, pp. 1–17.
8. Reference 5, p. 32.
9. *The Rocket Men*, Rex Hall and David J. Shayler, Springer-Praxis 2001, pp. 102–110; *Russia's Cosmonauts*, Rex Hall, David J. Shayler, Bert Vis, Springer-Praxis 2005, pp. 120–124.
10. *Exploring the Unknown*, Selected Documents in the History of the US Civil Space Program, Vol. 1 Organising for Exploration, pp. 394–403, NASA SP-4407.
11. *The Long Range Plan of the National Aeronautics and Space Administration*, 16 December 1959, as summarised in Exploring the Unknown Vol. 1, previously cited, pp. 403–407.
12. *Report to the President Elect from the Ad Hoc Committee on Space*, 10 January 1961, as summarised in *Exploring the Unknown* Vol. 1, previously cited, pp. 416–423.
13. Reference 10, pp. 423–454.
14. *Exploring the Unknown*, previously cited; NASA Summary Report, Future Programs Task Group, January 1965, pp. 473–489.
15. *Apollo: The Lost and Forgotten Missions*, David J. Shayler, Springer-Praxis, 2002.
16. *Deke! U.S. Manned Space: From Mercury to the Shuttle*, Donald K. "Deke" Slayton with Michael Cassutt, Forge Books, New York, 1994, p. 136.
17. *The Rocket Men*, previously cited.
18. *Project Mercury, NASA's First Manned Space Programme*, John Catchpole, Springer-Praxis, 2001, pp. 391–401.
19. *Gemini: Steps to the Moon*, David Shayler, Springer-Praxis, 2001.
20. Reference 15, pp. 111–153.
21. *Soyuz: A Universal Spacecraft*, Rex Hall and David J. Shayler, Springer-Praxis, 2003.

2

Scientists as Astronauts

"There is of course a certain risk to every space flight, specifically to the risky test flights of new craft. Mankind has had to pay dearly, not infrequently losing its best sons, for many of the achievements which have contributed to progress. Movement along the path of progress is unstoppable. Others will carry on the relay race of scientific success and go on further, true to the memories of their comrades." Yuri Gagarin[1]

On 12 April 1961, Yuri Gagarin became the first man to fly into space. A pilot, but not a test pilot, he voiced eloquent impressions of the view of Earth from space and though he never flew in space again (tragically being killed in an aircraft accident in 1968 during qualification to fly a new space mission), he was supportive of the cosmonaut profession. To him, this included the scientists and other specialists as well as the pilots. Though his own mission was essentially a "short" test flight into space, Gagarin recognised the importance that a space platform crewed by specialists would have in expanding our understanding of the cosmos. The early flights were accomplished by test pilots, but Gagarin knew that the era of scientists on space stations would come in time:

"We shall build a large orbital space station, a scientific research station, in order to be able to study outer space. It will make it possible for us to do different kinds of research, to study outer space, and to carry out experiments which we cannot do on Earth because certain conditions can only be achieved in space."[2]

After his flight, Gagarin became more interested in space research and science and often stated that "further space research would be impossible without cosmonaut scientists."[3]

Though the search for the first Soviet cosmonauts was limited to Air Force pilots, Korolyov, too, foresaw a time when civilian engineers and scientists would fly into space. Efforts to allow civilians to apply for cosmonaut training finally succeeded in

September 1961, when the medical commission authorised civilian applications, although nothing much would happen for a few years. In the United States at this time, the sub-orbital flights of Alan Shepard and Virgil Grissom had placed Americans "in space", but not into orbit. In light of Gagarin's success and his own frustrations with the situation in Cuba, President John F. Kennedy needed a challenging new goal to inspire the nation. As previously noted, his choice of the Moon landing programme made no provision for lunar science, but that goal was already being investigated by the scientific community in the United States.

AN ESSENTIAL PART OF FUTURE EXPLORATION

Between 1958 and 1961, the Space Sciences Board of the US National Academy of Sciences conducted a study into the scientific aspects of space exploration, including the role of man. A formal position was adopted during their 10–11 February 1961 meeting, and their recommendations were submitted to the Government on 31 March. A formal NASA news release on man's role in the US national space programme was released in August 1961, three months after President Kennedy's lunar landing commitment speech.[4]

The board recommended that: "Scientific exploration of the Moon and planets should be clearly stated as the ultimate objective of the US space program for the foreseeable future." Although the board also recognised that it was far too early (in 1961) to determine whether a human crew would be part of any early expedition to the Moon and planets, as many intermediate problems remained to be solved, they strongly emphasised that planning for subsequent scientific explorations should be developed on the premise that humans would be included. The board expressed little doubt that humans would be an essential part of future solar system exploration. Neither did they foresee that, when it became technically feasible to include astronauts on such flights, their judgement and discrimination in conducting scientific investigations of deep space would ever be superseded by instruments alone, no matter how sophisticated or complex they became. But they did feel that any expedition to the Moon or deeper into the solar system would require careful and detailed planning to ensure mission success over such vast distances. They also stressed that, in addition to establishing a clear and precise programme of objectives and goals, a new fleet of launch vehicles and spacecraft would have to be developed.

Perhaps most significantly, however, the board also felt that "Consideration should be given soon to the training of scientific specialists for spacecraft flight, so that they can conduct or accompany manned expeditions to the Moon and planets."

On 27 February 1962, just seven days after John Glenn had become the first American to orbit the Earth and six months after cosmonaut Gherman Titov had spent a full day in space, Professor James Van Allen testified before the House Committee on Appropriations with regard to the relative worth of instrumented unmanned satellites versus manned spacecraft. Unfortunately, he was misquoted in the media, to the effect of implying that the scientific value of manned space flight would be limited in the near future. What he actually stated was that a "man-in-space" programme would not be essential for scientific space exploration: "For the same investment of effort, we learn much more without the man ... A monkey made the first orbital trip and [he] made out alright."[5]

It was clear that not all scientists were in agreement that manned space missions were essential to space science. It is an argument that has continued ever since. To help clarify the situation, the Space Science Board of the NAS conducted, at the request of NASA, a study of the space agency's science programme between 17 June and 31 July 1962, at the State University of Iowa. This study would evaluate both the civilian NASA programme and the current and expected DoD programme, and would allow scientists to express their concerns to the agency's officials and to their peers. The study was aimed at maximising science within the programme, and the debate was not whether there should be a space programme or a space science programme (as that had already been decided over the previous few years), but to develop the science as a whole, whatever the programme. Since Apollo's brief was to meet "important national objectives" (specifically, beating the Russians to the Moon), it seemed logical that the scientific community should take advantage of this unique opportunity while ensuring that any scientific work conducted during Apollo was significant.[6]

In summary, the study suggested that scientist-astronauts should be included in the manned space flight programme. The board's most significant recommendation was that a scientist-astronaut should be included on the *first* Apollo landing crew (something that NASA, and certainly the astronaut office at Houston, was not likely to accept on the grounds of safety). It was also stressed that an institute for training scientist-astronauts should be created, administered by a university or by the department of NASA that was responsible for space science programmes. This again was something that NASA would not agree to.

During a news conference at MIT on 14 November 1962, Professor Van Allen reviewed the suggestions made by the panel of scientists, which had met during July and August 1962 to assess the NASA programmes. They suggested that scientists with NASA astronaut training should accompany pilot-astronauts on early lunar and planetary flights. The group also proposed establishing "an academy to train scientist-astronaut volunteers, located at Houston in Texas." The following month, on 26 December, Dr. Homer E. Newell, NASA Director of Space Sciences, made a speech before the American Association for the Advancement of Science (AAAS) suggesting that scientists should be among the next intake of NASA astronauts: "I have complete and utter conviction that we should take a scientist and make a flyer out of him, rather than the other way around."[7]

Deke Slayton, however, has been reported to have suggested that it would be far easier to train pilots to pick up rocks than to teach a scientist to fly!

Under careful study

Following the 1962 summer study, the drive to recruit scientists into the astronaut programme began to gather pace. In a letter dated 16 October 1962 from NASA's Deputy Administrator Hugh Dryden to Evan H. Walker of the University of Maryland, it was stated that the subject of scientists making flights to the Moon to carry out

A US Geological Survey photo showing Eugene Shoemaker with the first group of scientist-astronauts during a training trip to Meteor Crater (*http://astrogeology.usgs.gov*).

their own research, as well as participating in the scientific objectives of the Apollo programme, "has been under careful study by a NASA committee."[8] The letter also stated that, following the recommendation for a broad programme of scientific research and the direct participation of scientists in surface explorations, NASA was working towards obtaining the maximum scientific return possible from the Apollo missions. Dr. Eugene M. Shoemaker was brought into NASA in order to coordinate the planning of scientific research, including that of the manned lunar programme. Shoemaker, a geologist from the US Geological Survey, was also helping pilot-astronauts at MSC to prepare for Apollo missions to the Moon by training them in geology.

More than anyone else, Shoemaker left an indelible mark on planetary science before his tragic death in an automobile accident in the Australian outback in 1997. As his friend Jack Sevier put it, Shoemaker "was in it at the beginning when it was the domain of astronomers with their techniques of photogeology, geologic mapping, cratering analyses, and all the rest of the things that he helped to invent."[9] In 1961, Shoemaker created (and became the first Chief Scientist of) the USGS Astrogeology Research Program, which produced detailed maps of the ancient lunar surface. He secretly harboured a dream of becoming an astronaut himself one day and flying to the Moon to investigate the lunar terrain firsthand.

Early in 1963, NASA created a Manned Space Science Planning Group and a Panel On In-flight Scientific Experiments (which became known as POISE), replacing the *ad hoc* Committee on Scientific Tasks and Training for Man-In-Space. The new groups were formed to create closer ties between the astronaut office, MSC, the field centres, and the Office of Space Sciences at NASA HQ in Washington. This would ensure that the proposed experiments assigned to the Gemini programme would be managed more efficiently than those flown on Mercury.[10]

Taking immediate steps

The full report of the Space Science Summer Study, *Review of Space Research*, sponsored by the National Academy of Sciences, was transmitted to NASA Administrator James Webb on 6 January 1963.[11] Underlining the role of man in space exploration, the report concluded:

"Manned exploration of space *is* science in space, for man will go with the instruments that he has designed to supplement his capabilities ... to observe what is there, and to measure and describe the phenomena in terms that his scientific colleagues will clearly understand. A scientifically trained and orientated man will be essential for this purpose." The report urged that "trained scientist-observers be assigned to important roles in future US space missions," and asked NASA to take immediate steps to train scientists for space investigations. These steps included assigning a "scientist-astronaut" to each Apollo (lunar mission) crew to maximise the scientific return from each expedition (this only occurred on Apollo 17) and assigning meteorologists to co-pilot future manned orbiting space observatories in support of Earth resources and observation experiments, perhaps as early as the two-man Gemini missions planned for 1964. They also recommended training biologists to participate in the first manned flights to Mars (to assist in the search for life on the Red Planet) and preparing astronomers for the first use of space-based telescopes and astronomical platforms, and to assist in their maintenance and modification (as per the Hubble Space Telescope Service missions).

The report also urged the maximum possible participation of scientists in all space missions and outlined four areas of specific training for different categories within the scientist-astronaut programme:

Scientist-Astronauts would combine the experiences and resourcefulness of both professional scientists and trained astronauts (which, in the 1960s, included jet pilot qualifications for all NASA astronaut trainees).

Scientist-Passengers would include experienced and leading scientists in their fields, with adequate training in critical and emergency spacecraft operations, but principally focused on specific science mission objectives, payloads or experiments on a given mission (this would eventually evolve into the Shuttle payload specialist role over a decade later).

Astronaut-Observers would be career (pilot) astronauts with varying degrees of speciality training in scientific observations (similar to the pilot-astronauts, who trained in geology for the Apollo landing missions and in solar observation and Earth resources for the Skylab missions).

Ground Scientists would be leading scientists in specific fields, collaborating with the flight crew to accomplish specific mission goals. These could be non-astronaut support roles, but might also be a non-flight astronaut assignment (this evolved into the mission scientist concept for Apollo and Skylab missions in the 1970s).

Reasonably strong case for immediate selection

Dr. Robert B. Voas, the Human Factors Assistant to MSC Director Robert Gilruth and part of the team that selected the first NASA astronauts in 1959, had developed a comprehensive protocol for selecting the nation's first scientifically-trained astronauts. Though these were not official criteria, they were developed in conjunction with the In-Flight Experiment Panel and POISE groups to create a starting point for further development. These preliminary ideas were drafted on 25 April 1963 and sent to Gilruth on 6 May 1963, less than two weeks prior to the launch of the final Mercury mission (MA-9). Though it was suggested that the announcement for scientist-astronaut recruitment could be made during the MA-9 flight or press conferences, it would be a further eighteen months before any official announcement was actually made.

After reviewing Voas' proposals, William A. Lee, NASA's Director of Systems Studies, stated in a memo to Joseph Shea, the Deputy Director of Systems Engineering, "I believe a reasonably strong case can be made for the immediate selection of three or four physician-astronauts to be flown in long duration flight, if it becomes desirable to perform clinical procedures which would be dangerous when performed by a layman. In doing so, we would have to relax our present stringent restrictions for jet test pilot experience, thereby gaining training experience with non-test pilots."

Lee did indicate that this was not an urgent matter and suggested that a delay of a year or more would not be a serious issue. However, he also highlighted two considerations that weighed against a long postponement and which might affect the announcement to select scientists for the astronaut programme. Firstly, there was a growing and increasingly outspoken feeling of discontent among the more vocal sections of the scientific community, and secondly, there was the problem of recruiting qualified scientists who were prepared to move to Houston. Lee was suggesting that offering a "lunar flight as a carrot" might alleviate the shortage of scientific staff at MSC![12]

NASA issued a news release in June 1963[13] regarding the selection of a new (third) group of pilot-astronauts. Also included was a sentence stating that the space agency

An essential part of future exploration 31

Can you train a scientist to fly a jet and then fly in space? This was the question posed before the members of the two scientist-astronaut selections proved that you could. Here, Allen and Lenoir from the 1967 selection sit in a T-38 jet, used for proficiency flying and travelling between NASA facilities and contractors across the United States.

had begun discussions with representatives of the country's scientific community, with the objective of determining the best way of including scientists in the Apollo programme. The plan was not to include scientist-astronauts on early missions, but to do so on later ones, once the mission hardware had been qualified and the first landings had been achieved. It would be a balance between engineering and operational requirements (as well as scientific objectives) that would determine the selection process and the criteria that would be considered to narrow the field. The proposal to bring in scientist-astronauts seemed to point towards 1965, and a series of meetings was held throughout September 1963 to define the selection process.

On 4 September 1963, Eugene Shoemaker, Joe Shea, Robert Voas and George Low (NASA's Chief of Manned Spaceflight) met to discuss the main issues, which were itemised in a memo dated 13 September:

The appropriate mission to fly a scientist-astronaut: There was agreement that geologists and geophysicists were logical choices for lunar surface missions, but there was some disagreement over flying medically and psychologically qualified astronauts on extended duration orbital missions to evaluate the problems and phenomena such flights would encounter. The first intake of scientist-astronauts would therefore focus on geology and geophysics for lunar missions, and medicine and psychology for orbital missions.

Timing for Selection: Announcing the selection within the next year was generally agreed, but the size of this intake would be restricted to six.

Flight School Training: The MSC position required all scientist-astronauts to be flight trained, but there was also an argument in favour of selecting and training scientists with no flying experience. This might help to determine how applicable aircraft flying skills were to flying a spacecraft in a vacuum. It was agreed that this could be done, but it might create a high "washout rate" that would have to be factored into the selection process. It could also pose a recruiting problem, as the scientists would probably not volunteer if there was little actual prospect of making a space flight. The solution they proposed was to give candidates with flight experience priority, but those with little or no flying experience would also be accepted if they matched other criteria.

NAS Participation: The National Academy of Sciences would assist in publicising the process and would propose individuals to become consultants to NASA, to assist in assessing the scientific credentials of the applicants. The whole selection process would remain within the NASA organisation.

In summarising the meeting, Voas stated, "It appears to me that those in Headquarters are receptive to a proposal from MSC for scientist-astronauts and are ready to approve an appropriate program." By the spring of 1964, NASA was defining which outside organisations should be asked to assist in the recruitment and selection process. During the summer, the announcement was defined in detail and the proposal to select a further (fourth) group of pilot-astronauts was added.

A combination selection proved unworkable, so the scientist-astronaut selection was set for 1965, with the pilot-astronauts due to be selected early in 1966 to allow the scientist-astronaut candidates to graduate from pilot school prior to both groups completing the academic and survival training programme. Using the guidelines mentioned above, two categories of experience would take priority – one to support lunar landing missions and the other for Earth-orbital missions. Though it was clear that the plan was to send geologists to the Moon and to assign doctors to long duration space flights, NASA instructed the NAS not to limit their selections to just those disciplines, and to give consideration to other appropriate areas of speciality, such as meteorology and astronomy.

Selecting the selection board

By December 1964, the process of selecting the first group of scientist-astronauts was under way, but there remained the question of who would form the selection board to review the applications. NASA Headquarters proposed that representatives from the Office of Manned Space Flight (OMSF) and the Office of Space Science (OSS) should participate, along with representatives of MSC and the Astronaut Office. But MSC Director Robert Gilruth disagreed. He thought that the selection board should remain small and that if members from Headquarters were added, this would complicate the

An essential part of future exploration 33

Five flight surgeons/naval aviators attend a meeting at Houston's Manned Spacecraft Center in 1964 for training as aero-medical flight controllers for the Gemini programme. Left to right: Drs. Fred Kelly, Jeff Jeffries, Joe Kerwin (later a Group 4 scientist-astronaut), Bob Kelly and Ed Jacobs

exercise. In Gilruth's opinion, the field centres should carry out the selection, with later vetting by OSS and OMSF prior to final approval. The inclusion of a scientist was recommended and a seven-member NASA board was proposed, to represent management, programme, scientific, engineering, medical and operational viewpoints, as well as meeting political, congressional and scientific agendas. Eventually, a six-member NASA board was formed, with a further nine members comprising the NAS selection board that processed the applications prior to passing them on to NASA.

Table 1. Selection boards for the NASA Astronaut Class of 1965 (Group IV).

NAS Selection Board Members:	
Dr. Allan H. Brown	Department of Biology, Joseph Leidy Laboratory of Biology, University of Pennsylvania.
Prof. Loren D. Carlson	Department of Physiology, University of California Medical School, Davis.
Prof. Frederick L. Ferris Jr.	Educational Service Inc.
Dr. Thomas Gold	Chairman, Astronomy Department and Director, Center for Radio Physics and Space Research, Cornell University, Ithaca, New York
Dr. H. Keffer	Hartline Rockefeller University
Dr. Clifford T. Morgan	Department of Physiology, University of California.
Dr. Eugene M. Shoemaker	Astrogeology Branch, United States Geological Survey
Dr. Robert Speed	Department of Geology, Northwestern University
Prof. Aaron C. Waters	Department of Geology, University of California.
NASA Selection Board Members:	
Charles A. Berry, MD	NASA Manned Spacecraft Center, Houston, Texas
John F. Clark	NASA Goddard Space Flight Center, Maryland.
Maxime A. Faget	Director, Engineering and Development, NASA MSC
Warren J. North	Chief, Flight Crew Support Division, NASA MSC
Alan B. Shepard Jr.	Chief Astronaut, NASA MSC
Donald K. Slayton	Astronaut and Director of Flight Crew Operations, NASA JSC

The inclusion of astronauts in the selection board was logical, as it enhanced the general understanding of the requirements needed by each individual for current and future programmes, and would give the applicants a chance to meet someone with whom they would work and who had gone through a similar selection process. The inclusion of "active" astronauts on NASA astronaut-candidate selection boards has continued to the present day.

With all these issues resolved, the objective was not to select America's first group of scientist-astronauts immediately, but to refine the selection process to take into consideration additional academic qualifications and limited flying credentials.

Gene Shoemaker, a member of the NAS Selection Board, was among those who had lobbied hard for the inclusion of a group of scientists into NASA's astronaut corps. In fact, he was also one of the early leading candidates for selection before being diagnosed with Addison's Disease, a hormonal disorder characterised by muscle weakness, fatigue and low blood pressure, which precluded his involvement as a scientist-astronaut.

A CHANGE IN SELECTION CRITERIA

In 1959, and again in 1962, the primary criterion for NASA astronaut selections was the attainment of test pilot qualifications through military service or civilian institutions. During the 1963 selection, proven piloting skills were still a high priority, but the criteria had been widened to include pilots who had not achieved test pilot flying status, in the hope that this would attract applicants with scientific backgrounds as well as flying experience. It did. On 17 October 1963, fourteen new astronauts were announced to the world's media. Half of the selection held test pilot status, while Air Force Major Edwin ("Buzz") Aldrin had a PhD. Civilian applicants R. Walter Cunningham and Russell L. Schweickart had gained flying experience in military service and had also performed scientific research prior to selection.

Of the 1959 group, Gordon Cooper, "Gus" Grissom and "Deke" Slayton held engineering degrees, while Wally Schirra and Alan Shepard held Bachelor's degrees from the US Naval Academy. The remaining two (Scott Carpenter and John Glenn) had attended college but had not received degrees. All nine members of the 1962 group held degrees: Frank Borman, Elliot See and Ed White had been awarded Master's degrees in engineering; Neil Armstrong, Pete Conrad, Jim McDivitt and John Young had attained Bachelor's degrees in engineering; and Jim Lovell and Tom Stafford held Bachelor's degrees in science from the US Naval Academy. In the third group of fourteen, the academic qualifications were even more varied: Aldrin's PhD was in astronautics; William Anders, Gene Cernan and Theodore Freeman held Master's degrees in engineering; Cunningham's Master's degree was in physics; Schweickart and Dave Scott had attained Master's degrees in aeronautics and astronautics, and Donn Eisele held his in astronautics. There were also six Bachelor's degrees; in engineering (Charles Bassett, Alan Bean, Roger Chaffee and Clifton Williams), chemistry (Dick Gordon), and science (Michael Collins).

For the next intake, NASA would amend the selection criteria once again in order to expand the range of experience within the astronaut office still further.

A new breed of astronaut

On 16 April 1964, the National Academy of Sciences was requested to participate in identifying scientific criteria for the selection of "scientist-astronauts" by Dr. Homer E. Newell, now serving as NASA's Associate Administrator for Space Sciences and Applications.[14]

A series of meetings by an *ad hoc* Committee on Scientific Qualifications of Scientist-Astronauts began in May 1964. In the selection process, the NAS would screen the scientific qualifications of all the applicants, while the Office of Manned Space Flight in Washington and the Manned Spacecraft Center in Houston would assume all other responsibilities for selection and screening. The announcement of the opportunity came on 19 October, just one week after the Soviet Union had placed a three-person crew into orbit for twenty-four hours aboard the first Voskhod spacecraft, under the command of pilot Vladimir Komarov. Flying with him were physician Dr. Boris Yegorov and engineer Konstantin Feoktistov, the first "civilians" to enter orbit. All the previous cosmonauts and astronauts had been serving members of the military forces, including the 1962 Soviet Women Group, who had been specially inducted into the Air Force shortly after selection.

The NASA release stated, "A vast scientific frontier is being opened to direct scientific exploration by man. Observations made by scientist-astronauts will provide new information on the solar system and on man's ability to perform effectively in prolonged space flight." The scientist-astronaut programme was open to applicants from scientific, medical or engineering fields, or any combination of these specialisations.

To be eligible for consideration, each applicant had to:

- Have been born on or after 1 August 1930; be a citizen of the United States; and be no taller than 182 cm (six feet, limited by the Apollo spacecraft's internal volume).
- Have either a Bachelor's degree, a doctorate in the natural sciences, medicine or engineering, or the equivalent in experience.
- Have transcripts of their academic records sent directly to the Scientist-Astronaut Selection Board at MSC from all the institutes of higher education they had attended.
- Have scores in the graduate record examination sent directly to the Scientist-Astronaut Selection Board by Educational Testing Service (ETS) of Princeton, New Jersey. Applications for the examination (and the appropriate fee) had to arrive at the ETS by 31 December 1964, with the examination taking place on 16 January 1965.
- Submit Standard Form 57, the Federal Employment Applicant Form.
- Submit Standard Form 89, the Report of Medical History, signed by both the applicant and their physician.
- Submit Standard Form 78, the Certificate of Medical Examination, completed by the applicant and their physician.

All applications had to be postmarked by midnight on 31 December 1964 to be eligible for consideration. Following preliminary screening, certain applicants would be asked to submit additional material, including published or unpublished scientific or engineering reports; essays of field experiences, research activities, or hobbies related to space missions; and individual thoughts on the scientific objectives of manned space missions.

Going through the process

As part of the selection process, all the applicants would receive a thorough physical examination and were required to participate in a limited space simulation programme. This simulation would serve to familiarise them with the space environment and help determine (to a degree) their ability to withstand the stresses of launch, space flight and re-entry.

The announcement also made it clear that any applicant who was finally selected and who had not become a qualified pilot would be given "the individual flight training necessary to qualify them as pilots of high-performance aircraft and helicopters."

NASA had indicated that it was looking for between ten and twenty candidates.

By the cut-off date of 31 December 1964, a total of 1,351 applications or letters of interest had been received at a rate of about twenty per day. Just prior to Christmas 1964, as the applications process was drawing to a close, NASA issued another press release, which revealed that about 200 of the 900 applications that had been received to date had been rejected as ineligible.[15] This was mainly due to applicants failing to meet the stated requirements for either vision, age, height, or US citizenship.

The visual requirements demanded 20/20 uncorrected vision in each eye. As each applicant had to pass a Class I Military Flight Status Physical, this prohibited the wearing of glasses while flying. In addition, the full pressure helmet that had to be worn during space flight could not accommodate glasses, while contact lenses were deemed impractical for space flight conditions.

On 10 February 1965, the names of 400 applicants (including four women) were forwarded to the NAS to review their scientific credentials. The academy would then forward a short-list of between ten and fifteen candidates to NASA for final evaluation and medical examinations prior to selection.

NASA's astronaut selection process

In the 1959 group, the selection process was primarily a biomedical experimental study. No one had selected people for space flight training before, so there was no previous experience to draw upon. Originally tailored for the Mercury programme, the selection criteria were amended and broadened for the second and third groups as experience was gained. Those chosen for Group 4 (the scientist-astronauts) were expected to be assigned to later Apollo missions and follow-on programmes. As Mercury gave way to Gemini and then Apollo, the requirements for crew participation also changed, along with the volume inside the spacecraft, the mission profile and the scientific payloads and objectives. This further refined the selection requirements.[16]

Selection and training for Mercury had to allow for unknown or untested demands on each astronaut. As well as relying on their previous piloting skills, the Mercury astronauts had to be trained to operate the spacecraft systems, activate small, simple experiments, make scientific observations, and become the primary medical subject, all in an environment that no one had yet explored.

For Gemini astronauts, who would endure missions of up to fourteen days operating as a two-man team, the primary objectives were rendezvous, docking and station-keeping, and the first spacewalks (EVA – Extra Vehicular Activity). But Gemini also had a far more extensive programme of experiments and scientific observations to perform, all of which required the astronauts to develop additional skills in a range of academic subjects.

In the early planning for Apollo, it was suggested that the third man should be a scientifically-trained observer, as the objective of the flights was the scientific exploration of the Moon. But it was soon recognised that this third crew member would also have to be able to fly either the Command or Lunar Modules equally as well as his two pilot-astronaut colleagues. It was therefore clear that in addition to science, piloting skills would have to be included in their training prior to assignment to a flight crew.

For the predominantly military selections of 1959–63, medical testing was conducted at the Lovelace Clinic in Albuquerque, New Mexico, and the Wright-Patterson Aerospace Medical Center in Dayton, Ohio. For the 1965 selection, these evaluations moved to the USAF Aerospace Medical Center at Brooks Air Force Base in San Antonio, Texas.

By early 1962, with the experience of the first astronaut selection and the opening missions in the Mercury programme completed, NASA's basic training format for pilot-astronauts evolved into a process that would be followed, with few amendments, until the end of the decade. Based on USAF biomedical studies, a four-part evaluation programme was developed that could identify those candidates with the greatest potential for meeting the requirements of an astronaut. Those who failed to meet these requirements were medically disqualified from further participation in the selection process. The tests and criteria would then preclude applicants with:[17]

- Significant diseases or abnormalities, such as peptic ulcers, diabetes, gall bladder stones, and other conditions which could interfere with a prolonged space flight mission.
- Predisposition to disease or to limited performance capability, such as obesity in persons with borderline glucose tolerance tests.
- Complex evaluation of mental and character dynamics, including motivation, intellectual ability and learning aptitude, emotional adaptability and maturity.
- Physiological capacity under different loads and stresses, including maximum exertion, automatic control of the cardiovascular system, hyperventilation and breath holding.

For the 1965 selection of the first scientist-astronauts, the overall procedures varied only slightly from those used for the preceding pilot groups. The main difference was the involvement of the National Academy of Sciences in determining each candidate's professional scientific competence and academic qualifications. As with the pilots, past medical histories were screened and those candidates found to have "disqualifying defects" were no longer considered for selection. New medical examinations for the remainder would then be conducted at the USAF Aerospace Medical Center at Brooks AFB in San Antonio. Finally, the remaining candidates appeared before the Astronaut Selection Board, after which the new intake would be selected.

As the 1965 candidates did not have the extensive medical documentation that was mandatory for the military pilots as part of their service history, it was originally decided to repeat the medical evaluations, including the stress testing, at Wright Aerospace Medical Laboratory, following those conducted at Brooks AFB. However, as this procedure had already been deleted for the new pilot candidates that would be selected in 1966, Dr. Charles Berry, the medical member of the NASA Selection Board, concluded that such additional testing was unwarranted for the scientist-astronaut applications. He also supported the general NASA decision that extensive environmental stress testing was no longer required for new astronaut candidates, based on the experience of the first three groups and the flight results from the six completed manned Mercury missions.

As the 1965 candidates were all drawn from the nation's scientific community, their medical records varied considerably in frequency and depth, compared to those of the three previous pilot groups. In order to retain flight status, pilots underwent a regular annual physical that was far more extensive than most of the scientist-astronaut candidates had even considered, let alone been through before. The scientist candidates with previous military and flying service could obviously provide such in-depth medical records, but there remained a high proportion of civilian candidates with no such documentation, as they had never served in any branch of the US armed forces. Applications from some of the candidates revealed a litany of medical problems. The ensuing medical examinations revealed cases of myopia, nasal polyps, varicose veins, and inguinal hernias among the candidates, a major reason for the lower than expected number of finalists.

The characteristics, physical and overall abilities of the successful candidates were perhaps best summarised in the 1985 book *The Real Stuff* by Atkinson and Shafritz, which commented on all those who were selected to the programme between 1959 and 1969:

"Overall, the astronaut applicants who were accepted into the [astronaut] program were found, upon intensive physical and psychological evaluation, to be healthy specimens. They were well organised, pragmatic, concretely orientated, aggressive people of action; the successful candidates tended to handle sensitive interpersonal relationships rather distantly. They derived significant personal satisfaction from mastery and competence in increasingly complex flying vehicles and technical pursuits. They possessed self-confidence from realistically assessing their own capabilities, although they spent little time in introspection. The physical prerequisites for scientist-astronauts were relaxed slightly as the space vehicle environment became less physically demanding and more comfortable."[18]

By early 1965, the 400 applications under consideration for the first scientist-astronaut intake were being processed by the National Academy of Sciences. NASA's Director of Flight Crew Operations, Deke Slayton, was hoping the NAS panel might recommend ten geologists and ten doctors, which would allow the space agency to select three of each for pilot and astronaut training: "That way, we would end up with at least one geologist and one doctor for one of the earlier Apollo crews, which was all I planned to use."[19] When the selection options were expanded to include physicists, meteorologists and astronomers, had Slayton's desire for ten per category been followed, it would have led to a selection group of between fifty and sixty candidates. But the NAS was unable to find a pool large enough and nominated only sixteen candidates to NASA in April 1965. This was just as well, as NASA Headquarters was only looking for between ten and fifteen in the group.

SCIENTISTS AS COSMONAUTS

While NASA was debating and evolving the selection of its first scientist-astronaut group, a selection process was also under way in the Soviet Union. To help satisfy future demands, a group of cosmonaut trainees from the Soviet Academy of Sciences

(*Akademiya Nauk* – AN) was being created. Like their American colleagues, gaining assignment to any flights would prove to be a huge challenge for these candidates. In fact (unlike the astronauts), none of the scientist-cosmonauts selected under this plan ever made it into space.

Initially, the Soviet programme to put cosmonauts into space aboard the single seat Vostok spacecraft meant there was no need to select scientists for space flight. The spacecraft was developed by the designers of rockets and missiles, and automated systems were incorporated into the vehicle. Vostok's primary objective was to prove that a human could survive the rigours of launch and orbital space flight and return to Earth. Science was not high on the agenda.

The Soviet AN did develop a programme of scientific observations in the early years of the space programme. These included observations and investigations of the surface of the Earth, its oceans and atmosphere; near-Earth and interplanetary space; the Moon, the planets of the solar system and our sun; and other astronomical targets and phenomena. But, even on the later space station missions, research in these and other scientific fields was conducted by cosmonauts who were military pilots and engineers, civilian engineers, test pilots and a few physicians, and not those from the team of Academy of Sciences cosmonauts.[20]

Voskhod – the first opportunities

Once Vostok had completed the proving and development flights, the subsequent Soyuz programme would be based around a vehicle large enough to accommodate up to three cosmonauts on each mission. In addition to having a commander and flight engineer on each crew, the third place could be made available for research cosmonauts, including representatives from the Academy of Sciences. Before that, however, the introduction of the reconfigured, multi-seat Vostok as Voskhod in 1963 (purely designed to beat the introduction of the American two-man Gemini) gave the Soviets the opportunity to select, train and fly a non-pilot cosmonaut. The original plans for Vostok included missions that have subsequently been identified as "Vostok 7-13". These were high-altitude, ten-day duration flights (for biomedical and radiation studies), including the development of EVA equipment and procedures and proving flights for new spacecraft systems and hardware. Though these missions did not take place, some of these draft plans evolved into the series of Voskhod missions scheduled for 1964-6 (although in fact, only two manned missions in the series were actually flown). However, since Voskhod was essentially an "upgrade" of the automated Vostok, it would require no more than one "pilot", creating the opportunity to fly one or two cosmonauts from other organisations.

Voskhod: For the initial flight of this "new" spacecraft, the pilot-commander would come from the Air Force pilot and engineer group. Korolyov had previously suggested that space research scientists, physicians and engineers should be selected to train for future flights, with the engineers coming from his own OKB-1 design bureau since they had worked on spacecraft theory, design and fabrication for years. Initially opposed to the idea, the Soviet Ministry of Defence relented and screened up to thirty "passengers" and thirty "physicians", from which six would be selected for actual cosmonaut training. During April and May 1964, the Academy of Sciences and the Ministry of Health selected thirty-six candidates, but only fourteen passed the medical board examination. From these, ten candidates (eight physicians and two scientists) were reviewed by the Credential Committee on 28 May, with five nominated for training. Only one of these, Dr. (Eng) Georgy P. Katys, then aged 37, was from the Academy of Sciences. Fourteen OKB-1 engineers were also nominated, and on 17 May eight were medically qualified, but only one (Konstantin Feoktistov) was nominated to undertake space flight training for the engineer's seat. The training group for Voskhod consisted of nine air force pilots and engineers, a test pilot from the Ministry of Aviation Production, and four physicians, plus Feoktistov and Katys. After receiving "cosmonaut training" for only about six months, Katys was selected as back-up to Feoktistov for Voskhod on 9 October 1964. Shortly after the mission, Katys was stood down from his temporary role as "cosmonaut".

Katys. Georgy Petrovich was born on 31 August 1926 in Moscow. After graduating from school in the seventh grade, he entered the Moscow Motor Building School, followed by the Auto-Mechanical Institute, from which he graduated in 1949 as a mechanical engineer. Completing his post-graduate research work at the Baumann Higher Technical School, Katys received his Candidate of Technical Sciences (CTSc – his degree) before becoming involved in space research at the Academy of Sciences (AN) from 1953. He completed his doctorate studies at the Institute for Automatics and Telemechanics in 1962 and was subsequently employed at several industrial and defence institutes. He was suggested for cosmonaut training in 1962/1963 but was over the imposed age limit and was not considered. However, he was assigned to the Voskhod training group in 1964. During subsequent investigations into his background by the KGB, it was found that he had relatives in France and that his father, Petr I. Katys, who was employed by the Ministry of Post and Telegraph, had been executed during the repressive years of the Josef Stalin regime. Even though he was posthumously exonerated in 1957, his alleged anti-social history seems to have had a detrimental impact on his son's progression as a cosmonaut.

Voskhod 3 and 4: Following the success of the Voskhod mission carrying Feoktistov, Katys returned to the institute to prepare his own programme of in-flight experiments for space flight. These included an electro-optical device for Earth observations, which had both scientific and military application. To operate this instrument, scientist-cosmonauts would need to be assigned to future crews, and in April 1965, the experiment was approved for flight. Katys was himself considered for assignment to the fifteen-day Voskhod 3 mission in November 1965 that would carry his experiment and trained for some time alongside mission pilot Boris Volynov. However, as he was essentially only selected to operate both a gravity experiment and his own electro-optical device, when these instruments were seriously delayed in October 1965, Katys was replaced by pilot candidate Viktor Gorbatko. Katys was assigned to a back-up role for a while, with no chance of flying into space, until the flight itself was cancelled in late 1966. Prior to this, there remained some chance of flying his experiment on Voskhod 4 and Katys was briefly selected to fly with pilot Georgy Beregovoy. But their training had not started when that flight was also cancelled, along with the rest of the Voskhod programme, in favour of the more advanced and versatile Soyuz.

Academy of Sciences Cosmonaut Group

At the time of Katys' experiment being assigned to Voskhod, Mstislav Keldysh, the President of the Academy of Sciences, had decided to form a dedicated cosmonaut group at the academy and to authorise a selection of candidates from the fields of biology, astronomy and physics, organised by the AN's Gennady Skuridin. When Katys finally lost his chance of flying in space, he became involved with the academy's cosmonaut group and its plans to participate in both the 7K-OK Soyuz Earth observation programmes and the L-1 lunar programme.

The search for suitable, qualified scientists to fulfil the role began at the end of April 1966, and resulted in eighteen potential candidates, seven of whom were from the Institute for Terrestrial Magnetism, Ionosphere and Radio Wave Propagation of the AN (IZMIRAN). Medical examinations were completed at the Central Military Scientific Research Aviation Hospital (TsVNAIG), from which only four were passed for further consideration in November 1966. In May 1967, authorisation was finally given to allow civilian institutes to select their own cosmonaut detachments. The final selection for the AN team therefore came on 22 May 1967. A year later, in May 1968, Katys was named as commander of the group. The members of this initial (and, to date, only) AN cosmonaut group were:[20]

Gulyayev, Rudolf, 33, was born on 14 November 1934 in Izhevsk into a family of teachers. A graduate of the Astronomical Division of the Faculty of Physics at Moscow State University, he joined IZMIRAN in Troitsk, near Moscow, where he was assigned at the time of passing the cosmonaut selection process.

Kolomitsev, Ordinard, 33, was born in Tula on 29 January 1935. His father was serving in the military forces. A graduate of the Radio Physics Faculty at Saratov State University, he joined IZMIRAN directly. An experienced polar explorer, his pre-cosmonaut experience included three Soviet Antarctic expeditions, logging over four years and four months in total and working at the Vostok intercontinental station and at the Southern Geomagnetic Pole. For this work, Kolomitsev was awarded the "Sign of Honour" order, as well as the title "USSR Honoured Polar Serviceman".

Fatkullin, Mars, 28, was born on 14 May 1939, the son of a local executive of Tartar origin, in the village of Staroye Shaymurzino, Drozhzhanovsk Raion, Tartarian Autonomous Republic. A graduate of the Kazan State University, he joined IZMIRAN and gained his CTSc degree in 1965.

Yershov, Valentin, 40, was born on 21 January 1928 in Moscow. His father was an officer in the NKVD (People's Commissariat for Internal Affairs) and was supposedly killed by members of the same NKVD in 1945. After graduating from the Moscow Aviation Institute (MAI) as a rocket scientist, Yershov worked at the KB-1 design bureau in Moscow under Sergey L. Beriya on guidance systems for missiles. After a year, he moved to P.D. Grushin's KB-1 design bureau, where he was employed on surface-to-air missiles. He joined the Institute of Advanced Mathematics (headed by Mstislav Keldysh) in 1956, specialising in spacecraft navigation. His work in this field was recognised when he co-proved a theory in the specialised field of statistics of independent measurements, which became known as the Elwing–Yershov theorem.

The principal reason for the four men's selection was most likely their specialisations. After selection, Gulyayev, Kolomitsev and Fatkullin would work on the investigations of solar–terrestrial relationships, while Yershov would became involved in the development of navigation systems for the planned circumlunar missions.

Lack of assignments

In July 1968, the three IZMIRAN cosmonauts completed their initial OKP. Yershov was assigned directly to the L-1 group while he was still undergoing his initial training. A candidate for navigator-cosmonaut aboard one of the early Zond ("probe") circumlunar missions, he was assigned to development issues with the L-1 autonomous navigation system; one part of a system that included onboard computers, a sextant and "a crew console". Sadly, with constant delays to the lunar programme and an increasing shortage of funds, it became obvious even to the scientist-cosmonauts that there would be nothing for them to fly on.

Demise of the scientist-cosmonaut group

Despite their frustration at the lack of flight opportunities, the three IZMIRAN cosmonauts continued to maintain their physical condition. They took regular medicals in order to retain their cosmonaut status in spite of the increasingly slim prospects of flying in space. However, despite years spent helping to develop a scientific programme for space flight and the hope that the potential for scientific research on the new space stations (Salyut) would improve their chances of flying in space, none would ever do so.

Gulyayev and Kolomitsev left the team in 1968 to return to IZMIRAN, but they eventually failed their medicals which saw an end to their dreams of space flight. Fatkullin ceased attending his medical check-ups in 1970 when it became clear that he would not be assigned to a flight in the near future. He also returned to IZMIRAN where he worked towards his doctorate in physics and mathematics, which he was awarded in 1975. Fatkullin died on 16 April 2004, a month before his sixty-fifth birthday.

In May 1970, a further attempt was made to rejuvenate the Academy of Science team with the creation of the Zarya (later Salyut) programme of space stations. The rivalry of other teams, and the suspended status of the Academy of Science team, meant that this plan did not progress very far. If it had been implemented, Katys and Yershov would probably have been joined by four new candidates.

Katys eventually left the cosmonaut detachment in 1972, due to the lack of any firm flight plans, and joined the Scientific Research Institute for Automatic Systems (NIIAS). Yershov fared better, working on the L-1/L-3 programme until 1974. He was considered for inclusion in a proposed Salyut crew in 1973, but failed the medical and was eventually dismissed from the team in 1974 due to progressive deafness. He had also refused to join the Communist Party, which probably did little for his prospects. The lunar programme was itself terminated in 1974, formally ending all plans for an Academy of Sciences cosmonaut group

Waiting for the call

In the twenty-five years following the demise of the Academy of Sciences group, several scientists passed the Medical Commission, but none would receive the authority of the GMVK to begin cosmonaut training. During this same period (1972–97), the Soviet Union, and then Russia, successfully operated a series of space stations under the Salyut and Mir programmes and began cooperative work with NASA, ESA and other international partners to create the International Space Station. It is a great pity that, while numerous pilot-cosmonauts, engineer-cosmonauts, physician-cosmonauts, guest cosmonauts and "space tourists" have all journeyed to these space stations, no "pure" Russian scientist has ever made the grade.

Military scientists

In the 1960s, there was a programme of military manned space operations and experiments. These were initially planned for Vostok and Voskhod but never developed to flight status (although the unmanned military version of Vostok, called Zenit, flew for many years[20]). In addition, plans were formulated to fly military research missions on a military-class Soyuz spacecraft (Soyuz VI), a military space plane (Spiral) and on the military space stations (Almaz) that flew as Salyut 2 (1973), Salyut 3 (1974) and Salyut 5 (1976). The crews for these missions came from among the officers of the Air Force pilot and engineer cosmonaut classes selected in 1960, 1963, 1965 and 1967, and the in-depth military research would follow the initial evaluation flights of Soyuz VI and Almaz. This was similar to the USAF Manned Orbiting Laboratory (MOL) programme of the 1960s and the Department of Defense (DoD) Shuttle missions of the 1980s. To support this emphasis in science over engineering, three researchers were chosen with the air force military cosmonaut selection of 1967.[21]

The selection process for the 1967 group began on 28 February and the order to transfer three researchers to the Soviet VVS (*Voenno-Vozdushnye Sily*) Air Force selection was issued on 12 April 1967 (Order No. 0369). This was the sixth anniversary of the Gagarin flight, and right in the middle of the selection process for the second group of NASA scientist-astronauts. Engineer-Majors Vladimir Alekseyev, Mikhail

Burdayev and Nikolay Porvatkin all later became involved in the Soyuz VI and Almaz programmes. All three subsequently received their CTSc degrees (and Burdayev also went on to receive a Doctorate of Science – DSc – degree in military science), but all three finally left the programme after many years awaiting a flight assignment. They never flew in space due to a combination of cancellations, delays and changes to the original plans. The order dismissing the three researchers finally came on 20 April 1983, sixteen years after they had been selected. A similar selection was made in America, under the USAF Manned Spaceflight Engineer (MSE) programme in July 1987, when two civilian scientists working for the DoD were assigned to the Strategic Defense Initiative "Starlab" Shuttle/Spacelab mission planned for 1992. Dennis L. Boesen and Kenneth P. Bechis trained for three years, until the programme was eventually cancelled in August 1990.

Physician cosmonauts

Of the seventeen scientist-astronauts selected by NASA, five were qualified Doctors of Medicine. In the Soviet programme, medical doctors have been selected at various times since the 1960s, although only three have flown in space. Given the Soviet programme of long duration space flights over the past three decades, it may seem strange that so few doctors have made it to orbit, but of course delays and changes to the programme, budgets, politics, medical issues and other more pressing events have all contributed. The sheer number of hours that cosmonauts require simply to maintain the station, let alone perform experiments, has also been a significant factor. A lack of willing test subjects among fellow crew members is also one reason why the longest flight by a medical cosmonaut included a significant period of self-experimentation by the doctor concerned. The background to the selection of doctors (and scientists) to the Soviet/Russian programme has been detailed in a companion volume in this series[21] and is only summarised here.

Between 1961 and 2005, a total of ninety-nine Soviet and Russian cosmonauts flew into space, of which only five were doctors. The original plans included flying doctors (and scientists) on many missions, but this did not materialise. The original selection of four doctors was to support the Voskhod flight of October 1964. The four were Captain Boris Yegorov from the Military Medical Services, military doctor Aleksey Sorokin from the staff of TsPK (the cosmonaut training centre), Dr Boris Polyakov from the Institute of Medical and Biological Problems (IMBP) and Air Force test pilot and doctor, Vasiliy Lazarev. Yegorov flew the one-day mission with Lazarev as his back-up. Lazarev later joined the Air Force group in 1966 and completed his first two-day space flight in 1973. Two members of the medical staff of IMBP, Major of Medical Science Aleksandr Kisilyov and Dr. Yuri Senkevich, performed a five-day simulation of a Voskhod mission, along with Dr. Yevgeniy Ilyin who was also a Captain in the medical services. The reserve was Lt. Sergey Nikolayev of the medical services.

In 1972, a group of civilian cosmonauts was formed at IMBP and included Dr. Valeriy Polyakov, the group's first commander. He went on to fly two long duration missions on Mir totalling 16,312 hours (or 679.6 days) in space. There have been a further six selections to IMBP up to 2003 (and one special selection), but of the seventeen doctor candidates (including five female candidates in 1980), only two others have made it into orbit. In 1983, Oleg Atkov was recruited from Clinical Cardiology at the USSR Academy of Medical Sciences to fly as a cosmonaut researcher on a long duration flight to Salyut 7. This mission flew in 1984 and Atkov completed a 240-day flight. In 2000, Boris Morukov was a member of the American STS-106 crew that visited the ISS station for a twelve-day mission.

The desire to fly physician-cosmonauts on long duration missions was influenced by the need to obtain physical and mental data on the effects of long-term space flight on space station crews. In participating in ground simulations and monitoring the crew in orbit as well as flying on some of these missions, physicians were able to obtain data that could be used to develop new equipment, procedures and countermeasures. These could then support the first expeditions to Mars and the sustained occupation of both the Moon and the Red Planet. With the expansion of ISS operations after its construction is completed in 2010, there should be opportunities for Russian doctor-cosmonauts to continue the work begun on Salyut 7 and Mir. Whether that option is taken up remains to be seen.

ISS offers an excellent resource to evaluate and develop new procedures in dedicated facilities. In the 1980s one of the ideas, based on a study called Medilab, was for a specialised module attached to Mir 2. This would have been a biotechnology module in which more focused biomedical research would have taken place, presumably by doctor-cosmonauts or medical technicians assigned to a space station crew. Of course, Mir 2 was not funded and some its hardware was reassigned to the ISS programme, although the biotechnology module was not one of the ideas taken up.

Other selections

In 1970, Zyyadin Abuzyarov of the USSR Weather Service (*Gidromettsentr*) successfully passed all his medical examinations and was approved for "special (cosmonaut) training". Even though he also passed biomedical training at IMBP, he was put on the reserve list until 1974 when he was dismissed from the team, along with Yershov. Gurgen Ivanyan of the Leningrad State University, a professional geologist, also passed his medicals in 1971, but was never called up for cosmonaut training.

Science not a priority

During the 1980 selection of female cosmonaut trainees, women from a variety of agencies and ministries were considered and tested, including Irina Latysheva of the Academy of Sciences. Because there was no AN cosmonaut team at that time, she transferred to NPO Energiya and qualified as a civilian research-engineer cosmonaut. She was never called for mission training, but she remained a member of the team for several years, formally retiring on 25 February 1993. Other likely candidates for the AN (later, RAN – Russian Academy of Sciences) team existed in the 1980s, but although many passed the medicals, none were selected. In 1985, Arkediy Melua, a specialist in information processing at the St. Petersburg Department of the Institute of Natural Sciences and Technology History of RAN, passed his medicals but progressed no further. In 1988, Sergey Fursov also passed his medical examination and was authorised for cosmonaut training, but he never began the course.

> *Latysheva, Irina D.* was born on 9 July 1953. Employed by the Institute of Radiotechnics and Electronics of the AN, she was later seconded to the Space Research Institute (IKI), prior to selection for cosmonaut training.

Apparently, science was not a priority for crew assignments in the 1980s and 1990s, though in 1993 after the break-up of the Soviet Union, a decision was made to reform the Russian Academy of Sciences (RAN) cosmonaut group. Though no scientist-cosmonaut has flown in space, several other cosmonauts have gone on to earn CTSc and DSc degrees after their space flights. In the 1980s, two former engineer-cosmonauts with space station experience transferred to the AN, even though there was no formal team in place at that time.

Veteran civilian engineer cosmonaut Georgy Grechko (selected from OKB-1 in 1966) had flown two space station missions in 1975 (thirty days) and 1978 (ninety-six days) before transferring to the Academy of Sciences "team" on 6 July 1985. He flew his third and final mission in 1985 as a scientist, not a flight engineer, representing the Institute of Atmospheric Physics of the AN where he was employed as chief of a science laboratory. Leading up to this mission, Grechko prepared a broad programme of scientific experiments, which he conducted on the Salyut 7 space station during his week-long mission. He later expressed a desire to fly a fourth mission – perhaps to the Mir space complex – but this never materialised. Grechko retired from the AN team on 1 March 1992. The second experienced cosmonaut to transfer to the AN (on 4 November 1989) was former *NPO Energiya* cosmonaut Valentin V. Lebedev. He had been selected as an engineer-cosmonaut in 1973 and had flown two missions (in 1973 for eight days and in 1982 for 211 days). He transferred to a team working at the GeoInfo Centre of the AN, but he never flew again and retired from cosmonaut status on 25 February 1993.

Following the reactivation of the RAN team in 1993, former Air Force pilot-cosmonaut Anatoly Artsebarsky was seconded to the Centre for Applied Research of the RAN, acting as an advisor in the creation of a new RAN cosmonaut group. He was in line to head up the team when it was formed. His transfer occurred on 7 September 1993 and he officially enrolled in the RAN team in January 1994 as Chief of Sector for Information Technologies at the Laboratory of Large Scale Constructions (Space Stations). He began the difficult task of forming the RAN team, but by 1 April, only Artsebarsky himself was required to pass a medical as there were no other cosmonauts formally assigned to the team. On 28 July 1994, Artsebarsky formally retired from the RAN "team" and since he was its only member, the team again disbanded. On 20 March 1995, Yuri Stepanov, a medical research engineer, transferred from the IMBP group (to which he had been selected ten years before) to the RAN team, but left in February 1996.

Changes in selection

A new draft document on the status of cosmonauts of the Russian Federation was completed which predicted a unified cosmonaut selection under the new Russian Space Agency (essentially a counterpart to America's NASA). This, it was hoped, would eliminate the need for special cosmonaut teams and open the way for a single, large team that could also include scientist-cosmonauts. However, on 9 February 1996, a new cosmonaut selection board approved candidates from the Air Force Military Space Forces and RKK Energiya. No mention was made of a unified team and no scientists were considered. Specialist teams continued to be permitted for some institutes but the RAN was not one of them, and no scientist has subsequently been named for cosmonaut training.

A good career move?

To the casual observer, it would seem sensible to assign scientists as flight crew members. After all, astronomy is one of the oldest "sciences" and studies of space phenomena have been a high priority since the dawn of the space age. However, in reviewing the sometimes difficult process of recruiting scientists to space training in both America and Russia, this does not seem to have been the case. The American scientist-astronaut groups of 1965 and 1967 experienced this new challenge, but faced almost insurmountable hurdles and long delays before finally being accepted and flying into space. Their journey was far less frustrating than that of their Russian colleagues, but selection to the programme was only the first hurdle. Once they were assigned as a space trainee, these experienced academics had to go right back to school to learn brand new skills and overcome new challenges on their road to space.

REFERENCES

1. *Our Gagarin*, Progress Publishers Moscow, 1978, p. 12.
2. Reference 1, p. 215.
3. Reference 1, p. 283.
4. NASA News Release, Manned Spacecraft Center, 7 August 1961.
5. *Astronautical and Aeronautical Events of 1962*, p. 24, entry for 27 February; NASA report to the Committee on Science and Astronautics, US House of Representatives 88th Congress 1st Session, 12 June 1963.
6. *The Real Stuff: A History of NASA's Astronaut Recruitment Program*, previously cited, Chapter 4, Selection of Scientist-Astronauts (Group 4), pp. 54–86.
7. Reference 6, p. 240, entry for 14 November, and p. 277, entry for 26 December.
8. Reference 6, p. 69.
9. Lunar and Planetary Institute (LPI) online newsletter article "*Gene Shoemaker (1928–1997)*".
   Website: *http://www.lpi.usra.edu/publications/newlstters/lpib/lpib83/shoemaker83.html*
10. *Gemini: Steps to the Moon*, David J. Shayler, Springer-Praxis 2001, pp. 331–359.
11. *National Academy of Sciences Review of Space Sciences*, NAS Release, 6 January 1963.

12	Reference 6, p. 71, William A. Lee memo to Joseph F. Shea, dated 14 May 1963.
13	NASA News Release, MSC-63-95, 5 June 1963.
14	NASA News Release, MSC Houston, MSC-000 (not numbered), 19 October 1964.
15	NASA News Release MSC 64-195, 16 December 1964.
16	Reference 6, pp. 79–83.
17	US Congress, Senate Committee on Aeronautical and Space Sciences, *Scientists Testimony on Space Goals*, Hearings, 88th Congress, 1st Session, 10–11 June 1963, pp. 650–651.
18	Reference 6, pp. 81–82
19	*Deke! US Manned Space: From Mercury to the Shuttle*, previously cited, p. 152.
20	*Russian Scientist-Cosmonauts*, Igor Marinin and Igor Lissov, Spaceflight **38**, No. 11, November 1996, pp. 388–390, British Interplanetary Society.
21	*Russia's Cosmonauts*, Rex Hall, David J. Shayler and Bert Vis, Springer-Praxis, 2005, p. 133.

3

The Scientific Six

Soviet space planners enjoyed a seemingly unassailable edge over America's space agency in the first years of the space race to the Moon. While keeping plans for their own manned space flight efforts completely secret, they certainly used NASA's open information policy to great advantage. Armed with an almost complete knowledge of America's manned space flight programme, Soviet space designers and planners became ever bolder and increasingly more confident in their undisguised efforts to stay ahead of America in the space race.

A GAMBLE FOR GLORY

Then, in October 1964, they pulled off what was regarded at the time as a truly awesome feat that completely pulled the rug from under the feet of the Americans. By all measures it was an audacious move, but one that history now reflects as being filled with the most appalling and unnecessary hazards.

The Vostok spacecraft, which had emerged from Chief Designer Sergey Korolyov's OKB-1 design bureau, had proved itself many times over to be a reliable and functional vehicle. In April 1961, Lieutenant Yuri Gagarin (promoted to major during his flight) was the first person to fly aboard one, and in doing so became the world's first space traveller, on a single-orbit, 108-minute flight.

Success followed success when cosmonaut Gherman Titov flew Vostok 2 into orbit in August 1962, his mission setting a new record of more than a day in space. Then followed the concurrent flights of Vostok 3 and 4 in August 1962, each carrying a single crewman who would become jointly known as the Heavenly Twins – Andrian Nikolayev and Pavel Popovich. In June 1963, Valery Bykovsky flew into orbit aboard Vostok 5, and was followed a day after by the world's first spacewoman, Valentina Tereshkova, aboard Vostok 6.

While the American manned space programme toiled along at a carefully measured, but nevertheless spirited pace, the Soviets had begun to emerge in the eyes of most people as likely winners in the race to put the first man on the Moon. To a grudgingly admiring world, they seemed almost invincible, and certainly appeared superior in spacecraft and rocket technology.

A propaganda machine

But beneath the spotlight of success was another story that would not emerge for some time. The Soviet space programme, while brilliant in its execution, was becoming something of a sham, completely at the mercy of powerful, politically-driven manipulators, who were turning its very success into an orgy of propaganda. While flying headline-grabbing space spectaculars with ostensible ease, the Soviet Union was in fact stretching its technological capabilities and resources to the limit.

The safe return of three cosmonauts in their "new and spacious" Voskhod spacecraft would garner international acclaim and immense propaganda for the Soviet Union, but the history-making space flight would one day attract justifiable criticism. While words such as "reckless" and "hazardous" would later be used to address the first manned flight of the Voskhod spacecraft, it was a space mission that created history at a time when the world was in awe of each new space spectacular. It was, after all, a phenomenal era of national one-upmanship and governmental urgency in a rapidly accelerating effort to become the predominant space nation.

Seven years later, in 1971, the Soviets would try to emulate the incredible flight of the first Voskhod spacecraft, but this time it would lead to the tragic deaths of three cosmonauts. In 1991, Korolyov's deputy, Vasily Mishin, spoke about the hazards associated with the flight of Voskhod 1. "Was it risky? Of course it was. It was as if there was a three-seater craft and, at the same time, there wasn't. In fact it was a circus act, for three people couldn't do any useful work in space. They were cramped, just sitting – not to mention that it was dangerous to fly."

Just as Sergey Korolyov would recognise that there was a place for scientists and physicians aboard multi-person Soviet crews, there were rumblings in the United States by the scientific community for a presence on NASA flights. But as the world heard about the first physician and first aerospace engineer travelling aboard a three-man spacecraft, the National Academy of Sciences and the NASA Space Science Board were just a week away from announcing that they were seeking applications for astronaut candidates who were holders of a PhD, or the equivalent in natural science, medicine, or engineering. Yet again, with the successful but highly dangerous flight of the first Voskhod spacecraft, American technology and NASA had been figuratively pipped at the post.

TESTING THE CANDIDATES

On 2 May 1965, an early morning commercial flight departed San Francisco Airport on a flight to San Antonio, Texas. There would be intermittent stops in Los Angeles, Tucson, Phoenix and El Paso, and the passengers included a number of men who were travelling to San Antonio for medical and psychological evaluations in the process to find NASA's first group of scientist-astronauts.

Owen Garriott was one of those men, and he recalls that by the time the flight had left Phoenix, there were nine of the candidates on board. They were, he says, "mutually identifiable by concern over eating numerous candy bars," in order to raise their blood sugar levels before the tests.[1] He also affirmed that there were supposed to be sixteen candidates in total making their way to San Antonio, but one applicant from California had apparently changed his mind at the last minute and withdrawn. On arrival in El Paso at three o'clock that afternoon, there was a problem with a fuel pump on the left engine, causing a lengthy delay. It did little to ease the tension these men were feeling, and it would be some six hours before they were finally transferred onto another flight for the last leg of their journey. The fifteen who eventually arrived in San Antonio were:

- David J. Atkinson, from the Basic Geology Department of the Shell Development Company in Houston.
- C. William Birky, Jr., a zoologist from the University of California, Berkeley.
- Michael B. Duke, an astrogeologist with the USGS in Washington DC.
- Ernest J. DuPraw, a zoologist from the University of California, Davis.
- Owen K. Garriott, an electrical engineer from Stanford University.
- Edward G. Gibson, from CalTech, specialising in rocketry and plasma physics.
- Duane E. Graveline, an Air Force medical doctor.
- Don V. Keller, a physicist from the University of California, Berkeley, and with the Northrop Corporation in Camarillo, California.
- Joseph P. Kerwin, a medical doctor with the US Navy.
- F. Curtis Michel, a physicist from Rice University, Houston.
- Daniel J. Milton, an astrogeologist with the USGS at Menlo Park, California.
- Harrison H. Schmitt, a geologist with the USGS in Flagstaff, Arizona.
- William G. Tifft, an astronomer from the Steward Observatory, University of Arizona.
- Robert Woodruff, a doctor of psychiatry with the US Navy.
- Philip J. Wyatt, with the Defense Research Corporation in Santa Barbara, California.[2]

On arrival in San Antonio, the men made their way as scheduled to the Bachelor Officers' Quarters at Brooks Air Force Base. The following evening, they were picked up by bus and transported to the Aerospace Medical Sciences Division of the USAF School of Aerospace Medicine for a briefing. Here, each of the scientist-astronaut candidates received a typed note of welcome from Captain Lawrence J. Enders, Chief of the Flight Medical Evaluation Section, which was attached to a personal schedule. In the note, he bid welcome to the men, with the hope that, "your stay here will be pleasant and enlightening; you may also find it exacting at times."

Enders' note continued by stressing that they were about to undertake "the most thorough medical evaluation you will ever have." The men were told that there would be some thirty-five or more hours of actual medical surveillance and examinations on a rotating schedule, to allow for maximum thoroughness in evaluation by the medical staff, in the least number of days, and with the fewest non-productive hours. They were warned that certain portions of their evaluation would require advance preparation, such as fasting, abstinence from alcohol and "bowel preparation, etc." Each candidate would receive two runs on the centrifuge. The second run, they were told, "is not meant to test your maximum tolerance. The run to which you will be subjected will be similar to the re-entry 'g' profile, which might be experienced in one of the presently available spacecraft."[3]

The following morning, Monday, candidate testing began in earnest. The men were split into three even groups; the first consisting of Michel, Schmitt, Milton, Kerwin and Keller; the second of DuPraw, Wyatt, Woodruff, Tifft and Duke; and the third of Gibson, Atkinson, Graveline, Garriott and Birky.

Garriott's diary

Owen Garriott kept a daily record, and the extent of the tests, evaluations and interviews can be established from this. First of all there were blood sugar tests: 50 cc of blood was extracted, and then each subject had to drink a large cup of glucose, after which four smaller blood samples were taken at thirty-minute intervals to observe any variation of sugar concentration. Eighteen dental X-rays were taken, and then there was a brief discussion about the subject's medical history. As Garriott recalls, the doctor involved in this was in something of a hurry, "since just that morning he had been notified that he was leaving for Vietnam that afternoon."[2] After lunch there was another general examination, largely abdominal, and then the candidates took part in a "cold pressor" test, which involved immersing a hand in ice water for a minute while doctors watched their blood pressure and heart rate. After more blood pressure checks, there was a session of lengthy electrocardiogram recordings with electrodes strapped all over their bodies.

On Tuesday, vestor cardiograms were carried out in a shielded cage, followed by phono-cardiograms and then a session on the "tilt table" to conduct an evaluation of the way each candidate's body regulated blood pressure in response to some simple stresses. Stretched flat in a supine position on this table, the degree of tilt was adjusted from horizontal to almost vertical while the subject was asked to hold their breath and then hyperventilate while readings were taken. Each man was then subjected to four or five centrifuge runs without the aid of a pressure suit; first quickly up to around 4 G and back, and then quickly to 5 G.

Later, they experienced the simulated g-profile of an abort off the pad using dynamics associated with the "Little Joe" abort escape rocket that reached a peak loading of 10.5 G. "It felt like an elephant sitting on your chest," according to Gibson. Breathing could only be accomplished using the diaphragm, not the chest. During the entire test, heart EKGs were taken, as well as television pictures of their faces to monitor their physical and physiological status.

In the afternoon they endured "double masters," involving forty-eight trips over a turnstile while the doctors checked their blood pressure. This was followed by a session on the treadmill. The speed was set at 3.5 mph and the tilt was slowly increased by one degree per minute. Garriott was later told that the normal range was around twelve to eighteen minutes, although his group went for seventeen to twenty-seven minutes. His own "record" was twenty-two minutes, while Gibson hit twenty-seven minutes. After a short break it was time for more photos and abdominal X-rays.

The next day started on a sour note, with orders issued not to have breakfast before the next round of tests. Their first meal of the day was actually a "tritium cocktail," which comprised a small, 250-millicurie dose of Hydrogen-3 in a cup of water, after which further blood samples were taken to measure total body water. Then came the psychology tests: the standard Rorschach ink-blot quiz, making up stories based on seven pictures; mental arithmetic; word definitions; jigsaw puzzles; and IQ examinations. This lasted for about four hours. Following this they completed a neurological check-up and an electro-encephalograph (EEG). The latter included placing fifteen quarter-inch needles just under the skin, which were then taped parallel to the skin surface in order to make a good, subcutaneous electrical contact.

Thursday also began without breakfast, although the candidates had taken six tablets the night before to prepare them for their gastrointestinal "barium milkshake". They then underwent a fluoroscope, which tracked the barium through their stomach and duodenum. Thirteen new X-rays were also taken, followed by respiratory tests, with each man's vital capacity and residual volume measured. In part, this entailed breathing in pure oxygen and then checking for any nitrogen concentration in their exhalation. A visit to the centre's psychiatrist followed, but most of the men were only quizzed for up to thirty minutes.

On Friday the routine continued as before with no breakfast, but each candidate was required to drink a litre of water. Ocular tests were next on the list, with each candidate's eyes carefully checked during night vision, dilation and pressure evaluations. The latter test was developed to measure the eye's susceptibility to glaucoma and to check that the eye canals drained fluid away satisfactorily. In order to measure intraocular pressure, a five-gram weight was rested on each eyeball, which had been anaesthetised beforehand. Later that day came the measurement of their body weight and volume in a tank, to determine body density and fat percentage, followed by a full dental check-up. The day ended on a brighter note, however, with a cocktail party at the Officers' Club, compliments of the school's commander, Colonel Harold V. Ellingson. But they all took care to heed a notice they had been handed that day, which read: "Tomorrow you are going to receive a routine altitude chamber indoctrination flight. For your own comfort we advise you to avoid supper and breakfast foods which would promote excess gas formation in the intestines; such foods include cabbage, beans, green roughage and fried or greasy foods. Eat a light breakfast. Smile! Your friendly Aviation Medicine Department."[3]

Saturday entailed the promised trip to the centre's altitude chamber to test their susceptibility to hypoxia and any other difficulties connected with high altitudes and oxygen deprivation. Prior to the first test, they were required to breathe pure oxygen, and were then "flown" to an altitude of 5,000 feet and back to check that all their breathing canals, such as the Eustachian tubes, were open. The simulation then took them up to 43,000 feet at a rate of 7,000 feet per minute, then back down to 25,000 feet, at which point they were told to remove their oxygen masks to experience an hypoxic state. During this period they were asked to write their names, and Owen Garriott recalls that his signature became an illegible scribble within three to five minutes. After this, the pressure chamber was regulated to bring them down to ground level, then back up to 8,000 feet, where they underwent a rapid decompression before continuing up to 23,000 feet and then finally being brought back to ground level.

Sunday was a day off from the tests, so the men mostly slept in or sat around and rested. Garriott stated that there was nothing scheduled for Monday morning, "so I talked the testing crew into an extra Gemini re-entry profile in the centrifuge. Up to 8.2 G with no difficulty, except [for] pressure on chest." In the afternoon, they were given an exercise at the controls of an aircraft simulator, and then they were handed more written exams. Garriott says he had to stay up all night for an EEG repeat, and was very grateful to Joe Kerwin, who remained with him until 5:15 a.m. to help him stay awake.

It was back to the tests on Tuesday, with more ear, nose, and throat tests in the morning. When these were finished, the men were able to check on their psychological tests, and all seemed in order, with fairly consistent results across the group. That evening, the candidates were flown to Houston, where they were checked into the Rice Hotel.

Their first day in Houston began with an interview conducted by a medical panel. Among those doing the interviews were Drs. Clarence Jernigan, Charles Berry, and Bill Carpentier. They revealed any anomalies in the medical tests to the candidates, and then asked them about their motivation for joining the space programme. The next part of the day was given over to one of the most anticipated events – a flight in a T-38, some with NASA test pilot and Chief of Aircraft Operations, Joe Algranti, others with NASA test pilot, Bud Ream. The men sat in the back seat discussing manoeuvres and experiencing aerobatics performed by some of the best pilots in the skies. Garriott recalls Algranti, his pilot, performing dizzying loops and quick rolls of 360 and 720 degrees, and then initiating stalls with and without flaps. After this, he descended gradually toward the ground to accelerate up to about Mach 1.1, quickly manoeuvred and rolled, then climbed again on idle power at sixty degrees until the T-38 stalled, after which he made a hands-off recovery. After his flight, Garriott wrote that he thought he would find the T-38 "a very easy plane to fly and extremely easy to handle."

Late that afternoon, the candidates met for a general interview, this time with Deke Slayton, Alan Shepard, Max Faget, Warren North, Joe Shea and Jack Clark. They discussed space flight experiments it was felt the candidates could conduct, their motivation, experience and other general topics. With that, the tests were at an end. The following day, 13 May, they were taken on a tour of the Manned Spacecraft Center and local housing areas, had a meeting with Flight Director Chris Kraft and, that evening, celebrated the end of the tests and interviews with a cocktail party and dinner. Most of them seemed to feel they had done well, although a few had their doubts.

THE CHOSEN FEW

On 28 June 1965, NASA released a news bulletin, in which it revealed the names of its first cadre of scientist-astronauts; six men in all. They were Owen Garriott, Edward Gibson, Duane Graveline, Joe Kerwin, Curt Michel and Harrison "Jack" Schmitt. The men came from a diversity of backgrounds, and each could boast many varying inspirations and motivations along their path to a possible flight into space.

OWEN K. GARRIOTT

On 22 April 1889, right on the stroke of high noon, gunfire rang out along the borders of what would become known as Oklahoma Territory the following year. At the sound of the shots, twenty-five thousand hopefuls began a frenzied stampede into flat, dry Indian country, urging on their horses and wagons. Some even proceeded on foot, but all were engaged in a desperate race for newly released tracts of land in former Cheyenne–Arapaho territory. By sunset, thousands of 160-acre "quarter sections"

The six scientist-astronauts of NASA's Group 4 are introduced to the media. From left: Joseph P. Kerwin, MD, Edward G. Gibson, PhD, F. Curtis Michel, PhD, Duane E. Graveline, MD, Harrison H. Schmitt, PhD, Owen K. Garriott, PhD.

Answering questions from the world's media following the announcement of their selection on 28 June 1965.

and many of the better town lots had been staked and claimed by new settlers, who quickly created new tent cities. It was the first of five such land rushes, and these are certainly one of the most dramatised and colourful events in western history.

Among those who had set out carrying sledgehammers and claim stakes were the forebears of a man whose name is now synonymous with Oklahoma, and in particular his cherished hometown of Enid, in the northwest part of the state. His father's parents staked their claim in the land run of 1889, while his mother's parents settled in the so-called Cherokee Strip some time later, after buying another person's claim, reputedly for "a wagon and team."

In the footsteps of pioneers

Owen Kay Garriott can trace his family back over several generations; notably on his father's side to French immigrant brothers who first set foot in the colonies in the late eighteenth century. "Most have been farmers, until my father's generation. No criminals or cattle thieves that I've been told about!"[4]

The future scientist-astronaut was born in Enid, Oklahoma, on 22 November

Owen K. Garriott, PhD.

1930, the first child of Mary Catherine (nee Mellick) and Owen Garriott. While he was given his father's forename (he in turn had been named Owen after a much-admired local Welsh teacher), Kay was derived from an abbreviated form of his mother's middle name. Garriott speaks with tremendous pride of his forefathers in the days before Oklahoma achieved statehood in 1907, describing the way these rugged pioneers worked the land and "lived in partial dugout homes for some period, until crops were grown and sold and some money became available."

He also recalls being told how grandfather Garriott "freighted goods and supplies by horse and wagon from the nearest railroad station to their small community store – a distance of some twenty miles. It took a full day each way. Seventy years later, as I approached Vance Air Force Base for a landing in my NASA T-38 aircraft, I travelled across my grandfather's tracks, but covered the same distance in about four minutes. In 1973, some eighty years after my grandfather's trips, I covered the same distance in orbit in only four seconds!"

Childhood recollections of growing up in Enid with his younger sister, Donna Jean, evoke pleasant memories, unclouded by the sprawling impact of the Depression. "Perhaps this is because my father and most uncles always had jobs, and we never worried about material ownership. We played games, hiked with kids, and swam at the local pool, or occasionally in ponds. I had quite a few nearby cousins, friends and marvellous parents – very idyllic!" If anything, he found elementary education in the Enid Public School system to be disappointingly easy, but nevertheless enjoyable. "I still remember almost all my school teachers by name," he adds.

One teacher he particularly remembers from around eight years of age was Bessie Truitt, a poet laureate for the State of Oklahoma. "In her third grade class she showed us a simple orrery, which was a word whose definition I did not learn for a good many more years. It's a device that has the whole planetary system in it, with the sun at the centre, then all the planets and the moons of the Earth and Jupiter and so on. You turn a little crank handle and they all rotate. To see how the solar system moved was a real awakening for me, and quite fascinating. When I moved into the fourth grade the next year, I was asked to go back to the third grade class and explain its operation to them, which I considered quite a privilege."

His father, Owen, would also provide considerable inspiration. He held a Bachelor of Arts degree in geology, but when promised jobs in geology evaporated in 1930, he turned instead to his minor area of study for employment. For the next fourteen years, he worked as a chemist with Pillsbury Mills, and then spent a further thirty years as an oil and gasoline distributor. One evening in 1944, while the younger Owen was still in eighth grade, his father came home and mentioned that a friend at work would be teaching an adult class on radio theory, which involved instruction on electronics and how radios and transmitters worked. All this was in order for his father to become an amateur radio operator, more commonly known as a 'ham', so he asked his son if he would like to attend with him.

"I was pleased to participate in an adult activity with my father and we went to class three nights a week for four months or so. Next, we found that a 'code class' would be starting, so we both went to that together. When all this was finished we took our Federal Communications Commission (FCC) exam and we both passed, much to my father's relief!"

Due to existing war restrictions, ham radio operators were precluded at that time from on-air activities, but this situation would not last much longer. At the age of fifteen, Garriott was well on the path to his future, and he genuinely credits that study experience with his father as a major factor in his determination to enter the field of engineering.

After graduating from Enid High School in 1948, Garriott went to undergraduate school at the University of Oklahoma on a Naval Reserve Officer Training Corps (NROTC) scholarship, which paid about half of his college expenses. While still in high school, he had worked as a technician in the local radio station. The income from this work had been largely saved and would cover most of his remaining college education costs. Studies did not occupy all his time, however. After dating through their college years, he married his high school girlfriend Helen Mary Walker in 1952.

The following year, he was awarded his Bachelor of Science degree in electrical engineering and then, because of the NROTC scholarship, he was obliged to serve in the United States Navy. During three years on active duty, he served as a line officer and electronics officer, and was stationed aboard several destroyers at sea. Garriott then returned to graduate school, this time at Stanford University in Palo Alto, California, where he was selected to work as a research assistant in the Radio Propagation Laboratory with an eclectic group of ham radio operators, many of them faculty members.

By late 1957, Owen Garriott had completed his Master's degree in electrical engineering at the laboratory and was actively seeking an interesting research topic for his dissertation. On 4 October that year, the Soviet Union launched their first Sputnik satellite into orbit, so most of the graduate students and many of the professors in the laboratory converged on their field site, where a number of radio receivers and antennas had been set up for research purposes. Through this equipment, they were able to listen in unconcealed awe to the famous "beep-beep" signals emanating from the world's first artificial satellite. Very fortuitously this event, followed soon after by the launches of Sputnik II and III, handed Garriott his research topic on a platter.

"My dissertation used the radio signals from Sputnik III to study the electron content of the ionosphere. As these waves transverse the ionosphere some 300 to 400 kilometres above the Earth, various measurable effects were imposed upon them, such as a modified Doppler shift." Observation and interpretation of these changes formed the basis of his dissertation.

Having achieved his doctorate in electrical engineering in 1960, Garriott stayed on at Stanford, now as a member of the faculty involved in teaching electronics, electromagnetic theory, and ionospheric physics, as well as conducting his own research. At this time, he was honoured to receive a year-long National Science Foundation Fellowship at Cambridge University and the Radio Research Station, located at Slough in southern England, so the Garriott family took temporary leave of their American home.

By now, he and Helen had become proud parents to three of their eventual four children; Randall Owen, born on 29 March 1955, and Robert on 7 December 1956, with Richard patriotically arriving on 4 July 1961, albeit in England. Their daughter Linda would be born five years later, on 7 September 1966 after his selection by NASA, to complete the family. While they were living in England, Garriott was enthralled by the news that twenty-seven-year-old cosmonaut Yuri Gagarin from the Soviet Union had become the first human to fly into space. His epic flight was followed just three weeks later by that of American astronaut Alan Shepard.

An interesting proposition

From late 1961 to 1965, Garriott worked as an assistant professor, then Associate Professor, in the Department of Electrical Engineering at Stanford University. During this time, his interest piqued, he became even more absorbed in unfolding events as the United States and the Soviet Union engaged in what had become known as "the Space Race." While the Mercury and early Gemini flights had proved that humans could live, work and function well in space, more physiological and scientific research was going to be needed at some point. He began hearing reports that NASA was considering the possible introduction of scientists and physicians into the flight programme.

"One night over a casual dinner, a friend happened to ask if I'd heard that NASA was considering inviting people with research backgrounds to apply for the astronaut corps, as one of a new group to be called 'scientist-astronauts.' That's almost the first time I really ever thought about the possibility of actually participating personally in the flight programme.

"When I first began to think maybe NASA would accept somebody with an academic and research background, I asked myself 'what will they look for?' Not necessarily how many papers you have produced – although this would be important to the science reviewers – but pilot selectors would be interested if you could fly an aeroplane. Since I'd been meaning to get my flight license for a good many years, this provided the extra motivation to complete qualification for a pilot's license and instrument rating. Whether or not it factored in, I still don't know, but whatever the reasons were, it was a good thing to have from my standpoint."

This was added to his list of academic and research qualifications, which was eventually submitted to NASA on a standard government employment form, "along with those from a good many candidates." Then, as his application was considered, there came the first of many lengthy waits he would experience in his long association with NASA.

EDWARD G. GIBSON

Sharing his birth date with that of noted astronomer Edmund Halley, Edward George Gibson came into the world on 8 November 1936, the youngest of three children born to Geraldine (née Shannon) and Calder Alexander Gibson. The Gibson family lived in the quaint, tree-lined village of Kenmore, located in Erie County near the Niagara River just north of Buffalo, where his father ran the A.C. Gibson Company, a marking devices firm founded by and named after his paternal grandfather, Alexander Calder Gibson.[5]

Ed adored his older siblings; Calder, his brother, had been named after their father, and he also had a doting big sister named Helen. Growing up in Kenmore with its small town atmosphere and friendly inhabitants was a wonderful experience for any young boy, but during his early childhood Gibson suffered with osteomyelitis, an inflammatory process that particularly affected his right shinbone. This ailment would preclude his participation in athletics until he reached the sixth grade.

An inauspicious start

He admits he was far from a good student at Lindbergh Elementary School in Kenmore, where he began his education in 1942, and had an inauspicious beginning to his academic life. "I started out being president of my first-grade class two years in a row. They kept me around, [but] not because they liked me." Soon, however, a boyhood fascination with stars and planets began to develop, and this eventually grew into a fundamental interest in science and astronomy: "I used to draw pictures of the solar system and so forth." But his hobby was still not enough to entice him into any sort of enthusiasm for regular schoolwork. "I was young and had many other distractions, so this early interest was really the only thing I ever did that was academic in any way. My two older siblings were both A students, so I suppose I was the dunce of the family."[6]

Edward G. Gibson, PhD.

When asked how his interest in astronomy came about, the answer came quickly. "I was always fascinated with the questions: 'What's around the next bend? What's out there beyond Earth? Are there people somewhere in the stars?' Maybe I found my forced studies boring, and retreated to that in which I could find some stimulation."[6]

Having completed primary school in 1949, Gibson graduated to Kenmore High School on Highland Parkway, but he was still indecisive about what direction he wanted to take in life. However, the thought of flying with the Air Force had held a youthful appeal for him, although he admits he had not truly knuckled down to his studies at this stage of his life. "When I got to high school I improved my performance a little bit, and finally learned through several sad experiences that I had to study if I was going to get anywhere. I was still oriented toward science and math, but this did not translate itself into a firm commitment to excel in my studies. I liked football, astronomy, airplanes, girls and school – pretty much in that order. Because of all this, I barely got into college."

To his disappointment, and that of his parents – particularly his father – an application to attend Cornell University was turned down. Fortunately another application, this time to the University of Rochester in New York, was successful. "The University of Rochester is where I grew up scientifically. I'm forever in the debt of these wonderful people because they took me in and gave me guidance and an opportunity to improve myself. The quality of the education I received there was absolutely first class, and set me in the right direction."

Asked about any teachers who may have proved a particular influence or guidance in his scholastic courses and ambitions at this time, Gibson nominated Dr. Louis Contra (Rockets and Thermodynamics) and Dr. Joseph Frank (English Literature).

For the summer of 1958, Gibson went to work as a machinist at his father's marking devices company in Buffalo. Here, he learned how to operate such equipment as lathes, milling machines and punch presses. As he recalls, there was quite a bit of fatherly pressure to become involved in the running of the firm. "He made rubber stamps, steel dies and stencils among other things, and wanted me to go into business with him. It was his feeling that if I first learned the engineering side of things, I could always pick up the business side later. So, for lack of any real course direction in my life at that time, I went into engineering."

The decision would cause a greater determination and commitment in Gibson's life, but it also led him on to a new course. "Once I got into it, I found I really liked basic science more. I enjoyed studying and exploring physics, and combined that with my passion for astronomy. In this way I developed an interest in rocketry and space travel. In addition, I was a quarterback on the varsity football team, which kept me from becoming a one hundred percent bookworm! Although it was a lot of hard work to get my brain in full gear, college was a great experience."[6]

In 1959, Gibson graduated from Rochester University with a Bachelor of Science degree in engineering, and on 22 August that year, he married Julie Anne Volk from the neighbouring township of Tonawanda. She had been his girlfriend since the days when he was a senior in high school, and she was in her freshman year.

Gibson's earlier thoughts of becoming a jet pilot had dissipated due to his childhood bouts with osteomyelitis, and he realised that this was enough to preclude him from pursuing that particular ambition. "I thought that if I couldn't fly them, I might as well build them. So, much to my father's regret, I accepted a National Science Foundation Fellowship and the opportunity to earn a Master's degree in mechanical engineering with a jet propulsion option at CalTech." For the summer before going to California, he worked as a design engineer with Sylvania Electric in Buffalo, where he involved himself in conducting thermal, vibration and shock tests of electromagnetic countermeasure systems for Convair's B-58 Hustler.

During his time as a graduate student, America's space programme was on the move, and he followed each flight in the Mercury and Gemini programmes. "Like everybody else, I would stay up and watch the launches late at night or early morning, never thinking I'd have a chance to be involved in them. But I could see where it was headed and was just fascinated by it."[6]

His new bride Julie proved sublimely helpful to him, both in her moral support and by working to help make ends meet as he battled to get through his studies at CalTech. They also began a family, and became proud, excited parents of Jannet Lynn on 9 November 1960 – the first of their eventual four children. As they cared for their baby daughter his relentless studies continued. "After five hard years I got my PhD. I never would have anticipated that earlier on in life, but your self-image changes over time. It was during those five years of struggling that a certain realisation fully hit home – that if you work hard enough at almost anything, you can do well."

John Edward Gibson entered the world on 2 May 1964, and the following month, suitably armed with his PhD in engineering and a minor in physics, his father went to work as a senior research assistant for Applied Research Laboratories, an adjunct of the Ford–Philco Corporation at Newport Beach in California. Gibson's work in their Aeronautronic Division involved theoretical and experimental studies in laser pumping and optical breakdown of gases. In this he would utilise high voltage energy storage, rapid energy release and high vacuum technologies.

Changing careers

The job at Aeronautronics was pleasant, and sometimes challenging. The study of high-temperature gases was consistent with his background in plasma physics, he had a great boss, and he and Julie both loved Southern California. But while he felt the work was interesting, he also found it curiously unsatisfying. "I was kind of sitting on the fence, depending upon the type of project." One day while he was sitting at breakfast, Julie's attention was drawn to an article in the *Los Angeles Times*. She read aloud that NASA was looking for scientists who wanted to become astronauts. At first, Gibson thought his wife was playing a joke on him. "I really thought she was making it up, but then it went on and on and I knew she couldn't make it up quite that fast. It sounded too official, and it was. They were looking for scientist-astronauts, the first group. I thought long and hard about it – and at eight o'clock that morning, I applied!

"It was something I really wanted to do, and it offered a great opportunity to fly airplanes – which is also something I'd always wanted to do – and to be at the forefront in science and technology through personal space travel. I needed that type of challenge from a personal standpoint. Thankfully, Julie was with me, all the way.

"I had really been debating about whether I should apply, as the odds against me were quite extreme, and there was also my history of osteomyelitis. I thought they would blow me off pretty quickly, but then I felt I had nothing to lose."

Gibson completed the paperwork sent to him, and answered subsequent questions. He also had to go through medical examinations, and the question of his osteomyelitis came up. To his surprise, NASA's doctors finally decided that, as the disease had been dormant for twenty years, it would probably remain so, and should not be a factor in determining his suitability. He was then told that anyone who made it to the interview stage at Houston would be given a flight in a two-seat T-38 jet trainer to check his physiological reactions to strenuous aerobatics. If nothing else, he felt it was worth the effort just to get a ride in a T-38.

"So I kept on sending the paperwork back and forth and then I travelled down to the spacecraft centre in Houston for an interview. Actually, it was a physical first. They took us over to Brooks AFB (Air Force Base) in San Antonio, where they shook us and heated us and cooled us and vibrated us, and then sent us to the shrink to see what they could learn. I enjoyed the airplane ride, of course."

At the age of twenty-eight, Ed Gibson was selected in the first group of scientist-astronauts, and frankly admits he thought the testing process would have been far more demanding. "I was used to thinking and working in confined spaces, so it was no real problem for me when they put us through these exercises just to see how we would mentally respond. I felt very comfortable in all those things because of my background, and simply because I was highly motivated to do it. Eventually I thought, "Jeez – is this all they're going to have us do?" I truly expected a lot more, actually. So I was really surprised when they called me and said that I'd gotten in.[6]

"You never think of yourself on a national scale. You're always used to working on a local scale. I'd spent almost all my life up to then going through school, so I hadn't really had a chance to get out in the world and grow up or develop in any way. I was really just a kid thrown right into the thick of things in a big way. And then, to be involved in a national program like that – it was very daunting at that time."

JOSEPH P. KERWIN

On 19 February 1473, Nicolas Copernicus was born in the Polish town of Torun, nowadays called Thorn. Long recognised as the founder of the heliocentric planetary theory, he concluded that the planets, including the Earth, actually revolve around the sun. The "Copernican Revolution," as it came to be known, was a cornerstone in the development of modern science, in particular physics and astronomy.

Copernicus came from a middle-class Catholic family, which belonged to the Third Order of St. Dominic. He lived at a time when physicians made use of astrology, so he studied mathematical science and optics at Krakow University. While practicing medicine in Poland, and later as a canon at Frauenburg Cathedral, Copernicus became involved in studies of the sun, and would make celestial observations from

Joseph P. Kerwin, MD.

a high turret situated on the wall of the cathedral. As his theories evolved, he became an eminent astronomer, whose radical ideas would revolutionise perceptions concerning the order of the universe.

Just like Copernicus

On 19 February 1932, 459 years to the day after the birth of Copernicus, Joe Kerwin was born to Marie (née LeTourneux) and Edward M. Kerwin, a Chicago businessman. Like the famed astronomer, he took his early education at a Dominican Catholic school, studied medicine at university, and would later take part in space studies and experiments (in Kerwin's case, aboard America's first space station, Skylab). And just like Copernicus, his work would help revolutionise many aspects of space science.

Joseph Peter Kerwin was born in Oak Park, Illinois, as the seventh of eight children. His father had a strong work ethic, rising to become the senior vice president of confectioners E.J. Brach and Sons, who operated a large candy-producing plant on nearby Kinzie Street.

Oak Park, locally described with superlatives such as "The World's Largest Village", is situated nine miles west of the Chicago loop. It has a double cachet – both as the birthplace of renowned writer Ernest Hemingway, and as home to the largest gathering of Frank Lloyd Wright buildings and prairie-style dwellings. Originally built on a low sandy ridge, and just four-and-a-half miles in area, it remains stubbornly independent of Chicago, politically, socially and culturally. Around the time of Joe Kerwin's birth, Oak Park had a growing population estimated at 64,000 residents, but today that figure has dropped to around fifty thousand.

Like his older siblings, Joe went to elementary school in Oak Park, and then took his higher education at nearby Fenwick High School, a private Dominican Catholic college preparatory school in the Archdiocese of Chicago. It was here that Father Victor Feltrop, described by Kerwin as "my special mentor and example," taught him Latin and German. In 1949, he graduated from high school, still with no clear picture of where his future should take him. "I did have a fascination with astronomy, geology and science in general," he recalls when asked about his childhood interests.[7]

Despite the derision of his older brothers, Kerwin loved the adventure he found in reading science fiction novels, especially those by prolific authors such as A.E. Van Vogt and Robert Heinlein, while the C.S. Lewis trilogy became a special favourite. "I was fascinated by the prospect of space travel, at one time even joining the British Interplanetary Society. I didn't expect it to happen in my lifetime, and when it began I didn't expect I'd have the slightest chance of participating."

As Marie Kerwin once stated, her son's career path seemed of little interest to him at the time. "He started college really with no idea of what he wanted to become. He was in college for a while before he tried medicine."

In fact, he undertook his college education at Holy Cross in Worcester, Massachusetts. Here, he opted for a Bachelor of Arts philosophy major with a pre-med minor, thereby combining his loves for science and literature. "At the beginning of junior year, when I had to go one way or the other, I went for pre-med." It was a decision that would help shape his future. He had two admired mentors at Holy Cross; Father William Keleher, his philosophy teacher, and Vincent McBrien, a man he describes as "a brilliant math teacher." Both remained friends and occasional correspondents for years after he finished college.

In reminiscing about his days at Holy Cross, Kerwin recalled one unforgettable episode in 1951. "I wrote a sophomore English paper about life on other planets. It came back from Father 'Fuzzy' Foran with a B-minus, and with the comment in red ink on the front page: 'The grammar and paraphrasing are fine, but the subject matter is a bunch of garbage.' I still have that paper and I chuckle about it every now and again!"

After graduating from Holy Cross in 1953, clutching his Bachelor of Arts degree

The Scientific Six has become the Incredible Five. The formal NASA portrait of the Group 4 scientist-astronauts taken after the agency-enforced removal of Duane Graveline. From left: Curt Michel, Owen Garriott, Harrison (Jack) Schmitt, Ed Gibson and Joe Kerwin

and pre-med minor, Kerwin knew he wanted to continue in medicine and entered Northwestern University Medical School in Chicago. In his final year, he decided he wanted to study paediatrics, a science relating specifically to the care of infant children and treatment of their diseases.

While at Northwestern, he became casually acquainted with the Floor Supervisor at the university's hospital, an attractive young woman named Shirley Ann Good, known to one and all as Lee. They soon found an easy enjoyment in each other's company, and began dating. The romance managed to blossom even as he continued the travails of his medical studies, but those studies would prove worthwhile in 1957, when he was presented with his Doctor of Medicine degree.

Kerwin next applied for a residency in paediatrics, and was subsequently granted a military deferment. He took his internship at the District of Columbia General Hospital in Washington DC. During this time, the Soviet Union launched their first Sputnik satellite, but it did not have any memorable impact on the young intern. "I had no feeling of connection to Sputnik or the space effort at that time. I was merely a civilian guy going though an internship."

Eventually, Kerwin realised that paediatrics was not where he saw his future and decided to withdraw before he became too involved. He wrote a letter cancelling the residency, which also meant that his military deferment would be automatically revoked. "In the return mail I had a letter from Uncle Sam," was his wry recollection of the event.

Flight surgeon school

Drafted into the Navy, Kerwin was interviewed at the Bureau of Medicine and Surgery. One of the duties offered was the very last place in a flight surgeon school, involving a training programme in aviation medicine at the Naval Air Station in Pensacola, Florida. In addition to studying to be a flight surgeon, he would learn the basics of flying at pre-flight school, with about twenty hours of instruction. This appealed to Kerwin and he accepted, joining the Naval Medical Corps in July 1958.

From 1959 to 1961 he served as a medical officer in North Carolina, based at the Marine Corps Air Station at Cherry Point. He thoroughly enjoyed his tour with the Marines, and stated without hesitation: "That's when I got the bug for flying." In 1960, he and Lee were married and Kerwin then applied for a special Navy programme, in which a small number of flight surgeons could train to become naval aviators. He was accepted.

In 1961, Kerwin transferred from the naval reserve to the regular Navy, and the Illinois physician-turned-Navy doctor then entered advanced flight training at NAS Chase Field, located in Beeville, Texas. To his delight, he was named the outstanding student in his pre-flight class. Flight training appealed to him, and he would earn his coveted gold wings as a designated naval aviator at Beeville in 1962. Subsequently, he became one of only twelve Navy flight surgeons who served simultaneously as doctors and jet pilots.

On the domestic front, he and Lee had a daughter, Sharon, on 14 September 1963. Eventually, two more daughters joined the family – Joanna, born 5 January 1966, and Kristina, born 4 May 1968.

Kerwin's next naval assignment was on the medical staff of Carrier Air Group (CAG) 4, stationed at Naval Air Station Cecil Field, Florida. It was here that he met two men who would have a big impact on his life – fellow naval aviators Jim Lovell and Alan Bean. Both men had applied to be NASA astronauts and needed help with their medical documentation, and this would prove to be his first real link with the space agency. "It wasn't until I'd gotten my wings and met Lovell and Bean that I began to think I might have a chance. I was certainly ready!"

One night in 1964, he and Lee were watching television when a bulletin was read that would change his life forever. "The TV newsman was David Brinkley, who said, 'NASA has announced that it will hire scientist-astronauts – to go to the Moon.' Lee looked over at me and said, 'I'll bet you'd like to do that.' I replied 'Oh, they'd never pick me.'" At that time, he had plans to leave the Navy after his next tour and take an ophthalmology residency at Northwestern.

Despite his plans, the announcement had enthralled Kerwin. By this time he had accumulated over 2,000 hours of flying time, so he knew that with this and his medical background he was eminently qualified under the guidelines laid down by NASA for applicants. He also recalled the excitement he had felt in discussing the space programme with future astronauts Lovell and Bean. "I thought about it and applied through the Navy. Lee was initially hesitant, but she went along with my NASA ambitions with her usual grace and generosity."

He modestly shrugs off the many difficulties involved in being selected, including the rigorous battery of tests and interviews. "Since there weren't many physicians who could pass the physical and had 2,000 hours of time in a single-engine jet aircraft, that turned out to be enough, and I was accepted into the astronaut programme in 1965."

F. CURTIS MICHEL

Frank Curtis Michel was born in the Mississippi River town of La Crosse, Wisconsin, on 5 June 1934, the only son of Viola Olivia (née Knudsen) and Frank J. Michel.

F. Curtis Michel, PhD.

He was still very young when the marriage fell apart and his mother wed again, this time to Samuel Morris Bloom of San Francisco. As a result he did not have time to become attached to La Crosse, with all its history, charm and the romance of the steamboat era. The family now made their home in the California capital of Sacramento, and in 1942 his half-brother Otto was born. Just two years later his father Frank died.[8]

A career in science

Curt (as he prefers to be known) attended grade school at California Junior High from 1945-8, and over the following three years was a student at the C.K. McClatchy Senior High, where he met his future wife, Beverly Muriel Kaminsky. Having a desire to pursue a career in science, he applied for a position at the California Institute of Technology (CalTech), was successful, and began his studies in 1951, living at the popular undergraduate lodgings of Blacker House in Pasadena, one of four on-campus residences attached to the institute. In order to assist with his college fees, he worked for a short time as a disc jockey with Sacramento's KPBK radio, and then as a clerical assistant for Hale Bros.

Michel went on to receive a Bachelor of Science degree in physics from CalTech, graduating with honours in 1955. In June that year, he took up a position as a junior engineer with the Firestone Tire and Rubber Company's Guided Missile Division in South Gate, California, working on the Corporal missile programme. Then, on 6 December, he began his military service as a Second Lieutenant in the US Air Force Reserve, with an introductory posting to Lackland Air Force Base for elementary training.

In February 1956, as an Air Force Reserve Officer Training Corps (AFROTC) graduate, Michel travelled to Marana Air Base in Tucson, Arizona, where he would spend the next six months learning to fly single-engine trainers, before his reassignment to Laredo AFB in Texas for more advanced training on jet aircraft, and then Perrin AFB, also in Texas, where he learned to fly the F-86D all-weather interceptor jet aircraft.

In October 1957, Michel was given overseas duty at Bentwaters AFB in the English county of Suffolk. Here, he continued flying F-86 aircraft until his squadron's reassignment to Sembach Air Base in Germany in April 1958, situated amid rolling farm country near the vineyards of the Rhine Valley. He would remain at Sembach for the next four months, before returning to the United States in September 1958 and seeing out the few remaining weeks of his Air Force reserve service at Travis AFB in Vacaville, California. In October 1958, he received his honourable discharge.

When he left the Air Force, Captain Michel had accrued five hundred hours of flying time, mostly in the F-86, and his experience with this particular aircraft would later serve him well in his astronaut application with NASA. In a 1963 interview with the *Los Angeles Times* newspaper, he said of the F-86: "This is a radar bird, similar to the type of guidance an astronaut uses. There is very little visual contact."

While serving in his country's Air Force, he and Beverly Kaminsky had become engaged, and they were married in Sacramento on 4 October 1958, immediately following his return from Germany. He then returned to CalTech as a graduate student. "Over the next four years, I worked on an experimental thesis project under Professor Thomas (Tommy) Lauritsen and, at the same time, a theoretical thesis under Richard P. Feynman." In 1965, Feynman would win a shared Nobel Prize for Physics in recognition of his work in quantum electrodynamics. In 1986, as part of a commission set up to investigate the loss of Space Shuttle Challenger and her crew, he was the first to discover and publicly identify the cause of the explosion.

In June 1962, Curt Michel was awarded a PhD for his work in the experimental project. He stayed on at the institute to finish his work under Professor Feynman as a Research Fellow in Physics, and then did some theoretical work on astrophysics of interest to Professor William Fowler. In 1983, "Willy" Fowler (as he was universally known) would also share the Nobel Prize for Physics, as a result of his research into the creation of chemical elements inside stars. Michel certainly enjoyed an impressive array of teachers and colleagues in his time at CalTech.

By this time, a national effort to accelerate scientific research programmes essential to America's space efforts had begun. A wide range of academic programmes were undergoing intense evaluation by NASA, together with the National Science Foundation and the Department of Defense. Of particular interest to Michel was the fact that government bodies were proposing the establishment of a programme for training scientists as astronauts on Project Apollo and beyond.

When NASA began recruiting for their third intake of astronauts, Michel noticed that the test pilot requirement had been lifted, although the mandatory jet flying and education-to-degree-standard requirements remained. As well, advanced degrees or scientific research could be substituted for operational flying experience, which resulted in a large number of applications from experienced pilots with advanced academic skills and experience. He completed his application on 20 June 1963 and submitted it with high hopes that he might be one of the ten to twenty candidates NASA required.

Unfortunately, he did not make the cut of thirty-four from which the final fourteen were selected, but when he read the academic qualifications of many who had made it, he felt NASA was moving into an area which would soon encompass those whose skills were more scientifically based, rather than judging them chiefly on their flying ability. He was disappointed, but resolute; he was still building his résumé, and another opportunity would soon come along.

Rice University

In September 1963, now a theoretical physicist, Michel took on a new job as an assistant professor at Rice University in Houston. This was virtually right next door to the Manned Spacecraft Center, and many of its personnel came to Rice for graduate training. Michel began work under Professor Alexander Dessler, who would be instrumental in the formation of America's first college-level department of space sciences, which Dessler – an early supporter of the scientist-astronaut programme – would chair. According to Michel, "Alex Dessler was instrumental in getting me to come to Rice and to continue my efforts at joining the astronaut corps." For his part, Alex Dessler, now with the University of Arizona in Tucson, was delighted to welcome Michel to Rice, recalling a good friend and colleague from those days:

"I first met Curt Michel as a new faculty member in 1963. He arrived at Rice University just a few months after I did, and he was one of the founding members of the new Department of Space Science (now a part of the Department of Space Physics and Astronomy). He immediately showed himself to be an 'ideas man', who almost daily came to work with a new, refreshingly different, but always logical point of view on how things ought to be done. He thus brought to the creation of the new department a sense of adventure in what otherwise might have been a dreary administrative task.

"Curt's ability to think deeply and creatively has earned him the admiration of his fellow physicists. I have often heard him described as 'one of the smartest guys I know'. I share this opinion, and I have said the same. However, I should add that this opinion does not extend to all areas. For example, I would hesitate to take his stock-market advice.

"Because Curt is interested in just about everything to do with the physical universe, figuratively covering it from A to Z, he can be counted on to have something interesting to report, ideas that extend to even the intricacies of the latest biological findings. The area that I most enjoy seeing his dazzling display of brilliance is in explaining some new application of Einstein's Theory of General Relativity. He is able to talk about it in a way that makes you think that, for the first time, you understand relativity. I know of no one else who can do this.

"Curt Michel's varied career has been filled with remarkable achievements, and he had fun doing it. An interesting guy, a loyal friend, and a great physicist, I consider myself lucky to have spent time in his company."[9]

Rice's president in those days, Dr. Kenneth Pitzer, was also excited about having Michel at the university, and knew all about his ambitions of one day being accepted as an astronaut. In welcoming Michel to Rice, he described him as "a possible candidate as a scientist-astronaut," and said he was one of the very few men in the United States with the necessary qualifications. For his part, Michel responded by cheerfully declaring: "I'm available – and *how* I'm available!"

When reporters asked him if he had any regrets at turning down a job in industry to take the post at Rice, Michel said that it was a position he wanted very badly. "First of all, it's the kind of work I like best. I expect to teach some graduate courses and possibly do some research with the reaction of elementary particles in magnetic fields. But I'll have to admit that the proximity of Rice to the Manned Space Center was not exactly a deterrent!"[10]

Asked about his hopes of entering the space programme, he became suitably thoughtful. "I don't want to be put in the position of telling NASA what it should do. NASA knows the requirements for the first space flights much better than I do. But I personally believe it can't be very long before we must send scientifically trained men into space. We are going into space because we expect to find the unexpected, and only a man trained in science will know what he's seeing when the unexpected comes along. I certainly think a scientist should go to the Moon, even if it's someone else. I think most scientists would want to go. It's a place where man has never gone before, with many things man has never seen before.

"I mean no discredit to pilots, because I'm a pilot myself, but it's easy to visualise significant findings which an untrained mind would overlook. I'm sure there won't be much agreement at first as to what kind of training the first space scientist should have – whether he should be a physicist, an astronomer, a chemist or a geologist. I know we'll want a man with the widest training possible. I know something about physics and chemistry, and I'm studying hard to be stronger in the geological sciences."[10]

On 19 August, he and Beverly celebrated the birth of their first child – a son they called Jeffrey Bryan. A few weeks later, as part of his geological studies, Michel travelled to Australia as a member of an American survey party on a month-long expedition to map and study thirteen meteor craters near Henbury, eighty miles south of Alice Springs. The expedition was headed by Dr. Dan Milton, a crater expert with the US Geological Survey, which was mapping the Moon's surface in co-operation with the US Air Force. The area was chosen because it was located in a semi-arid region where erosion is slow, as it is on the Moon, and because of the "rays" or lines of rocky debris spread like wheel spokes around the craters, just like those observed through telescopes around their lunar equivalents. Michel described the effect as "similar to the stream of dirt thrown up when you kick the ground."

On his return to Rice, Michel began studying a theory concerning the effects of solar winds on the Moon's surface. These "winds" are a thin blast of atomic particles thrown out by the sun, and he performed research on the interaction of solar winds and the lunar atmosphere. His aim was to discover if solar winds were constantly sweeping away any atmosphere, however infinitesimal, that clings to the Moon. He was also interested in the shock waves created by solar winds, theorising that the first lunar explorers could even find the surface of the Moon composed of charged dust particles. Around this time he published a paper called *Collapse of Massive Stars*, resulting from his earlier theoretical research under Willy Fowler, which caused a considerable amount of excitement in scientific circles as it examined the notion that vast amounts of unexplained energy in space might come from solar bodies, millions of times larger than our sun, which had finally become excessively hot and collapsed inwardly.

Although his wife had been what he described as "a little negative at first," in regard to his ambitions of becoming a scientist-astronaut, she had become increasingly enthusiastic as time passed. When NASA officially announced it was seeking suitable candidates and began recruiting in October 1964, Curt Michel was already well ahead of the chase, having submitted his application for astronaut selection and security clearance documentation in June the previous year. He now updated the application with his work and studies at Rice, and to his previous involvements with the American Physical Society and American Geophysical Union, he added membership of the American Astronomical Society.

HARRISON H. SCHMITT

One inescapable reality faced by those who walked on the Moon during the Apollo programme was the comprehension that, as relatively young men, they had already carried out what would be the pinnacle achievement of their lives. They were there at the culmination of humankind's greatest-ever scientific accomplishment, so in that respect they were in the right place at the right time, and in possession of the right qualifications. However, none of them could count on doing anything to eclipse those historic hours they spent traversing the lunar surface. In some well-documented cases, it would prove to be a difficult and frequently dominating obfuscation, but for Harrison Schmitt, going to the Moon would provide the key to many new challenges and experiences, which continue to this day. He could easily, and correctly, refer to himself as the last man to step on the Moon; yet it is a mantle he seems to shun, preferring instead to be known as a trained geologist (and later, pilot) who happened to be in the right place at the right time.

Hereditary interest in geology

Harrison Hagan Schmitt certainly inherited a deep and enduring love of geology from his father, but it was defined by the rich mining history and intriguingly rugged character of the area in which he grew up. Santa Rita, where he was born on 3 July 1935, was a New Mexico mining town located fifteen miles east of Silver City. It no longer exists. Once covered by vast oceans where limestone, shale and sandstone rocks piled up into a thick, sedimentary layer over millions of years, and then changed by igneous intrusions, Santa Rita has been consumed by the very industry for which it was created, falling victim to open pit mining and a huge expanse known as the Chino Pit. It was here, tucked against rolling mountain foothills, that the techniques of bulk tonnage copper mining were first developed in the southwest, and where burrowing, digging, loading, hauling, milling and smelting have been ongoing activities for centuries, beginning with prehistoric Indians. Mining then, as today, fed the needs of many societies.

His father, Harrison Ashley Schmitt, was born in Mankato, Minnesota in 1896, and grew up in an area famous for the Farr/St. Clair Fissure, which created one of the largest areas of natural hot springs in the world. He served with distinction in the Marine Corps during World War One and according to his son, this is when his father first became interested in geology. Post-war, he worked as chief geologist for the New Jersey Zinc Mining Company in Mexico and in the United States. He gained his doctorate in geology from the University of Minnesota in 1926 after thesis work in Mexico.

On 19 October 1929, Harrison Schmitt married Ethel Malissa Hagan in Nashville, Tennessee. Ethel, one of nine children, was a pretty young teacher and amateur botanist he had first met in Silver City, New Mexico. According to Schmitt the younger, his mother came from "Giles County, Pulaski Post Office, Tennessee, where she grew up on a farm, became a teacher, took a Master's degree from Peabody College (now part of Vanderbilt), and moved to Silver City in 1928." During this

Harrison H. Schmitt, PhD.

period, she was attached to the Education Department of New Mexico State's Teachers' College, working mostly with college students who were planning careers in teaching. Following their wedding, the couple made their first home in Hanover, where the New Mexico offices of the mining company were based, and later worked together in El Paso, where Harrison started up his own geology consulting business. He quickly became an internationally recognised leader in the evaluation of ore deposits. During the hard times of the Great Depression, he would often write scientific articles for mining magazines.

April 1932 proved a happy time for the couple when their first child, a daughter named Alexandra (who became known as Sandra), was born. A year later, the couple decided to quit their El Paso office and moved back to the mining camp house they had left two years before in Hanover.

In 1936, after their son Harrison was born in Santa Rita, the family moved to the Silver Heights area of Silver City, at that time little more than a college and cow town that also served the local mining camps. It may have been a small, seemingly insignificant town, but it boasted a rich and colourful history replete with infamous wild-west characters such as Butch Cassidy and his Wild Bunch. For a time, it was even home to the notorious outlaw, Billy the Kid. It was here that another daughter, named Paula, was born to Harrison and Ethel. Sadly, she died from pneumonia in January 1939 aged just fifteen months, but the following year another daughter named Armena came into their lives, and the young family was complete.

In 1943, the Schmitts finally moved into a beautiful homestead on Cottage San Road in Silver City, nestled in the foothills of the spectacular Pinos Altos mountain ranges, on the edge of a forest leading up to the Continental Divide. As Schmitt recalled, "I could wander in the hills anytime I felt like it and I enjoyed doing that. My parents encouraged reading and knowing things about the world around you, even though we were living on the outskirts of a very small town."[11] His parents would maintain their consultancy business at this house until his father died in 1966, and his mother Ethel would remain in Silver City until her own death in 1999, aged 95.

In 1949, following his elementary education at Silver City public schools, Schmitt attended Western High School (now Silver High) and would spend many happy summer breaks working alongside his father as a field assistant. On these protracted trips, they would be actively engaged in mineral exploration across New Mexico and Arizona. At other times, his father's work for the US Defense Minerals Production Agency meant he was seldom home, taking him abroad on long trips to such countries as Peru and Yugoslavia, so young Harrison treasured those precious times he and his father spent living and working together in the great outdoors. When asked if anyone in particular provided inspiration or guidance to him at this time, his response came quickly. "Both my parents had profound influences on me in very different ways."

Following his graduation from Western High in 1953, Schmitt undertook further studies at the California Institute of Technology, receiving his Bachelor of Science degree in geology in 1957. As the winner of a Fulbright fellowship, Schmitt then travelled to Norway, where he studied geology at the University of Oslo during 1957. While he was in Norway, Russia launched the first Sputnik satellite, and Schmitt credits this event with not only capturing his imagination, but also as the occasion on which he first became interested in space exploration as a human activity.

The next step in his scholastic career took him to Harvard University, where he received his Doctorate in 1964. When asked if he could recall anyone who gave him specific guidance or inspiration at this time, he responded, "No ... I followed my instincts, but I did have many outstanding CalTech and Harvard professors." Along his scholastic way, he picked up a Society of Kennecott Fellows scholarship in geology (1958–59), a Harvard Fellowship (1959–60), a Harvard Travelling Fellowship (1960), a Parker Travelling Fellowship (1961–62), and a National Science Foundation Post-Doctoral Fellowship from the Department of Geological Sciences at Harvard (1963–64).

From 1955 to 1961, while studying for his advanced degrees, Schmitt had also spent productive time with the Norwegian Geological Survey back in western Norway. Then followed work for the US Geological Survey at sites of interest in his old stomping grounds in New Mexico and Montana, and for the US Steel Corporation in south-eastern Alaska. As a teaching fellow at Harvard in 1961, he had even assisted in teaching a course in ore deposits.

After his time at Harvard, and now clutching his brand-new PhD, Schmitt (or "Jack" as he was commonly known) was eagerly recruited by the renowned planetary scientist and geologist Eugene Shoemaker and joined the US Geological Survey's (USGS) Astrogeology Center, located at Flagstaff, Arizona. It was a definite turning point in his life, as he recalls: "Gene Shoemaker was then putting together a team of scientists to work on space-related problems, and specifically how the astronauts should explore the Moon when they successfully landed there."

Looking at the Moon

In Flagstaff, he participated in putting together photographic and telescopic maps of the Moon. Additionally, as an instructor, he began training many of NASA's pilot-astronauts to be geologic observers and field workers on their forthcoming lunar missions. He became Chief of the Lunar Field Geological Methods Project, developing techniques for sampling, description and photography on the Moon. Like Shoemaker, Schmitt knew that anyone interested in geology also had to be interested in physical activity. "You have to enjoy it, or you're not going to be a very good field geologist. There's no other way but to get your hands dirty."

Gene Shoemaker was keen to become a lunar geologist and fly to the Moon on an Apollo mission but, to his keen disappointment, was found unfit for scientist-astronaut selection after tests revealed he had Addison's Disease, a rare hormonal disorder. Nevertheless, he urged Schmitt to apply when the opportunity came along four months after he'd begun working with the Geological Survey. "I did not really plan to be an astronaut; not until NASA and the National Academy of Sciences asked for volunteers for the scientist-astronaut program back in 1964. So I thought about it for perhaps ten seconds and decided to volunteer."

Still a bachelor when he applied to NASA, it was an easy decision for Schmitt, who was prepared to devote whatever time and energy was necessary in order to participate in this great national effort and, if possible, to get to the Moon and explore its geological features. Both he and Gene Shoemaker, who actually chaired the committee that would recommend the final candidates to NASA, knew it would be the trip of a lifetime for any geologist.

Meanwhile, Schmitt's work continued for the USGS, and his colleague Don Wilhelms recalled in his book, *To a Rocky Moon*, the work Schmitt did in processing photographs from the Ranger 8 unmanned lunar probe. "The most novel new stroke was by Jack Schmitt. Drawing on his mission-planning work in Flagstaff, Jack used a high-resolution photo as the basis not only for a geologic map, but also for a simulated manned mission. Traverses meander from the landing site of the LEM across features of geologic interest, just as they would on the maps packed in lunar modules a few years later."[12]

Because of Shoemaker's need to avoid any conflict of interest as part of the selection committee, Schmitt had to make the grade and qualify on several fronts through his own efforts. In fact, two of his colleagues at USGS, Dan Milton and Mike Duke, fell by the wayside as a result of the intense physical testing and evaluation, and were eliminated. However, as Wilhelms so rightly noted in his book: "Shoemaker's committee nominated only sixteen survivors of the testing. Among all us USGS astrogeologists, which he had been for a year, only Jack had the required combination of good health, good eyesight, good hand–eye coordination, and the potential ability to fly a jet."[12] So, in July 1965, the one geologist who would both prepare and execute a lunar landing mission transferred from the USGS to NASA and began fifty-three weeks of jet pilot training at Williams Air Force Base in Arizona.

DUANE E. GRAVELINE

For a young Air Force medical trainee named Duane Graveline, the defining moment in his life occurred one day in 1957 when he heard that the Soviet Union had launched a satellite named Sputnik into Earth orbit. "All things are supposed to have a beginning," he recalled, "and I guess Sputnik started it for me. From that moment on I did my best to guide my path towards space."[13]

Early influences

Duane Edgar Graveline was born on 2 March 1931 in the small border town of Newport, Vermont, located ten miles south of Canada on the southern tip of the spectacular Lake Memphremagog. His parents, Edgar and Tina, imbued in their older son a strong, lifelong pride in his French Canadian heritage. Before they met, his mother Tina Lamere had been a renowned ski jumper, while his father Edgar would become a sporting goods store owner and marina operator with a grand passion for flying.

Young Duane spent much of his early childhood living with his beloved grandparents on their 300-acre dairy farm, and took his education at a nearby one-room schoolhouse, which his father had attended years before him. He still recalls with fondness the day he found his father's name etched deeply into one of the classroom seats.

Today he credits his grandmother, affectionately known as Mamere, with giving him guidance on the road to becoming a doctor. "Even before I started school, my Mamere always told me I would be a doctor. It did not seem to be a matter of choice. I willingly walked the path she set for me from my earliest years, never doubting even for a moment that it was the right path."

Sadly, tragedy would strike him early in life. He has strong recollections of his grandfather, his wonderful Papere, "holding me by the hands as he jiggled me up and down on his foot while singing ribald French Canadian songs of women and college days in his past. At other times, he would just hug me and dance around the living room to imaginary music. Although it was Mamere who provided the guidance, it was Papere who gave me love and the most heart-warming memories. He was a great bear of a man with enough affection to go around for everyone." Duane was only six years old when he witnessed a kerosene heater in his grandparent's house explode into flames, killing his adored Papere. The horror of that day remains a brutally vivid memory, and the thought of death by fire still haunts him.

Life continued after the accident, and Graveline remains eternally grateful to his grade school teacher, Iris Wheeler, for developing and nurturing his interest in science. In one such instance, she was able to spike her young student's curiosity by "the simple act of placing a jar of frog eggs on my desk so that I could more intimately watch the almost hourly change from egg to tadpole. This probably more than any other thing pointed me along the path of scientific enquiry."

As he grew up, other teachers influenced his career path. His high school science teacher Arlene Cushing was one who "got my undivided attention with a demonstration of the explosive qualities of a beaker of hydrogen gas," while another, Lyman Rowell, later astounded the young college biology student with his "chalkboard dexterity using two hands simultaneously as he sketched the process of gastrulation. It still impresses me."

The University of Vermont had long held a strong appeal for him, and seemed a logical college choice. It was also not too far from where he lived. "Getting home easily seemed to have been a high priority in those early days, when the apron strings were still tight and your mother looked forward to doing your laundry." It was 1948, and the college was jam-packed with older war veterans recently out of the military, all taking full advantage of the government's free college education programme.

The University had a compulsory ROTC programme, and Graveline soon grew to love the disciplines involved. He became so accomplished in close order drill that he was made squad leader. "They later offered me the officers' training programme, which I would have accepted, but it was designed for a four-year college curriculum and I knew that when my third year of college ended I would have all my graduation credits and would be starting medical school, so I could not accept the programme. However, my introduction to the military was pleasant and had a very important role in my future career decisions."

As he was trying to cram four years of studies into three, Duane had little time for extra-curricular activities, although he did play second trumpet in the university band in his first year. In order to raise a little extra spending money, he also took on a part-time job as an assistant instructor in biology, conducting class work and special tutoring for freshmen students.

At the beginning of his junior year in undergraduate school, he made the unprecedented move of applying for medical school. The school's initial response stated that his application was invalid because he was not yet a senior, but when they checked the records it was clear he had met, or would soon meet, all the preliminary requirements. As he would have sufficient credits in a few months for graduation with a Bachelor of Science degree, his application was finally accepted and approved. Following this, he started medical school in 1951 as a member of the university's June class. He also married his high school sweetheart, Carole Jane Tollerton, in Newport, Vermont. Twelve months later Graveline received his Bachelor of Science degree, and in 1955 was awarded a Doctor of Medicine degree from the university's medical school.

Flight surgeon

By now he was keen to join the Air Force as a flight surgeon, no doubt influenced in this ambition by his father, who had been a pilot in the Civil Air Patrol. Graveline could never understand his younger brother, Norman, not sharing this sublime passion. Following a two-month assignment as an assistant to a family doctor in

Vermont, he drove down to Washington DC to take up a USAF Medical Service internship at the Walter Reed Medical Center.

In June 1956, Graveline attended the primary course in Aviation Medicine, Class 566, at Randolph AFB in Texas. He was later assigned to Kelly AFB, also in Texas, as Chief of the Aviation Medicine Service. His job as flight surgeon required him to become familiar with the flight environment, and he loved every minute of this training, during which, "I flew in everything they had with more than one seat."

Reasoning that a pilot might become incapacitated during flight, requiring an accompanying flight surgeon to assume control, the Air Force trained its flight surgeons to a "ready-to-solo" level of experience. In this way, he got to fly such aircraft as the C-45, KC-97, C-124, XC-99 and C-119, while he racked up thirty hours of instruction time, "the unforgettable frosting on that delicious flying cake," in the sleek T-bird jet trainers, particularly the TF-100 and TF-102. "I actually used to feel somewhat guilty when I picked up my pay check with its extra cash for flight pay. Where else but the Air Force was this kind of thing readily available to you and they paid you for it?" When asked, his favourite recollection of that time is of the many hours he spent flying as second pilot aboard the nimble, two-seater T-33 jet trainer. "During my instruction in this bird, I still remember the thrill of rolling her over into a split S and holding her on 'burble' during pullout. I loved it."

In October 1957 Graveline was attending Johns Hopkins University as part of his Aerospace Medical residency, and studying the effects of prolonged weightlessness on the human body, when he learned that Russia had launched the world's first artificial satellite into orbit. His excitement at this event was only compounded the following month when canine passenger Laika was sent aloft on a one-way journey aboard Sputnik II. "Little did I know then that only a few years later at Wright-Patterson AFB I would be studying Laika's bio-readouts in the top-secret labyrinth of FTD, our Foreign Technology Division. Through Foreign Technology, I learned the true scope of the Soviet bioastronautics programme and it was impressive. By the time we launched John Glenn, the Soviets had given us a tremendous amount of information about space flight."

His time at Johns Hopkins came to an end in 1958 when he departed with a Master's degree in public health. He then took up the Aerospace Medical residency at the Air Force School of Aerospace Medicine, completing his residency training at Brooks AFB, San Antonio, in July 1960.

Because of his interest and research into the biological effects of weightlessness, Graveline (now known to one and all as "Doc") became a medical flight controller for NASA's Mercury programme. In this capacity, he was involved in events leading up to the orbital mission of John Glenn, especially the biomedical studies surrounding the precursor space flight of chimpanzee Enos. At this time, he was temporarily based on Canton Island in the South Pacific. He then continued working for NASA as a flight controller in the early phases of the Gemini programme. During this period of his life, he was also still working with the FTD as a medical analyst, attempting to deduce what the Soviets were doing in their bioastronautics programme.

Eventually, he carried out two tours at Canton Island, and two tours aboard the tracking ship *Rose Knot Victor*. While conducting his studies on one of these seaborne tours in March 1965, he was engaged in some particularly rewarding work during the flight of cosmonauts Aleksey Leonov and Pavel Belyayev aboard their Voskhod 2 spacecraft. "I was able to direct our entire worldwide tracking network to monitor the bio-data emanating from a Voskhod multi-manned spacecraft, coincidentally launched by the Soviets during our deployment. All we needed were our high frequency receivers and simple antennas – standard items at every tracking station. The frequencies and orbital parameters were carried in my head."

The year before, Graveline had read that NASA was accepting applications for the position of scientist-astronaut, and fired in his application – one of 1700 eventually received by the space agency. It was a long, anxious wait of almost a year for the applicants, as their numbers were slowly culled down to four hundred through a process of elimination, and the remaining applications were forwarded to the National Academy of Sciences for further perusal. Their review eventually recommended that he be one of a group brought to the School of Aerospace Medicine at Brooks AFB for evaluative testing.

"The screening process took several weeks, and included the usual assortment of medical tests, prodding and probing into every nook and cranny, personality and emotional evaluation with all the usual inkblot, Rorschach and MMPI (Minnesota Multiphasic Personality Inventory) tests, and even intelligence testing with all the latest methods." When asked about the rigorous testing to which he was subjected in the latter stages of the selection process, he recalled his experience pulling 10 G on the centrifuge as the most difficult. "At that time I was thirty-five, but even then, transverse Gs at those levels were very uncomfortable. Perhaps as a doctor I was aware that the discomfort was coming from the stretching of tissues in my body not designed for this kind of thing."

NASA flight surgeon and Navy pilot Dr. Fred Kelly (himself an unsuccessful applicant for the same group) had previously worked with "Doc" Graveline, and was watching his application with interest, as he later wrote in his book:

"Duane Graveline – now there was a fierce competitor. He was real competition. He was firmly established as one of the leading researchers in aerospace medicine, had performed most of the basic research into the physiology of weightlessness, and was the first to use underwater simulation of weightlessness. He was now assigned by the US Air Force to monitor the biological data from every space flight the Russians had put up. He probably knew more about the Russian space programme than the Russians. In addition to all of his professional qualifications, he was young, articulate, ran three miles every day, and certainly had the motivation. I would be disappointed in the system if Duane were not selected. Duane was definitely my first choice – after me."[14]

With Graveline's selection as one of six Group 4 scientist-astronauts, there was another decision to be made. "After my acceptance as a NASA scientist-astronaut was announced, I made my decision to leave the Air Force. I was a major at that time. My reasoning was that I wanted to be permanently with NASA in a civilian capacity, not as an Air Force 'loaner.' I thought my career in NASA would be better if they 'owned' me." It was a decision that would soon have an enormous impact on his future life.

84 The Scientific Six

A time of devastation

Within weeks of NASA announcing the names of the six scientist-astronauts, a furore of monumental proportions erupted. In July, Carole Jane Graveline filed for divorce in San Antonio, accusing her husband of "harsh, cruel and tyrannical treatment" and saying he had an "uncontrollable" temper. While understandably devastating for her husband, it created a massive media headache for NASA, who at that time was desperate to maintain the pristine, clean-cut, all-American image of its astronauts. Surprising as it may now seem, reporters, editors and other media people had cooperated. They meekly complied with this carefully created image of the squeaky-clean,

Duane E. Graveline, MD. Graveline said this NASA portrait was taken only hours after his return to Houston after being dismissed from the astronaut corps. "You can see I look the picture of complete misery," he later reflected on the photograph.

untarnished hero, and chose not to report many of the wilder activities and affairs readily ascribed to many of the early astronauts. Now there was mounting confusion. But the newly-selected scientist-astronauts continued as normal with their preparations to go to supersonic flight school at Williams AFB in Arizona in August. Following this, they would undertake their initial astronaut training in the summer of 1966.

The uproar quickly descended like a black cloud upon the Astronaut Office, and Chief Astronaut Deke Slayton was unimpressed. "The programme didn't need a scandal. A messy divorce meant a quick ticket back to wherever you came from – not because we were trying to enforce morality, which was impossible, anyway, but because it would detract from the job."[15]

Fuelled by adverse publicity, the story of an astronaut's wife suing for divorce caused growing concern for the space agency, who could no longer keep a lid on the situation. Faced with this dilemma, NASA director Dr. Robert Gilruth had to make a quick decision, and his verdict was soon despatched to the Astronaut Office for prosecution. The news hit a bewildered Graveline like a blow; he was out – end of story. As Deke Slayton later related, "Graveline was the first to get it. He was back in the Life Sciences Division so fast he never even made it to the group photo." It was also a salutary lesson for the other astronauts, as evinced by Walt Cunningham from the previous group, selected in October 1963: "That was as dramatic an example to the rest of us as a neck attached to a swinging rope was to horse thieves in the Old West."

The whole sordid business outraged Graveline's friend and fellow NASA flight surgeon Fred Kelly, and he later wrote of his disgust at the way the whole affair had been handled. Like other insiders at the space agency, he knew that many of the astronauts were far from innocents themselves, and were living perilously close to the marital edge. It was well known that two of the Mercury astronauts, for instance, had only maintained their failed marriages for the sake of keeping up appearances, and to remain part of the space programme:

"Astronauts were supposed to be immune from social and marital ills. They were all fair-haired boys living in vine-covered cottages in perfect harmony. Duane's wife called a press conference to deliver an indictment so devastating that NASA had asked for his resignation. It was more than a request; he was out! Here was a man I considered head and shoulders above all the other scientist-astronaut selectees. He was wasted because his wife had filed for divorce. If this was going to set a precedent, they would have to select more astronauts."

Looking back, "Doc" Graveline can afford to be philosophical and forgiving to an extent, and he says he cannot help but "deeply envy" those of his colleagues who realised their dreams of living and working in space. "I had no problem with NASA's reaction. This was the first publicised divorce of one of their shining knights. Their reaction was entirely reasonable and even predictable. What else could they do? With respect to Dr. Gilruth, it was his decision that day, in his office at NASA. I bear him no grudge. My divorce publicity was anything but appropriate for a newly-appointed astronaut. He did what he had to do at that time. I have to admit that when the publicity came out I felt like someone who had just received a lethal dose of radiation – I knew I was dead but was just not sure of when it would happen. Five months after the ruinous publicity, and well into the flight-training programme, I resigned from NASA officially 'for personal reasons' and took up life again with my family. My resignation was the hardest decision of my life."

"It was a reconciliation doomed to failure, for every time an Apollo mission occurred I was pulled more deeply into self-doubt and troubled thoughts. A year passed and finally, like Thoreau, I moved alone into my little cabin in the woods of northern Vermont and gradually regained my footing."

The divorce went ahead, but it was a muted affair after the early, scandalous publicity that had ruined his career as a scientist-astronaut. He will not speak publicly of Carole's motivation except to add that she later admitted her timing could have been better. "I consider that a masterpiece of understatement," was his final comment on the matter.

With a return to civilian life, Duane Graveline once again began to practice medicine as a family doctor in Burlington, Vermont. During this time he also served as a flight surgeon for the Vermont Army National Guard.

Prior to the first Space Shuttle flight in 1981, Dr. Fred Kelly invited his friend to make a temporary return to NASA as Director of Medical Operations. Happy to be involved, he took a six-month leave of absence from his family practice to assist Kelly on the first four Shuttle missions. "Otherwise my only NASA contact has been as a regular participant in the Longitudinal Study of Astronaut Health, getting my annual physical check-up at Johnson Space Center."

Other roads to travel

Following his retirement from medical practice at the age of sixty, 'Doc' Graveline has become a prolific author of medical and science-fiction thrillers, with nine published novels to his credit. His website[13] gives a revealing and far more comprehensive look into his impressive career, with detailed information on the groundbreaking bioastronautical work in which he participated.

He now lives a contented but prodigious life, with a supportive wife at his side. Suzanne Gamache had lived in California, Hawaii and Florida before they met. "Both her parents were born in Canada, making this French Canadian beauty the perfect match for me. We were married in Newport, Vermont." When asked if he would change any facet of his life, the smile and answer came quickly and easily. "I would want to turn back the clock to 1955 and meet my present wife Suzanne just as I started my USAF internship at Walter Reed Army Hospital. We are a perfect pair – we just found each other thirty years too late. I would love to restart my life with her and travel my same path to astronaut selection."

These days, Duane Graveline owns a second home in Cape Canaveral, where he and Suzanne happily spend each winter. It must be a very wistful, introspective sensation for him to live within sight of a place where he might have achieved his now unrequited dream of a lifetime. "My entire career has been devoted to the space programme. The Cape seems to fit me."

"Doc" Graveline remains philosophical about the many and varied directions his life has taken him, but he remains passionate about space and aerospace medicine.

Duane Graveline addresses a forum on the misuse of statin drugs, Sydney, Australia, November 2004 [Credit Colin Burgess].

Every year he returns to the Johnson Space Center for his astronaut physical, writes about the space programme in his many science-fiction novels, and has prepared a poster display for the Space Walk of Fame near the Kennedy Space Center, as an educational resource for science teachers. He is still actively involved with KSC aerospace researchers on medical support for space crews.

In late 2004, following the publication of his latest non-fiction book *Lipitor, Thief of Memories: Statin Drugs and the Misguided War on Cholesterol* (republished as *Statin Drug Side Effects and the Misguided War on Cholesterol*), Graveline began working with KSC to establish liaison biomedical research projects with a hospital located near Merritt Island in Florida. The outcome of this endeavour was a KSC funding package that was submitted to NASA Headquarters. It contained four proposals, uniquely his, and for which he will be principal investigator.

"So, after forty years, I am back in the research saddle in space medicine. My current projects are (1) the use of short radius arm centrifuge to prevent bone demineralisation, (2) the effectiveness of anti-diuretic hormone to assist astronaut rehydration prior to return to Earth, (3) use of intermittently inflated waist tourniquets during space flight to prevent post-mission orthostatic intolerance, and (4) 2 G-adapted mice (centrifugation requirements for maintenance of dense bones). In many respects, I have never left the programme."

Despite the travails of the past, however, there will always be a certain quiet comfort and satisfaction in having been able to ascend through his own endeavours from a one-room schoolhouse education to the position of a USAF research scientist, and now a space medicine researcher. And of course, albeit briefly, being selected as one of his nation's very finest – a NASA astronaut.

THE "ALMOST" SCIENTIST-ASTRONAUTS (1965)

As the six new scientist-astronauts became known to the world, the other nine applicants who had attended the medical and psychological evaluations at San Antonio returned to their institutes and, for many, anonymity. Perhaps, if this selection was successful, then they might get the chance to apply again and receive the call from NASA to commence training as astronauts. Perhaps ...

The unsuccessful nine were:

David J. Atkinson: No other information known.

C. William Birky: Born on 5 June 1937 in Champaigne, Illinois. His father Carl William Birky and his mother Pauline Elizabeth (formerly Livengood) were from farming families, although his father was a sociologist and his mother a home economist. For five years he grew up on the farm in Illinois before the family moved to Bloomington, Indiana, where his father served on the faculty at Indiana University for three years. His father then became a faculty member of Colorado State University, moving his family to Fort Collins, Colorado.[16] Birky completed undergraduate work at the Indiana University, Colorado State University, the American University of Beirut in the Lebanon, and Woods Hole Marine Biological Laboratory. He gained a BA in 1959 and his PhD in 1963, both from Indiana University. From 1964 to 1970 he was an instructor, Assistant Professor at the Department of Zoology and a member of the interdepartmental genetics programme at the University of California at Berkeley. He was the Associate Professor, Department of Genetics, Ohio State University between 1970 and 1976, and a member of the Interdepartmental Molecular, Cellular and Developmental Biology Programme at Ohio State from 1970 until 1986. In 1976 he became Professor of the Department of Genetics (now Molecular Genetics) at Ohio State until 1997. He moved to the University of Arizona in 1997 as Professor of the Department of Ecology and Evolutionary Biology and Chairman of the Genetics Committee/Graduate Interdisciplinary Program in Genetics. He has authored over sixty papers and book chapters and given over seventy-five seminar presentations, as well as public service lectures in his field.[17]

Michael B. Duke: After receiving his PhD in Geochemistry from the California Institute of Technology (his thesis was on basaltic meteorites), Duke joined the Astrogeology Studies Branch of the US Geological Survey. Subsequently his work on micro-mineralogical analysis led to an interest in lunar regolith, which in turn would see him selected as a principal investigator in the Apollo 11 Lunar Sample Program. Duke became Curator at the Lunar and Planetary Institute near the Johnson Spacecraft Center, where he helped develop the procedures and facilities that house the Apollo lunar collection. His enthusiasm for human exploration of the Moon and Mars was evident during his term as Chief of Solar System Exploration at JSC, initiating a series of symposia and publications dealing with the human exploration of the Moon and eventually Mars, as related to the Space Exploration Initiative. He continues to participate in the planning activities for robotic and human exploration of the Moon and Mars and works at the Colorado School of Mines as a research professor.

Ernest J. DuPraw: Published several papers and books in the fields of cell and molecular biochemistry.

Don V. Keller: A Berkeley-educated physicist who holds a PhD in high-energy particle physics, he founded Effects Technology Inc. in 1969, conducting experiments simulating the effects of nuclear weapons on various materials. Two years later he founded Ktech Corporation in Santa Barbara, California, as a three-man weapons testing contractor. Two years later the company moved to Albuquerque, New Mexico, where it now operates the world's largest ion beam fusion accelerator. He retired as chairman of Ktech Corporation in January 2006. He and his wife live in a desert home in the Albuquerque area.[18]

Daniel J. Milton: No other information known.

William G. Tifft: Born 5 April 1932 in Connecticut, he gained his BA in Astronomy in 1954 from Harvard University before earning his PhD in 1958 from the California Institute of Technology. His research interests have been in galaxies, super clusters and red shift problems. Tifft's early career included a Harvard undergraduate education, CalTech graduate work, and post-doctorate work in Australia. He also worked at Vanderbilt and Lowell Observatory before becoming involved in space astronomy in 1963. His involvement included work in the manned space programme, and in particular in Apollo Applications (Skylab). Tifft worked on the UV sky survey camera and other projects with NASA between 1963–65, including working at Marshall SFC and participating in a 1964 meeting to coordinate manned and unmanned space programmes. His involvement with the Space Telescope dates back to a 1965 meeting at Woods Hole at which he made suggestions for managing and observing time in space astronomy and discussed a Moon-based observatory plan. His interest in using the Moon for astronomical research conducted by astronaut "space-scientists" was explained in a 1966 article:

"The lunar observatory will come progressively. Various investigations and design studies could lead to a pilot telescope of perhaps 50-inch aperture on a LEM truck or shelter in the mid-1970s. A permanently manned large telescope could come by 1980 ... Several small outlying sites may be developed for continuous access to all parts of the sky ... Instruments will be operated from one or more pressurized control centers by a crew that will probably rotate on about a yearly basis. A typical crew might include 'space scientists' ... Leading observatories on Earth making major use of the facility might provide 'observatory representatives,' who would carry out assignments for their home staff ... Once in a while uniquely talented men could be brought in as 'passengers.' Engineering scientists ... would be needed to maintain the operating facility and basic 'technical staff' for general support. A variety of other scientists, including geologists, biologists, and physicists, would logically utilize the same basic location for their operations. The major astronomical center on the Moon could well emerge as the basic center for permanent lunar operations. It will come."[19]

After applying for the scientist-astronaut programme, he decided to return to ground-based astronomy.[20] In 1970 he performed comparisons of red shifts within a single galaxy cluster and found that individual red shift differed by multiples of 72 km/sec. He is an Emeritus Professor of Astronomy at the University of Arizona and Principle Scientist for the Scientific Association for the Study of Time in Physics in Cosmology (SASTPC).

Robert Woodruff: No other information known.

Philip J. Wyatt: Received undergraduate education in liberal arts, physics and mathematics at the University of Chicago and Christ College in Cambridge, England. He gained a BS degree from the University of Chicago, an MS degree from the University of Illinois and his PhD from Florida State University. He became an industrial physicist and during his career, founded the first commercial instrument company to incorporate a laser and an on-board microprocessor. The author of over fifty articles and co-author of three text books, Wyatt has had over thirty foreign and domestic patents issued relating to laser light scattering. In March 1982 he founded the Wyatt Technology Corporation, which was formed around his patents, ideas and inventions for industrial, military and medical domains. The work of WTC has included developing instruments capable of measuring the multi-angle light scattering characteristics of macromolecules (and particles in solution) and refractometers, as well as instruments for airborne aerosol samples.[21]

REFERENCES

1. Email from Owen Garriott to Colin Burgess, 23 Oct 2002.
2. From papers supplied by Owen Garriott to Colin Burgess, 5 Mar 2003.
3. F. Curtis Michel papers, Rice University, Houston.
4. All information supplied and checked by Garriott in emails to Colin Burgess; 23 Sep 2002, 25 Sep 2002, 22 Jan 2004, 28 Jan 2004.
5. Most information supplied and checked by Gibson in emails to Colin Burgess, 2 May 2003, 5 May 2003, except where noted.
6. Quotes from Gibson's NASA Johnson Space Center Oral History interview with Carol Butler, 1 December 2000. Used with permission of Ed Gibson.
7. Family information supplied and checked by Kerwin in emails to Colin Burgess, 25 November 2004, 1 December 2004 and 12 January 2005.
8. All information and quotes supplied and checked by Michel, except for quotes otherwise noted.
9. Quote taken from an email from Alex Dessler to Colin Burgess, 7 December 2002.
10. This quote and others taken from the *Los Angeles Times*, 7 July 1963.
11. Family information supplied and checked by Schmitt in emails to Colin Burgess, 3 December 2002, 27 December 2002 and 3 January 2003.
12. *To a Rocky Moon*, Don Wilhelms, University of Arizona Press, 1993.
13. All of the quotes and verification in this story come from Duane Graveline's website, *www.spacedoc.net* (used with his written permission) including responses he gave in a telephone interview with David Cannetti; and from emails from Graveline to Colin Burgess, 28 & 29 Jan 2004, and 22 May 2004.
14. *America's Astronauts and Their Indestructible Spirit*, Dr. Fred Kelly, TAB Aero Books, 1986.
15. *Deke! US Manned Space: From Mercury to the Shuttle*, previously cited.
16. Email to Dave Shayler from Bill Birky, 4 June 2002.
17. C. William Birky, Jr., Curriculum Vitae, University of Arizona website: *http://eebweb.arizona.edu/faculty/birky/CV.html*
18. *UPI Science News*, 17 January 2006, "Successful physicist-businessman retires." Online at *http://science.monstersandcritics.com/news/printer_1076701.php*
19. Excerpt from "Astronomy, Space, and the Moon," *Astronautics & Aeronautics*, Dec 1966, pp. 40–53.
20. Interview abstract from the oral history of Tifft, William G., 13 December 1984. Interviewer: Joseph Tatarewicz. Auspices: Space Telescope History Project, NASM.
21. *www.wyatt.com/company/ourhistory/* and *www.wyatt.com/company/executiveteam/Philip.cfm*

4

School for Scientists

In selecting the first group of scientist-astronauts, NASA required the successful applicants to complete a period of academic study and survival courses. While this was a logical and acceptable condition, there was another, particularly rigid requirement; any person selected into the programme would have to undertake training to qualify as a military jet pilot – if they had not already done so in earlier military service. While it may have seemed a daunting prerequisite to some other scientist applicants, Owen Garriott says that in fact, it was "an eagerly anticipated portion of the training" for him and his successful colleagues.[1]

FLIGHT TRAINING

Ed Gibson had once nurtured a dream of flying for the United States Air Force, but a youthful bout of osteomyelitis, a bacterial inflammation of the bone, had put an end to this particular ambition.[2] Back then, even when successfully treated, it was still regarded by the Air Force as a basis for disqualification. Suddenly, with his acceptance as a scientist-astronaut, Gibson was no longer precluded from taking to the skies and flying solo in some of the country's finest jet aircraft. This, combined with his acceptance as a NASA scientist-astronaut, brought an excitement he found hard to contain.

First however, there was much to be done, and a great deal to learn. "I felt like an imposter," he said of his first few days with NASA. "People treat you like you know everything about the space program and you don't. So it takes a while to make that adjustment."[3]

Two of their number, Joe Kerwin and Curt Michel, were already qualified pilots with jet experience, so they were exempted from this aspect of the group's astronaut training. Because of this, NASA made the sensible decision to defer the scientist-astronauts' academic and survival training until the four requiring military qualification returned from flight training. They would then combine forces with the next astronaut selection group (Group 5), in order to fulfil these phases of their training.

Kerwin learned of the decision to recruit more astronauts when he attended his first Monday morning pilots' meeting at MSC's Building 4, in the third-floor conference room. "Deke [Slayton] and Al [Shepard] were at the front of the room. The guys are sitting there, and I'm in the back, and they introduced me. Then Shepard says, 'Headquarters has okayed the selection of another group of astronauts next year.' And Dick Gordon said, 'Are they going to be pilots?' Shepard said, 'Well, I certainly hope so!' So I realised that I was here, but I didn't have a distinct role yet." Kerwin also recalls asking Shepard if he should keep up his medical proficiency by going to the clinic once or twice a week. Shepard responded that he didn't think it was a very good idea, as the new astronaut would have more than enough to occupy his time. "You get your priorities in order that way," Kerwin admitted, "and realise that clinical medicine is not going to be your job in this program; that keeping on top of medical research and medical issues in space flight was very important. He encouraged me to train as an astronaut. I was going to sit in that cockpit, and they had to be able to rely on me." As he later discovered, there were so many medical issues involved in flying into space "that it was a very easy profession to reconcile with the duties of an astronaut."[4]

Curt Michel says that, in his opinion, there was a little uncertainty in their group about their actual role within NASA, and the way in which they were being portrayed as Apollo astronauts. "Among the various items of evidence we were asked to supply was a statement of what scientific objectives we would perform on the lunar surface," he wrote in his journal at the time. "Newsmen asked us why we wanted to go to the Moon. Everyone evidently thought of us as going on a lunar landing mission, except the Astronaut Office, which kept a discreet silence beyond letting it be known that the first three landing missions would be manned by regulars."

On 29 July 1965, while Kerwin and Michel remained in Houston, their four colleagues were despatched to Williams Air Base in Chandler, Arizona, under a special agreement between NASA and the Air Force for what that service called their Undergraduate Pilot Training Program. Owen Garriott was a private pilot, while Ed Gibson had soloed small tail-dragger aircraft ("with the instructor hiding behind the barn," Gibson once quipped) in San Clemente. Jack Schmitt was the only one who had never flown an airplane.

Duane Graveline had run up plenty of logged hours in light and fixed-wing aircraft, and while under pilot instruction in the T-33 jet airplane, but it was not considered sufficient experience under NASA's guidelines. "Despite my thirty hours of IP (Instructor Pilot) time in the T-bird (to the ready-to-solo status, routine for most USAF flight surgeons at that time) and some 1,500 hours overall in dual-seat T-birds, I could not transition into a T-38 without additional training. I suppose you could quibble that I did not need the full training year required for my more academic peers, but I was so delighted to be selected as a scientist-astronaut that quibbling never entered my mind."[5]

Now the four scientists faced up to some *real* flying in powerful jet aircraft, but first there were certain preliminaries. Flight training for the four men would entail 300 hours of classroom work and around 240 actual flying hours – all but about thirty of which would be in T-37 and T-38 jet trainers. Over the next fifty-one weeks they would work twelve-hour days beginning at 05:30 a.m. and attend classes in navigation, meteorology, radar, aviation physiology and other flight-related subjects. Like their sixty-four fellow students, they would also be required to fly many missions in a Link trainer as well as in actual aircraft. Apart from some military officer training that the other cadet-pilots had to go through, the scientist-astronauts would participate in all phases of student activity, which included a course in self-confidence. There would also be callisthenics, and supervised sports.

Assuming all four men graduated from flight training, the newly-qualified jet pilots would then link up with Kerwin and Michel back at MSC to complete their specialised astronaut training. In the meantime, Kerwin and Michel would occupy their year performing routine support assignments in the developing Apollo Applications Program.

"I loved it," Gibson recalled of the year's flight training, although he quickly added, "The military service is a bureaucratic process, and so they didn't know how to treat us. One day we'd be out there as sub-airmen picking up cigarette butts – a part of one day, as I recall – and the next day we'd be out there meeting dignitaries coming through. But we just looked like anybody else, had standard Air Force flight suits on (albeit without any kind of military insignia), and went through the classes like anybody else."[3]

Gibson and Garriott were assigned to the same squadron, Graveline and Schmitt to another. Schmitt recalls that most of his group, the Class of 1967A, were Air Force Academy graduates. "So, it was a really remarkable group of young men that were going through pilot training, all of whom were about ten years younger than we were." Not only did the quality of the young men wanting to become pilots in the Air Force impress Schmitt, but also the staff that were part of Air Force Training Command. "They did a really remarkable and highly professional job. You can understand why American pilots do as well as they do in combat as well as in peacetime, because of the quality of training they're getting."[6] In fact one of the men in their group, Eugene Habinger, would end up as a three-star general and Commander-in-Chief of the Strategic Command some thirty years later, responsible for all the US Air Force and US Navy strategic nuclear forces.[1]

Schmitt also believes the Air Force had difficulties in "putting up with" the four civilians on their base. "If I remember correctly, we were told we were the first civilians ever to receive Air Force pilot wings. The Air Force seemed to tolerate us. We were probably a bit of a thorn in their side, because we could not be disciplined in the same way as the military pilots. They could just tell NASA, 'We don't want them anymore,' I guess, and then deal with NASA at that point. I think everybody was satisfied with the relationship with the Air Force. From our perspective, I think we fit in very well."[6]

"Of course we were really old folks at that time," Garriott laughingly recalled. "I was thirty-four years old when we went to Williams, and I think I was about the oldest one that ever went to pilot training and graduated; that is until Karl Henize joined NASA as a scientist-astronaut candidate two years later and took on similar flight training. Most of those in our training group were in the twenty-one and twenty-two-year-old range."[1]

Then, about three weeks after their flight training began, came a bombshell. NASA unexpectedly recalled Graveline from flight training. As he prepared to leave Williams AFB he had no inkling of what was to come. "In my enthusiasm and naïveté, I seriously entertained thoughts that it might be for a special assignment, possibly to do with the Soviet space programme because of my background in intelligence." It was, in fact, quite the opposite; his wife had announced she was suing for divorce. It was going to get ugly, and he was forced into an immediate resignation from NASA, effective 18 August. "My impending divorce was the last thing on my mind," he mused, looking back with considerable regret to that difficult time. "But it proved to be the first thing on theirs. Definitely not my best day."[5]

There was a clear rationale and importance behind the flight-training requirement for the scientist-astronauts. They would have to become familiar with certain capabilities and conditions such as life support systems, acceleration, communications, vestibular activity and vertigo. As well, there was a man-to-machine integration phase to be experienced. This not only included aircraft controls, but checklists and procedures discipline, systems analysis and operation, and the relationship between the operations manual, vehicle performance, and vehicle performance specifications. Additionally, there was information and training on attitude control, lift, thrust and drag, and fuel management.

Also under scrutiny were functions associated with time dependency, which meant not only completing operations or actions, but also comfortably performing them within a given time frame. "Although unspoken," states Garriott, "I believe it was also vital to demonstrate to the earlier astronauts that these new 'science types' could work in their environment of high performance jet aircraft; that we could remain cool under pressure and would not falter in critical situations. Solo flight in T-38s in weather, at night and with very limited fuel is one of the few ways to demonstrate this to others and to gain the self-confidence that you, too, can operate in their professional world."[1]

The three men knew it would prove a challenge beyond anything else they had known. Their training consisted of many hours of lectures and tutoring, including subjects such as airmanship, aviation physiology, T-41, T-37 and T-38 systems and procedures, and parachute familiarisation. They would also study the principles of flight, techniques for navigation (both electronic and by dead-reckoning), the influences of weather, flight instruments, communications, aerodynamics, aural and visual codes, flight safety, survival and physical conditioning.

In order to achieve these goals, each of them would be expected to spend fifty hours in aircraft simulators, and thirty flight hours in the short-range, propeller-driven T-41 Mescalero, a high-wing, low-performance military version of the Cessna 172 Skyhawk. They would then graduate to the twin-engine, dual-control Cessna T-37, a jet aircraft whose flying characteristics would give them the feel for handling the larger and faster T-38A Talons they would fly as NASA astronauts. Ninety hours were assigned to mastering the T-37 jet trainer. Eventually they would graduate to the supersonic T-38, in which they would fulfil 120 hours of flight training.

In-flight training itself would consist of putting each aircraft type through its performance envelopes, including ground reference manoeuvres, take-offs, patterns and landings, aerobatics, instrument flight, navigation, and formation flying. From the outset of the training, the steepness of their expected learning curve surprised Gibson. Even though he had done a little light airplane flying he quickly learned that flying Air Force style was a lot more challenging. "When you first start out," he reflected, "you think, 'I'm never going to get this.' It's like rubbing your head and patting your stomach and touching your nose with your tongue while you're shining your shoes with the back of your cuff, and trying to do all those things simultaneously. It takes a while before you get to where it becomes second nature, and then you feel comfortable doing it. But at the beginning you realise this is not a pushover."[3]

Schmitt agrees that it was a difficult undertaking. "I had a harder time learning how to fly these things, particularly on instruments, than the others did – for whatever reason that was. Eventually, I succeeded, but it was a bit traumatic for me to try to work my way through that."[6]

As if things were not difficult enough for Schmitt, he broke his elbow during a game of basketball, just as the group transitioned from T-37s to T-38s. "So, I had to sit down for a few weeks while that elbow healed, and then try to catch up, which I ultimately did. But it meant an awful lot of flying, awfully fast. Which was fine. That's the best way to learn, I think, to get all your flying in at once."[6]

It was a tough year, and it was extremely hot out at Williams, but eventually the three men graduated. Garriott recalls that it was actually a pleasurable experience, and all three "would like to have had even more time in flight, but their syllabus did not allow it." All finished high in their class, with standings well above the average, and were now military-qualified jet pilots. All three would continue to fly T-38s for their entire NASA careers, even on some occasions as Instructor Pilots. Later, they would also qualify in solo helicopter flight, although this new dimension to their aeronautical skills was eventually terminated as an unnecessary expense and risk.[1] "Most of us got helicopter training," Kerwin pointed out. "I knew I wouldn't use it to go to the Moon, but I enjoyed it anyway!"[7]

But for now, it was time for the three flight school graduates to return to Houston.

Screaming Purvis

Owen Garriott has some interesting recollections of his time during flight training, and of one instructor in particular.

"As already noted, being sent to jet training felt just like Br'er Rabbit must have felt upon being thrown into the Briar Patch! That's just what we would want. The first primary jet that we saw was the T-37, a small side-by-side seating aircraft, not too fast, but very noisy.

"Although only thirty-four years old at the time, the USAF apparently thought I might be too old to learn how to fly fast jets. They even thought I might have difficulty being the student of some young, Second Lieutenant flight instructor – or at least that was the story told to me by my assigned T-37 instructor, Captain 'Purvis' (name slightly changed for reasons that will become apparent in a moment!). Purvis was also in his mid-thirties, the oldest Instructor-Pilot (IP) in the squadron. He considered himself an excellent instructor, but had had the misfortune to have been involved in an earlier mid-air collision and had then been passed over for promotion to major. As a result, any rapid head motion from the student sitting next to him, such as a quick turn of one's head to the left, could easily initiate an excited 'Where? Where?' as if the student were about to involve him in another mid-air collision.

"Captain Purvis was known by another nickname throughout the Training Squadron, acquired from his personal technique for instructing his students – it was 'Screaming Purvis.' He had no favourites; everyone got the same treatment, literally screaming at them all. However, for particularly egregious errors (perhaps failing to set in an 'out bound heading' correctly for example, or any number of others), the screaming could continue on for quite a number of seconds. Then when back on the ground, he was usually quite calm, even complimentary, about the student's overall flying technique and if one dared inquire about the excited nature of his instruction, he was informed that it was for the purpose of teaching the student to 'concentrate.'

"Upon occasion, it exceeded what could even be reasonably tolerated. One of my fellow students had made some simple mistake which apparently drove Screaming Purvis into a rage. He reached across to the student's oxygen hose (we always flew breathing through an oxygen hose and face mask), grabbed the hose cutting off all air flow and jerked the student's head up and down, all the while screaming some kind of correction, never understood by the student! They made it back, with a somewhat shaken student-pilot, but almost no mention was made at the post-flight debriefing of the now-forgotten cause of the incident.

"But this did provoke some serious thinking on my part. What should I do in a similar circumstance? The first thought was to simply return the favour with a sharp blow across his face mask. But a fight in the cockpit would almost surely end both of our careers, and I had a lot more to lose than he did! I finally decided that in the same situation as my fellow student, I would simply fold my hands in front of me and declare 'Captain Purvis, you have the airplane.' (I *would* take back control, however, if the ground appeared to be approaching too fast! High speed 'chicken,' I suppose.) I was convinced that his superiors on the ground would see the wisdom in my action.

"I could very well have been mistaken in this, however, because he was finally promoted to major later in the year. Vietnam required all the pilots the Training Command could supply in 1965–66. And I was fortunately never faced with this decision, only my share of the 'screaming.'

"An interesting sequel to this story was provided by another 'elderly' student. Twenty-six-year-old Lieutenant John Fabian had been scheduled to be in our class starting in July 1965, but he was notified by the AF that he would be delayed about five months because the AF had to accept four NASA students in the earlier class. But when he finally arrived at Williams, he was placed right in the original slots where we had been. In fact, his T-37 instructor turned out to be the infamous Screaming Purvis as well. He endured all the screaming and the polite debriefings as before, except this time he also encountered a grabbed oxygen mask and head jerks to the point of his face mask striking the control stick of the airplane! After graduation, John completed a PhD program in the AF, flew ninety combat missions in Vietnam, and was selected as a NASA astronaut in 1978, flying two Space Shuttle missions thereafter. As close friends now, we can almost joke about these experiences but neither of us would attribute any of our later successes, whatever they might be, to the flight training of Captain 'Purvis.' We should also add that we believe the AF IP ranks no longer hold any of these outdated, bizarre examples of misguided instruction."

Technical assignments and the AAP Office

At the time the new group began their training, many of the other NASA astronauts had begun receiving Astronaut Office (CB) technical assignments on future programmes. This included early studies in extending the range of Apollo flights to include more scientifically orientated missions.

At a meeting held on 6 August 1964, Astronaut Chief Alan Shepard outlined several technical assignments that would be conducted prior to any further assignments to specific flight crews. Most of these were associated with the Gemini and Apollo programmes, but Russell ("Rusty") Schweickart from the third astronaut group was also assigned to "future programmes and in-flight experiments" within the Apollo branch office, which pointed to the extended Apollo flights under serious discussion at that time. Fellow Group 3 astronaut Walt Cunningham also picked up an assignment to non-flight (ground) experiments in the same branch office.

In the absence of their three scientist-astronaut colleagues, Kerwin and Michel were handed technical assignments on 23 September. Kerwin was given technical duties in pressure suits and EVA in the Operations and Training branch, and would also play a key role in the thermal vacuum tests of the Apollo Command and Service Modules. Michel found himself with an assignment in experiments and future programmes within the Apollo branch.

The preparation of hardware and experiments for any flight into space consumes a considerable amount of time. It involves many evaluations and mock-ups before a design is even considered qualified for assignment to a flight. The technical tasks given to the astronauts involved reviewing the ideas and proposals and evaluating their suitability for flight in areas of operational use and crew safety. As a consequence, hundreds of plans, proposals and ideas passed through the office. Many would never reach the stage of design configuration, let alone be assigned to a flight. The Gemini and Apollo programmes were also consuming so much of people's time and energies that no one in the Astronaut Office would give more than scant attention to paper studies on plans for future extended Apollo missions. That would change when the Apollo Applications Program (AAP) Office was finally set up at MSC in late 1965.

The AAP Branch Office was established once proposals for the development of a series of extended Apollo missions had progressed sufficiently to warrant a serious examination of all future plans and possible hardware. It became responsible for crew activities within the programme, and several astronauts were assigned to help carry out initial work studies.

Six months later, on 3 February 1966, Shepard issued a CB memo announcing the creation of yet another new branch office within the CB structure. This would be known as the Advanced Programs Office. It would be associated with the CB Apollo Program Office and headed by former Mercury astronaut Scott Carpenter, who had recently returned from a temporary assignment to the USN Sealab underwater habitation project. This office would assimilate Kerwin, assigned to pressure suits and EVA development, and Michel on experiments. Shepard's memo went on to state that Garriott, Gibson and Schmitt would be joining the branch in September, following their return from jet pilot school. On 23 March 1966, NASA unveiled its first AAP schedule, and it was a staggering concept. It projected a total of forty-five missions, made up of twenty-six Saturn IB and nineteen Saturn V launches, in both lunar and Earth orbit phases of the programme, by the mid-1970s. AAP would be an autonomous programme, separate from the mainstream Apollo lunar landing missions. Among the launches then envisaged would be three Saturn S-IVB Spent Stage Experiment Support Modules, otherwise known as "wet" workshops, three Saturn V-launched orbital laboratories, and four Apollo Telescope Mount missions. The first AAP launch was anticipated to take place in April 1968, although this depended on progress within the Apollo lunar landing programme and assumed minimum modifications to the hardware and launch schedules.

There was some early astronaut input into AAP development, as outlined in a memo dated 6 May, with Kerwin and Slayton expressing the concerns of the Astronaut Office over the lack of experiment planning and hardware operational safety in the Saturn IVB wet workshop configurations then under consideration. Then, in August, after completing his backup work on Gemini 10, Alan Bean was reassigned from the astronaut flight line to become the first Chief of the AAP Branch in the Astronaut Office. The following month, the three newly-qualified jet pilot scientists returned to Houston.

Work begins in earnest

As a naval aviator prior to becoming a scientist-astronaut, Joe Kerwin was asked if he found his reception among the astronaut corps any different to that of the other members of his group. "I think Jack was well accepted, even though he was a geologist," he responded. "I think Owen Garriott and Ed Gibson were pretty well accepted, even though they were physicists and engineers, whereas Curt Michel, who was a paper-and-pencil astronomer was not so well accepted. But I think that was the degree to which we merged and subordinated our scientific interests to those of the program."[4]

With the return of the three scientist-astronauts, further technical assignments in the AAP Branch would take place, as outlined by Shepard on 3 October 1966. Bean would remain at the head of the group, with astronauts Bill Anders, Joe Engle, Jack Lousma, Bill Pogue and Paul Weitz assigned to him. Owen Garriott became the Chief of the Experiments Branch, with Ed Gibson, Don Lind, Bruce McCandless, Curt Michel and Jack Schmitt under his administration. By 5 December, Joe Kerwin had also joined the Experiments Branch under Garriott, together with Ed Givens. Both offices would continue to change members several times until they finally amalgamated into the AAP Branch Office on 4 April 1967. The office now consisted of twelve astronauts, all under the administration of Alan Bean.

During his tenure as a scientist-astronaut, Curt Michel kept a journal, in which he recorded his impressions of the training, and the prevailing philosophies of NASA and the Astronaut Office. Early in his training he set down his impressions of what was then known as the Apollo Applications Program, or AAP.

"My interest naturally heightened in AAP and in particular the Apollo Telescope Mount (ATM) that was to carry a suite of instruments to study the sun. The basic plan of all the experiments was to record data on film which was to be returned to Earth prior to development; consequently no scientific analysis or direct observation would be required of the crew. Indeed, the experiment had been transferred almost directly from the defunct Advanced Orbiting Solar Observatory, which had been an unmanned satellite project. The metamorphosis into experiments requiring man had not been entirely successful. At least one experiment initially required only pointing at the sun and throwing a few switches in a predetermined order. The manned requirement was produced by effectively replacing an automatic function with a switch."[8]

The Advanced Orbiting Solar Observatory (AOSO) satellite of which Michel wrote was intended as a larger-scale version of NASA's Orbiting Solar Observatory satellites. These dated back to 1962, when they had first been built and launched. NASA's Office of Space Science and Applications then set in motion plans for a much larger satellite which they proposed as the "Helios." This would evolve into the AOSO.

The AOSO, 3.2 m long by 1.5 m in diameter with eight fold-down solar panels, was originally devised as a free-flying, unmanned satellite system that would maintain a 300-mile high polar orbit, continually monitoring the sun and near solar environment through the use of an array of detectors and electronic imaging devices covering a broad frequency band. The array of instrumentation built into the AOSO would include a high-resolution X-ray telescope, a white light coronagraph, ultraviolet and hydrogen-alpha spectroheliographs and an ultraviolet scanning spectroheliometer.

The concept of building an advanced solar observatory linked to the manned space programme was first mooted at a NASA-sponsored summer study at Woods Hole, Massachusetts, in 1965. The study was convened to discuss options for post-Apollo programmes, which would include planetary and solar physics missions. One of the many suggestions proposed was to fly a number of large solar telescopes that would be operated from an Apollo Service Module. The Apollo Telescope Mount is said to have derived from this discussion.

Before the programme was cancelled in 1969, the AOSO hardware went no further than the construction of a full-sized mock-up, which is now held at the National Air and Space Museum's Paul E. Garber Restoration Facility. However, many of the scientific instruments that had been designed for the AOSO underwent re-evaluation. Apart from working on photographic detectors to achieve greater resolution, one major modification was to convert the major telescopes to film, as the data could be physically returned to Earth from a man-tended vehicle.[9]

GENERAL TRAINING

When they returned from the USAF pilot training course in the summer of 1966, Garriott, Gibson and Schmitt embarked on a fifteen-month period of general NASA astronaut training. They were joined by Kerwin and Michel, who had remained at Houston completing their technical assignments in the Astronaut Office while the other three were away at jet pilot school. In April 1966, NASA had selected a new group of nineteen pilot astronauts who were to be trained to support the main Apollo lunar mission effort and crew later Apollo-type missions to the Moon and to space stations under the Apollo Applications Program. To save time, costs and administration headaches, NASA decided to include the new scientist-astronauts in with the pilot-astronaut academic and survival training programme. With twenty-four new astronauts undergoing basic training, this was by far the largest group NASA had trained in one go. There would not be such a large group again until the first Shuttle-era selection (of thirty-five candidates) began their training in July 1978.

General training plan – 1966

The class of 1966 had to report to NASA Manned Spacecraft Center, Houston, by 1 May 1966. Their training would commence from 9 May with an orientation programme at MSC. The three scientist-astronauts attending flight school would not graduate until the end of July and would not return to the MSC until mid-August, so they would have a few courses to catch up on from September. Meanwhile, Kerwin and Michel received the training schedule they would join in May of 1966.

General training overview

The fifteen-month plan for the new astronauts was designed to prepare the group for assignment to a specific mission, with the training directed towards the Block II Apollo spacecraft and mission profiles (lunar and Earth orbital). However, in order to enhance their knowledge of spacecraft and space flight operations, the group would also monitor the remaining Gemini missions (Gemini 9 to 12) and the early Apollo Block I and Block II missions. Following assignment to a flight crew, their training would be covered under specific mission training plans.

Over a period of four months, a series of ten science and technology summary courses initiated the general training programme, combined with Gemini and Apollo operational briefings. This was followed by eight weeks of detailed systems briefings on the Apollo Command and Service Modules and the Lunar (Excursion) Module. After six months of training, the group would undergo the environmental familiarisation, survival training, control task training and a continued programme of geology training, which would be completed by the middle of August 1967.[10]

The duration of the overall course was dictated by the seasonal requirements of the survival and geological field trips and to allow the participants to become familiar with both the flight hardware and operational procedures prior to any flight assignment.

The science and technology summary courses were devised to provide the astronaut trainees with "a means of attaining a common level of understanding in the prescribed subjects." The courses were chosen to provide sufficient background information about the design and operation of the spacecraft, launch vehicles and mission experiments and were fundamental in nature. However, the course on guidance and navigation included a functional description of the Apollo G&N system, as well as covering the basic components of inertial guidance systems.

Each week, courses were scheduled for Monday to Wednesday, with the rest of the week devoted to operations, project briefings and field trips. At least one of the scientist-astronauts (Michel) also used the spare days to maintain his connections in the academic world (in Michel's case, the nearby Rice University in Houston).

Science and technology summary courses

Geology I (56 hours of instruction): This covered basic terrestrial mineralogy, petrology and geological processes, identifying basic rock structures and geological mapping techniques (a total of fourteen sessions in geological processes).

Geology II (56 hours of instruction): Terrestrial analogues of lunar geographic features, geological mapping, geophysical studies and appropriate sampling techniques (fifteen sessions), plus fourteen sessions of mineralogy and petrology.

Geology field trips (an integral part of the geology course)
1. Grand Canyon, Arizona (2–3 June 1966).
2. West Texas (Marathon Basin and Santa Elena Canyon) (23 June 1966).
3. Bend, Oregon (Newberry Crater and Lava Butte) (27–29 July 1966).
4. Katmai, Alaska (Valley of The Thousand Smokes) (22–26 August 1966).
5. Los Alamos, New Mexico (Valles Caldera) (21–23 September 1966).
6. Pinacate Volcanic Area, Mexico (Cerro Colorado and Elegante Craters) (30 November–2 December 1966).
7. Hawaii (Island of) (week of 13 February 1967).
8. Flagstaff, Arizona (sunset Crater Area and Meteor Crater) (26–28 April 1967).
9. Medicine Lake Area, California (22–23 June 1967) and Iceland (Askja Caldera and Laki Fissure Area) (week of 3 July 1967).

Astronomy (15 hours): Astronomical terminology, solar system and celestial sphere (five sessions); 9–10 June 1966 – field trip to Morehead Planetarium.

Digital computers (8 hours): Digital computer components, operation, and programming techniques (four sessions).

Medical aspects of space flight (17 hours): Physiology of the human body as affected by the space environment (six sessions).

Flight mechanics (24 hours): Earth orbit, lunar, mid-course and entry mechanics (twelve sessions).

104 School for Scientists

Gibson (right), Joe Engle (left) and Ron Evans (second left) study rhyolitic ash flow from Mount Trident with NASA Geologist Dr. Ted Foss during an August 1966 geological field trip to the Valley of Ten Smokes, Katmai National Monument, Alaska.

Meteorology (4 hours): Meteorological considerations on space flight operations, global weather system observations (two sessions).

Guidance and Navigation (34 hours): Apollo navigation techniques, functional description of the Apollo guidance and navigation system (seventeen sessions).

Rocket Propulsion (8 hours): Rocket performance parameters, liquid rocket engine operation, solid propellant rocket operation, reaction control system operation (four sessions).

Communications (10 hours): Basic communications concepts, radio ranging, radio telemetry and Apollo telecommunications performance (five sessions).

Michel studies rock samples during a November 1966 geological field trip to the Pinacate volcanic area in Sonora, Mexico.

Physics of the Upper Atmosphere and Space (12 hours): The environment of the interplanetary medium and the sun's effect on this environment, the Earth's upper atmospheric condition and associated phenomena (six sessions). Michel was one of the lecturers on the topic of Upper Atmosphere and Space Physics to the rest of his group.

Operational briefings

Apollo Project Familiarisation: This consisted of three Apollo orientation briefings:

- Mission Profile Briefing (approximately 4 hours) – covering the Apollo mission objectives, the proposed launch schedules, a general description of the spacecraft and a definition of the lunar profile (19 May).
- Launch Vehicle Familiarisation (3 days) – Visiting Marshal Space Flight Center in Huntsville, Alabama, where briefings were conducted on the Saturn 1B and Saturn V launch vehicles and systems. The stay at Marshall included a tour of centre facilities, and a third day spent at the Mississippi Test Facility to familiarise the astronauts with these facilities and associated test programmes. A static test

firing of one of the Saturn boosters was planned as part of this visit (25–27 May 1966).
- Spacecraft Familiarisation (30 hours) – North American Aviation instructors presented eighteen hours of general briefings on the Command and Service Module (5–8 July 1966). This was followed by twelve hours of briefings by Grumman instructors on the Lunar (Excursion) Module (14–15 July 1966). Following the formal academic programme, complete and detailed systems briefings on each of the spacecraft were also delivered. There were also two days of familiarisation with the CM and LEM mock-up controls and displays (21–22 July).

To facilitate their understanding of the CM and LEM controls and displays, the group was divided into two teams to use the mock-up held in Building 5 at MSC. Kerwin received CM training – in Block I configuration – on 8 July, and Michel on 21 July. Kerwin's LEM briefings were conducted on 21 July, with Michel on 22 July.[11]

Space Flight Operations Familiarisation: A programme of tours and briefings to instruct the astronauts in aspects of manned space flight operations:

- Launch Operations – A visit to the Kennedy Space Center in Florida to receive briefings on the spacecraft and launch vehicle, Apollo launch preparations and countdown operations. The astronaut group toured the Apollo launch complexes, the Vehicle Assembly Building, Apollo Checkout Buildings and Launch Control (4–5 August 1966).
- Mission Control Centre Operations – Flight Operations Directorate staff presented a full briefing on the MCC-H facility and its operations. This included an overview of each Flight Control console position in the Mission Operations Control Room (MOCR), as well as the function and operation of each of the Staff Support Rooms. The astronauts were briefed on the network data flow and real-time operations organisation of the control room during a detailed tour of the Building 30 facility at MSC (18–19 August 1966).
- Recovery operation: The Recovery Operations Branch gave detailed briefings on the planned recovery for the Apollo programme (11 August 1966).

Spacecraft systems training

At the completion of the academic programme, a series of detailed spacecraft system briefings were scheduled. These systems encompassed four days a week, six hours a day for approximately eight weeks. Instructor personnel from both primary spacecraft contractors (North American Aviation for the CSM and Grumman Aerospace for the LEM) used the systems trainers where applicable to support their course work.

Command and Service Module Systems Training (95 hours): A series of system briefings presented by NAA instructors (12–20 September, and 26 September–11 October 1966, either side of Geology Field Trip #5).

System Course Title	Duration (Hours)
1. Structures	6
2. Electrical Power System	12
3. Crew Systems	6
4. Communications	8
5. Environmental Control System	12
6. Sequential Events Control System	15
7. Propulsion Systems	12
8. Stabilisation and Control System	24

Lunar (Excursion) Module Systems Training (82 hours): Instructors from Grumman Engineering Aircraft Corporation presented a series of systems courses on the lunar landing vehicle (13 October–3 November 1966).

System Course Title	Duration (Hours)
1. Structures and Mechanical Systems	6
2. Electrical Power Systems	10
3. Instrumentation	6
4. Crew Systems	6
5. Communications	6
6. Environmental Control	12
7. Propulsion	12
8. Guidance, Navigation and Control	24

Environmental Familiarisation: For experienced aircraft pilots (as well as for the three recently qualified scientist-astronaut "pilots"), the environmental conditions of space flight are unique in that they are more extreme or of longer duration than nominal aircraft flying. These environment conditions also have an effect on the performance of the astronaut relative to their familiarity with the condition. To help prepare them for this, each astronaut was exposed to the environmental conditions of space flight: weightlessness, launch and re-entry accelerations and decelerations and pressure suit familiarisation.

Weightlessness (4 days): Using a US Air Force KC-135, each astronaut was exposed to approximately thirty seconds of "zero-g" per parabolic trajectory with between eighteen and twenty parabolas flown on one flight. There were twenty-four astronauts flying in groups of three and two flights per day for four days, so eight flights were scheduled. (24–27 January 1967 – one day per man).

Launch and Entry Acceleration: Using the centrifuge at MSC, four to six familiarisation runs for each astronaut (in "crews" of three to simulate an Apollo flight profile) were used to simulate expected acceleration profiles of the Saturn 1B Earth-orbit launch, as well as selected launch aborts and orbit re-entry. These runs gave the astronauts familiarity with the forces encountered during these mission periods and an understanding of the accelerations, as well as an appreciation for their operational capability during these phases of the mission. As they used an early Apollo centrifuge crew configuration, it did not have control capability, so crew station angular positions and acceleration profiles were pre-programmed into the centrifuge central computer. As there were also no active controls or instrumentation, except for a "g" meter and an event timer, a running commentary on the mission events and crew tasks was given to the crew over the radio link as the run progressed. The crew were then required to make the proper motions to simulate these tasks on the basis of the commentary. Each crewmen received two sessions separated by at least twelve hours, with three training periods (09:00, 11:00 and 13:30 hrs) per day scheduled (7–15 November 1966).[12]

Session 1
Run 1 – Normal launch
Run 2 – SPS entry
Run 3 – Pad abort
Run 4 – High altitude Launch Escape System (LES) abort

Session 2
Run 1 – Normal launch
Run 2 – RCS entry
Run 3 – SPS abort – minimum altitude
Run 3 – SPS abort – maximum G (15)

Those relating to the Group 4 scientist-astronauts (and the Group 5 pilot-astronauts who rode with them) were:

Date	Session	Run	Time	Position	Crew
7 November	1	B	1100	1	Kerwin
				2	Garriott
				3	McCandless
	1	C	1330	1	Lind
				2	Michel
				3	Gibson E.
8 November	2	A	0900	1	McCandless
				2	Kerwin
				3	Garriott
	2	B	1100	1	Michel
				2	Gibson E.
				3	Lind
14 November	1	C	1330	1	Givens
				2	Swigert
				3	Schmitt
15 November	2	C	1330	1	Swigert
				2	Schmitt
				3	Givens

Pressure Suit Indoctrination: As a way of introducing Apollo pressure garments, the group were given an indoctrination on the Block II pressure garment by representatives of the Crew Systems Division through a programme of briefings and demonstrations. These included:

- Briefing on the suit design and construction.
- Demonstrations of suit donning and doffing.
- Demonstrations of mobility at 3.5 psi differential pressure.
- Briefings on miscellaneous crew equipment with demonstrations as required.
- Experience of suit donning, mobility and doffing.

During later phases of training, in mission support back-up roles as well as flight crew assignments, the astronauts would wear (training) pressure garments to become familiar with its operation in different phases of the mission (5 December 1966).

Wilderness and survival training

Nominal and launch abort recovery for Apollo missions was in the ocean, but a mission could be terminated at any point and entry and landing could result in an emergency terrain or out-of-range ocean landing. Therefore, the astronauts were given three types of basic survival training in tropical, desert and water environments. Since the orbital inclination planned for Apollo did not normally cover polar flights, Arctic wilderness training (which the Soviet cosmonauts underwent) was not part of the programme. The purpose of this phase of their training programme was to provide the astronauts with the confidence and ability to survive in an emergency landing environment until rescue. In each case, the training was divided into three phases: lectures and briefings on survival techniques; demonstrations of the survival methods; and field experience to apply the academic and practical knowledge gained from the first two phases.

Water survival (1.5 days): The Recovery Operations Division delivered an academic presentation which encompassed the requirements for human survival at sea, food and water requirements and sources at sea, progressive aspects of water survival, and the effects of drinking sea water. This was followed by practical experience of water survival at the Water Safety and Survival School at the Naval School of Pre-flight in Pensacola, Florida. Here, the astronauts worked in an enclosed water tank, with and without the pressure suits, on basic swimming strokes, underwater escape from a cockpit, life raft boarding, and helicopter rescue by sling and seat. Water egress training during subsequent specific mission training would add to the astronauts' experience in handling survival equipment and procedures. (Lectures – 17 November 1966, one half-day per man; water survival 8–9 December 1966, one day per man.)

Tropical Survival (5 days): With the support of the USAF Tropical Survival School at Albrook Air Force Base in the Panama Canal Zone, the astronauts were given a

110 School for Scientists

Curt Michel (centre) during jungle survival training in Panama with pilot-astronauts Paul Weitz (left) and Jim Irwin (right).

split programme of academic and demonstration training and in-field experiences. The first two days encompassed lectures in the major types of tropical rain forests, tropical plants and animals applicable to survival, terrain, travel, self-first-aid, the use of kit equipment and contact with indigenous people. The demonstrations included the construction of shelters, improvising equipment (such as recovery parachutes), building animal snares and traps and signalling.

The next three days were spent in the field, with one day devoted to travelling to and from the field area and two days spent at the field site. Once at the field site, the group was split into teams of three men each, as would be the case in an Apollo landing, and then assigned an area for their campsite which was remote to the other teams. One instructor was assigned to two teams to monitor the activities and, where necessary, give advice. The three men would have exactly the same equipment as they would have in a real survival situation, essentially what they would have aboard the Apollo CM. In the field training, the "crews" received first-hand experience in procuring food other than on-board rations, establishing a camp, improvising equipment and clothing and signalling rescue aircraft in the tropical environment (week of 12 June 1967).

General training 111

Michel (right) and Jim Irwin relax during a break in jungle training in Panama.

Desert Survival (5 days): This five-day course was similar to the tropical school, but this time the Air Force Survival School, 3635th Flying Training Wing, based at Stead Air Force Base in Nevada provided the instruction, tailored to space flight mission requirements. The characteristics of world desert areas and survival techniques were the subject of a one-and-a-half-day academic programme, followed by a one-day demonstration programme given at the field site on the proper use and care of survival equipment and in the use of parachutes for constructing clothing, shelters and signals. Again teamed into "crews" of three, the astronauts spent two days at remote sites putting the theoretical studies and demonstration information into practical use (week of 7 August 1967).

Control task training

As Apollo featured side-arm rotation and translation controllers, and unique techniques were necessary to use them safely and effectively, a certain amount of time was required on part-task trainers to acquaint the astronauts with a variety of space flight manoeuvres. These simulators incorporated partial crew station displays with "out-

Jack Schmitt (centre) helps Al Worden (left) and Ed Mitchell (right) gather jungle leaves and branches to construct a lean-to shelter during jungle survival training in Panama.

of-the-window" infinity optics displays of the Earth, stars and a variety of rendezvous targets.

Gemini Part-Task Trainer (approximately 12 hours): Though the astronauts selected in 1965 and 1966 were not going to fly the Gemini spacecraft, the programme's part-task trainer had the proven capability to simulate orbit attitude control, manoeuvre thrust control, retrofire control, terminal rendezvous, and (planned) re-entry control in three modes (pitch, roll and yaw), as well as rate command, direct (acceleration) and pulse control. As the astronauts would not be using this simulation until the latter months of 1966, there were fewer Gemini crews in training, but as the Apollo simulators were taken up by the Block I and Block II crews preparing for 1967 missions, it was sensible to utilise the Gemini trainer in this time frame before progressing to the Apollo simulator later in their training flow.

The simulator featured out-of-the-window displays that presented a star field, with an occluding disc to produce a horizon. For rendezvous control practice, the out-of-the-window display was an electronically generated image of an Agena docking target. Each astronaut was to complete the following training simulations:

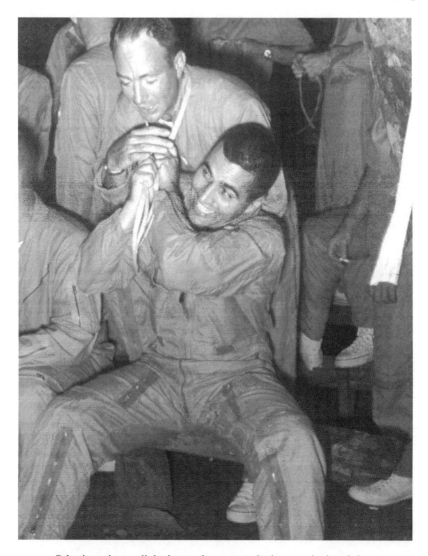
Schmitt enjoys a light-hearted moment during survival training.

- *Orbital attitude and manoeuvre control:* Using all three control modes, practice in attitude control and thrusting manoeuvres was conducted, initially using the cockpit attitude reference system, then the out-of-the-window display. Particular attention was given to the problem of controlling the cross-coupling from the manoeuvring thrusters in the direct control mode.
- *Retrofire Control:* Using the 8-ball attitude indicator and the out-of-the-window horizon display, attitude control of the engine misalignment torques was practised using the rate command and direct control modes. The training instructor had the capability to vary misalignment torques.

114 School for Scientists

- *Re-entry Control:* The astronauts practised damping the oscillations in pitch and yaw and controlling the roll to the commanded position in both the rate command and direct control modes.
- *Terminal rendezvous:* This manoeuvre was practised using the electronically-generated image of the target display under different initial conditions. To accomplish the manoeuvre, the astronauts used the cockpit display instrumentation of range and range rate and the flight direction indicator.

Translation and Docking Simulator (multiple 2-hour sessions): A series of sessions were scheduled on this simulator so that the astronauts could receive practice at manoeuvring their spacecraft during the final docking phase. This manoeuvre was practised using rate command, direct and pulse control modes with different initial conditions.

Launch vehicle abort training

An appreciation of manual abort requirements was gained by using the Dynamic Crew Procedures Simulator (DCPS). This simulator combined the crew station displays and physical cues for simulation of a wide variety of booster malfunctions and could run several abort situations in rapid succession. The types of run and their approximate number each astronaut experienced were:

1. Normal runs – 2.
2. Engine/propulsion system failures – 6.
3. Propulsion failures – 3.
4. Staging/sequential failures – 3.
5. Control failures – 8.
6. Structural failures – 2.

This training took place over 14–22 November 1966 and featured the astronauts working in pairs in four "crews" (A–D) during each day.[13]

Those featuring the Group 4 astronauts and their Group 5 colleagues were:

Date	Crew	Time	Position 1	Position 2
Nov 21	A	0800	Lind	Gibson E.
	B	1000	Garriott	McCandless
	C	1300	Kerwin	Michel
	D	1500	Irwin	Schmitt
Nov 22	A	0800	Gibson E.	Lind
	B	1000	McCandless	Garriott
	C	1300	Michel	Kerwin
	D	1500	Schmitt	Irwin

Aircraft flight programme

Flight proficiency and "spacecraft flight readiness" was maintained through the use of T-33 and T-38-type aircraft assigned to MSC and based at Ellington AFB. These aircraft were utilised for cross-country, as well as local flying requirements. In addition, a two-week course in helicopter familiarisation (though intended for the Group 5 pilot astronauts, the Group 4 scientist-astronauts also participated in this phase of flight training) was provided by the Naval School of Pre-flight, Pensacola, Florida, with a continuation programme at Houston. The use of helicopters provided initial familiarisation with lunar landing trajectories.

A HECTIC DIARY

This training programme was developed primarily for the 1966 pilot-astronaut selection, with the members of the 1965 scientist-astronaut group participating in two phases. As they were already jet qualified, Kerwin and Michel participated in the programme from the beginning (May 1966), while the other three (Garriott, Gibson and Schmitt) joined in August 1966, three months into the syllabus. A review of Curt Michel's files for the period shows a hectic diary for the period 2 May–29 August 1966. In a scheduling matrix of these first seventeen weeks of training, Michel records only ten free days (excluding weekends), of which five were scheduled for him to work at Rice University. In the following forty weeks (5 September 1966–27 August 1967), there were only seventy-seven days free (excluding weekends), from which he had to fit in his work at Rice University and the seasonal holidays during an academic year.

In his 2000 Oral History, Garriott recalled: "When we came back [from flight school], the Gemini program was nearing completion, so we began to work on whatever needed to be done to provide assistance to the flight crews getting ready for Apollo. Then the early stages of the Skylab (then called Apollo Applications Program) came into existence, so we began to spend time thinking about how AAP should be run, the configuration that should be used and how it should be flown. We had a number of really fascinating trips ... a lot of geology training that just had to be done, and survival training. Occasionally over the course of the next ten years, there would be refresher courses that needed to be taken. There were a number of very useful geology courses all over the world related to the geology of the Moon and they took up a fair amount of time; all of which we certainly enjoyed and I felt was to our personal advantage to have the opportunity to participate in all that."[14]

At the same time as the scientist group participated in the academic and survival training, they received technical assignments in the CB that they hoped would lead to being named to a flight crew, though it soon became clear that this would not be for some time.

REFERENCES

1. Owen Garriott email correspondence with Colin Burgess, 31 January 2004, 2 February 2004.
2. Ed Gibson email correspondence with Colin Burgess, 2 April 2003.
3. Ed Gibson, JSC Oral History transcript, interview 1 December 2000.
4. Joe Kerwin, JSC Oral History transcript, interview 12 May 2000.
5. Duane Graveline email correspondence with Colin Burgess, 28 and 29 Jan 2004.
6. Jack Schmitt, JSC Oral History transcript, interview 14 July 1999.
7. Joe Kerwin email to Colin Burgess, 12 January 2005.
8. Curt Michel's Rice University papers, previously cited.
9. *Living and Working in Space: A History of Skylab* by W. David Compton and Charles D. Benson, NASA SP-4208, Washington, DC, 1983.
10. *Project Apollo Flight Crew General Training Plan*, prepared by Raymond G. Zedekar, Assistant Chief for Crew Training, Flight Crew Support Division, NASA MSC, Houston, Texas. MSC Internal note no 66-CF-1, undated (1966), Curt Michel Collection, Rice University, Houston, Texas, copy on file AIS Archives.
11. Undated CB Memo – CM and LEM Controls and Displays Familiarisation, Curt Michel collection, Rice University, copy in AIS Archives.
12. Centrifuge Training, CB Memo from Raymond Zedekar, CF3-6M-173, dated 27 October 1966, Curt Michel collection, Rice University, copy on file AIS archives.
13. DCPS schedule undated CB Memo, Curt Michel Collection, Rice University, copy on file AIS Archives.
14. Owen Garriott, JSC Oral History transcript, 6 November 2000, pp. 10–11.

Additional information on the Group 5 training syllabus that included the Group 4 scientist-astronauts was supplied by Jerry Carr.

5

The Excess Eleven

As members of the first group of scientist-astronauts were completing their training, plans for the Apollo Applications Program were expanding in scope, but being cut back financially.

Over the following fifteen years, Apollo hardware was expected to be involved in at least ten manned landings in the mainstream programme, before being used to implement extended lunar surface explorations that could lead to a manned lunar research station. Redesigned Apollo hardware was also planned for use in developing the first space stations in Earth and lunar orbit, while new space logistics systems would be developed to support and eventually replace Apollo. This would lead to high space complexes, permanent lunar bases, and many more Americans venturing to the Moon. At least, that was the plan.

A SECOND SELECTION

In order to crew such grand visions for the future, it was quite evident that a corps of less than thirty-six astronauts was inadequate for the task. Therefore, a two-part astronaut selection process was initiated to begin expanding the Astronaut Office. These new recruits would support operations after the first manned Apollo landing, whatever they may be.

Deke Slayton was concerned, however, and stressed caution before overloading the programme with too many astronauts at a time of uncertainty. It was clear that mounting rumours of budgets cuts could threaten the Apollo lunar effort, let alone any endeavour beyond, so for any astronaut selected after 1966 there was less guarantee of flying than in the early days. Nevertheless, after nineteen pilots were selected in May 1966, a call for a second group of scientist-astronauts was issued on 26 September 1966. Once again, this included the active participation of the National Academy of Sciences and National Research Council (NAS-NRC). These eminent bodies would conduct the initial and intermediate stages of screening and selection, and NASA would then make the final selection from those applicants recommended by the selection committee of the NAS-NRC.

Applications were invited from those with a doctorate (or equivalent in experience) in the natural sciences, medicine or engineering. With NAS-NRC expecting relatively early flights for those selected, a strong emphasis was placed on the fields of astronomy, biology, engineering, geology, medicine and physics. The first stage in the process was based on information supplied by the applicants, which included their background, education, training, special capabilities, skills or qualifications, teaching, and research and clinical experience, as well as their involvement in professional associations and societies. Supporting documentation was also required, and this had to include multiple references, a research bibliography, college transcripts, aptitude tests and a medical history.

Accompanying the press release was a message from Gene Shoemaker, who would chair the Academy's selection panel. In it he wrote: "Scientific investigations from manned space platforms and direct observations on the Moon will initiate a new phase in man's quest for knowledge. While such missions call for daring and courage of a rare kind, for the scientists they will also represent a unique adventure of the mind, requiring maturity and judgment of a high order."

Geologist Donald Beattie could not resist the challenge; he knew Gene Shoemaker as a colleague, and he was also known to several other members of the selection panel. He had applied for the first scientist-astronaut group but had missed out, being one inch over the height limit and nine months over the age limit. In his 2001 book, he said that he felt he stood a good chance of selection.[1] "The Academy had been somewhat disappointed by the number of applications received from the first selection, although the six chosen had excellent qualifications, and thus the selection criteria were a little more relaxed the second time. The age and height limits had not been changed, but this time the press release stated that 'exceptions to any of the ... requirements will be allowed in outstanding cases.' Perhaps now I had a chance. Could I qualify as an 'outstanding case'?"

The screening process

By January, some 923 applications had been received for an expected twenty to thirty openings. Slayton had asked for that number due to the high attrition rate he expected as a result of several factors. These included flight school, the training programme, or even accidents. However it soon became clear that the number of qualified applicants would once again fall far short of what had been envisaged.

The next stage of the NAS-NRC screening then took place, which involved looking through reprints of published scientific papers that were representative of the applicant's work, and determining its relevance to the role of scientist-astronaut. The screening board also read essays on investigations the candidate would like to conduct in Earth orbit or on the lunar surface, or both. At the completion of this stage the NAS-NRC committee made its selections, and these names would in turn be passed on to NASA. Eventually, only sixty-nine names were forwarded to NASA for final screening.

The next step in the process was a background and security check conducted by the United States Civil Service Commission. They would examine each candidate's life history and carry out extensive interviews with family members, work colleagues, instructors, neighbours, friends and other sources of information from every period of the candidate's life. Once this was completed, the CSC would issue a clearance for the applicant's involvement in such a sensitive and classified position.

As with the first group of scientist-astronaut applicants, the candidates were then required to present themselves to the School of Aerospace Medicine at Brooks Air Force Base in San Antonio.[2] Here, they would undergo a full week of tests and examinations lasting about ten hours per day. These were designed not only to qualify the candidate medically for space flight status, but also to provide extensive baseline data on parameters and systems that might be altered by space flight or other parts of the operational environment experienced by an astronaut. While a thorough systems review and examination was performed, emphasis was placed on aero-medically important areas such as biochemistry, haematology, neurology, psychiatry, ophthalmology, the vestibular system and cardiology. There were also several operational tests, including cardiovascular and performance responses to acceleration forces of 9 G for one minute, and cardiovascular response to exercise on a slowly-accelerating treadmill (which also determined maximum oxygen consumption).

On the final day, there came what would be the highlight for many – an acrobatic flight in a supersonic aircraft to measure their reactions to a new and relevant environment. This exercise would also give them an early and dramatic insight into the nature of the career for which they were now applying.

Following this battery of tests and interviews, each of the remaining candidates was asked to make their way to Houston in June and present themselves for a final array of tests and a probing interview with members of the NASA selection committee, who would make a final determination.

Candidate Don Beattie had made it through to this testing process, and recalls enduring "a week of prodding, blood work, and spinning, IQ, and many other tests, some of which were vividly shown in the movie 'The Right Stuff,' though not with the same comic detail. While I was tilted upside-down with my stomach filled with a barium solution, they discovered that I had a slight hiatal hernia: the muscles in my oesophagus couldn't hold all of the solution in my stomach."

Beattie's other results had seemingly been good, and as it was only a minor ailment, he was sent to the Walter Reed Medical Center in Washington DC for further evaluation. The doctors there concluded that it was not a disqualifying ailment, and could be easily overcome by taking an antacid tablet. He was therefore one of the twenty-one candidates who presented themselves for the last selection interviews. "Our backgrounds included almost all scientific disciplines, but as I read the list I saw I was the lone geologist, along with one geophysicist. Only two Earth scientists! Most of the post-Apollo science activities we were planning had some Earth science connection; I thought my selection was in the bag!"

120 The Excess Eleven

After the twenty-one candidates had checked in they were gathered at Ellington AFB, where they would be taken for a spin in one of NASA's T-38 jet aircraft to check their level of comfort during aerobatic manoeuvres. Next, they were strapped into the Spacecraft Center's centrifuge and whirled around at 6 G while they tried to perform some simple tasks with light switches. Then came the final interviews, as Beattie relates:

"I recall only four people in the room: Al [Shepard], Deke [Slayton], Bill Hess and Charles Berry, who was head of the medical sciences office – the astronauts' doctor. All the questions were rather innocuous. Berry asked about the hiatal hernia. The only question that stands out in my mind was the one Deke asked: 'Don't you think you're too old to be an astronaut?' I was thirty-seven at the time and not the oldest of the final twenty-one candidates, but I knew I was over the advertised age allowance, so I had done a little homework. I answered, 'I don't think so; after all, I'm younger than Wally Schirra, and he's still flying!' I thought my selection was now only a formality. That afternoon I did some preliminary house hunting in the neighbourhoods around NASA."

The official Group 6 portrait. Standing (l to r): Joseph Allen IV, PhD, Karl Henize, PhD, Anthony England, PhD, Donald Holmquest, MD, Story Musgrave, MD, PhD, William Lenoir, PhD, Brian O'Leary, PhD. Seated: Philip Chapman, SD, Robert Parker, PhD, William Thornton, MD, and John (Tony) Llewellyn, PhD.

Unfortunately, Beattie did not make the cut. Early in August, Alan Shepard called to say he hadn't been one of the eleven selected. The other Earth scientist, Anthony England, was one of the lucky ones.

THE GROUP SIX SELECTION

On 4 August 1967, the names of eleven new scientist-astronauts were announced. Unlike the first group of scientist-astronauts, none had previous pilot training, and all eleven would have to complete the fifty-three-week jet pilot course. With academic studies and survival technique training as well, they were not scheduled to report for active flight assignment at NASA until the summer of 1969.

They were a diverse collection of talented men who would ultimately have mixed space careers, but who joined the programme for a variety of reasons. Their collective qualifications were outstanding: all were PhD (or D.Sc) scientists or MDs, and the group comprised three astronomers, two physicists, one chemist, one geophysicist, one electrical engineer, two MD-PhD physiologists, and one MD/physicist. Not all of

Members of the sixth astronaut group and second scientist-astronaut group in a more informal gathering shortly after being announced by NASA. Left to right: (at the back) Llewellyn, England, Thornton, Lenoir, Chapman, (kneeling) Parker, Henize, O'Leary and Allen. Missing from this photo are Holmquest and Musgrave.

them were expecting, or necessarily wanted, to fly to the Moon. In the event, it would be many years before any of this group even reached orbit. On the very day they reported to Houston, Deke Slayton informed them in no uncertain terms that they would definitely not get flight seats in the shrinking Apollo programme, and that if any of them wanted to leave immediately, it would not be held against them. None did, at least on that day.

As well as Slayton's less-than-welcoming words, it was soon made abundantly clear to these men that they were actually surplus to NASA's requirements. In a defiant countermove, they had soon devised and adopted the prophetic sobriquet of the "Excess Eleven", or "XS-11".

It was an eclectic, eager group of highly talented and qualified professionals who eventually made their way to a new career in Houston, little realising the enormous problems that would soon beset them.

JOSEPH P. ALLEN IV

While many honours and deserved plaudits have been heaped on America's astronauts, perhaps the most incongruous of all is one awarded to Dr. Joseph Allen: since 1984 he has enjoyed honorary membership of the Indiana Wrestling Hall of Fame. Not only that, but in 1998, Joe Allen was inducted into Oklahoma's National Wrestling Hall of Fame, as a distinguished incumbent in their Outstanding American category.

A distinguished heritage

Standing at just 168 cm (five feet six inches), one would have to admit that the diminutive adventurer resembles neither a former wrestler nor an astronaut, yet he was involved in some of the most accomplished science carried out in space during his two Shuttle missions in the early 1980s, logging a total of 314 hours in space, including close to eleven hours of EVA, and participating in the only space salvage mission to date. These days, he is highly amused when reminded of his wrestling past and resultant accolades, but says the sport was good for both his physical and personal development.

"Wrestling pitted me against people my own size and it gave me a great deal of confidence," he stated. In high school he wrestled at 44 kg, in college at 58 kg, and his win–loss record in his final two years while wrestling for the DePauw University team was 16–2. "I don't think I was really that good a wrestler, but my team mates were remarkable. Eight of the ten of us became physicians, lawyers or PhDs."

Joseph Percival Allen IV was born on 27 June 1937 in Crawfordsville, Indiana. Crawfordsville is a small farming and industrial town located in west central Indiana, about forty-five miles from Indianapolis. His father was Joseph III (also known as Perk) and his mother was Harriet Elizabeth (née Taylor). They had graduated from DePauw University in 1928 and 1930 respectively, and Joseph III later became a professor on the university's staff.[3] In fact, Joe Allen came from a long line of

Joseph P. Allen IV, PhD.

academics associated with the university. His great-grandfather, the original Joseph Allen, did not attend college, but he grew up in Greencastle, Indiana where DePauw, founded as a teacher's college in the mid-1800s, is located. On a corner of the town square, he and his brother established and ran the Allen Brothers dry goods store, later sold to J.C. Penneys. Joe's grandfather, Joseph Jr., graduated from DePauw in 1897, followed just two years later by his great-uncle, the noted humanitarian Dr. Percy Hypes Swahlen, who was also a football letterman at the university. A rich academic heritage and a loving family environment linger as integral elements of Joe Allen's upbringing and development.[4]

"I was, and am, lucky in the extreme that I grew up as the older of two boys in a very caring and supportive family. Dave is my only sibling and we have always been the best of friends. My mother was the daughter of the Methodist minister in Greencastle. The Methodist church sits directly across from the office of the president of DePauw and it is an important church to the university, originally associated directly with the Methodist church. My mother and father encouraged us to pursue any discipline which intrigued us, including studies of any kind, hobbies, sports ...

anything which exercised our mind, or body, particularly if focus and discipline were required."

Joe Allen began his academic days at the Caleb Mills School, where he undertook elementary education from 1942 to 1948. Then followed seventh and eighth grade education at Crawfordsville Junior High, and finally Crawfordsville High School, graduating with high distinction in 1955. "Actually, I was second in my class behind my good friend Nancy Wells, whose father was the physics and chemistry teacher of the high school. Mr. Wells was a wonderful teacher, and much of my later success in science I credit to him. I suspect that on our graduation day, Mr. Wells was very proud of both Nancy and me!"

Another teacher who would have a profound effect "in all subjects involving life's lessons" was Bob Hauck, his wrestling coach. "My dream all through high school was to become a state champion, a goal I nearly accomplished in all four years of the state tournament, but never totally accomplished, to my deep disappointment. Many years later, I realised that knowing how to deal with disappointment in a gracious manner is invaluable. My failed spacewalk on STS-5 is a good example." He also credits teachers who were "sticklers on fundamentals" for furthering his education and desire to succeed. "Although coming from a relatively small school in a small town, I was nonetheless well prepared in the basics when I arrived at age eighteen at the university."

Deciding on a future

On admission to DePauw, Allen became a Rector Scholar, a programme that recognises students who have demonstrated significant academic achievements on their high school transcripts and college entrance exams. At this time he still had no clear thoughts about what he wanted to do later in life, although science would obviously be predominant.

"When I grew up there were no astronauts. The idea of going to space was crazy – science fiction. I wound up in a profession that had not been invented when I was in high school.[5] Living in a rapidly changing world as we do, particularly with regard to science and technology, one cannot prepare in school by studying all the processes and technologies that will possibly be needed in your years ahead. Rather, one must learn the basics of science and the scientific method and learn, in addition, 'how to learn'. When one does that as a young person in school, then one is safeguarded against becoming obsolete."

Both he and his brother David, later to become a physician, were involved in varsity wrestling and were made members of the Phi Beta Kappa fraternity. Joe had also met an attractive fellow student named Bonnie Jo Darling from Elkhart, Indiana, who had entered DuPauw as a freshman when he was a junior. They dated briefly in his senior year.

After graduating in mathematics and physics in 1959, Allen travelled to Germany under a grant as a Fulbright post-graduate scholar to continue his studies. This programme was established in 1946 under legislation introduced by Senator J. William Fulbright of Arkansas, and was intended to increase mutual understanding between the people of the United States and those of other countries. It was only available to students who had achieved a Bachelor's degree or the equivalent. Allen was readily accepted by fellow students at the Christian Albrechts University in Kiel, but only on the challenging proviso that he would speak nothing but German, even though many of his new classmates spoke excellent English. It was the best possible way to learn German, and he consequently became fluent in the language – as they had intended. "Every journey changes a person. Part of the way a person is changed by a journey is in being away from home and gaining a new perspective and a new appreciation for home." During that year he formed many friendships of lifelong tenure, and would later host several German high school students in the United States.

On his return home, Allen continued his studies at Yale University. He had also been regularly corresponding with Bonnie Jo, now a senior at DePauw. "We were good at writing letters and meeting occasionally during school breaks, usually at her parents' home in Elkhart, near the border between Indiana and Michigan. She graduated from DePauw in June 1961, and we were married in Elkhart in July 1961. We then drove to a small apartment we had in New Haven, Connecticut. I continued my physics studies at Yale, and she taught second grade at a nearby school."

In 1961 he was awarded his Master of Science degree at Yale, and became a guest research associate at the Brookhaven National Laboratory from 1962–1965. During this time he completed his doctoral dissertation, entitled 'Studies of Quantum Shapes in the S-D Shell.' This, he says, involved "using nuclear particle accelerators known as 'van de Graaff accelerators'. I bombarded oxygen nuclei with proton, deuteron and helium beams, causing the oxygen nucleus to absorb energy; then, nanoseconds later, releasing that energy by gamma-ray decay, thus revealing properties of the excited quantum states of the oxygen nuclei. Such studies provide data giving credence to (or casting doubt on) the validity of various theoretical nuclear models. One such model of the nucleus is known as the 'shell model,' hence the term 'S-D Shell'."

After receiving his PhD in nuclear physics from Yale in May 1965, Allen became a staff research physicist for the following year at the university's Nuclear Structure Laboratory. During the period 1963 to 1967, he also served as a guest research associate at the Brookhaven National Laboratory. He then took up an appointment at the University of Washington in Seattle, where he would become a research associate and instructor on a post-doctoral fellowship in their Nuclear Physics Laboratory.

Joe Allen did not apply for the first scientist-astronaut group in 1965, as he was simply "totally unaware of the opportunity." However, in the fall of 1966, and just before relocating to the University of Washington, he saw a notice on a Yale bulletin board. It said that NASA was recruiting scientist-astronauts. "I responded by sending in an inquiry and expressing interest in further communication from NASA – if appropriate in that I had just finished a degree.

"On 30 January, I received a letter from a NASA official asking if I was still interested. This letter had been sent to me at Yale in early January, and then forwarded to me in Seattle. Reading it for the first time when I did gave me a macabre feeling to say the least, in that the Apollo fire had occurred just three days earlier. For this reason, I decided not to share with my dear wife Bonnie the fact that I had applied and was going to pursue the application. It was not unlike telling one's spouse of spending hard-earned money on buying a lottery ticket. The tricky part of this story comes if you win the lottery!"

When asked where he thought his career path might have taken him had he not become a scientist-astronaut, Joe Allen said that was "just about impossible to know. However, when selected by NASA, I had a very enjoyable and responsible job as junior faculty at a major university. I quite possibly would have continued as a research physicist and physics professor."

PHILIP K. CHAPMAN

Philip Kenyon Chapman was born in Melbourne, Australia, on 5 March 1935, the son of Colin Robison and Phyllis (née Kenyon) Chapman. When he was four, his journalist father joined Sydney's *Daily Telegraph* newspaper, and the family moved north to New South Wales. The Chapmans now made their home on Spit Road in peaceful Mosman, a northern Sydney suburb with magnificent, sweeping views of Sydney Harbour. It was a splendid place to grow up and he soon learned to swim in the seawater baths at the Spit, as well as exploring the rocky cliffs and beaches.[6]

He doesn't remember how it happened, but when he was eight, Phil Chapman met and made friends with Ethel Turner, author of the beloved children's classic, *Seven Little Australians*. Then in her seventies, she had a house with what Chapman describes as "a wonderful, mysterious garden, overlooking Balmoral. I often went there after school and spent a lot of time reading on her veranda. I also remember using her telescope to inspect departing ships, wondering what lay out there across the ocean, beyond the Sydney Heads."

Growing up in Australia

Chapman, who had learned to read when he was three, was excited by Ethel Turner's large library of paperback novels. She introduced him to the thrillers of John Buchan and Dornford Yates, and to the science fiction of H.G. Wells, Jules Verne and Edgar Rice Burroughs. "I think it was John Carter's adventures under the hurtling Moons of Barsoom that first made me a space buff," he reflects.

Initially he attended Mosman primary school, but when he was nine, he moved over to Fort Street Opportunity School, established a century before as the first model public school in Australia. Teachers and friends at those schools, and later at Parramatta High, often teased him about his fascination with rocket ships and space travel. In 1951, he won a book prize at school, and asked for Arthur C. Clarke's *Interplanetary Flight*. "At the presentation, the headmaster joked about wasting my time on fantasies. Many years later, Arthur Clarke and I became friends, and he autographed my tattered copy. While I was an astronaut, I returned to Parramatta High for a tree-planting ceremony, bringing my book with me. My old headmaster had retired, but he was there, and I was glad to make him eat his words."

Philip K. Chapman, SD.

His youthful zeal even led Chapman to write a letter to the Australian Prime Minister, Robert Menzies, in which he pointed out that the proposed rocket launch facility at Woomera in South Australia was pointed in the wrong direction for launching satellites. "Oh, yes, Menzies wrote back. He thanked me for my interest, but told me I should not worry, since his experts assured him that launching a satellite would always be impossible!"

Despite this response, Chapman remained convinced that people would fly in space during his lifetime, and set about maximising his chances of being one of them. Learning to fly seemed an obvious step, but he could not afford the lessons. Like all young men in Australia at the time, however, he knew that when he was eighteen he would have to spend six months in National Service doing military training. A very small number of trainees – thirty from each intake, nationwide – were taught to fly by the Royal Australian Air Force (RAAF). To make himself a better candidate for that selection, he joined the High School Army Cadets when he was fourteen, and rose to become the senior cadet officer by his final year at Parramatta High.

After passing the Leaving Certificate (with the highest grade in the school), he enrolled in physics and mathematics at Sydney University. By joining the university's

Squadron of the RAAF Reserve, he made sure he would do his National Service in the Air Force rather than the Army. The plan worked; he spent two summer vacations as an Aircraftsman Recruit (Minor) at Bankstown airfield, learning to fly Tiger Moths. The result was a private pilot's licence and many happy memories of flying a biplane, with his head out in the slipstream.

After graduating from university in May 1956 with a Bachelor of Science degree, Chapman joined the Sydney branch of Philips Electrical Industries as an electronics engineer. In mid-1957, he applied for a job as a physicist with the Australian National Antarctic Research Expeditions (ANARE), run by the Antarctic Division of the Australian External Affairs Department, but he was late in responding to their advertisement. He was told that the expedition teams had already been chosen, but that he could apply again the following year.

In early December, Chapman was working in his lab at Philips when ANARE called, to say that a physicist who was scheduled to spend the next year at Mawson base in Antarctica had just broken his leg. Could he join the expedition in the man's place – that day? Chapman quickly went to see his boss, who was very understanding, and gave him ten minutes' notice that he was leaving. He was on a plane to Melbourne that evening, and a week later, found himself in the *MV Thala Dan*, ploughing through seventy-foot waves in the Southern Ocean.

International Geophysical Year

The worldwide scientific community had designated 1958 as the International Geophysical Year, a period of intense study of the planet. Chapman's principle job was to observe the *Aurora Australis* – delicate, luminous curtains that light up the night sky in Antarctica. One of the objectives was to take simultaneous photos of the aurora, using a camera mounted on a theodolite at Mawson, and another one at a camp beside an Emperor penguin rookery at Taylor Glacier, some eighty kilometres to the west. The display is typically at an altitude of around one hundred kilometres, and the two photos formed a stereo pair that would allow researchers to plot its position in space. These studies required that Chapman spend most of the Antarctic winter at Taylor. There was always one other man with him, "but I was a resident and my companion was always a tourist taking a vacation from the hurly-burly metropolis of Mawson – population twenty-nine, all male!"

Chapman was twenty-three, and he found life at Taylor an exercise in self-reliance, and a formative experience. "If an emergency arose when the weather was fine, a de Havilland Beaver aircraft from Mawson could be there in less than an hour. But the weather was appalling much of the time, especially in the depths of winter. The temperature was often below −40 degrees Celsius, the wind above 100 knots, and the visibility zero in blowing snow so cold and dry that it would sandblast any exposed skin in seconds. In any case, for several weeks around midwinter, there was too little daylight for flying."

The first hut intended for Taylor blew away in a blizzard, so they lived in one made from a packing crate, two-and-a-half by two metres in area, and two metres high. Their toilet was a tide crack where the sea-ice met the land; and the weather soon taught them to be quick about it. Most of their food was in cans, stored in a stack of wooden crates that blew away one night. The crates all broke open but they managed to recover many of the cans, which were spread out for miles across the sea-ice. "All the labels had come off, so thereafter, dinner was always a surprise."

The first satellite, Sputnik 1, was launched on 4 October 1957, shortly before Chapman went to Antarctica. While he was there, America's first satellite, Explorer 1, discovered the Van Allen Radiation Belts, showing that the aurora is caused by high energy particles from the solar wind that are trapped in the Earth's magnetic field. "By early 1959, when the *Thala Dan* relieved the expedition, the aurora was understood by physicists everywhere – except for those who had been out of touch because they were busy studying it in Antarctica!"

The space age was now definitely under way, and Chapman knew it was time to move to the United States. Engineers and scientists were in great demand as competition between America and the Soviet Union in space intensified, and he had no trouble getting a job offer from the Massachusetts Institute of Technology (MIT). That put him on the first-preference list for a US immigrant visa, but getting one would still take time, because the Australian quota was then only fifty per year.

In December 1959, Chapman married Pamela Gatenby from Herberton, Queensland. They sailed the next day on a journey that took them eventually to Montreal, Canada, where he found a job as an engineer working in flight simulators for Canadian Aviation Electronics Ltd. The main attraction of Montreal was that he could run down to MIT every few months, to make sure he would still have a job there when his immigrant visa finally came through.

At long last, in April 1961, Phil and Pamela moved to Boston with their first child, Peter Hume Chapman, who had been born on 20 November 1960. At MIT, Chapman became a staff physicist in the Experimental Astronomy Laboratory of the Department of Aeronautics and Astronautics.

This was a momentous time in space flight. On 12 April came Yuri Gagarin's historic first flight, followed just three weeks later by Alan Shepard's sub-orbital ballistic mission. President John Kennedy followed that with a stirring speech in which he committed his nation to the goal of landing a man on the Moon and returning him safely to the Earth before the decade was out. It was an incredible undertaking, and MIT was heavily involved.

The Experimental Astronomy Laboratory was an offshoot of the MIT Instrumentation Laboratory, which was responsible for building the inertial guidance system that would take Apollo to the Moon. It was one of the best possible places to begin a career in the new field of astronautics. Phil's office colleague was Rusty Schweickart, who later became one of the third intake of astronauts and flew in Earth orbit on Apollo 9. Buzz Aldrin, the second man on the Moon, and Dave Scott, commander of Apollo 15, had both earned degrees through that laboratory.

In 1963, Chapman and his wife travelled to Washington DC, where the Australian Ambassador invested him with the British Polar Medal for services in Antarctica. His job allowed him to take courses at MIT, which led to a Master of Science degree in Aeronautics and Astronautics in 1964, and a doctorate under the joint auspices of the Departments of Physics and of Aeronautics and Astronautics in 1967. His doctoral dissertation was entitled *Theoretical Foundations of Gravitational Experiments in Space*. By this time, the space race between the United States and Russia was at its peak. Russia had not only sent the first man into space, but also the first woman, and the first three-man crew. In 1965 Alexey Leonov had become the first person to "walk" in space, outside his Voskhod 2 spacecraft.

But 1967 would become a tragic year for space exploration. On 27 January, astronauts Gus Grissom, Ed White and Roger Chaffee suffered a horrifying death when the interior of their Apollo capsule caught fire during a training exercise on Launch Pad 39B. In April, Russia's experimental Soyuz 1 spacecraft crash-landed following in-flight problems and a parachute failure. The sole cosmonaut on board, Colonel Vladimir Komarov, was killed instantly when his capsule slammed into the ground at high speed and exploded into flames.

Soon after receiving his doctorate, Chapman applied for American citizenship, which he realised was a requirement for any astronaut applicant. He completed the necessary paperwork, and on 8 May 1967, became a citizen of the United States. By this time, as he had hoped, NASA had announced that it was looking for a second group of scientist-astronauts, and Phil mailed in his résumé. "I was fortunate to know several astronauts, who had told me a lot about the programme. There is also no doubt that my year in Antarctica was a great advantage, because it suggested that I could survive in isolation and under stress. I met all the requirements, but I also knew there were 1,100 applicants." NASA wanted only the best, and only one in every one hundred applicants would make the grade.

The selection process quickly sorted out the dedicated applicants from those who took the whole exercise with less enthusiasm. The selection panel questioned them endlessly about their qualifications, and why they wanted to join the astronaut corps. Because they were not pilots, or had limited flying experience, the applicants were taken on dizzying flights aboard jet trainers, to test their reaction to heavy G-loads and unusual attitudes. They were also told that, as NASA astronauts, they would be required to learn to fly those same jets themselves.

At the time, NASA was undergoing a profound re-examination of its policies and procedures, trying to ensure that preventable accidents such as the Apollo fire never happened again. Chapman was quite aware of the risks involved in space flight, but he was also confident that the country's best scientific and engineering brains were working on solutions. In any case, exploration demanded that people resolve to press ahead, despite losses that might occur along the way. It was something he accepted, and then set aside.

ANTHONY W. ENGLAND

Quite surprisingly, Dr. Tony England still holds a minor record with NASA. Aged just twenty-five when selected in 1967, he remains the youngest candidate ever to be chosen for training by the space agency in six decades of selecting astronauts. He was also a mission scientist for Apollo 13 when the flight ran into serious trouble on the way to the Moon and, together with the Crew Systems Division at Mission Control, helped devise a makeshift lithium-hydroxide canister that prevented the crew from expiring of carbon dioxide poisoning. Assembly instructions were meticulously transmitted to the imperilled astronauts by radio, and after their safe return, England noted with justifiable pride that the unit they had patched together in space was an exact replica of the one he and his team had constructed.

A family on the move

Anthony Wayne England was born in Indianapolis, Indiana, on 15 May 1942, the first child of Betty (née Steel) and Herman Underwood England. "My mom's side were English and Irish, but on my dad's side we're not really sure – we think it was Scottish. My ancestors got lost in the Kentucky/Tennessee area." He would eventually have three younger siblings; the first was named Ethan, and then came Alice and finally another brother, Michael (who died in the spring of 2002). Tony, as everyone knew him, took his first tentative steps in education at Indiana Public School 49, also known by the less austere title of William Penn Elementary School, situated southwest of Indianapolis in Rhodius Park. A community swimming pool was located in the park and attracted exuberant hordes of neighbourhood kids on hot, sticky Indianapolis summer days.[7]

His father spent forty-three years as an agent for Hartford Life Insurance, insuring farms, livestock, and livestock transit – both train and truck. "He worked primarily with farmers, feedlot buyers, packing house buyers and livestock truckers, so his knowledge of farming was extensive." Herman England also did carpentry work by night and weekends, and he would buy up and renovate older houses on speculation, which the family would sometimes occupy until they were completed and sold. As a consequence, the family always seemed to be on the move. As England recalled: "When I was sixteen I'd lived in sixteen different houses. About half were temporary quarters when a house was sold before another was purchased, or when a house that we were to move into was not yet habitable." Armed with the fascination of a small boy for such things, he loved watching his father and the carpenters he had hired at work, and there was a meaningful lesson for him in his father's work. Everything Herman England built was built well, and the studious way he approached any project or task was something his son would emulate.

With another family move also came a change of school, and he continued his early education at Public School 82, located west of the city in Christian Park. When he was eight years old, the family moved to the country, twenty miles south of Indianapolis and it was here, at a small community school, that he took his education from fourth to sixth grades. But it was also a time of pre-pubescent rebellion, and he remembers being "a little unruly" at school. Looking back, he says he was one of the boys who would sometimes receive a well-deserved paddling from the teacher.

In 1953, Herman England was promoted to Hartford District Manager of North and South Dakota and part of Minnesota, and that winter set up a home for them in Fargo, on the eastern edge of North Dakota. The rest of his family followed in the summer of 1954. The impact on twelve-year-old Tony England was almost immediate. "Oh, I really enjoyed growing up in North Dakota. I loved it because it was a very

132 The Excess Eleven

Anthony W. England, PhD.

small pond where everybody could try anything – and we were all encouraged to try everything – so it was a great place to be a kid."

With the move to Fargo came a new enthusiasm for science in seventh and eighth grade at Agassiz Junior High School, where he also studied carpentry. "I was really very fortunate that my parents pushed education. I probably didn't really appreciate that until about the seventh grade. But that, and I think the freedom to explore different things in North Dakota, were the two aspects of growing up that made the biggest difference to me." By this time, he was also becoming, in his own words, "a fairly decent carpenter" and helped his father build a new family home a mile south of West Fargo along the Cheyenne River. Betty England was adamant that this would be their last move – she was tired of living in half-finished houses. "So dad sold his tools and never worked on houses again."

West Fargo back then was a farming community, and he would often walk out of his door with a single-shot .22 rifle to hunt rabbits and squirrels along the river. "The Red River Valley is absolutely flat and very fertile. Except for some trees along the Cheyenne River, a person could start walking away from our house and be seen until they dropped over the horizon several miles away. The only hills were man-made road or railroad overpasses, and the bluffs along the Cheyenne and Red Rivers. The stockyards and local packing plant were in West Fargo, as was Dad's office."

Tony England continued his education at West Fargo High, which was little more than a small country school six miles out of Fargo. In the summer of 1955, he hired out to a farm in southern Indiana. "After that, until I left for college, I always lived at home, though I always worked either in a lumber yard or as a carpenter during those summers."

A real turning point

His only academic problem now rested with the fact that, by the end of his junior year, he had devoured all of the courses then available at the school. In his ninth grade year he went to see the superintendent of schools in West Fargo, a man named Leonard (known as L.E.) Berger. He wanted to graduate early, and was seeking some advice. Happily, Berger saw that there was little point in holding the keen youth back, and his advice was that if he was willing to work for it, they would help him achieve this. England recalls that as a real turning point in his life – so much so that he stayed in contact with the Bergers, and is still in touch with Berger's widow, Dorothy.

His secret ambition at this time was to become a jet pilot, but he also realised that his less-than-perfect eyesight would likely preclude him from this particular goal. Little did he realise that in later years he would achieve his ambition, but not quite in the way he expected. Nevertheless, he had developed an interest in aeronautics and began attending the local Civil Air Patrol. On one momentous occasion, he and some other young people involved in the CAP went on an encampment at Ellsworth Air Force Base in Rapid City, South Dakota. By this time, he had begun flirting with girls, and was looking forward to taking one in particular to an organised dance. However his date fell ill, and so he took her girlfriend instead – a pretty young Fargo girl also in the CAP, named Kathi Kreutz. "She was fourteen, I was fifteen, and that did it, I guess!"

Meanwhile, as a distraction from his studies, England would often take to the football field for his school, and he also played trombone in the school band. Sometimes, there was a combination of the two. "I played football, and at homecoming I would still play in the band at half time. Another time, the football team got the 'flu badly enough that we actually had to scrimmage the band – there weren't enough to make up the team!"

England now had to decide on a college, but one of his friends told him he wouldn't get into a good college by graduating early. Undaunted, he took up the challenge by applying to both the Massachusetts Institute of Technology (MIT) and Harvard. To his complete surprise he received acceptances from both colleges, but he finally chose MIT as it "gave the best scholarship." With his college plans now laid out, he graduated from high school at the end of eleventh grade in May 1959, one of a senior class of just forty-six students.

While living in West Fargo, England had also acquired a growing interest in amateur radio. Even though he lived in a fairly remote area ("it was a little isolated, perhaps, from the rest of the world – everybody thought we lived in igloos and such up there!"), he found immense enjoyment in building radios, which allowed him to communicate with other ham radio operators all around the world. "It was a big feature in my growing up there," he emphasised. What began as a hobby and curiosity would later become a notable diversion for him when enjoying some free time during his Shuttle flight in 1985.

England's first three undergraduate years at MIT were in physics, but he had discovered a particular fondness for doing fieldwork, and during his senior year, switched to a five-year honours course in Earth sciences. This course included geophysics, which combines the principles and methods of physics and geology and applies them to Earth science. By now, he and Kathi Kreutz had been an item for quite some time and they were married on 31 August 1962.

In October 1964, NASA began advertising for a new breed of astronauts, whose skills were more inclined towards science than in test flying jet aircraft. "They sent the information, but it was clear I wasn't qualified yet; I was still in graduate school."

At the end of summer 1964, Tony England achieved his Bachelor of Science degree in Earth sciences, and Master of Science degree in geology and geophysics, which were awarded the following January. His parents were particularly proud of him earning these degrees. "It meant so much because I was the first in my family, except for one of my grandmother's sisters, who was a teacher, to earn a college degree. Dad had attended two years at Indiana University while on leave from Hartford, but had quit to support his family. He actually went to night school at North Dakota State University in Fargo in his late forties and early fifties to complete his BA in economics."

England stayed on at MIT, where he worked on developing theories that would predict the electrical properties of the Moon and the planets. As part of his earlier graduate studies, he'd written a paper on what to expect from the influence of lunar electromagnetic propagation. Another field of study involved the structure of glaciers in the western United States.

Much to his delight, in late 1966, NASA eventually called for applicants for a second group of scientist-astronauts. Now, as a graduating PhD in geophysics, he felt he was qualified to complete and submit the form. "I was particularly interested in trying to go to the Moon to do my own physics, and that's why I was anxious to apply. I'm not sure even now whether I'd be any more insightful about whether I was qualified or not, but you just try and see how it works out." This time, it certainly worked. Confirming his successful application by letter, Director of Flight Operations, Deke Slayton, asked him to report for duty in Houston on or before 18 September.

There would be one unexpected difficulty for Tony England following his selection. "I thought my PhD research was completed when I left for NASA, but while writing the dissertation during the fall of 1967, I found that some of the experimental work was inconsistent and had to be redone. My first opportunity to repeat the experiments was during the summer of 1969, after flight school. I spent most of the summer at MIT and then wrote the dissertation at night during that fall, while working on Apollo. I defended the dissertation that winter and was awarded the degree in May 1970."

KARL G. HENIZE

On the evening of 15 September 2000, a small but well-attended dedication ceremony took place at the Harper College Observatory in Palatine, Illinois. Located at the northern end of the Harper Campus, the building houses the historic Peate Telescope. On this particular Friday, the observatory was being renamed to honour the life of a modern-day astronomer, not only for his accomplishments, but also for his dedication to fulfilling his dream of reaching and working in space. That scientist, who died suddenly while ascending Mount Everest in 1993, was Dr. Karl G. Henize.

A close friend and colleague, Dr. Loren Acton, gave the dedication address. He had flown into space with Henize as a payload specialist aboard Spacelab 2 on Shuttle flight STS 51-F in July and August 1985. Interviewed for this book, he recalled a man for whom life was an astonishing adventure, but all too short.[8]

"Karl and I were on the Spacelab 2 'red team' with Roy Bridges. The red team served alternate twelve-hour shifts from the 'blue team' of Tony England, John-David Bartoe and Story Musgrave. It was a joy to work with Karl because of his total enthusiasm for the space flight experience, even when he was feeling rotten from space adaptation effects. Karl was the one who was at the window taking Earth photos at every opportunity. He was the one who chose to sleep floating free in mid-deck rather than in his more constraining bunk. When I was tending to overdose on 'responsibility', Karl suggested I not miss the opportunity to look out the window and enjoy the ride; sound advice for which I am eternally grateful."

Even before he was selected as a scientist-astronaut in 1967, Karl Henize's name was already well known and respected within NASA. As a professor in Northwestern University's astronomy department, he specialised in stellar spectroscopy, testing the amount of light emitted by stars in order to determine their composition. In this capacity, he had personally supervised experiments carried on Gemini missions that allowed stars to be photographed using ultraviolet light. He had certainly come a long way from the hills and farmlands of Plainville and Mariemont in Ohio.

Just like Daniel Boone

Karl Gordon Henize was born on 17 October 1926, the third son of Fred Raymond and Mabel Henize. Fred Henize, raised on a farm in Ohio's Brown County, was a third generation American, whose paternal great grandfather Jacob had emigrated from Germany. Karl's mother, born Mabel Claire Redmon, had also grown up in Brown County, where she and Fred met and eventually married. Her side of the family went back several generations on American soil, and Karl could boast ancestors who had fought in the war of independence from Britain.[9]

As a young man, Fred Henize had taken up a job with the post office, first working an urban route and later, one covering rural areas. Along the way, he and Mabel managed to acquire some land, and became property owners and farmers in Ohio's Mariemont–Plainville area, some ten miles east of Cincinnati. Here, they worked a twenty-acre spread situated atop a plateau overlooking the magnificent Little Miami

Karl G. Henize, PhD.

River. The rear boundary of their land butted up against a vast estate once owned by former President William Howard Taft, who had sold the estate back in 1889. An industrious and loving couple, Fred and Mabel ran a small dairy and icehouse business on their property, in addition to raising dogs and ferrets.

Their happiness was boundless when their first son Wilson was born in 1918, but tragedy would strike the family a few years later when their second son Claire, then two years old, found and swallowed some insect poison. They rushed him to the doctor, but he passed away soon after. For the rest of their lives, Fred and Mabel would maintain that their doctor had mistakenly given Claire the wrong antidote. The birth of baby Karl in 1926 was treated as a blessed event, and he was raised in an environment of love and security.[10]

In the days before suburban development began to flourish, Karl grew to love the character and history of the local area. It was a vast, tree-filled countryside, presenting endless challenges and enthralling possibilities for any young, adventurous boy. He was told that legendary frontiersman Daniel Boone once used to hunt and trap along the Little Miami River, so he and his friends would similarly spend untold hours on expedition forays, especially around Indian Hills to the north, where relics from earlier Native American occupation could sometimes be unearthed.

In their free time, Karl and Wilson loved nothing better than to go swimming or bank fishing in local swimming holes, especially on those long summer days when they'd meet up with friends also keen on seeking relief from the relentless heat. During the Depression years, it was not uncommon to see carloads of adults and children heading out to these pools, which were actually tributaries of the Little Miami and Great Miami Rivers. Favourite excursions included shady pools in Remington, Shademoore and Bass Island.

When Karl Henize was just eight years old, he lost his father to a combination of pneumonia and kidney infection. He and his brother not only found themselves saddled with the daunting responsibility of maintaining the farm, but also with assisting their mother through the latter days of the Depression, which would only end with the declaration of war. Fortunately, the established family business of selling and delivering milk and ice helped to ease their burden.

After completing his elementary and primary school education in Plainville, Karl went on to attend Mariemont High School, situated on tree-lined Pocahontas Avenue in the quaint English garden community of Mariemont. But he would never complete his schooling there, as a result of America's entry into the Second World War. When Wilson volunteered for submarine duty in the Pacific, Karl also became keen to serve his country. In 1943, he decided to quit high school and enter the Navy's revolutionary V-12 Program. Spread across 131 college and university campuses, V-12 was not only designed to prepare large numbers of men for the Navy's Officer Candidate Schools, but also to increase the war-depleted student bodies of many campuses. By joining this programme, Karl knew he could receive college credits over some seven semesters, which would eventually qualify him for officer commission.

He first attended Denison University in Granville, Ohio, before relocating to the University of Virginia in Charlottesville. While he studied, he also fell in with a dedicated group of cave explorers, known as spelunkers. The Blue Ridge area of Virginia boasts the highest number of caves of any area in North America, so with his friends, he found endless enjoyment making his way through the complex recesses of many caves, particularly those in the Shenandoah National Park. His agility and cautious daring in the pitch-black mazes would soon earn him the nickname Monk, short for 'monkey.' These experiences in working as part of a team under extreme, remote and often hazardous circumstances would also bode well for him in his later application to NASA.

The skies and a thesis

The war came to an end before Henize received his commission, so he decided to stay on and become a member of the Naval Reserve. In 1947, he received his Bachelor of Arts degree in mathematics from the university, to which he added a Master of Arts in astronomy the following year. He then took up employment with the University of Michigan, and travelled to their Lamont-Hussey Observatory in the South

African town of Bloemfontein, where he spent the next three years working on a spectrographic survey of the southern sky for stars and nebulae showing hydrogen emission lines. When not occupied with his work, he took time out to play rugby, organise a small baseball league, and enjoy long cross-country hikes. Following his return to the United States, he became a candidate for a doctorate in physics at the University of Michigan, and his survey plates of the southern skies not only formed the basis for his thesis, but of his life's work beyond that.

Karl Henize met his future bride, Caroline Rose Weber, at Michigan in 1952 and they were married on 27 June the following year. A few months later, they made their way across to California, where Karl had taken up a Carnegie post-doctoral position at the renowned Mount Wilson Observatory in Pasadena.

In 1955 their first son, Kurt, was born, and the following year an opportunity came along to work as a senior astronomer at the Smithsonian Astrophysical Observatory in Cambridge, Massachusetts. Here, Henize was given responsibility for setting up a series of Baker–Nunn wide-field photographic satellite tracking stations during the International Geophysical Year, even though no satellites had been launched at that point. Despite this, high priority had been given to establishing and operating a global network of twelve stations for photographic tracking of future satellites. On 3 October 1956, their daughter Marcia was born and just one year later, the Russians sent their first Sputnik artificial satellite into orbit, and the reason for haste in setting up the system became evident.

Three fulfilling years passed at the Smithsonian before his next appointment, as an associate professor at Northwestern University's Department of Astronomy in Evanston, Illinois. As well as his teaching responsibilities, he also carried out extensive studies into phenomena such as peculiar emission-line stars (classified as Be stars), S-stars, T-associations, and planetary nebulae.

Their second daughter, Skye, was born in June 1961 and soon after, he and Caroline gathered up their young family and flew over to Australia for a year, during which time he served as a contributing guest observer at the Mount Stromlo Observatory on the outskirts of Canberra. His work involved the support and development of new research directions in stellar and galactic astronomy, using instruments that included a 20/26-inch Uppsala Southern Schmidt telescope. Henize was particularly interested in Stromlo's 74-inch parabolic reflector; just two years earlier, this telescope had made observations of 30 Doradus, yielding the first information ever obtained on the chemical composition of anything beyond our Milky Way. He also participated in studies of ultraviolet optical systems and other research suited to the manned space flight programme, developing a stellar spectroscopy experiment later flown on the last three manned Gemini missions. When their posting came to an end late in 1962, the family returned home to Evanston, Illinois.

When NASA announced that it was seeking more scientist-astronauts in September 1966, Karl talked over their options with Caroline, and she finally convinced him to apply. Empirically realistic about his chances, his only major concern was his age. Now forty, he was already five years beyond the given limit. He wrote a letter to Chief Astronaut Deke Slayton, asking if this might preclude his selection. Happily, Slayton's reply said the age limit could be waived for especially qualified individuals, and Henize certainly fitted into that category. Given this assurance, he allowed his application to stand and continued to work as he waited to hear back from NASA.

DONALD L. HOLMQUEST

Oak Cliff, Texas, was an exciting place for any youngster to grow up, and Don Holmquest was no exception to this hometown appreciation. Situated two miles south of downtown Dallas on the south bank of the Trinity River, Oak Cliff was an ethnically and architecturally diverse community with a rich heritage stretching back to the first European settlers in the mid-1800s. His forebears had been among many robust pioneers who first settled the area, originally making their homes in hand-hewn cabins and tents on the river's west bank, across from John Neeley Bryan's dynamically growing city, Dallas.

Graced by gentle, rolling hills, the area young Don lived in was filled with sparkling creeks, parks, and massive oak trees – all of which he and his boyhood friends explored with utmost vigour. "I had a great childhood, spending most of my time in the woods and in tree houses."

A strong educational discipline

Donald Lee Holmquest was born on 7 April 1939 to Lillie Mae (née Waite) and Sidney Browder Holmquest. At some point, his mother's family had anglicised their original German surname from Weitz to Waite, and she told him that her grandfather had been a gun maker in Germany. She had elected to leave school and find work at the tenth grade level so she could attend the funeral of her older brother, who was killed while working on the railroad in Chicago. Lillie never returned to school, as it was not considered all that important at that time, especially for a woman. However, she would always maintain a strong educational discipline as her own son grew up, making it clear that "doing second best was not an option."[11]

His father Sidney, whose parents were Swedish emigrants, had also left school early, giving up his education after the eighth grade so he could help support his family. He became an electrician with the local power company and, as a person with a reputation for meticulous care in anything he did, was very proud of the work he carried out for them. This love of crafting and repairing things had come in turn from his own father, who had been a cabinetmaker.

Sidney Holmquest taught Don very early in life to make and fix anything – family values that would always stick. As he grew up, Don would learn the attributes of persistence and patience when working on any task, and take immense pride in completing a difficult job, knowing he had done well.

When not playing with his friends, he also enjoyed the many simple pleasures associated with hanging out with his father, who was often more like a friend to him. "We camped and fished and hunted together, and he taught me to love the outdoors and be self-sufficient. His attachment to building things probably steered me into math and physics, and ultimately into electrical engineering."

140 The Excess Eleven

Donald L. Holmquest, MD.

Early schooling for Don was taken at Roger Q. Mills Elementary, a public school half a mile from his home. Depending on the weather, or his inclination, he would either walk to school or ride his bicycle, along the way passing many magnificent houses constructed of distinctive Austin stone, a popular post-Depression building material. The area encompassing Oak Cliff had once been an elite residential and vacation community, but as more and more lots were sold off to middle- and working-class families, bungalows and economical wood-fronted houses had begun to proliferate.

When asked what he remembered of his public school and teachers, his answer came after a little reflection. "My only recollection is that it was pretty, had a great playground, and had teachers who did in fact motivate me to do my best. I recall absolutely no science in elementary school, but I did realise a serious fondness for math."

His learning did not end when the school bell rang, however, as his mother always took care to see that his education was going well. "On my first report card in the fourth grade, I had one B. She sat me down and made it clear that there was no reason whatsoever for that grade. She made sure I reviewed my spelling words each night and my multiplication tables until they were perfect. I guess I would have been some sort of neurotic person if it had actually been difficult. As it was, I was lucky in that it didn't really cut too much into my playtime. I always found school easy, as well as the tests, which probably explains much of my success. I never took books home for homework from public school because it wasn't necessary to making top grades."

After graduating from the W.H. Adamson High School in Dallas at the age of eighteen, Holmquest decided to undertake a degree course in electrical engineering at the Southern Methodist University, also in Dallas. He worked his way through engineering school by taking on part-time jobs in industry as a student engineer. In 1958, he joined the Dallas-based firm of Chance Vought Aircraft, which would become Ling-Temco the following year. In 1961, his final year with the firm, they merged with Ling Electronics to become Ling-Temco-Vought Electronics. "Some of the time [I] was at Chance Vought Aircraft where I developed an interest in fighter aircraft," he stated. Ling-Temco-Vought would re-enter his life later on, as principal designers and suppliers of the Gemini and Apollo spacesuits.

Holmquest was able to achieve a 4.0 average at Southern Methodist, despite adding on some pre-med courses in his senior year, and graduated in 1962 with his Bachelor of Science degree. He then began working at Texas Instruments, also in Dallas, and recalls that this was where he "designed computer chips and became fascinated with computers."

Some time before his graduation, he had become interested in medical science, but it was certainly not his primary focus at that time. "My entry into medicine was due entirely to a close friend, who explored medicine as a means of pursuing an interest in psychology and psychiatry. Sadly, I got into medical school easily and he did not."

While his studies at Houston's Baylor University now occupied the better part of every week, Holmquest would always set aside some leisure time to pursue his favourite outside activities. "Most of my close friends in my teens were from church rather than school. While I was not all that serious about religion, our church group was very close. It was my main outlet for athletics in that we had great baseball and basketball teams and lots of social events."

It was in this group of his peers that Holmquest first met his future wife, a pretty young girl of Czech heritage named Charlotte Ann Blaha. Asked during the interview if she was related to Shuttle astronaut John Blaha, he said it was not the first time that had been asked. "While she had Czech parents like John, they are not related to my knowledge. I don't know John's roots, but there are very large communities of Czechs in Texas, and Blaha is a rather common name."

While studying for his doctorate in medicine, and later in physiology, Holmquest came to admire the work of Chairman of Surgery at Baylor, Michael DeBakey, MD. When he took up an internship at Baylor's affiliated Methodist Hospital in Houston studying internal medicine, he was able to continue his studies of Dr. DeBakey's work. In 1965, DeBakey introduced the use of telemedicine, when he performed open-heart surgery that was televised live to another hospital in Geneva. It was a ground-breaking event in cardiac surgery procedures, and Holmquest felt fortunate to be studying at Methodist Hospital at such an historic time, when his twin passions of electronics and medicine combined in a most innovative and practical way. Later, in 1968, Dr. DeBakey would direct the world's first multiple-organ transplantation procedure at the hospital. Meanwhile, Holmquest had begun a dissertation for his PhD, having decided that the subject would be telemetry studies of the thermal rhythms in laboratory rats.

Applying to NASA

Late in 1966, and prior to completing his doctorates and his internship, Holmquest discovered that NASA was seeking applications from qualified scientists for their second intake of scientist-astronauts. The requirements for the second group were not as rigid as those for the first, and he noticed that there was no need to have prior jet pilot experience, or even any flight time. Certain physical requirements were also a little more relaxed than before, and such things as the age limit could even be waived if the applicant was particularly highly qualified.

It was obvious that NASA and the NAS had been badly burned the first time around by their insistence on far too many inflexible standards, resulting in just 1,400 applicants. This time around, the easing of the jet pilot qualification would ensure a much larger pool from which they could draw their finalists. To sway the undecided, the National Academy of Sciences and NASA issued several bulletins, in which they described future plans for an indeterminate number of orbital and lunar flights, which would require the crew participation of scientists and medically qualified astronauts. They suggested that in around three years, the selected applicants could find themselves flying to the Moon, and perhaps on to Mars in a decade or two.

The whole idea appealed to Holmquest and he filled out an application, careful to mention that he had yet to be awarded his doctorates in medicine and physiology. The launch-pad tragedy that took the lives of three Apollo astronauts in January 1967 made him doubly aware of the dangers he could face as an astronaut, but he remained resolute. Early in March, he was informed that his application had passed the NAS screening and it would now be passed to NASA, who would resume the process of elimination and final selection. He would later discover that he was one of sixty-eight scientists to undergo this further evaluation, and was one of the eleven who made the final cut. He was now a NASA scientist-astronaut.

In 1967, Don Holmquest was awarded his doctorate in medicine, and the following year achieved a second doctorate in physiology.

WILLIAM B. LENOIR

In August 1967, the state of Florida was enjoying a mild celebration. Home to America's bustling spaceport, the state could finally boast its own native-born astronaut. Group 2 astronaut John Young had previously been "adopted" by Florida, having been raised in Orlando, but he had originally come from California. On the other hand, William Lenoir was born in the city of Miami and took his early schooling in Coral Gables, so he was perfectly happy to be associated with the south-eastern state when he joined the second group of scientist-astronauts in Houston. It was a high point of his life – a true pinnacle of achievement – and he was ready to fly to the Moon or into orbit following his training. He was not to know it back then, but Bill Lenoir would not achieve his first and only flight into space until 1982, by which time he had been with NASA for more than fifteen years.

William (Bill) Benjamin Lenoir was born on 14 March 1939, to Iona (née Yann) and Samuel S. Lenoir. His mother was born in New Glarus, Wisconsin, in 1915, but had moved to Florida with her family when she was quite young. His father, born in 1910, came from a family well associated with Tennessee; in fact he was born in a place called Lenoir City. Originally known as Lenoir's Station, the 5,000-acre tract of land along the northern bank of the Tennessee River was named for family patriarch General William Lenoir, who, as a junior officer, had served under Colonel Benjamin Cleveland during the Revolutionary War. His eldest son, Major William Ballard Lenoir (1775–1852), had been bequeathed the land by his father, and gave the burgeoning town its name in honour of his father.[12]

When he was a child, Sam Lenoir's parents moved to Harriman, Tennessee, and subsequently to St. Louis, Missouri. According to Bill Lenoir, his father's family then moved to Miami in the 1920s, "where he endured the Depression as a bookmaker (gambler). He served in the infantry during the Second World War, returning home at its completion." In 1943, Sam's father, Dr. Benjamin Ballard Lenoir, was killed in an accident. It was from his paternal grandfather that Bill Lenoir had inherited his middle name. His first name derived not from his famous ancestor, William Lenoir, but from his maternal grandfather, William R. Yann, a barber by trade who doted on Bill and his younger sister, Barbara (who was born in December 1940). "In the 1950s, my father and a friend set up a swimming pool maintenance business, where my father managed the store."

A natural-born engineer

As a youngster, Bill Lenoir attended public school at Coral Gables in Dade County, South Florida. He began his high school education in 1953 at Coral Gables Senior High, which had been inaugurated the previous year. When he graduated as a member of the Class of 1957, he was just one year behind Janet Reno, who would become the longest-serving (and first female) United States Attorney General, in the Clinton administration.

"I don't remember deciding to go into engineering. Since junior high school, I had known that I would be an engineer or scientist, based on my affinity for mathematics and sciences. In a way, high school was a breeze. I studied some (not a lot), got excellent grades, and partied a lot. My strengths were math, science and English. The English foundation that I built in high school is often lacking in recent engineering graduates, much to the dismay of all who communicate with them."

When asked if any of his educators had had a significant effect on him, two names came readily. "The two teachers I remember most are Mrs. Joanne Woltz (English) and Mr. James Newmeyer (math). Mrs. Woltz was a no-nonsense 'old world' teacher;

William B. Lenoir, PhD.

Mr. Newmeyer was an MIT graduate who helped me navigate the MIT admission and scholarship application processes."

Following his graduation from senior high school, and with Newmeyer's considerable guidance, Lenoir took on studies of electrical engineering at the Massachusetts Institute of Technology. "I began at MIT in September, 1957 as a Sloan Scholar, a position of respect at MIT. This was a full scholarship that paid for tuition, room, board, and incidental expenses. As seniors, Sloan Scholars were flown to New York for a banquet and overnight at the Waldorf Astoria with Alfred Sloan presiding. Mr. Sloan was the CEO of General Motors and an MIT graduate.

"As an undergraduate, I was a member of Sigma Alpha Epsilon fraternity and lived in the chapter house on Beacon Street in Boston, Massachusetts. I studied more (still not a lot), did well, and maintained my interest in parties as a diversion from the day-to-day stress of school. The fraternity was an independent entity from MIT, fully responsible for its own actions. This self-reliance of the group was an integral part of the overall education process, especially when I served as treasurer and president in my junior and senior years."

Lenoir majored in electrical engineering and participated in the Institute's co-op programme, where he alternated terms at school with terms in a work environment at the General Radio Company. He was awarded his Bachelor of Science in electrical engineering in June 1961.

One year later, he was awarded his Master's degree in electrical engineering. "My Master's thesis was a report on my design and development of a thermoelectric temperature control device for an upcoming General Radio product. It was supervised by Professor Paul Gray, the recently-retired chairman of the MIT Corporation." While pursuing his doctorate in electrical engineering under the tutelage of Professor Emeritus Alan H. Barrett, a noted radio astronomer, Lenoir continued to work at the institute, first as a teaching assistant, and later as an instructor. He won the institute's Carleton E. Tucker Award for Teaching Excellence in 1964.

The dissertation Lenoir researched and wrote for his doctorate went by the title *Remote Sounding of the Upper Atmosphere by Microwave Measurements*. It set the theoretical basis for performing space-based measurements to infer temperature distribution in the upper atmosphere. "In addition, I supported and led a team doing related experimental work from unmanned balloons at 100,000–130,000 feet. I was also a principal investigator and a co-investigator on several proposed space experiments."

On 4 July 1964, prior to the awarding of his doctorate, Lenoir married Elizabeth May Frost of Brookline, Massachusetts. They had met in the summer of 1959 while he was doing several part-time jobs in order to augment his standard of undergraduate living. "I was shovelling coal and tending a furnace in an apartment building on Beacon Hill in Boston when I met Liz. Her parents were substituting for the caretakers and she came down to give me my pay cheque. I was a sorry, black, dusty mess." The young couple would have two children; William Benjamin Jr., born 6 April 1965, and Samantha Ellen, born 20 March 1968.

Research for Apollo

Lenoir was awarded his doctorate in 1965. He became a Ford Post-doctoral Fellow and an Assistant Professor of electrical engineering at MIT, and continued teaching and researching as before. His work now included teaching electromagnetic and systems theories, as well as carrying out absorbing research into remote sensing of the Earth and its atmosphere. At the same time, he took on a role as a researcher and investigator in several Apollo Applications experiments, which included work on developing experiments intended for use on what became the Skylab programme. Later, following his selection as a scientist-astronaut, he would continue with this work as a co-investigator with his colleagues at MIT and JPL.

For two years, starting in 1965, Lenoir carried out graduate research with MIT's Research Laboratory of Electronics (RLE) under Professor Barrett, who had pioneered the microwave spectra measurement of high-temperature diatomic molecules. Barrett had probed the atmosphere of Venus using theoretical microwave radiation studies, and was the first to recognise that the planet's thick carbon dioxide atmosphere contributed to its extremely high temperatures. He would also design microwave detection equipment used on NASA's unmanned Mariner missions, which conducted flybys of Venus and Mars.

As part of his research with RLE, Lenoir helped design a 60-gigahertz atmospheric sensing microwave receiver to remotely sense the temperature profile of Earth's atmosphere at different altitudes.[13] The instrument package was deployed on board a high-altitude helium balloon launched from the National Center for Atmospheric Research in Palestine, Texas. As a result of these experiments, Lenoir's research group, as well as groups led by former students such as Dr. Joe Waters at JPL, developed new instruments that were used in NASA's Nimbus series of advanced global meteorological satellites, forerunners of the Landsat satellites and today's weather forecasting system, operated by the National Oceanic and Atmospheric Administration (NOAA).

How did he come to apply to NASA? "In the fall of 1966, I decided that I needed to broaden my experience base by spending some time – several years – somewhere other than MIT. I had been at MIT for the entire ten years of my professional life. I answered inquiries from the University of Michigan, the NASA Goddard Space Flight Center, and Drexel University. Later that year I came across a small notice in *Science* magazine that NASA was accepting applications for scientist-astronauts. Given my field, this seemed to make a lot of sense, so I returned the small clipping with my name and address. Like magic, this small clipping became a one-inch thick stack of paperwork to fill out and return. Then, the physical; the interview in Houston; the phone call from Alan Shepard."

JOHN A. LLEWELLYN

At the time of his selection by NASA, Dr. Tony Llewellyn was asked what he hoped to achieve as a scientist-astronaut. His response was typically straightforward but eloquent. "My ambition is simply to make a successful flight, do some good experiments and get some good first-class science out of it. You know, Isaac Newton said he had walked on the edge of the sea picking up pebbles. I think that all of science has the chance to walk on the edge of a brand-new sea – space – and pick up the pebbles."

Most people rightfully envisage Wales as an ancient country steeped in Celtic culture and bearing a rich heritage, although the spectacular green hills and valleys of South Wales still bear the scars of an industrial past, when massive excavation and mining yielded millions of tons of coal and iron. Situated in the county of South Glamorgan is the sprawling city of Cardiff, since 1955 the capital of Wales. In its boom years, Cardiff had been the largest coal-exporting port in the world, and this thriving industry was reflected in an expanding and rapidly growing population.

In the early 1930s, when Tony Llewellyn was born, the population of Cardiff stood at just under a quarter of a million, and the once-great industrial city was in the merciless grip of the Depression. Despite this, Cardiff remained resolute, and was an exciting, fiercely proud place for any youngster to live in. The massive working docks on the River Taff were the ultimate draw for any young boy's curiosity, as were the brightly-painted buses and electric double-decker trams that made their way past

John A. (Tony) Llewellyn, PhD.

throngs of people down busy Cardiff streets, with its multitude of narrow, covered Victorian arcades. For the older boys there was also the sheer joy of crawling over such ancient local sites as Cardiff Castle, in the central part of the city – a sprawling gothic structure built by the Normans in the eleventh century – and the magnificent "Red Castle", Castell Coch, just five miles north of town by bus.

Tony Llewellyn was born in Cardiff, and he would later achieve a deal of minor renown in Britain when, at the age of thirty-four, he was selected as one of the first two NASA astronauts born outside the United States. He was born John Anthony Llewellyn on 22 April 1933 to John and Morella (née Roberts) Llewellyn. Llewellyn is a common enough surname in southern Wales, and is perhaps best attached to the famed writer Richard Llewellyn, whose richly evocative book about the travails of coal mining communities in the area, *How Green Was My Valley*, became a classic of modern literature, as well as a celebrated Hollywood movie. Tony Llewellyn has often been asked if he is related to the writer, but he has never established a family link.[14]

His father, John Llewellyn, was an engineer who worked with the large Guest, Keen and Baldwin Iron and Steel company, situated near the docks, which was one of the town's major employers. He imbued in his young son a fascination with science and mechanics, and both parents saw to it that Tony and his younger brothers David and Roger took an active interest in their education. He would begin that education at the Adamstown Public School on System Street, and today fondly recalls an energetic curiosity, being "always interested in the balls and springs of the universe; how everything worked."

Early influences

His early fascination with all things mechanical caused John and Morella to give their older son a set of red-bound Arthur Mee's *Children's Encyclopaedias*. In them, he found endless pages of subjects that provoked and sometimes satisfied his childhood quest for answers. These days, he credits the books with being at the very origin of his interest in how things worked. He recalls spending countless hours curled up in a comfortable chair with his beloved encyclopaedias, absorbing information.

Towards the end of June 1940, Luftwaffe bombs occasionally rained down on Cardiff, but war truly came to the city on 15 September, when massive German air strikes were carried out on several large cities on a date that came to be designated as Battle of Britain Day. Fortunately, the Luftwaffe bombing was not very accurate, due to a determined and spirited defence by Royal Air Force (RAF) fighter pilots, but Cardiff still suffered extensive damage and a number of civilian casualties that day, together with London and other major cities across Britain.

Harassing night attacks by German bombers continued however, now with an emphasis on major ports and harbour installations, which made Cardiff a prime target. John Llewellyn also worked as a civilian for the Royal Navy, and this caused the family to move around often during the war – and afterwards – to cities that also came under bombardment, such as Bristol, Plymouth and Southampton. The war years were a deeply concerning time for the civilian population, and Tony Llewellyn remembers those times: "It seemed that the Luftwaffe followed us, since bombing always seemed to speed up after we arrived!"

With peace declared, the family returned to Cardiff and life slowly resumed its course. Tony now attended Cardiff High School, graduating in 1949. He recalls one teacher named Whittle as being an early inspiration to him in that time. He was also active in the Boy Scout movement and would eventually achieve the honour of becoming a King Scout – the British equivalent to an American Eagle Scout.

A future in science now seemed to lay before him, and with this conviction in mind, Llewellyn registered for a Bachelor of Science degree at the University College of Cardiff, which he eventually achieved in 1955. During the course of his degree studies he had become excited about the specific field of chemistry, and once again an educator was to provide the inspiration: "I had read about the 'magic bullet' idea of specific molecules to cure everything and thought chemistry was interesting, but found (in college) that organic chemistry was messy ... physical chemistry was much more logical. At the university, A.G. Evans' lectures convinced me that physical chemistry was where I belonged, especially his stuff on statistical thermodynamics." It would be a determining revelation in his life, as statistical thermodynamics plays a vital linking role between quantum theory and chemical thermodynamics, yet most students find the subject difficult to grasp and therefore unpalatable.

While studying for his degree, he had found a little diverting relaxation and the thrill of competition as a member of the college's swimming team, where he soon found himself drawn to another distraction – a young lady also from Cardiff named Valerie Mya Davies-Jones. "She was a Physics/Math major, but that wasn't a handicap!" They were married in 1957, and their first child, a son they named Gareth Roger, was born on 30 October that year.

Working in Ottawa

Upon gaining his Bachelor of Science degree, Llewellyn signed up for extended courses at Cardiff University, and in 1958, was the proud recipient of a Doctorate degree in chemistry. Following the award of his doctorate, Llewellyn emigrated from Wales with his wife and son to a new life in Canada, where he had decided to take on a position as a post-doctoral fellow at the National Research Council (NRC) in Ottawa, which enjoyed a reputation as a place to perform leading edge, world-class research. Here he would work alongside such luminaries in the field of chemistry as Gerhard Herzberg and Keith Ingold. During this exciting period, the NRC published numerous scientific papers, such as those on the rates of free radical exchanges. Many of these papers would become classics in chemical literature.

In 1960, Dr. Llewellyn was on the move again, this time taking up a post in Tallahassee as a research assistant in Florida State University's Chemistry Department, where faculty member Dr. Charles Mann had recently developed a graduate course in chemical instrumentation. The following year, he also took on the role of research associate in the university's Institute of Molecular Biophysics, and would subsequently be appointed an assistant professor.

On 20 April 1962, a daughter named Sian Pamela made the Llewellyn family a foursome. By now, Llewellyn had seen astronauts and cosmonauts taking those first pioneering steps out into the cosmos and, being so close to Cape Canaveral, he was swept up to a degree in the excitement of space flight. But he could not see a future for anyone such as himself in this bold venture. Still, a further eminent advancement in his academic career came in 1964, when he was appointed as an associate professor in both the university's School of Engineering, and in their Department of Chemistry.

The Florida lifestyle appealed to the young family, with the Llewellyns continuing their love for water sports through swimming, scuba diving and boating, and on 17 February 1966, Dr. Llewellyn swore the Oath of Allegiance and became a United States citizen. In September that year, he spotted an advertisement from a local newspaper pinned onto the university's notice board. It was headed *Opportunities for Scientists as Astronauts*, and said that NASA was looking to recruit a new intake of scientist-astronauts as the space agency's sixth astronaut group. He noted with interest that the flying proficiency attached to the first group of scientist-astronauts had now been dropped, and he felt he could meet many of the necessary qualifications. All applicants had to be American citizens, and he had fortunately achieved this mandatory qualification just seven months earlier. As with many others who would also apply, he was hoping the stated possibility of flight training might not be required, and after consulting with Valerie, he filled out his credentials and posted an application off to Houston.

On 7 February 1967, Tony and Valerie celebrated the birth of another baby – a second daughter they christened Ceri Eluned, a strongly Celtic name, as with their other two children. Within days of Ceri's birth, Llewellyn received a telegram from the National Academy of Sciences, requesting further information to back up his application. All of a sudden, his impulse to apply had taken on a whole new dimension, and he had to ask himself once again if he really wanted to become an astronaut. Much as he tried to find logical arguments against it, the answer always came back the same way: Yes, he did.

F. STORY MUSGRAVE

In many respects, Story Musgrave's life and accomplishments were shaped by a troubled youngster's desperate affinity with nature. It was to nature he fled in body and spirit to seek a calming distraction from a life riven by conflict and torment. Just as he would later turn to the elevating works of Thoreau and Emerson, the young boy found comfort in nature, and inspiration in the night sky.

There are many who would describe picturesque Stockbridge, Massachusetts, as a sublime sort of place in which to live and raise children. The hometown of Norman Rockwell, the famed artist once characterised this quaint, picture-postcard village as "the best of America, the best of New England." His painting, *Main Street at Christmas*, achieved renown for its portrayal of a classic New England town stamped with elegance and history – a comfortable place wreathed in old-fashioned charm, gracious living, and a spirit of patriotic bonhomie.

A childhood filled with despair

A few miles west of Stockbridge, an old English-style stone mansion bearing the name Linwood still nestles peacefully in a quiet valley beneath the Berkshire Hills. Once situated on a thousand-acre dairy farm overlooking the gentle Housatonic River, the five-storey dwelling, set amidst formal gardens and hedges, is now part of the magnificent Norman Lindsay Museum estate. At one time, it was the boyhood home of future scientist-astronaut Franklin Story Musgrave, born on 19 August 1935 to Marguerite (née Swann) and Percy Musgrave, Jr.[15]

Musgrave's ancestry certainly laid an academic path for him to follow, boasting nine straight generations of doctors on his father's side. This included his paternal great-grandfather and great-uncle, both of whom were professors of surgery at Harvard. In addition, his paternal grandfather was a noted physician who studied the effects of troops' exposure to poison gas in the First World War. Story, who had three ancestors on the *Mayflower*, received his preferred first name from his mother's side of the family, which included such luminaries as Joseph Story, an early Supreme Court justice, and William Wetmore Story, an eminent 19th century sculptor.

F. Story Musgrave, MD, PhD.

Despite the cherished image of Rockwell's Stockbridge, it was never a place of peace or tranquillity for young Story and his older and younger brothers. Both parents were alcoholics, but while their mother's drinking merely caused her to slide into meekness and acquiescence, their father was often brutal towards the three boys. "I came from an extraordinarily dysfunctional family full of abuse and alcoholism. Dad was very violent, very harsh, exceedingly malicious."[16]

They lived a life of isolation on the farm, and rarely had visitors: "They either weren't permitted, or didn't dare come into that environment." The unhappy situation would often cause young Story to flee his home by night, making his way into the embrace of a nearby forest, where he would lie on his back, look up, and marvel at the stars. He recalls doing this when aged only three, but the darkened forest held no fears for him. "Nature became my world. Even as a three-year-old I could go out in the forest at seven or eight o'clock at night. I was totally at home in the fields, the woods and the rivers from the earliest age. I was totally immersed in nature. Lying in a damp, cool, freshly ploughed field just after sunset, and looking out into the heavens – that became my world.

"It's hard to say what drives a three-year-old, but I think I had a sense that nature was my solace – a place in which there was beauty, in which there was order. My parents knew I was out there, but I always came back. Even on the darkest of nights I could never get lost. I would just feel the trunk of a tree and I could tell north or south by how it felt."

He had to learn independence at an early age. "By five or six I had built my own raft and I was on the rivers. And so I was, as Emerson would say, very self-reliant in terms of being out in nature, and that was very important. I learned things on the farm that I was to use later on. I drove tractors at the age of ten, and was soon fixing farm machinery, because I was in remote fields and if a tractor broke, I either had to walk home or fix it." In light of his later achievements, it is quite remarkable to note that Musgrave hardly read as a child. He would read books in school because it was required, but never outside the classroom.[17]

By the time he was ten years old, his mother had finally decided she could no longer tolerate her husband's ways, so while the other two boys stayed on the farm, she took Story and went to live in Boston. Later, they returned to live with relatives back in Stockbridge, and then moved on to other homes in Lee, Cheshire and Pittsfield.

Horrendous tragedy loomed large on both sides of the Musgrave family. Story's great-grandfather and grandfather had committed suicide, and both his parents would also take their own lives. His older brother, Percy III, later died in an aircraft accident while catapulting off a carrier in the Atlantic, and his younger brother Tom fatally shot himself. Through it all, Story's self-reliance was of crucial importance. "You learn to associate with the good, and even though you suffer, you do get enough distance psychologically from what is going on in order to form your own ground. Those unbelievable tragedies are what built me. I look back on them as my Rock of Gibraltar, strangely enough."

His fascination with repairing and running anything mechanical eventually led him to aeroplanes and visits to a neighbouring farm, where he learned to fly at the age of sixteen, albeit "in a very informal kind of way. I drove them like I drove tractors, and then one day just leaped off!"

In 1947, Story began attending St. Mark's High School in Southborough, and here he became vitally interested in the science of biology through one of his teachers, Frederick Avis. He discovered an intrigue in researching the transplantation of fertilised eggs, and felt that in his own way he was part of a pioneering effort in this area of elementary biology. Despite this interest, he was still not driven to be more than just a competent scholar. At the age of seventeen he was badly injured in a car accident, which caused him to miss a substantial amount of vital pre-graduation exam schooling. Unfortunately, his school chose to make no allowance for this missed education, and so, with mixed feelings of regret and bitterness, he decided to move on and dropped out of high school. He joined the Marines and went off to Korea to expand his horizons.

Settling into the Marine Corps

The Marine Corps quickly introduced Story to the world of military aviation and technology. After training, he became an aviation electrician and instrument technician, and for a time worked as a plane captain while completing duty assignments in Korea, Japan and Hawaii, and aboard the carrier *USS Wasp* in the Far East. Musgrave's love of flying had remained with him, and he resumed his studies in order to get a pilot's licence. He began reading any technical manuals on aircraft that he could get his hands on, and this would lead to a lifetime appreciation of books and fine literature.

Eventually, Musgrave left the Marines in order to study mathematics and statistics at Syracuse University in New York State, and this is where he says he "got into computers." As with biology in high school, his curiosity was aroused. "This was in the early fifties, so it was vacuum tubes and ancient stuff back then, but computers got me interested in the nervous system, the brain, and how it works. I was in graduate school when Sputnik went up, and that's how I was introduced to space." It was at Syracuse that he met his first wife, Patricia Van Kirk from Patterson, New Jersey, who had recently transferred from Connecticut University to study nursing.

In 1958, Musgrave was employed as a mathematician and operations analyst for the Eastman Kodak Company in Rochester, New York, while he continued his studies at Syracuse. That same year, he was awarded his Bachelor of Science degree – the first in a string of many higher academic achievements. An MBA in operations analysis, business administration and computer programming from the University of California followed in 1959, then a BA in chemistry from Marietta College in 1960, and his Doctorate in medicine from Columbia University in 1964.

At Columbia's Presbyterian Medical Center, he had begun his own research into the nervous system under the tutelage of Dr. Dominick Purpura, and Musgrave now decided he would become a surgeon, "not just to heal people, but because of a curiosity of what a human is, and what it means to be human."

That same year he began a one-year surgical residency at the University of Kentucky Medical Center in Lexington, eventually staying on for two years as both an Air Force post-doctoral and National Heart Institute Fellow, working in aerospace and space physiology, temperature regulation, exercise physiology and clinical surgery. He gained his Master's in physiology and biophysics from the university in 1966.

Deep interests in computers, physiology and the human brain all came together for Story Musgrave when he discovered that NASA was calling for applications from potential scientist-astronauts. A new door was being opened for him, and he applied, fully prepared to trade his ambitions of becoming a neurosurgeon for the prospect of one day working as a mission physician on flights to the Moon and Mars.

"Everything I'd ever done, I realised – every unknown path I took – was leading me to this. As soon as NASA expressed an interest in flying scientists, people with a formal education as well as being military pilots, that was an epiphany that just came like a stroke of lightning. I saw that everything I had ever done in life could be used in that endeavour. It just fit, and felt right."

The numerous qualifications he had accrued nearly caused an early end to his plans. "They almost didn't take me, because they said I was so over-trained that I might not be comfortable. Apart from everything else that I was doing in life I had about six earned degrees at that time, an active laboratory and a surgical practice, and was a commercial pilot, flight instructor and parachutist."

He may have been over-trained and over-qualified in his opinion, but NASA recognised Musgrave's definite potential as a scientist-astronaut. His name eventually ended up on their list as one of the eleven selected.

BRIAN T. O'LEARY

Perhaps it was because he was small for his age that Brian O'Leary had a childhood fascination with things that he found extreme in any way: "Lofty, large and long marvels, such as the highest mountains, the tallest buildings, the longest roads, and the biggest animals." It is a fascination with superlatives and the unexplained that still dictates his life's many directions, accomplishments and goals.[18]

Mary Mabel O'Leary was the first to convince her son that if he was looking for grandness, there was no greater immensity for a young lad than the night sky. Here, he would find massive planets, far distant galaxies, and gargantuan rotating stars billions and trillions of miles from Earth.

His schoolteachers at Chenery School, already impressed with his rapid proficiency at mathematics, happily encouraged him as he began to ask elementary questions about the solar system. In particular, his fourth grade teacher, Miss Perault, was an educator of considerable influence. "She was an adventurer, pilot, etcetera, and it was just about that age I was getting turned on to space, which she was also excited about. She was a real inspiration."

The influence of the heavens

The greatest influence of all, however, came one Tuesday night in 1948 when his parents took him to an open house at Harvard Observatory. It was the evening of 2 November, and the nation was gripped in a convoluted electoral fever. Early returns that night resulted in premature headlines declaring Thomas Dewey the winner of the presidential election – an announcement later rescinded in favour of Harry S. Truman, but the race for the White House paled into insignificance for one small boy. "I was eight years old. I never had a telescope, but looking at Mars and Saturn through the one at the Harvard Observatory ... it thrilled me. I wanted to go to Mars, and from that point on it appeared inevitable I would become a planetary astronomer."

In this way, a small boy from suburban Massachusetts was set on a determined course to those stars. Less than two decades later he would become NASA's only planetary scientist-astronaut in the heady days of the Apollo programme.

Brian Todd O'Leary was born in the residential suburb of Belmont, seven miles west of Boston, on 27 January 1940. The youngest son of Fred and Mary O'Leary had an older brother named Fred (Junior), then six, and a three-year-old sister, Judith. His father was a radio and electrical appliance sales manager for a distribution company.[19]

By the age of nine, the young enthusiast was trying to absorb all he could about astronomy and space travel, and took delight in reading about those subjects in *Life* and *Colliers* magazines. Anything written by the erudite German-born rocket

Brian T. O'Leary, PhD.

engineer, Dr. Wernher von Braun, was eagerly sought and lovingly pored over many times.

O'Leary's interest in astronomy, as well as his academic strengths, went right along with him as his education continued at Belmont High School. He recalls that, as a senior at Belmont High, he was required to write a long essay in history class on an important contemporary issue. While others concentrated on such issues as national security, communism, and the evils of tobacco, he did not have to give the subject matter much thought at all:

"I wrote my essay on space satellites. This was in 1956, one year before the launch of Sputnik I. At that time, the United States had started the Vanguard project, a programme intended to launch basketball-sized satellites into orbit. This gave me enough material for the essay, but I still recall the puzzlement with which the history teacher and class reacted to my choice of topic."

The following year, Williams College presented fresh challenges, but O'Leary began to find the curriculum patently mundane. Despite his father's mounting concerns, he somehow managed to successfully follow through on a physics major. Unfortunately, he was now on a perilous educational slide. His previous love of mathematics had become moribund in the wake of more social activities; there was no longer any enthusiasm for his physics courses; and he began to rebel against authority. "My attitude was carefree, in contrast to that of the majority of my fraternity brothers, who were seriously preparing for careers in law or medicine." Even the much-anticipated announcement that Russia had launched Earth's first artificial satellite failed to excite him.

Happily, his interest in planetary science and space technology was rekindled following his graduation from Williams in 1961 with his Bachelor of Arts degree in physics. A job had come his way at NASA's Goddard Space Flight Center near Washington DC, and though this work was otherwise mundane and routine, the massive juggernaut of manned space exploration was inexorably under way. The public was swept up in all the excitement of this new technological age, and he was thrilled to simply be a part of it. "I plotted graphs which predicted models for the response of a pressure gauge to the thin atmosphere surrounding a satellite – but I welcomed any opportunity to immerse myself into the space business, and there was no place better to do it than Goddard."

His poor grades at Williams College now came back to haunt him as he endeavoured to gain entry into an astronomy graduate school. To his relief, he was finally accepted at Georgetown University, and over the next three years, his grades and his interest in astronomy soared. In July 1962, he took a short break from his studies to fly overseas and fulfil a long-held dream, as part of a small group that successfully climbed to the peak of one of his childhood superlatives – the mighty Matterhorn. He also took part in another gruelling personal challenge, finishing a twenty-six mile Boston Marathon with badly blistered feet.

Despite his high grades, he began to have serious fallings-out with the faculty, and a final confrontation led to his questionable and unfair dismissal from Georgetown. While disappointing, it was not an unexpected development, but it would ultimately prove extraordinarily beneficial to his life. O'Leary's dismissal from Georgetown only served to make him more determined than ever to continue his studies. While he wrote his Master's thesis on astronomy he also took on part-time work teaching high school mathematics, and in the midst of this, managed to find time to meet his future wife, Joyce. His hard work and long hours finally paid off in 1964 when he was accepted into graduate school at the University of California.

That summer, after he and Joyce were married, O'Leary took his new bride to the West Coast. Here, he found the university curriculum and teaching staff at the Berkeley campus a sublime change from those he had endured at Williams and Georgetown. This was quickly reflected in his superior grades and research, and in his published papers. These days, O'Leary credits much of his ultimate success to his thesis advisor and colleague, Donald Rea. "He was a young research chemist, recently converted into a planetary scientist at Berkeley's Space Sciences Laboratory. He was brimming over with ideas and had plenty of NASA money to spend." Under Rea's guidance and mentorship, O'Leary became a professionally recognised planetary astronomer.

Overcoming the obstacles

In September 1966, O'Leary was reading some pamphlets pinned to a notice board at Berkeley, when he saw one issued by the National Academy of Sciences, urging qualified scientists to apply for selection in a second NASA scientist-astronaut group. The final selection would not be made until the following summer, and with his doctoral thesis on the planet Mars scheduled to be completed in the next few months, he considered the timing to be perfect. "My childhood dreams had told me that this would be the most exciting thing a man could do in this age, and that other things society offered were relatively extraneous. I was convinced space science was my calling, so what better way was there to answer that calling than to become an astronaut?"

Despite his enthusiasm, O'Leary realised he had several slight physical impairments that could easily go against him, which he and Joyce discussed prior to his application. He was nearsighted, with 20/40 vision in his left eye and 20/100 in the right, and he held deep concerns about his lifelong inability to sufficiently distinguish reds and greens. He knew he would eventually have to face an Ishihara test for colourblindness, which consisted of a pattern of coloured spots, and his failure to correctly perceive several numbers had earlier prevented him from qualifying for the Naval officer programme. It was an obstacle he would have to face when he came to it, and with his wife's backing he filled out the form and sent it off.

It was not until the following February that he finally received a telegram from the National Academy of Sciences, requesting further information. They also wanted five reprints of his published papers and an essay of 250 words or less on experiments he would carry out in orbit or on the Moon. In his usual meticulous way, he completed this using exactly 250 words, and sent off everything that had been requested. On 3 March, he was advised that he had passed initial screening and was "scientifically qualified" for the programme.

Early in May 1967, O'Leary was asked to report to the School of Aerospace Medicine at Brooks AFB in San Antonio, Texas, for a week of physical and mental examinations. The Ishihara test was still preying on his consciousness every day, and in desperation, he even resorted to memorising the numbers embedded into patterns on each page. In the end, all his concerns were dispelled on the first afternoon of testing, when he passed the colour vision test without a problem. "Oddly enough, it was not the Ishihara test which I had memorised. The technician giving me the test was lax and paid no attention to my answers, although many were probably correct anyway. I passed the same test that had eliminated me from the Naval officer program."

ROBERT A.R. PARKER

The anticipated letter of confirmation arrived amid great excitement at the Madison home of Dr. Robert Parker, an assistant professor of astronomy at the nearby University of Wisconsin (UW). The letter, date-stamped 2 August 1967, was from

Robert A.R. Parker, PhD.

the office of Deke Slayton, NASA's Director of Flight Crew Operations. The style was quite formal, but it was written verification of the news Parker had received in a call from Alan Shepard just two weeks earlier, on 21 July. In part, it read:

"I am happy to confirm your selection into the Scientist/Astronaut Training Program of the Manned Spacecraft Center. Because of the heavy demand for flight training in the Air Force, you will not be attending flight school until next March. Consequently, we intend to start your training with the ground-based aspects of the program here at the Manned Spacecraft Center. Your first move, therefore, will be a permanent one to Houston. We ask that you report for duty on or before September 18. Representatives of our Personnel Division will be in contact with you to arrange details and a specific reporting date which will be convenient. Congratulations, again, on your selection and we certainly are pleased to have you in the program."[20]

Robert Allan Ridley Parker was born in New York City on 14 December 1936, the son of Allan Elwood and Alice (Heywood) Parker, and twin brother to Peter. When the twins were aged five, the family moved to historic Shrewsbury, a suburb of Worcester, Massachusetts. Shrewsbury had earlier been home to both of his parents, and before that to his paternal grandparents, Vernie (née Elwood) and Samuel Ridley Parker.

Parker's grandfather, who preferred to be called Ridley, was born in March 1868 to Sarah (née Ridley) and John Parker, and was brought up in the small village of Waterbury in the northwest corner of Washington County, Vermont. Having attended the state's prestigious St. Johnsbury Academy, Ridley Parker became a high school teacher, while continuing his studies at Amherst College in Massachusetts. Following his graduation he became a librarian and in 1901, he moved to New York City with his wife Vernie in order to take up a position in the magnificent Astor Library. They would later return to Massachusetts and settle in Shrewsbury, located some thirty-four miles west of Boston. It was a peaceful town, and in those pre-Depression days, Shrewsbury enjoyed a thriving agricultural economy based on a profusion of apple orchards.

Astronomy beckons

Robert Parker took his early education at local primary and secondary schools. It was in fourth grade that he says he first developed "a determined desire" to become an astronomer – long before there were any thoughts of a space programme. "Astronauts were no more than a Flash Gordon movie then," he later reflected. His graduation from Shrewsbury's Beal High School took place on 16 June 1954.

Parker then continued his combined studies in astronomy and physics at Amherst College, where he also did some laboratory teaching in his junior and senior years, and became a member of Phi Beta Kappa. Among his college honours and scholarships, he achieved the distinction of Magna Cum Laude when graduating with his Bachelor of Arts degree in astronomy and physics on 8 June 1958, joining his grandfather on the alumni roll.[21]

His interest in science, and astrophysics in particular, certainly seemed to have its origins in being part of a science-oriented family. His father Allan was chairman of the physics department at Worcester Polytechnic Institute, while his twin brother Peter taught physics at Yale. His younger brother, Allan Jr. (born in 1938), became a computer systems analyst in Boston.

Six days after graduating, Bob Parker married Joan Audrey Capers, a registered nurse from Waynesville, Pennsylvania, whom he had first met when they were both seniors in high school. They would have two children; Kimberly Ellen, born on 7 February 1962, and Brian David Capers Parker, born on 8 March 1964. He and Joan would eventually divorce in 1980 and Parker later married again, this time to Judy Woodruff of San Marino, California, who already had three children by a previous marriage.

After receiving his undergraduate degree from Amherst College, Parker continued his chosen career path in science by attending the California Institute of Technology in Pasadena. At CalTech, he achieved many deserved distinctions, including: Woodrow Wilson Fellow 1958–9; Danforth Fellow 1958–62; Graduate Teaching Assistant 1959–60; National Science Foundation (NSF) Cooperative Fellow 1960–1 and 1961–2; Woodrow Wilson Summer Fellowship 1961; and NSF Summer

Fellowship 1962. In 1960, on an NSF travel grant, he also attended the NUFFIC (Netherlands Organisation for International Cooperation in Higher Education) Summer Institute on Galactic Structure, held at Breukelen, near Utrecht in the Netherlands.

He received his Doctorate in astronomy from CalTech on 7 June 1963. His PhD dissertation, under the guidance of Professor Guido Münch, was entitled *Physical Conditions in Some Possible Supernova Remnants*.

From 1962 to 1963, Parker was an NSF post-doctoral fellow on the Madison campus of the acclaimed University of Wisconsin. He became an assistant professor of astronomy on the university's Badger faculty from October 1963 to 1967, and taught the introductory course in astronomy for non-majors, as well as other courses, including those at the graduate level. Quite often, more than two hundred students would attend his lecture sections.

Parker was also on the subcommittee of the Student Life and Interest Committee (SLIC), and for recreation on campus joined a group of faculty members working out at the old Armory Gym. He also ran two miles on the university track every other day.

For relaxation at the family home on Tomahawk Trail, he enjoyed nothing more than the simple pleasures of gardening. He also took great pride in building a stereo unit for his family, and loved tinkering under the bonnet of an old Triumph sports car he'd bought soon after he and Joan were married. Fond of many outdoor activities, he especially enjoyed ice-skating on Madison's frozen lakes in winter, and had once camped and hiked in the majestic Rockies.

By 1964, Parker was a full-time staff member of Washburn Observatory on the Madison campus, and that year he became a supervisor of Washburn's major research facility, in charge of the day-to-day operation of their country observatory at Pine Bluff – the chief research facility for university astronomers. Here, and up to the time of his astronaut selection, Parker became engaged in advanced astronomical research, specialising in interstellar matter – phenomena lying between the stars such as supernova and nova remnants, dust clouds, extragalactic H II regions, neutral hydrogen clouds, and planetary nebulae.

Reasons against selection

In October 1966, he submitted his details when the National Academy of Sciences asked for applications from suitably qualified people wanting to become scientist-astronauts with NASA. With unnecessary modesty, he recollects turning in his application form "for kicks, just to see if I could make it," but says he did not really think he stood a chance. After all, he had never been a pilot, a military man, nor even a varsity athlete – three attributes everyone naturally associated with astronauts at that time.

Following his selection, reporters asked thirty-year-old Parker what set him aside from other applicants: "I saw twelve or fifteen of the other candidates when we were taking physical and psychological examinations. On the surface, there's nothing to distinguish us. I never saw such a homogenous bunch of men." Did he indulge in any extreme sports such as mountain climbing or flying? "Until now, I've been strictly an

Earthbound astronomer. I've never climbed a mountain – it's too dangerous – or flown in anything other than a commercial airliner. Now that I'm accepted, I'm just waiting for the chance to get up there. The training and ground work should be fun, but the ride into space is the payoff."[22]

Bob Bless, an emeritus professor in astronomy who worked alongside Parker at UW, says in praise of his former colleague:

"He was a very good teacher; articulate and enthusiastic. He continued the research begun as a graduate student on the physical properties of supernovae remnants. This interest led him to design and supervise the construction of a photoelectric spectrum scanner for use in this research. It was no surprise when he applied for the astronaut programme since it was obvious that he was captivated by the 'romance' of space."[23]

Vivien Hone, when reporting on Parker's selection for the *University of Wisconsin News*, stated that the professor seemed more amused than impressed with the excitement the announcement had caused on campus.[24] She asked him what had led someone who studied what has been called "the oldest science" to apply for a post on the newest and highest science frontier, and what he saw as his major goals in the programme: "It's a natural extension of ground-based observations. A bigger and better observation site up there. And beyond that, it looks like an interesting and exciting job. The kinds of experiments I'll probably be carrying out will probably be similar to the ones we've been trying to do here from the ground, or from rockets, X-15s, and an orbiting astronomical laboratory."

When selected, Parker was a month away from being promoted to Associate Professor of Astronomy, at the beginning of the new school term. However, he would be accorded this honour a few weeks before he and his family left for Houston, as it was expected that the scientist-astronauts would be permitted an average of one week per month to work on their own research. He would continue in this role up until 1972, even as he undertook training and other astronaut duties.

WILLIAM E. THORNTON

In Brian O'Leary's book, *The Making of an Ex-Astronaut*, there's a photo of O'Leary and six other members of the second scientist-astronaut group standing a little self-consciously in front of a Saturn F-1 engine at the Marshall Space Flight Center in Huntsville, Alabama. It's a rather unremarkable picture, except that Bill Thornton, at 185 cm, has been imprudently positioned next to Joe Allen, who stands at just over 168 cm tall.

The man they used to call "Moose" in his younger days, because of his height and bulk, certainly dominates most photos of the "Excess Eleven." In writing about Thornton, O'Leary impishly states that: "His weight, plus Joe Allen's, divided by two would come out normal for an astronaut," but he also talks about a man with "a sensitive though determined manner; he is soft-spoken in a deep, North Carolina accent."

Faison, North Carolina, was one of several small trackside towns founded when the Wilmington and Weldon Railroad, then the world's longest, was completed in 1840. Today, some of the town's 700-strong population are descended from plantation owners who settled in the rich agricultural area. William Edgar Thornton was born there on 14 April 1929. His mother was Rosa Lee (née Blanchard), and while he was given the same forenames as his father, he never attached the suffix Junior. He would be the couple's only child.[25]

"Apropos my father, he was proficient and worked in many areas, including accounting. One job was timber 'scaler' – measuring the timber before and after cutting and calculating lumber content for a major lumber company. In later years, he directed farm production on land we owned."

A fascination with anything aeronautical

A bright and inquisitive boy, the future astronaut loved building devices and putting broken things back together – the more complicated, the better – and he had a

William E. Thornton, MD.

fascination with anything aeronautical. A lifelong friend, Anna Stroud Taylor, once said that he "was always doing electronics ... and science, even as a child."

Bill Thornton was just eleven years old when he lost his father, who himself had been orphaned as a teenager during the American Civil War reconstruction. Despite this, and at the cost of his own education, Thornton's father had raised two younger sisters who went on to become college professors, and a brother who became manager of the export division of an American cartel. Recalling his father, Thornton says: "He undoubtedly was a determining factor in my interest in science and electro-mechanics, as well as an intense interest in nature, biology, and the universe." Assuming a mantle of responsibility, the young boy took on some odd jobs to help things along and make things a little easier for his mother.[26]

Other influences now began to shape his future. "My education in space began when I discovered a series of science booklets in the third grade that included rockets and space travel, gravitational attraction and so forth." In 1943, after his public school education ended, he began attending a local combined school known as Faison High School. Here, he would find further influence and encouragement in his education. "The teachers were real teachers in spite of Depression-limited resources. I began work in the sixth grade with paper routes, and by eighth grade I was also a full-time motion picture projectionist with responsibility for the maintenance of sound and pictures equally in several theatres over a two-year period."

By eleventh grade, Thornton had opened and began operating a radio electronics repair shop that would help finance his university education. In his last year of high school, and throughout college, he was supported entirely by weekend and vacation work in electronics. This included the design, installation and maintenance of entertainment, industrial and communication electronic systems. He later stated in his application to NASA that: "informal study during this period has proved to be more than equivalent to a degree in electronic engineering."

In 1948, Thornton undertook further studies at the University of North Carolina (UNC), where he gained his Bachelor of Science degree in physics on 2 June 1952. He also became a Reserve Officer Training Corps (ROTC) graduate with the USAF. College was, in his own words, "a balance between survival, education, Air Force ROTC, and ill-considered participation in collegiate football without a scholarship."

Following his graduation, and having completed his ROTC training with the rank of second lieutenant, Thornton served as Officer-in-charge of the Flight Test Photo Optics Instrumentation Laboratory at the Air Proving Ground, Eglin AFB. He would later become a consultant to Air Proving Ground Command on instrumentation for airborne fire control systems.

At Eglin, Thornton was involved with in-flight testing of all-weather interceptors, and he developed the first successful airborne target and evaluation missile scoring systems that were eventually standardised and used in all Free World air forces. During this period, he chalked up over a thousand hours in various crew functions during aircraft flight tests, and was proficient with a variety of aircraft systems, including airborne radar and armament systems. For this work, he was awarded his first patents and the USAF Legion of Merit.

Introducing electronics into medicine

After spending more than two years in the USAF in secure laboratories, aircraft, or in desolate test stations, Thornton decided to introduce electronics into medicine, rather than fly. His work began at the UNC's Memorial Hospital, where he worked, briefly, in their Departments of Anaesthesiology, Cardiology and Neurology. It was here, in July 1955, that he first met Jennifer Fowler from Hertfordshire in England, who was part of a medical exchange programme, and a chase ensued. "I now needed money to pursue a medical education – and Jennifer."

In 1956, with these issues to spur him on, Thornton developed and ran the Electronics Division of Del Mar Engineering Laboratories in Los Angeles, where for the next three years he designed and produced military electronics under his patents. During this time, he was able to convince the president of the company to put some of its profits into a medical electronics subsidiary, which he designed and ran. It subsequently became Del Mar Electronics, an international leader in medical electronics.

On 14 June 1958, Thornton and Jennifer were married in London, England, after which he returned to UNC's medical school, "where my medical education was shared with continued medical electronics development." Bill and Jennifer would have two children; William Simon, born 15 March 1959, and James Fallon, born 4 January 1961.

In his first year back at UNC, Thornton designed and established the first clinical anaesthesia monitoring of patients and in his second year, the first automatic, online analysis of electro-cardiograms (EKGs). This system was patented and later applied to Holter EKG monitoring. The basic principles are in worldwide use today.

A symposium at the USAF Aerospace Medical Division in San Antonio, Texas, turned Thornton's attention to space medicine. "Al Shepard had just made his flight and it was too good a show to miss. Though I didn't tell anybody, by then I had decided I wanted to fly, including space flight."

Thornton was awarded his Doctor of Medicine degree from UNC on 3 June 1963. Shortly after accepting an appointment to the university's faculty, where he became an instructor and worked on medical applications of his developments, he decided to actively pursue a career in space flight medicine. At that time, the Air Force seemed to offer the best chance and in 1964, now returned to active service with the rank of captain, he completed a rotating internship at the USAF Wilford Hall Hospital at Lackland AFB in San Antonio, and a primary flight surgeon's course in aerospace medicine at nearby Brooks AFB.

It was at Brooks, in their Aerospace Medical Division, that Thornton's primary responsibility became the development of medical experiments and instrumentation for the Air Force's Manned Orbiting Laboratory programme. This included the design and development of personal EKG/EEG systems with automated analysis, non-invasive blood studies and exercise equipment and programmes, evaluated in conjunction with Dr. Kenneth Cooper.

Another major project, urgently required by NASA, was the development of the first successful non-gravimetric mass measuring system ("weighing in weightlessness") for a manned Earth-orbiting programme that would later evolve into Project Skylab. Developed and patented in 1965, the system was eventually flown on Skylab and some STS flights, and would remain a standard NASA instrument. Thornton subsequently became a principal investigator for the civilian space agency.

By 1966, the MOL programme was crumbling. Thornton called Deke Slayton, asking if there was any possibility of becoming a candidate for their scientist-astronaut programme. He told Slayton he had not applied for selection in the 1965 group, as his age even then was a few months beyond the established and rigidly applied maximum. Slayton's advice was to hold out for a while longer, as NASA would shortly initiate another scientist-astronaut selection. He gave a further assurance that previously disqualifying size and age limits would be relaxed significantly, "in exceptional cases."

In a follow-up letter dated 18 August 1966, Slayton seemed to be on Thornton's side. "You have nothing to lose by trying. The Academy will provide us with a recommended list from which we will make final selection, so your first chore is to impress the Academy with your scientific competence." In light of his later selection, it would seem he did just that.

THE OTHER "ALMOST" SCIENTIST-ASTRONAUTS (1967)

Of the 1965 selection, both Birky and DuPraw thought about trying again in 1967. As Birky recalled in 2002: "I was invited to apply again. I talked to DuPraw at the time, and he was also invited. It seemed likely that I'd be selected the second time, because there was reason to believe that I was rejected because of inferior stereoscopic depth perception, and NASA was going to become less prickly about physical requirements. (Also at my interview, one of the interviewers pointed out that the pioneering pilot Wiley Post had only one good eye and hence no stereoscopic depth perception at all.) But I didn't apply again because the budget for Skylab and lunar missions had been cut and it seemed likely that I'd never get a mission. I believe Ernie also did not apply."[27]

Other short-listed candidates for this selection are not so well documented but have included:

Donald A. Beattie: Born in 1931. Became a geologist and served in the US Navy as a jet pilot, and worked as an exploration geologist for Mobile Oil before joining NASA as a lunar aerospace technologist, Office of Manned Spaceflight, Manned Lunar Mission Studies at NASA HQ in Washington DC. He worked at NASA from 1963 to 1973 in a variety of management positions, finally as programme manager of the Apollo Lunar Surface Experiment Package (ALSEP). He has also worked at the National Science Foundation and Department of Energy, and later as a private consultant. He authored *Taking Science to the Moon* in 2001.

Peter Eltgroth: Attained a BS in physics from CalTech in 1962 and a PhD (also in physics) from Harvard University in 1966, joining the Lawrence Livermore

National Laboratory in 1967 after failing to be selected to the space programme. His expertise lies in computational and mathematical physics with an emphasis on numerical algorithms. He has published a number of papers and worked in a variety of fields during almost forty years at Livermore. These fields included relativistic fluid dynamics, plasma physics, massively parallel computing and innovating algorithms. He is also a qualified flight instructor. His initial work at Livermore was in the original T (Theoretical Physics) Division, and prior to taking up his current position he was Group Leader for Computational Physics in the Center for Applied Science Computing. In October 2001 he became director of that organisation, which is part of the Computing Applications and Research Department in the Computation Directorate at Livermore. He applied for both the 1965 and 1967 selections, but in both cases, despite self-financed corrective operations, reoccurring kidney stones was an insurmountable problem in his application. He was at the Manned Spacecraft Center as an NRC post-doctoral fellow during the selection process. He remembered England, Allen and Musgrave at the time of his application.

Donald T. Frazier: Currently Professor and Director of the Science Outreach and Career Opportunities Center, University of Kentucky College of Medicine. He gained a PhD from the University of Kentucky in 1964. Frazier wanted to fly high performance jets, but was disqualified for medical reasons. His current research focuses on how information emanating from respiratory afferents is taken in the central nervous system and used to control the drive to breathe.[28]

Everett K. Gibson, Jr.: Gibson received his BS and MS degrees in chemistry in 1963 and 1965 respectively at Texas Tech University, and his PhD in geochemistry from Arizona State University in 1969. He visited Eugene Shoemaker at USGS and saw the prototype of the MOLAB (Mobile Laboratory) being developed for extended lunar exploration missions. He received a "no thanks" letter from Dr. Charles Berry informing him that his vision was a disqualifying factor. Gibson subsequently went to work at the Lunar Receiving Laboratory at MSC on 24 July 1969, just in time to participate in the Apollo 11 "splashdown party." NASA Road 1 was closed to cars and a wall-to-wall party spilled over into it! In 1970 he accepted a position in the Geochemistry branch at NASA, and worked in the LRL for Apollo 14. Since 1983 Dr. Gibson has been a principal investigator in NASA's Planetary Biology Program, and in March 1993 was promoted to the Senior Scientist position at NASA-JSC.[29]

R. Thomas Giuli: Born in 1936 in Lansing, Missouri, he received a doctorate in Astronomy and Astrophysics from the University of Stockholm, Sweden. He was an astronomer at JSC. On 26 June 1974 Giuli, then of the Science and Applications Directorate, was appointed Apollo–Soyuz Test Project Program Scientist, responsible for coordinating all scientific aspects of the joint mission.[30] In 1977 he was nominated as one of eighteen potential American payload specialists for

Spacelab 1 (along with Bill Thornton) [NASA News 77–83 5 December 1977] but was not selected.[31]

Albert R. Hibbs (1924–2003): Known worldwide as the "voice of JPL" Albert Roach Hibbs was born on 19 October 1924 in Akron, Ohio, and died aged 78 on 24 February 2003 from complications following heart surgery. A 1945 physics graduate of CalTech, he received his Master's degree in mathematics at the University of Chicago in 1947 and a PhD in physics at CalTech in 1955. He studied under and wrote with his friend Richard Feynman. Hibbs joined the Jet Propulsion Laboratory operated in Pasadena by CalTech in 1950. Following a range of technical assignments he became a systems designer for Explorer 1, America's first successful satellite launch in January 1958. He became the NASA spokesman who explained the intricacies of space flight in layman's terms, describing the missions of the early Ranger and Surveyor missions to the Moon and Mariner missions to Mercury, Venus and Mars. He became host of NBC's children's programme *Exploring* in 1960 and wrote over seventy scientific and technical papers as well as two text books. After failing to be selected to the astronaut programme he continued his work at JPL until he retired in 1986 as Director of Space Science.[32]

John A. O'Keefe: Born 13 October 1916 and died 8 September 2000 aged 83. In 1937, he earned his BS degree in astronomy from Harvard and in 1941 his PhD in astronomy from the University of Chicago. Rejected for military service during WWII, he joined the US Army Corps of Engineers as a civilian and began a new career as a geodesist. From 1945 until 1958, he was Chief of the Research and Analysis Branch, Army Map Service. In 1958 he joined NASA's Goddard Space Flight Center as Assistant Chief of the theoretical division, where he remained for the rest of his career. After O'Keefe and his colleagues analysed the orbit of Vanguard 1 in 1959, he realised that the Earth's gravitational field affected the satellite's trajectory, especially in the Southern Hemisphere. This led him to coin the phrase "the pear-shaped Earth".[33] The champion of the idea that tektites originated from the Moon, O'Keefe was supported in his nomination for the scientist-astronaut programme by Egbert King of NASA, despite the fact that he favoured the Earth-impact theory. In a 29 November 1966 memo King stated "my comments should be regarded in light of my contact with O'Keefe which has been in lunar and geosciences only. Geosciences is not O'Keefe's strong field ... I am certain that he would perform with dedication any task that was assigned to him as a member of a space flight team. I think he would be most valuable as an astronomical observer in Earth orbit or under possible lunar base conditions. His qualifications as an astronomical observer are confirmed by the opinion of other astronomers. He has an outstanding command of the entire field of space science."[34]

Gerald K. O'Neill (1927–1991): Gerald Kitchen O'Neill was a physicist with a specialisation in high-energy particle physics and its application to navigation and space colonisation. A Professor of Physics at Princeton University, his 1997 book *The High Frontier* popularised the idea of looking at near-Earth space "not as a void but as a cultural medium rich in matter and energy." O'Neill argued strongly that the colonisation of space was an obvious solution to many of Earth's problems such as over-population, fossil fuel depletion and pollution. He was a supporter of creating space colonies equidistant between the Earth and the Moon known as Lagrange 5 or L5. A pilot, inventor, author, advisor, entrepreneur and teacher, his ashes were aboard one of the Space Service Inc. memorial space flights launched in 1997.

George C. Pimentel (1922–1989): George Claude Pimentel completed high school in 1939 and gained his chemistry major at UCLA in 1943. After joining the Manhattan Project in 1943 and realising the scope of the project, he enlisted in the US Navy and served on submarines, returning to Berkeley for graduate work in infrared spectroscopy in 1946, and gaining his PhD in Chemistry in 1949. He then joined the faculty at Berkeley, where he remained an active member until his death in 1989. In the 1950s, he had developed the matrix isolation techniques to trap free radicals. In the 1960s he completed studies of fast reactions, unlocking the secret for converting chemical energy directly into laser light. At the age of forty-five, he applied to be a scientist-astronaut and underwent the demanding physical and intellectual tests. The National Academy of Sciences ranked him first among its thousand or so applicants, but a minor abnormality in one retina precluded him from further consideration. When Pimentel was asked his reaction to a two-year, high-risk trip to Mars he instantly replied: "Where do I sign up?" In his lifetime he would receive many awards as an outstanding teacher and champion of science education.

Richard Paul Von Buedeingen, USN Retired: Born 14 September 1938 in Rochester, NH, he gained his Doctorate of Medicine from the University of Wisconsin and subsequently served in the Medical Corps.

These applicants never got the chance to apply again as this was the final selection under NASA's "scientist-astronaut" programme. In 1969 seven former MOL pilots transferred to NASA, and the next astronaut selection was not until 1978, over a decade later, to provide crews for the Space Shuttle.

REFERENCES

1 *Taking Science to the Moon: Lunar Experiments and the Apollo Program*, Donald Beattie, John Hopkins University Press, Maryland, 2001.
2 *Academic Training Program for Group Six Astronauts*, prepared by Flight Crew Support Division, Manned Spacecraft Center, NASA Houston, Texas, 1 September 1968.

References 169

3   Beta Theta Pi magazine, DePauw University, Winter 2001, article *Joseph P. Allen IV: Astronaut Scientist, Businessman, Author, Technology Pioneer.*
4   Family information supplied and checked by Allen, emails to Colin Burgess 9 January 2003 to 13 May 2004.
5   DePauw News, article *Astronaut Joe Allen.*
    Website: *http://www.depauw.edu/pg/news/db_article.asp?id=372085385300926*
6   Family information supplied and checked by Chapman, emails to Colin Burgess 12 and 15 August 2002.
7   Family information supplied and checked by England, emails to Colin Burgess 2 March 2003 to 1 April 2003.
8   Information supplied to Colin Burgess by Loren Action, email correspondence 2 August 2002 and 21–22 August 2002.
9   Family information supplied and checked by Carolyn Henize, emails to Colin Burgess 24 and 30 August 2002, 5 November 2002 and 9 January 2003.
10 Information from *Astronaut Karl G. Henize: A Personal History*. Webpage by Vance Henize *http://spacsun.rice.edu/~karlhenize.html*. Used by permission Vance Henize.
11 Family information supplied and checked by Holmquest, emails to Colin Burgess 11 and 27 July 2002.
12 Family information supplied and checked by Lenoir, emails to Colin Burgess 31 July 2003 and 31 August 2003.
13 From Massachusetts Institute of Technology *RLE History* webpage, *http://rleweb.mit.edu/groups/g-radhst.HTM*
14 Family information supplied and checked by Llewellyn, emails to Colin Burgess 5 November 2002 and 9 January 2003.
15 Information from Musgrave homepage *http://spacestory.com/newtexas.htm*. Used by permission Todd and Story Musgrave.
16 Family information supplied and checked by Musgrave, emails to Colin Burgess 15 March 2003 to 24 June 2004.
17 Lenehan, Anne, *Story: The Way of Water*, TCA Publishers, New South Wales, 2004. Used by permission Anne Lenehan.
18 O'Leary, Brian, *The Making of an Ex-Astronaut*, Houghton Mifflin, Boston, 1970. Used by permission Brian O'Leary.
19 Family information supplied and checked by O'Leary, emails to Colin Burgess 24 and 30 July 2002, 1 August 2002.
20 Archived letter from D. K. Slayton to Parker, 8 January 1967.
21 University of Wisconsin Faculty Information Sheet on Parker, released 4 August 1967.
22 Milwaukee Journal newspaper, article *UW Astronomer Named Astronaut*, issue 4 August 1967.
23 Bless email to Colin Burgess 1 July 2003.
24 Interview conducted by Hone for UWS News and Publications Service, 15 August 1967.
25 Family information supplied and checked by Thornton, emails to Colin Burgess 18 March 2003 to 16 June 2003.
26 Thornton's Curriculum Vitae and other private papers supplied to David J. Shayler. Extracts by permission of Thornton.
27 Email to Dave Shayler from Bill Birky, 3 June 2002.
28 Email to Dave Shayler from John B. Charles, 31 May 2002; U of C College of Medicine website
    http://www.mc.uky.edu/physiology/people/faculty/frazier%20research.asp

29 Email letter to Dave Shayler from John Charles 20 November 2002; also
    http://www.ares.jsc.nasa.gov/People/gibsoneverett.html
30 *The Partnership: A History of the Apollo–Soyuz Test Project*, Edward Ezell and Linda Ezell, 1978. Appendix E, p. 518, NASA SP 4209.
31 Emails to Dave Shayler from Rex Hall, 21 September 2002, and David C. Fowler, 7 October 2002.
32 Email to Dave Shayler from David C. Fowler 7 October 2002; also
    http://www.voy.com/87202/3/742.html Albert Hibbs, 78, scientist and Voice of NASA missions dies.
33 John Aloysius O'Keefe, Obituary, by David P Rubincam and Bernard Chovitz, Physics Today Online, 2001, *http://www.aip.org/pt/vol-54/iss-6/p76.html*
34 Email from Mark Bostick posted on *astronauts@yahoogroups.com* 29 March 2005.

6

"Flying Is Just Not My Cup of Tea"

The academic training programme for the second group of scientist-astronauts was implemented in October 1967, and was completed the following February.

The training schedule was aimed at developing a level of competence in each of the men across a variety of scientific and technical subjects relevant to the space programme. Experience gained in these subjects would also prove valuable in reinforcing in each of the scientists the scope of problems related to space flight. The selection of course material was critical, due to the overall diversity and level of academic background in the group, and two basic requirements were imposed: course matter involving the least repetition; and materials selection at a level of complexity and interest commensurate with the men's abilities and backgrounds.

Table 2. Group 6 Academic Training Programme October 1967–February 1968.

Following a similar format to the Group 4 academic and wilderness training (1966–7), this programme for the 1967 scientist-astronaut group (with Holmquest joining later) commenced on 2 October 1967. It started with a six-hour briefing on physiological training and was completed on 28 February 1967 with the final two-hour session on meteorology. In total, there were 333 hours of instruction, broken into separate sessions to cover various topics or fields. The basic academic course covered the following disciplines:

Academic courses

- Physiological Training   6 hours
- Space Sciences   28 hours
- Astronomy   23 hours

One of the instructors in this programme was Karl Henize, who covered the stellar and galactic astronomy fields.

(continued)

Table 2 (*cont.*)

• Life Sciences	26 hours
• Earth Resources	22 hours
• Planetology	36 hours

The latter was a new course that was aimed at more in-depth training for future assignments as observers in the Earth Resources Program during AAP.

• Rocket Propulsion	8 hours
• Communications	10 hours
• Space Fight Dynamics	20 hours

Phil Chapman expressed some initial concerns over this programme when he first saw it. There did not seem to be a review of preliminary mathematics, which Chapman felt was essential for some of his colleagues, otherwise the instructor would have been forced to deliver at "too low a level to develop a sound understanding of the subject." Chapman suggested supporting the lectures with briefings on elementary classical mechanics "aimed at developing an understanding of the effects of gravity–gradient forces and of gyroscopic effects in rotational motion." This was important for understanding the dynamic effects observed inside the S-IVB workshop and understanding the use of the CMGs on the Apollo Telescope Mount during AAP missions, which at that time, the new group were expected to fly on.[1]

• Computers	6 hours
• CSM Familiarisation	18 hours
• LM Familiarisation	12 hours
• Mercury Project Briefing	2 hours
• Gemini Project Briefing	4 hours
• Apollo Mission Profiles	4 hours
• MCC Operation	4 hours
• AAP Mission Plans	4 hours

Additional notes in the Michel collection at Rice University indicate that AAP mission plans actually increased to fifteen hours and was delivered by AAP Branch Chief Astronaut Al Bean, AAP Branch assigned astronauts Jack Lousma, Bruce McCandless and Paul Weitz, and scientist-astronauts Owen Garriott and Ed Gibson. In addition, Jack Schmitt instructed the new astronauts on Advanced Lunar Exploration mission plans under AAP.

• Recovery Operations	2 hours
• CSM Systems Briefings	50 hours
• LM Systems Briefings	25 hours
• AAP Hardware	10 hours
• Bioscience Training	16 hours

The latter was another new course of eight two-hour lectures, covering terrestrial biology, terrestrial ecology and exobiology. It was designed to provide basic information in bioscience disciplines, in particular, an "understanding of the complexities of microbial life forms [to] foster an appreciation for the relationships of micro flora to their environment." The most current data from the Biomedical Research Office was used to study the effects of space environments on microbial ecology and the possible resultant effects on the health of future Apollo crews.[2]

Tours and Visits (8 days)
- Morehead Planetarium (University of North Carolina); 2 days (12–14 November 1967).
- Launch operations, KSC, Florida; 2 days (22–24 January 1968).
- Launch vehicle familiarisation, Marshall Space Flight Center/Michoud Facility; 3 days (13–15 February 1968).
- Acceleration familiarisation, 1 day (27 February 1968).

Additional courses included briefings on the various lunar and planetary probes programmes (2 hours) and the Soviet space programme (2 hours).[3]

Survival Training

This was attended by Allen, Chapman, England, Henize, Holmquest, Lenoir, Musgrave and Parker. Thornton never completed survival training due to delays in his qualification from flight school.

- Desert – Fairchild AFB, Spokane, Washington. The actual site was in the Oregon desert just south of Pasco in Washington State; summer 1969.
- Jungle – Albrook AFB, Panama Canal Zone. The site was near the Chagres River; Three days of instruction and two days of field exercises, 25–29 August 1969.[4]

Joe Allen recalls this period more for what happened when he came home after a week in the jungle. Even after several showers, the smell of the tropics lingered and his wife demanded that before Allen was allowed back into the house, all his clothes, bags and mementoes from the trip had to be left outside, thrown away or burnt. Even the family cat took swipes at him with its paw from the top of the refrigerator for some days. Allen was relieved that he had only had to endure this challenging part of astronaut training once ... and so was his wife.[5]

- Water survival - at Perrin AFB, Texas, during the summer of 1969

KNUCKLING DOWN TO THE TASK

As there would be only five months in which to fit in the academic training prior to their Air Force undergraduate pilot training, the schedule was carefully constructed. Lectures were generally limited to two per day, each of around two hours' duration. When possible, afternoons and alternate Fridays were kept free in order to allow the scientists the opportunity to pursue any personal projects.

A total of 330 hours would be devoted to scientific and technical lectures, while orientation field trips were conducted on the subjects of launch vehicles, launch operations and star recognition. In addition to the many NASA and contractor personnel contributing to this training programme, some thirty-five lecturers also participated, including some of the scientist-astronauts themselves; namely, Karl Henize and Bob Parker (astronomy), Jack Schmitt and Brian O'Leary (planetology), Curt Michel (solar winds and the Moon and planets), Phil Chapman (space flight dynamics) and Owen Garriott (aeronomy). They were even given a talk on exobiology in space

Seven members of the Excess Eleven examine a model of the Saturn IVB workshop cluster at the Marshall Space Flight Center in Huntsville, Alabama. From left: O'Leary, Chapman, Thornton, Llewellyn, Parker, Musgrave and Allen.

exploration by the eminent scientist, astronomer, educator and author Dr. Carl Sagan.[6]

Back to school

Story Musgrave likened the initial orientation and academic training phase of the group to the first year of medical school: "It is an introduction to a new organization, a new discipline and a new career. The precise information which an astronaut must have or which will contribute to his effectiveness is, as in medicine, not precisely known, so that the education and training in the early phases is very broad in scope and narrows as a mission approaches. What is attained in this phase might be titled, 'the management, science, technology and hardware basic to manned flight.'"[7]

Musgrave and his ten new colleagues discovered the learning atmosphere to be both unusual and dynamic. They found themselves privileged to be lectured by international authorities, explaining their special fields to a handful of students who, as Musgrave states: "have doctorates and have proven themselves in the lab; are uninhibited and urgently motivated, and are forever asking questions such as 'Why?' and 'How do you know?' As in surgical training programs, examinations are not needed. Everyone is aware of the progress that he and the others are making and, if motivation is needed, the launch pad can be looked upon as the final exam."[7]

The seven astronauts (without hard hats) stand in front of a mighty Saturn F-1 engine while touring the Marshall Space Flight Center.

Areas covered during the initial orientation and academic training included an introduction to NASA, field trips and briefings on the three NASA centres dealing primarily with manned space flight, and detailed briefings and demonstrations of the functions of all the major divisions within the Manned Spacecraft Center in Houston. Every day, all eleven men would be collected by the longest limousine they had ever seen and driven to specific buildings or facilities, where they would sit through a talk given by a division head, usually accompanied by a lengthy slide show to complement the oral presentation.

In addition to the grand tour of NASA's facilities, the men attended numerous lectures on space sciences, astronomy, planetology and physiology. The latter involved cardiovascular and pulmonary studies, human engineering, haematology, clinical pathology, neurology, vestibular investigations, vision, and a summary of physiological and medical observations on previous missions. In addition, there was bioscience – including biology, the origin of life, evolution, ecology and exobiology (a scientific study of life outside Earth), space flight dynamics, Earth resources, meteorology, computers, rocket propulsion, communications theory, summaries of past programmes, the Apollo Program, and post-Apollo plans. They also attended

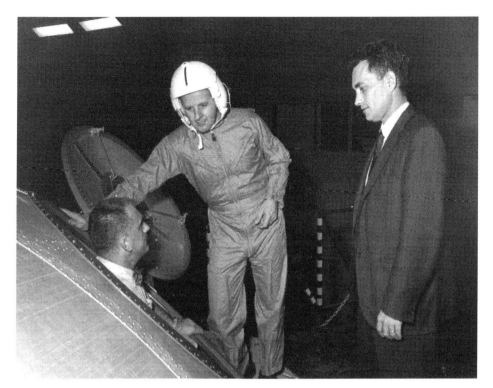

Chapman enters the centrifuge for an evaluation of g-stresses on his body

Apollo spacecraft systems briefings, Saturn launch vehicle systems and operations briefings.

As with their more formal training, the eleven men were encouraged to hold individual conversations and group discussions with instructors, managers, other astronauts, scientists and engineers. This informal approach outside of the classroom or laboratory environment was greatly appreciated, and would contribute tremendously to the men's education and development.

According to Brian O'Leary, many of the lectures were quite stimulating and worthwhile, "but we paid for our fun in the form of the lectures on spacecraft systems. With a few exceptions, they were extraordinarily dull – duller even than the 'come-see-our-computer' tours. Towards the end of the course work on spacecraft systems, the jargon and block diagrams were getting so tiring that we would rotate so that only five of the Excess Eleven attended any one lecture. We figured that any less than that might get back to the bosses, and then we would be conscripted to compulsory attendance."[8]

O'Leary was apparently coming to the realisation that this was not the easy ride to glory he'd imagined. He was rapidly falling behind those with better attitudes and more dedication, who possessed a "big picture" understanding of what was required of them. In a later book, he would offer a surprisingly frank insight into that period of his life:[9]

"I was a PhD and astronaut appointee on the ground floor of an exciting career in planetary science and astronautics. Soon, I felt myself to be somehow 'special', and my ego had grown to a point of seeming invincibility. I felt rather as if I were a young, tenured Roman senator during the reign of Caligula or Nero, or perhaps the Catholic Church's most junior cardinal during the Inquisition. As such, I would have been one of the least likely to point out that the emperor has no clothes."

As one highly critical scientist-astronaut later wrote about O'Leary: "He was quite immature and unrealistic in his expectations as to what the tasks, duties and responsibilities of a scientist-astronaut would be, and must have 'hidden his cards' very well during the selection process. In my opinion, he should never have applied or been selected. I cringe every time the term 'scientist-astronaut', to which I have devoted most of my professional career, is used to describe this egotistical and very doubtful team player."

In order to simulate accelerations associated with launch, launch abort, orbital re-entry and lunar re-entry, several of the scientists would train on a centrifuge, more formally known as the Aviation Medical Acceleration Laboratory, located at the Naval Air Development Center in Johnsville, Pennsylvania. However Bill Thornton, for one, does not recall being involved in this particular aspect of his astronaut training. The interior of the centrifuge's gondola could be configured to simulate a cockpit inside an actual spacecraft, and the engineers at the facility could install internal mock-ups of different spacecraft from the Mercury, Gemini and Apollo series into the cab. These cabs sat on long arms attached to a giant motor, weighing 180 tons, which swung the cab around at the end of the arm at tremendous speeds, accelerating from zero to 180 miles per hour in just seven seconds, sufficient to exert 40 G on the occupant. While the scientists and their astronaut colleagues trained at speeds and pressures far less than this, they found that a force of just 2–3 G produced only slight limitations to their performance, while 9–10 G was enough to cause a total inability to move or talk, and only a limited ability to breathe. They would later train on another centrifuge gondola in the Flight Acceleration Facility in Building 29 at the Manned Spacecraft Center in Houston.

FLIGHT TRAINING

One of the most exciting days for the new astronaut group occurred on the morning of 22 January 1968, when they were present at Cape Kennedy for the launch of the unmanned Apollo 5 flight. The Saturn IB rocket used for the launch was actually five times less powerful than the mighty Saturn V, which was still under development, but it would carry a prototype Lunar Module on its first test flight. This flight would verify the ascent- and descent-stage propulsion orbital performance of the Lunar Module.

Phil Chapman could hardly believe the noise and power of the launch, and he urged the huge rocket on as it cleared the gantry and thundered off into space. Following his selection the previous year, he had found himself under an especially bright media spotlight, because he and fellow candidate Tony Llewellyn from Wales were the first NASA astronauts to have been born outside the United States, so it was

Phil Chapman gets a close look at the 1 G mock-up of Lunar Module-3 during familiarisation training at the Manned Spacecraft Center in Houston. On the left is Louie Richard of Flight Crew Support Division's Crew Station Branch.

Chapman inside the Lunar Module Mission Simulator in Building 5.

In March 1968, Phil Chapman stands in the centrifuge gondola of the Manned Spacecraft Center's Flight Acceleration Facility.

Brian O'Leary and Joe Allen are strapped into the gondola of the MSC centrifuge, where they will experience forces up to 9 G.

good for the Australian scientist to get away from that for a while and become involved in watching real Apollo hardware at work. One day, he knew, he might be sitting on top of a rocket like that, soaring out over the Atlantic Ocean on his way into orbit, and perhaps even beyond.

Strapping on the jets

After the six-month academic and familiarisation training programme, the XS-11 were ready to begin their year-long pilot training course. The new group had been selected in the aftermath of the Apollo 1 pad fire in January 1967, and the space programme was changing as a result. In addition, they were out of the research laboratories and many were falling seriously behind in their own science work. Adding to their frustration, the new group of scientist-astronauts had tried to convince Air Force authorities to conduct their training at a single, suitable base, but this was denied as impractical. They would be sent to military air bases already stretched to the limit by the growing war in Vietnam.

Apart from Phil Chapman, only one of the group, Story Musgrave, had ever piloted an aircraft before, and none had flown a jet aircraft. With the exception of Don Holmquest, who would join the group a little later, they were sent to flight schools at several USAF bases all over Arizona, Oklahoma and Texas, beginning their training in March or April 1968.

After completing their Air Force training, the astronaut pilots would then be expected to maintain their flight proficiency in NASA's fleet of nimble T-38 aircraft. This proficiency would include acrobatic, instrument, formation, cross-country and experimental flights. They would also receive periodic and individually-requested performance honing from NASA's pilot-instructors, and they would undertake written examinations on aircraft performance and systems, procedures and regulations. Oral and flight examinations would take place every six months.

Brian O'Leary and Bob Parker were sent to Williams Air Force Base (AFB) in Phoenix, Arizona, where they would begin their flight training with nearly eighty air force officers, many of whom would be shipped over to Vietnam once their training was ended. The two men felt conspicuous and awkward in their civilian suits, especially O'Leary, the only man wearing spectacles. He was extremely out of place and very uncomfortable with the whole process, and had already entertained the idea of quitting the astronaut corps. But he'd decided to give himself a chance, reasoning that he might even surprise himself and become a good pilot.

Their introduction to the training was hardly reassuring, according to O'Leary. "The briefing consisted of a thousand forms to fill out and a few inspirational speeches from colonels, such as: 'Now you guys aren't going to find it easy here. You'd better believe we're going to work you hard, to the bone, but we guarantee you'll be damned good pilots by the time you leave. But if you don't work hard, we'll wash you out. This is not an outfit for amateurs. Do I make myself clear?' "[8]

In O'Leary's case, these words would certainly ring true. To make matters worse, he and Parker were separated and placed into two classes of about forty men. Each morning, beginning at 7:00 a.m., one group would receive flight instruction while the other group attended classes at the Casa Grande airport, fifty miles away. In the afternoon they would change places, and in the evenings they spent several hours studying what they had learned, knowing they could be quizzed at any time and had to know the right answers. On average, each trainee flew one hour a day, four days a week.

Over the first six weeks they would learn to fly in the T-41. Then they would move on to the jet-engine T-37 for a further twenty weeks, and during the second half of their training, they would fly the T-38. O'Leary was concerned: "The thought that I would be soloing the T-41 within two weeks scared me, because at the outset I couldn't even distinguish an airplane instrument panel from an automobile dashboard."

Chapman was fortunate enough to draw Randolph AFB, in San Antonio, Texas. This was the headquarters of USAF Training Command, and it was only 330 kilometres from Houston, which made it easy for him to keep in touch with the work on Apollo at the Manned Spacecraft Center. It was a very tense but exhilarating time for all of them.[10]

Flight training began with six weeks in a version of the relatively uncomplicated Cessna 150, a single-engine high-wing aircraft that is also commonly used in civilian flight schools. The principal purpose of this phase was to identify, at minimum cost and risk, any students who had little chance of completing the course. Chapman had no difficulty, especially as the Cessna was not very different from aircraft he had flown before. The surviving students then moved on to a subsonic T-37 jet trainer, affectionately known as the 'Tweety-Bird' because of its whistling engines, and finally to the supersonic T-38 Talon, a two-seat training jet aircraft. NASA owned a fleet of these T-38s, which were used by all existing astronauts.

The complete USAF pilot training programme took fifty-three weeks. The students were fresh out of university, so Chapman, at thirty-three, was the "old man" in his class. He was also the only civilian. "I normally wore a flight suit around Randolph, but I had no rank insignia. Even worse, I had no hat, which was a serious infraction for the military students. Occasionally, some colonel I passed in the street would tear a strip off me for being out of uniform, but of course he always apologised when I explained that I was a NASA civilian. I soon learned that if I made firm eye contact with an approaching officer, he would usually leave me alone. Sometimes, confused by seeing this rankless, hatless person in flying clothes, he would play it safe by actually saluting me. My best effort in this game was a salute from a two-star general![10]

"I really loved flying jets, especially the T-38. It is an extraordinarily agile machine, and it needs a very light touch on the controls. If you push the stick hard over, it will complete an aileron roll in less than two seconds. My first few solos were quite scary, but it became very easy to fly once I was used to it. Then we moved on to formation flying, which was scary all over again. It is quite intimidating to see your wingtip only a meter away from the lead's wingtip, when you are moving at 500 knots – but formation is easier than it looks. The slipstream from the lead aircraft pushes you away, so colliding is unlikely."[10]

Story Musgrave was another who really took to the training, and later wrote that:

Chapman has a coffee break in the lounge of the 3516th Pilot Training Squadron at Randolph AFB. Pouring the coffee is 2nd Lt. Paul Cunningham, while behind Chapman is instructor pilot Major Charles Herning.

"The entire process is superbly organised, efficient and an excellent example of an operational application of educational philosophy and learning psychology."

Bill Thornton had to take a little longer to complete his flight training. On 10 June 1969, NASA announced that the scientist-astronaut's graduation from jet pilot school would be delayed by an eye problem. An evaluation of his vision indicated that he should be fitted with specially designed eyeglasses which would allow him to achieve the degree of stereoscopic vision required for certain piloting manoeuvres. Wearing this type of glasses required four to six weeks' adaptation to use them effectively. Though temporarily grounded, Thornton would resume flight training at Randolph AFB in San Antonio, Texas. After the adjustment period, he completed the course without further delay as he was in the T-38 phase of the programme. When grounded he was close to completing the course and returning to NASA.

Eleven becomes ten, then nine

After graduation, the military students were allowed to select their own USAF assignments, with first choice going to those with the highest grades. Both Musgrave

"Flying Is Just Not My Cup of Tea"

Bob Parker gives a thumbs-up signal from the rear seat of a T-38 in 1983. Having graduated from the AF flying school in 1969 the scientist regularly used the T-38 for transportation across America.

and Joe Allen topped their particular classes[11] and if Chapman had been in the Air Force, he, too, would have had a wide choice, because he graduated second in his class. This was at the height of the Vietnam War, so he was a little surprised that the best students all wanted to be fighter pilots. "I admired their spirit, but dodging SAMs over Hanoi didn't seem like the best possible career move. I would have selected something a little less lethal. Test pilot school, perhaps."

Others, however, were finding it hard going. One month after flight training began, Brian O'Leary came to the conclusion that he simply couldn't handle the flying part of being an astronaut. He also had an overall ambivalence to classroom studies, and was quickly being left behind. He therefore decided to tender his resignation from the astronaut corps, which did not really surprise anyone, even though in his book he blamed the attitudes of the test pilot astronauts to the scientists and a perceived lack of future flight opportunities.

On 22 April 1968, O'Leary called Deke Slayton to say he was resigning from the astronaut corps, saying that he had flown fifteen hours and had soloed, but after much soul-searching he had decided to call it a day. "I guess flying isn't my cup of tea," he weakly admitted. Slayton was said by some to be understandably furious (although one of O'Leary's group suggested to the authors that there was "hearsay evidence of just the opposite"), and in such a competitive field, there was very little sympathy or support from many of his peers. Another scientist-astronaut said that O'Leary should not have applied for selection if he didn't think he could fly: "What the hell did he think astronauts did?"

Chapman, on the other hand, was far more committed and decided to ride it out, just to see what NASA had in store for him. He knew that the national objective at this time was just to get men to the Moon and back, and only the best, most experienced pilots would be on the first flights. Once NASA had demonstrated that men could land on the Moon and return safely to Earth, there would be a need for scientists to follow in their path.

On 23 August 1968, NASA announced in a news release that another of the Excess Eleven had withdrawn from the Astronaut training programme. Tony Llewellyn had been undergoing flight training at Reese Air Force Base in Lubbock, Texas, since 4 April. He had completed the first phase of training involving thirty flying hours in the propeller-driven T-41A, and had completed his required solo flight in the light aircraft. He had then begun approximately forty hours of dual flight training in the T-37 jet trainer, but he had already begun to harbour qualms about his ability to complete this assignment. He said that it had become apparent to him that he was not progressing as well as he should, and he held discussions with NASA and Air Force officials prior to his scheduled solo flight in the T-37. Based on their advice and his own doubts about his ability to go any further, the thirty-five-year-old scientist decided to tender his resignation and withdraw from the programme. His departure from the Astronaut Office effectively reduced the total number of NASA astronauts to fifty-two, and the Excess Eleven down to just nine members.

Following their return to NASA from the US Air Force, the nine remaining scientist-astronauts from the second group maintained their flight proficiency in NASA's Northrop T-38s. Instruction from NASA instructor-pilots was received periodically, as well as any time it was requested, and part of their proficiency included the oral and flight examinations every six months.

Training and conditioning to the zero-G environment was also started. The only true zero-G experience available was derived from parabolic flights carried out in NASA's KC-135 jet aircraft, a military version of the Boeing 707 jetliner. A simulation of orbital weightlessness could be sustained for up to thirty-five seconds during a series of parabolic dives and ascents. One moment the participants would be calmly floating around the padded interior of the KC-135, and the next they would be pulling two-and-a-half G as the aircraft was dragged upwards for the next parabola.

Neutral buoyancy was another reasonably effective method of simulating the weightless environment of space, and remains an important EVA training tool for astronauts to this day. In the early Apollo days, neutral buoyancy training was undertaken using traditional scuba gear, a Kirby–Morgan type of helmet which provided both air and communication, or in a pressurised spacesuit weighted to be both neutrally buoyant and also rotationally stable in the water. With this in mind, the scientists undertook a preliminary scuba training course at the Navy

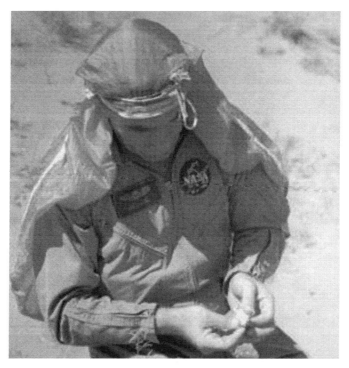

During Group 6 desert survival training, Karl Henize takes a few moments to examine a small toad he has caught.

Underwater Demolition School at Key West, Florida. Their course included swimming instruction, underwater physiology, snorkel and scuba equipment familiarisation, malfunction procedures, depth diving, rapid ascent from depth, underwater navigation and some underwater ocean experience.

Spacesuit training was also started early in their programme, requiring a basic knowledge of the suit and its system hardware, as well as the inter-relationships between total pressure, differential pressure, suit flow, inlet and outlet gas temperatures, oxygen and carbon dioxide, water flow to the liquid cooling garment, inlet and outlet temperatures of the water, metabolic rate and thermal balance. However, they would not wear the bulky suits at this time for, as Bill Thornton pointed out to the authors, all of the Apollo suits were individually custom-made units that wouldn't have fitted most of them.

Since ejection from an aircraft or an off-nominal de-orbit and entry in a spacecraft could bring astronauts down anywhere in the world within around fifty degrees of the equator, survival training similar to that carried out by all previous astronaut groups also fell to the scientist-astronauts. Desert survival training was conducted at Fairchild AFB in Pasco, Washington, while jungle survival techniques were studied outside Albrook AFB in Panama. For water survival, the men trained at Pensacola Naval Air Station in Florida. At Perrin AFB in Texas, vital training was given for ejection, free

Phil Chapman and Joe Allen undertaking Panamanian jungle survival training at Alsbrook AFB.

fall and parachute canopy manipulation procedures, as well as parachute landings on land, in trees, or in water. To complement all of this survival training, the men were given instruction by NASA experts on the contents and use of all survival equipment carried in their T-38 jets, and in the spacecraft they might fly. These courses followed a standard routine, comprising two days of lectures and laboratories, two days in the field with instructors, and finally two days alone in the field with the appropriate spacecraft survival kit.

Story Musgrave felt that their survival training was "excellent," but it was time to move on in their programme. "Further, and more, briefings and schematic reviews on the Apollo spacecraft and the Saturn booster vehicle were conducted at this point and initial instruction on the Skylab space station was also begun at this time. Study of the Apollo spacecraft included [the] electrical power system, communications, environmental control systems, sequential electronics control systems – automated launch aborts and Earth landing systems, for example – propulsion, and attitude control system, guidance and control systems, structures, hatches and docking mechanisms. A similar division and part-task approach was taken to the study of the Skylab space station. Part-task systems training in the Apollo spacecraft simulators was begun following the completion of the briefings and schematics reviews."

188 "Flying Is Just Not My Cup of Tea"

As part of their jungle survival tasks, the astronauts had to pair up and construct a shelter. Here, Bill Lenoir and Story Musgrave prepare to spend the night in their shelter.

NASA wanted all astronauts to maintain their proficiency by flying thirty or forty hours each month, but it was not easy to find the time, as the work on Apollo was building to a crescendo. When Phil Chapman's family returned to Houston early in 1969 after his Air Force training, he had become a fully qualified jet pilot, and he soon found himself flying all over the mainland United States attending technical meetings and giving public lectures about the space programme. Unlike other astronauts, however, he spent most of his time helping with the development of equipment for the lunar landing.

As an example, the Apollo Command Module was designed to enter the Earth's atmosphere under automatic control. If the guidance system failed, the pilot could re-enter by flying manually, using a display that showed the deceleration as a function of time. If the deceleration was too great, the descent path was too steep and the capsule would burn up; if the descent path was too shallow, the capsule would bounce out of the atmosphere. It was not clear that a pilot could handle the controls under the stress of atmospheric entry, so Chapman flew many simulated manual trajectories, pulling seven or eight G in a Command Module mock-up mounted on a centrifuge. The conclusion was that manual re-entry was feasible, but only if the pilot had practised it beforehand.

Members of the Group 6 astronauts practise their skills (or lack of) with life preservers and rafts in Panama. In the photo are Joe Allen, Don Holmquest, Phil Chapman, and Story Musgrave.

LOOKING TO THE FUTURE

The five remaining scientist-astronauts from the first group, meanwhile, were involved in several work assignments for NASA, but still without a clear indication as to their operational future. In September 1966, shortly after the formation of the Apollo Applications Branch in the Astronaut Office, Curt Michel was given an assignment to monitor progress on the Apollo Telescope Mount (ATM) project, working with specialists at the Marshall Space Flight Center in Huntsville, Alabama. He would also work with groups concerned with lunar atmospheres, manned space science atmospherics, and planetary atmospheres. As well, he took part in ongoing astronaut training and actively maintaining his flight proficiency.

Jobs on the line

In 1967, NASA recognised the need for the scientist-astronauts to also maintain and update their scientific proficiency. The agency established a review procedure to ensure that the scientists had adequate time for study and research in their areas of expertise, although this was initially limited to one day per week and one week per month. This was finally the fulfilment of the twenty-five per cent research time members of the group had been promised. However, some of the men would still find this very limiting, and Michel was one of those who expressed concerns about being able to meet the requirements of being an astronaut, as well as taking on study and research.

The war in Vietnam was proving a vast drain on the American economy, and governmental budgets were being drawn ever tighter. NASA, even in the throes of accomplishing a national goal, was not exempt. Budgetary problems would cause NASA to decrease the number of planned Apollo missions, which meant that there were suddenly less crew assignments available and far too many astronauts in training. Time and money were running out on their dreams. In light of this, Michel requested that he be given a year's leave of absence to return to Rice in the fall of 1968, in order to resume his teaching and studies and to conduct some research. The initial reaction was that he could quit the astronaut programme if he didn't feel he could devote his time to it, but eventually his bosses relented and he returned to Rice.

One thing that continually astounded the scientist-astronauts was the manipulative game of office politics played within NASA. Curt Michel recalls that during the Gemini IX mission, Gene Cernan complained that some "mysterious force" seemed to be pushing him away from the Gemini spacecraft during his curtailed EVA, and Michel was asked to identify the source of the problem. "Slayton told me to 'look into' the possibility that it was some 'gravity gradient' effect," he recalled. "I told him that it was simply action and reaction: every time Gene touched the surface he pushed himself and the spacecraft away from each other – mainly him of course. And even if each was a tiny effect, they all add up. The next day Max Faget was in the *Houston Chronicle* explaining this to the reporters as his theory!"

Both Karl Henize and Joe Allen seemed to feel that they were able to maintain an even balance between their astronaut and scientific careers, and both agreed that even the fifty-three weeks spent on the jet pilot course had not interfered with their ability to maintain proficiency in their particular fields. In fact, Henize felt that the Apollo programme was of considerable benefit to his work as an astronomer by enabling him to gain access to the programme's large spectrograph. He was also able to spend time at Northwestern University and would be able to publish two papers. Similarly, Joe Allen was able to keep up with nuclear physics by simply reading the literature and by working with a cosmic-ray physics team at NASA: "Any scientist would like to have more time for research, but we didn't have all the time we desired even when we were teaching."[12]

Everybody understood that Apollo was John Kennedy's answer to the initial Soviet lead in space. Now it seemed very probable that the first people on the Moon would be Americans, not Russians. The problem NASA faced was that beating the Russians too decisively might lead the Congress to decide that they could cut back the funding for human space flight.

According to Chapman, "Soviet cosmonauts visited Houston several times while I was there and we usually put on a party for them at the home of one of the astronauts. They claimed to speak little English, and were always accompanied by 'translators,' who were actually *zampolits* (political officers). Our standard procedure was to have some pretty girls on hand, in miniskirts, who would distract the *zampolits* while we fed alcohol to the cosmonauts. Their English improved remarkably after a few scotches! I remember one of them urging us to do something spectacular in space, and soon, because otherwise the Politburo would cut their funding. The space programmes of both countries depended on the Cold War.

"A few months before Apollo 11, I was sitting in a bar with a few other astronauts, when the conversation turned to what we could do if the USSR quit the space race. Somebody suggested that we should create a fake alien artefact, and give it to Neil Armstrong to 'find' on the Moon. Evidence that an alien civilisation had visited the Moon in the past would surely stimulate a major effort to find out more.

"To be convincing, the artefact had to be something subtle. Our solution depended on the fact that sugar molecules exist in two forms, called isomers, which are mirror images of each other. Sugar produced chemically contains an equal mix of both isomers, but biological sugars contain only one, and it is the same one for all life on Earth. Our plan was to obtain some goat urine, remove the sugar, and replace it with the other isomer. Neil would contaminate a soil sample, let it bake in vacuum until he was ready to leave, and bring it home. Then the chemists would discover that somebody or something not of this Earth had taken a leak on the Moon. Of course this was a joke – almost – and we never did anything about it. I rather wish we had: we might now be much farther ahead in space."[5]

Prior to the Apollo 13 mission, Chapman had been handed a choice assignment, as Mission Scientist for Apollo 14. He was not on the prime or back-up crew, but he became an essential part of the team, helping to organise the scientific training of the crew, coordinate lunar experiments and provide the interface between the scientific team and the crew on the Moon. After the mission, he served as Chairman of the Editorial Board for the Apollo 14 Preliminary Science Report.

In January 1971, Alan Shepard and his crew returned America to the Moon in style. America's first man in space had now made it to the lunar surface, and he even managed to hit a couple of golf balls a considerable distance in the Moon's low gravity, using a specially improvised club.

While Phil Chapman was glad to have had an active role in a lunar mission, his work on Apollo 14 strengthened a growing concern about the attitudes of NASA management. In particular, he often disagreed strongly with Deke Slayton, one of the first Mercury astronauts, who was Head of the MSC Flight Crew Operations Directorate which included the Astronaut Office. His position made him responsible for selecting the crews for each mission, and according to Chapman, he used that power ruthlessly to force the astronauts to obey his slightest whim.

"Problems with Deke became apparent soon after I joined the programme in 1967, when John Glenn sent a memo to all of us complaining that Deke had revoked his access to Mission Control. Glenn had left NASA, but he was still in great demand by the media. Cutting him out of the picture, instead of using him as a spokesman, was an act of breathtaking stupidity. It was hard to believe, but it really seemed that Deke was motivated by nothing more than envy because Glenn was a hero enjoying the limelight, while he had been grounded by a heart murmur.

192 "Flying Is Just Not My Cup of Tea"

Chapman photographed wearing an Apollo pressure suit.

"Removal from flight status was of course a great disappointment for Deke, so nobody minded much that he was usually irascible. Everybody knew he had been given his senior desk job as a consolation prize – but I didn't think a heart murmur was an adequate qualification for such a responsible position. To put it plainly, it seemed to me that Deke had no understanding that leadership is a two-way street, and no vision of space flight beyond keeping it as his own little fiefdom. Furthermore, he apparently thought that the only legitimate purpose of a space mission was flight-testing a vehicle; that science in space was a worthless distraction, and that scientists were inherently unacceptable as astronauts, regardless of their flying skills."

(On reading these comments, Bill Thornton strenuously pointed out that Slayton was not suffering from a "heart murmur" but from a condition known as an intermittent arrhythmia, and finds Chapman's recollection of these events bewildering. He says that Slayton's condition "was over-reacted to by consultant physicians, and certainly not Bill Douglas, who was in fact the flight surgeon for the Original Seven. Deke reluctantly took an important, crucial job and generally did it well in spite of obvious personal and system prejudice. He did it superbly compared to his successors who made a farce of arguably the most important office in the NASA system.")[13]

Chapman continues: "I believed then, and still do, that pilot training is very useful. Like space flight, flying is an activity that is not normally dangerous, but it is intolerant of mistakes. If you do the wrong thing, you die. Flying thus teaches you to be calm under time-critical stress – or, rather, to postpone the panic attack until you are safely on the ground. I had nothing but admiration and respect for the coolness shown by pilots such as Neil Armstrong, and by the entire crew of Apollo 13."

While Chapman admits that he was not sure that he could have matched this coolness given the circumstances, he did not believe that the scientists would endanger any mission. They had all been tested under stress, he emphasised, and would not have been there if they were unstable.

"Deke made his attitude very clear while I was working on Apollo 14. As in all lunar landings, the Command Module Pilot, Stu Roosa, would wait in orbit while the other two crew members went down to the surface. Stu would have very little to do, and he asked me if I could suggest useful activities or observations he might undertake. I talked to various friends in the space sciences, and we came up with an interesting list. None of our suggestions involved any special equipment or had any impact on other mission tasks.

"For example, a Polish astronomer named Kordelewski, using an Earthbound telescope, had reported seeing very faint clouds at places in the plane of the Moon's orbit called L4 and L5, which are located so as to make equilateral triangles with the Earth and Moon. These are points of stable gravitational equilibrium, and it was possible that dust or larger meteors had collected there. This was of great scientific interest, since it offered the possibility of obtaining meteoritic material that had not been altered by passage through the Earth's atmosphere. There were times in lunar orbit, on the night side of the Moon, when the Kordelewski clouds (if they existed) would be within the line of sight and illuminated by the sun. All Stu had to do was to point a camera in the right direction at the right time, and he might make a major scientific discovery.

"Unfortunately, Deke heard about this list, and he carpeted Stu and me. He pointed out that scientists whose proposals had been rejected would be angry if we undertook experiments that had not been through the formal selection process. I agreed, but said that formal experiments were only accepted if there was a good chance of completing them. This conservative approach meant that there would always be some spare time, and we should not waste it just to avoid upsetting a few individuals."

Slayton refused to listen, and told Roosa that he would be removed from the flight crew if he did not get rid of the list. He also warned Roosa that he would never fly in space again if he made any observations that were not specifically shown in the official flight plan. As a result, Chapman said, scientists still do not know if the Kordelewski clouds are real.

"Later, Deke sent a memo to all the astronauts, saying that TV commentators always judge the success of a mission by the percentage of the mission objectives that have been achieved. The obvious way to maximise that number, he said, was to reduce the number of mission objectives. In future, therefore, the Astronaut Office was to do everything it possibly could to eliminate scientific experiments on every mission. I was dumbfounded by the idea that the way to increase interest in space flight was to minimise the useful results, and, insubordinate Australian that I am, I told Deke what I thought of his new policy."

Losing the Moon

The final straw came when Slayton announced that he would never assign any scientist-astronaut to a lunar mission – or to any other mission – if he could help it. Chapman had no expectation of going to the Moon himself, but geologists Jack Schmitt and Tony England were clearly the astronauts best qualified to investigate lunar geology. They were both very competent pilots, and Chapman stated that excluding them for no better reason than blind prejudice was tantamount to sabotaging the space programme.

"I had always tried to be a team player, disagreeing with Deke only in private, but this was much too much. After some soul-searching, I discussed the issue of scientists on the Moon directly with Jim Fletcher, the NASA Administrator. I don't know whether what I said influenced his decision, but in any case Deke was overruled, and Jack Schmitt went to the Moon on Apollo 17, the last Apollo mission."

Chapman's efforts to maximise NASA's science potential also led to success on other fronts. Shortly after the conclusion of the Apollo 14 mission, he gave a talk about space flight at a meeting of the Ninety-Nines, the international association of women pilots. He shared the podium with Sheila Scott, a well-known British aviatrix who had just arrived from Fiji in her single-engine Piper Comanche. She told him that she was getting ready for a flight from equator to equator, over the North Pole.

"I knew that NASA had developed a transponder that gave a position report to the Nimbus weather satellite. It was intended to be hung around the neck of a caribou – a North American reindeer – to track their migratory patterns. I persuaded NASA to lend one to Sheila. This meant that the Goddard Spaceflight Center in Washington would get a fix on her every ninety-six minutes, when Nimbus passed over the Pole. If she came down anywhere in the Arctic, we could tell rescuers where to find her."

Chapman was at Goddard with Dick Hoagland, another friend of Sheila Scott, when she left Norway in a twin-engine Piper Aztec, heading for Point Barrow in Alaska. It was a strenuous flight, which she later described in her book *On Top of the World* (to which Chapman contributed a foreword). At Goddard, between Nimbus fixes, he and Hoagland discussed a request by Dave Scott, Commander of Apollo 15. According to Chapman, he wanted something interesting that he could do before leaving the Moon, "to be one up on Al Shepard's demonstration of lunar golf!" The suggestion that Chapman and Hoagland came up with turned out to be one of the most memorable but simple exhibitions of science ever carried out in the Apollo programme.

Towards the end of his lunar visit, Dave Scott told a live television audience back on Earth that he was hoping to demonstrate how Galileo Galilei had been right. Galileo, an Italian physicist and inventor born in 1564, had stated that objects of different weight always fall with the same acceleration under gravity. Scott produced a hammer and a falcon's feather (taken from the mascot of the US Air Force Academy), held them out, and dropped them at the same instant. On Earth, the ultra-light feather would have floated down, but on an airless Moon, both objects fell together. Because lunar gravity is only one-sixth that of Earth, they fell quite slowly and hit the lunar soil simultaneously. The video of this simple experiment is now widely used in schools to demonstrate the nature of gravitation.

By the time Apollo 15 flew, the American people were beginning to lose interest in the programme. President Richard Nixon did nothing to stem this decline, or to help NASA find the best direction for the future. Chapman's lack of enthusiasm for the man is quite evident. "Despite the fake enthusiasm he displayed during his phone call to Neil Armstrong and Buzz Aldrin while they were on the Moon, he had no real interest in space flight, and he was already sinking into the paranoid preoccupation with his political enemies that would lead to the Watergate break-in and his eventual resignation." Lacking strong support from the White House, the NASA budget was decreasing each year. The overwhelming enthusiasm when Apollo 11 landed had been replaced by uncertainty about the future.

Crippling budget cuts led to the cancellation of Apollo missions 18, 19 and 20, and also to a protracted debate within NASA about its future programmes. Skylab A, the first space station, was scheduled for launch in 1973, but it was not designed for re-supply of consumables such as oxygen, and would cease operating when initial stocks were exhausted. Three crews would spend months aboard the workshop, but after that it would be abandoned. Each crew included one scientist-astronaut, but none of them were from Chapman's group.

A second workshop, Skylab B, was planned for 1975. Chapman was a member of a committee, headed by Apollo 8 astronaut Frank Borman, advocating simple changes that would permit repeated re-supply and refurbishment, to give the station an indefinitely long life on orbit. It would be supported initially by Saturn rockets, using Apollo Command Modules to carry crew, but a series of incremental changes were suggested that would, over time, replace them with reusable vehicles that were much cheaper to operate.

There was, however, another influential faction in NASA, which urged the development of what became the Space Shuttle. They claimed that a space station was unnecessary, because the Shuttle would be so cheap to fly that astronauts could sleep at home and commute to orbit each day. Chapman had no faith in these claims, because in his judgement the estimates used to support them "were simply fraudulent" and, in any case, the radically new technologies needed by the Shuttle precluded any meaningful prediction of costs. He believed that NASA should work on advanced technologies as relatively low-cost research projects, and not commit to using them in an operational vehicle until they had been proven. "Economy could best be achieved by incremental improvements, rather than by another 'Giant Leap for Mankind'." While the necessary research was under way, NASA could use Skylab to establish a low-cost, permanent space station. That would permit work to begin in areas such as free-fall agriculture, aimed at enabling large numbers of people to live and work off-Earth.

Chapman suspected that the real problem with a space station, as seen by Deke Slayton and his supporters, was that it just sat there in Earth orbit. "No flying had to be done, and therefore there was no compelling reason to staff it exclusively with pilots. If it were supported by Saturn technology, the role of the pilot would essentially be reduced to monitoring the automatic guidance systems that controlled launch, docking with the station, atmospheric re-entry and descent on parachutes. The Shuttle was much more appealing, because it had wings and could actually be flown manually, at least during the final descent to a landing. The space station was a step towards a future in which all sorts of people would be needed in space – engineers, scientists, high steel workers, doctors and nurses, even farmers and cooks. That prospect was very threatening to those who were more interested in retaining control of the programme than in a growing human enterprise in space.

"It seemed obvious that extraterrestrial operations could not become a significant part of our civilisation if they were funded only by NASA. The NASA budget will always be determined by political considerations, in competition with other demands on government revenues. Substantial growth would be seen as an unacceptable burden on taxpayers, and it simply cannot happen unless some dire national emergency demands it. On the other hand, exponential growth in the private sector is highly desirable: that's what economic progress means. Like the early American colonies of Great Britain, outposts in space can grow until they dwarf the motherland – but only if we can engage the engines of free enterprise."

Putting things in perspective

In Chapman's view, NASA's proper roles in human space flight are to develop technologies needed by private enterprise in space, to reduce the present barriers to private investment in facilities off-Earth, and to provide those services that can best be handled by the government, such as space traffic control and search and rescue. Most astronauts should be employed by corporations working in space, he feels, and not by NASA.

"It is very unfortunate that the Shuttle proponents won this argument, because it has proven a catastrophe for NASA, and for human space flight. The Shuttle was supposed to be at least ten times cheaper to fly than the Saturn vehicles, but in fact it is ten times more expensive. In other words, the Shuttle programme missed its cost goals by a truly incredible 10,000 percent! As a result, the average cost of a Shuttle flight is an astounding US$600,000,000, so NASA can afford only a few missions each year.

"To put this in perspective, suppose General Motors announced that it was developing a new, greatly improved Cadillac, and the price would be only $20,000 each. What would happen to GM stock, and to GM management, if it was found when the new car reached production that building each one cost $2,000,000? This is exactly what happened to the Shuttle. Why didn't we see a mass dismissal of everybody in NASA who had anything to do with this fiasco?"

Faced with a declining budget and rising costs, NASA was soon forced to cut back in other areas. The Borman committee recommendations about making Skylab B permanent were rejected, and then the whole second Skylab project was cancelled.

This was a major blow for Chapman, as he had hoped to fly in that workshop, and was in fact developing two experiments in general relativity that he could undertake there. "The Skylab in which I hoped to live is now a tourist attraction in the Smithsonian Air and Space Museum. I sometimes visit it when I am in Washington, but it is very sad to see it wasted."

Chapman was disgusted with what he calls "the monumental stupidity of the course NASA had chosen." Moreover, it was now clear that none of his group of scientist-astronauts would get into space for at least another decade. With his views diverging so sharply from the course NASA was taking, he reluctantly decided, in July 1972, that he needed to find a better way to contribute to space flight than simply hanging around in Houston, working on that so-called "Giant Leap" in which he had no confidence.

REFERENCES

1. Memo from Phil Chapman to Curt Michel, 16 October 1967, Curt Michel Collection, Rice University, copy on file, AIS archives.
2. Memo (DB3/11/19) from Charles A. Berry MD, Director of Medical Research and Operations to Deke Slayton, Director of Flight Crew Operations, 27 November 1967, subject *Bioscience Training for the Scientist-Astronauts*, Curt Michel Collection, Rice University, copy in AIS Archives.
3. *Academic Training Program for Group Six Astronauts*, Mission Training Section, Mission Operations Branch, Flight Crew Support Division, NASA MSC, Houston, 1 September 1968, MSC Internal Note MSC-CF-D-68-16, from the Curt Michel Collection, Rice University, copy on file AIS Archives.
4. MSC Roundup, Vol 8, #23, 5 September 1969 p. 1 photo and caption.
5. Notes of conversations between Joe Allen and Dave Shayler, June 1994, England.
6. NASA MSC Internal Note MSC-CF-D-68-16, *Academic Training Program for Group Six Astronauts*, previously cited.
7. Paper by Story Musgrave called *The Selection, Training and Activities of a Surgeon Astronaut*. Part of I.S. Ravdin Lecture Series, American College of Surgeons, Chicago, Illinois, 17 October 1973.
8. *The Making of an Ex-Astronaut*, Brian O'Leary, Houghton Mifflin, Boston, 1970.
9. *Miracle in the Void*, Brian O'Leary, Kamapua'a Press, Hawaii, 1996.
10. All of the information about Chapman, and all of the quotes from him in this chapter, come from a lengthy email from Chapman to Colin Burgess on 5 October 2002.
11. Email from Story Musgrave to Colin Burgess, 7 Mar 2004.
12. *Scientific Research* magazine, issue 4, August 1969.
13. Email from Bill Thornton to Dave Shayler, 22 May 2006.

7

A Geologist on the Moon

Barring human intervention, the footprints of two men will remain undisturbed at a place called Taurus-Littrow for many millennia to come. One set of those footprints belongs to Lunar Module Pilot Harrison "Jack" Schmitt, who walked his way into one of mankind's most exclusive clubs when he became the last person to set foot in the Moon's ancient dust, on 11 December 1972.

The same feat three years earlier made Neil Armstrong a household name, yet for the most part, Schmitt enjoys a relative anonymity shared by almost all of his moonwalking colleagues. It was an historic task for which he had worked hard; in his dual capacity as a scientist and astronaut he had provided other potential Apollo astronauts with detailed instruction in geology, lunar navigation and feature recognition.

While not referring specifically to Schmitt, cosmonaut Yuri Gagarin had predicted in a 1966 press conference that: "Soon man will walk with a geological hammer on the mysterious surface of the Moon." Six years later, Jack Schmitt had his hammer ready.

SUPPORTING APOLLO

While Harrison Schmitt was the only one of the scientist-astronauts to set foot on the lunar surface during Apollo, the others all had roles to play in support of the programme to ensure that Schmitt and his eleven fellow moonwalkers successfully achieved their missions.

On 23 September 1965, three months after being selected as astronauts, Joe Kerwin and Curt Michel received technical assignments within the Astronaut Office (CB) while their colleagues were at flight school. For Kerwin, the only physician in the group, it became very clear that he would probably receive an assignment that was something to do with medical research on long-duration space flight rather than a

On 6 January 1966, Joe Kerwin and Rusty Schweickart were inside the crew compartment of a Lunar Excursion Module (LEM) simulator at Houston's Manned Spacecraft Center during stowage tests.

short mission. When the Apollo Applications Program (AAP) evolved to support such long-duration missions, it dawned on Kerwin that "I had a role and a mission to shoot for."[1]

He was assigned to environmental control systems and then to pressure suit and EVA issues in the Operations and Training Branch, while Michel worked with experiments and future programmes within the Apollo Branch. Most of this work would be reassigned to the Apollo Applications Office in late 1965. The two men received additional assignments over the Apollo years in support of the developing Apollo lunar programme.

Kerwin asked Astronaut Chief Al Shepard if he should keep up his medical skills at the clinic for a couple of days each week. Shepard informed the new astronaut that he would have more than enough to do without clinical medicine proficiency, and he was right. Assigned initially to the environmental control systems (taking over from William Anders who had been reassigned to back-up Gemini XI), he found himself dealing with life support, spacesuit and cabin atmosphere issues. But it soon became clear to Kerwin that his medical background would be useful in understanding such issues. He had to pick up on the engineering aspects of his assignments, and in this he was guided by Anders, who told him to stop learning every nut and bolt, reduce the tasks to the simplest things and keep notes on small file cards.

When Ken Mattingly received a support assignment in the early Apollo missions, Kerwin was given the technical assignment for pressure suits, participating in the Source Evaluation Board for the Apollo suits examining prototypes from Hamilton Standard and David Clark. This included evaluating their design, material and mobility. This assignment also involved technical assignments in AAP relating to suits and EVAs, running concurrently with his Apollo assignments. It is not unusual for NASA astronauts, even today, to receive two, three, four or more concurrent technical assignments while between flight crew assignments.

VACUUM TESTING APOLLO

Manned flights in Apollo would, like Mercury and Gemini before it, be preceded by a series of unmanned flights to qualify the hardware, systems, procedures and infrastructure. As part of this test programme, NASA planned a series of simulated missions in vacuum chambers, which would reproduce the characteristics of the space environment without leaving the ground. Though engineers from the prime contractors could and would participate in some of these tests, having astronauts aboard the simulations would enable the CB to be represented in such critical developmental tests of the spacecraft. It would also give rookie astronauts some experience of working as a crew and in the confinement of a spacecraft under as close to space conditions as possible.

Located in Building 32 at the Manned Spacecraft Center, Houston, the Space Environment Simulation Laboratory comprised two separate vacuum chambers, one of 17 m diameter by 36 m in height (Chamber A) and the other 14 m by 13 m (Chamber B). As well as being able to reproduce the vacuum of space to a simulated altitude of 240 km, there was a battery of carbon-arc lights that were used to reproduce the intensity of solar heating to 95 degrees Celsius, while cryogenic panels in the walls of the chambers could reduce temperatures to −140 degrees Celsius. With air in the chamber, the door could be opened to insert or remove the spacecraft being tested and evaluated. The two chambers were completed in 1965 and Chamber B was first used during January 1966 for testing Gemini pressure suits and EVA equipment. Later that year, a Block I Apollo CSM (planned for Earth-orbital missions only) was installed for qualification tests in Chamber A.

Chamber testing the Block I CSM

Initial vacuum tests of a Block I Command Module took place in a smaller chamber at the primary manufacturer, North American Aviation, in Downey, California, during the summer of 1966. The full CSM test conducted in Houston's Chamber A covered a period of eighty-three days, with a ninety-two-hour unmanned test followed by a 163-hour manned test (almost seven days). Astronauts Ed Givens (recently selected in the Group 5 intake) and Joe Kerwin were chosen for the test, along with USAF Captain Joe Gagliano, who was on assignment to the Flight Crew Support Division at MSC.[2]

The purpose of the test was to evaluate procedures and hardware prior to committing the Block I to manned orbital operations with Apollo 1 early in 1967. As a result of the test, a number of issues had to be addressed prior to clearing the Apollo 1 mission for flight. During the August 1966 review of the CSM-008 test programme, Director of Flight Crew Operations, Deke Slayton, put forward a suggestion that in future, a flight surgeon should be assigned to the next CSM chamber test crew scheduled to evaluate the Block II CSM for lunar missions (using CSM 2TV-1) to help define medical requirements for Apollo and to provide baseline data that could be used in developing crew training and test objectives. At that time, the only medical doctor in the astronaut corps was Joe Kerwin. Having already experienced the CSM-008 test, it was logical to assign him to the next one.

Kerwin recalls this period in his 2000 Oral History: "I had a fun assignment in there as a member of the crew. The first one was to precede Apollo 1. Three of us were assigned to spend a week in Chamber A in Building 32 (which was brand new at that time), testing the spacecraft out, and we had a good time doing it. We utilised equipment and procedures which were the same as those that killed the crew of Apollo 1, so we considered ourselves, in retrospect, very fortunate that we didn't have a spark [in the 100 percent oxygen atmosphere of the CM]."

Chamber testing the Block II CSM

With the Block I spacecraft tested and about to be man-rated, in early 1967 NASA scheduled a similar test for the Block II spacecraft that would support lunar missions. Originally scheduled for February 1967, the test was delayed as a result of the Apollo 1 pad fire of 27 January 1967, and the necessary changes resulting from the inquiry into the fire. The changes incorporated into the mainstream Apollo missions and hardware also applied to the chamber test programme. On 22 June 1967, a ten-person committee was set up to review the changes made in the chamber for future simulations. Kerwin represented the CB on this committee due to his experiences with CSM 008 in 1966.

The test crew were assigned later in 1967, but delays in preparing the hardware pushed the test well into 1968. The crew would be Joe Kerwin (commander), Joe Engle (Lunar Module pilot) and Vance Brand (Command Module pilot), the latter two from the 1966 pilot selection. "After the Block I test, I went back into Skylab for a while, and then out again to command my first mission, which was the second hanger queen."[1] Kerwin was probably chosen as the commander because of his medical background, his previous experience and his seniority in astronaut selection. This was the *only* time a scientist-astronaut would officially receive the Mission Commander designation for a crew (albeit a ground crew). Scientist-astronauts would later be lead astronaut for a "payload crew" in simulations of Spacelab-type missions, but never in overall command.

In the era of a pilot-orientated Astronaut Office, the idea of a scientist-astronaut not only commanding a crew but also two pilots was a bitter pill to swallow, but neither Engle (who had a bigger pill to swallow some years later in losing a lunar landing seat to another scientist-astronaut) nor Brand ever showed hostility, and knuckled down to ensure the test was a success, like the professional astronauts they both were. Kerwin was well aware of their experiences as pilots, especially Engle, who was a former X-15 pilot and had earned his USAF astronaut wings by flying over fifty miles high in the rocket research aircraft. Kerwin later remarked that it must have irked them "to be outranked by a stupid staff officer, but they never showed it."[2]

The crew visited the CM plant at Rockwell in California for two or three weeks to assist in getting the CSM though the contractor test programme prior to delivery to NASA, before returning to Houston for the real test. After a dry run in early June 1968, the three men entered the CSM on 16 June wearing full Apollo pressure garments. Once they were sealed inside the spacecraft, the chamber closeout crew left, the access platform surrounding the spacecraft was removed and the chamber was sealed and its pressure lowered to simulate the vacuum of space. Initially, the cabin bore a sixty–forty per cent oxygen–nitrogen atmosphere mix, a change implemented as a result of the Apollo fire, while the crew breathed pure oxygen through their suit system. When the cabin pressure was switched to pure oxygen at five psi, the crew removed their flight suits and remained dressed in lightweight coveralls, as they would on a real mission.

For 177 hours, the three men sat in the CM performing their duties as they would on a real lunar mission, but keeping the centre couch folded for additional room. The test programme included new launch procedures that avoided the use of pure oxygen, evaluating the new material inside the spacecraft, and rotating the spacecraft to simulate the "barbeque mode" to evenly distribute temperatures across the surface of the spacecraft (as would happen during the trans-lunar and trans-Earth portions of the Apollo mission to the Moon). The crew communicated with "Mission Control" via the radio link and were monitored by a TV camera inside the spacecraft. By following the flight plan of a real mission to the Moon, the crew would evaluate the structural integrity of the hull and inner pressure vessel, and verify the heat shield structure under conditions of extreme cold, the performance of the environmental control systems (one of Kerwin's early technical assignments), the radiators' ability to dump excess heat generated by the fuel cells, and the habitability systems. Though not as busy as during a real space flight, the crew occupied their time between test objectives by making notes, reading books and playing cards, or keeping a diary of their activities and findings. They also ate space food rations to evaluate them for the Apollo missions. The spacecraft was exposed to 55 degrees Celsius for the first fifty-five hours of the test, −100 degrees Celsius for the next fifteen hours and ambient (room) temperature for the remaining 117 hours.

On 24 June, the three bearded astronauts emerged having overcome a number of problems, including an unexpected rise in chamber pressure and a leaking 1 G chemical toilet that spilt urine all over the floor of the CM. They were obviously unable to open the hatch in the vacuum inside the chamber, so they had to endure and get used to the awful smell, despite efforts to clear it up. Kerwin fondly recalled the three of them walking into Apollo Program Manager George Low's office with a weeks growth of beard and "dressed up like hippies" to report that the spacecraft would be ready to support a crew in space. On 2 July 1968, a fourteen-page crew report listed their findings and, with a few issues to resolve, cleared the way for Schirra and his crew to launch in Apollo 7 in October 1968. That maiden flight of a manned Apollo has often been described as 101 per cent successful, starting the breathtaking pace of missions that culminated in July 1969 with Apollo 11 achieving the first Moon landing. The success of Apollo 7 and the progress of the other missions were due in no small part to the work of Kerwin and his colleagues on these and other ground-based simulations.[2] "We felt like we were breaking some new ground and doing some necessary testing to enable the Wally Schirra flight of Apollo 7. So that was good. And then it was back to Skylab again."[1] Before putting all his attention into Skylab, Kerwin still had some work to do on Apollo lunar preparations in 1968.

Qualifying the Lunar Receiving Laboratory

Though Apollo was conceived in 1960 and the lunar mission commitment initiated in 1961, it was not until 1964 that NASA recognised the need for a suitable facility to process lunar samples, reducing any associated risks (from the rocks or the astronauts) of contamination. Initially, a very modest facility was considered, whose origins date back to 1959. In fact, it was nothing more than a clean room in which lunar material could be packed in a vacuum, leaving the more complex experiments and investigations on the samples to principal investigators, who would apply for a load of lunar material for their research.

This plan for a "simple" Lunar Sample Receiving Laboratory evolved over the next few years into something much more complex, where initial data would be collected and experiment work collated, while protecting it against potential "back contamination" from the human crew and human-built machinery and equipment sent to the Moon. Construction for the Lunar Receiving Laboratory (or LRL) as it was later called, began in the summer of 1966 and was completed in late September 1967. It was designated Building 37 and would eventually house a geology laboratory, a biological laboratory, the astronaut quarantine facility and associated support rooms and services. Built and certified in time for Apollo 11, it supported the preliminary sample examination from all the Apollo missions, although the astronaut quarantine rules were relaxed for Apollo 12 and terminated after Apollo 14 in 1971. The LRL had to pass a lengthy certification process before it was deemed safe and usable. As an early step in this certification process, which did not always run smoothly, the Director of MSC, Robert Gilruth, established an Operations Readiness Inspection Team in October 1968, which was headed by John Hodge. Joe Kerwin served on this team, representing the Astronaut Office, until the formal review of the facility on 3 February 1969. During that review, and a six-week practice session by the LRL staff that followed it, problems were discovered that required additional committees and work before it could be declared operational and ready to receive the first landing crew.[3]

AN EXPERIMENT PACKAGE FOR THE MOON

On 16 March 1966, the Bendix Systems Division of the Bendix Corporation won the contract to design, manufacture, test and provide operational support for an initial four Apollo Lunar Surface Experiments Packages (ALSEP). This was designed to be deployed by the landing crew and left on the Moon to continue scientific investigations of the lunar surface and environment long after the astronauts returned home. As the company developed these instruments, the astronauts continued working on the other aspects of Apollo. For the first year after coming back from pilot training, the Group 4 scientist-astronauts spent their time completing academic and survival training, as well as becoming familiar with the Apollo spacecraft and its associated hardware, procedures and support facilities. As everybody was already working long hours, the new astronauts took on assignments overseeing certain aspects of the engineering operations of Apollo that were instrumental to the success of the programme. One such assignment was the qualification of the ALSEP experiment package, working out procedures to allow an efficient and successful deployment sequence while wearing the full Apollo EMU pressure garment.

Jack Schmitt worked on AAP but also had major assignments on Apollo. He had a technical responsibility for the LM descent stage and for what could be stored on the stage for use on the surface (including the ALSEP). When Schmitt and Bill Anders visited Bendix for the Preliminary Design Review Stage, they were astounded that the company had designed the instruments to *maximise* the astronauts' involvement instead of minimising it. Anders promptly suggested his "Big Red Button" concept, in which an astronaut could kick a big red button and ALSEP would automatically deploy itself. They never quite managed to achieve that, but astronaut involvement from the early stage of development was critical to ensuring the package's smooth deployment operations when attempted on the lunar surface.

Schmitt also worked with Eugene Shoemaker on the development of the geological tool kit used by the astronauts to collect samples, and with Don Lind on developing the flight plan for lunar surface activities. They (along with members of the Flight Crew Support Division) developed a two-EVA timeline for the first landing mission that included a full ALSEP deployment and geological sample collection, conducting a full EVA dress rehearsal to iron out the procedures. It soon became evident that the constraints imposed for the first landing would not allow such a "heavy" payload (it was over 140 kg), so Schmitt coordinated the minimum requirements for an experiment pack on the first mission in order to save weight and, with Lind, worked out simple procedures to deploy them. The abbreviated science package became known as the Early Apollo Surface Experiment Package (EASEP) and was successfully deployed on the Apollo 11 mission by Armstrong and Aldrin thanks to the work carried out behind the scenes by Schmitt and Don Lind.

AFTER APOLLO?

By 1969, the Apollo programme was gaining pace, leading to the Apollo 11 landing in July. At that time, ten lunar landings were planned, with the faint possibility of further landings, but signs were already becoming apparent that Apollo would end with Apollo 20 some time in 1972 or 1974. NASA needed to finalise the programme that would follow Apollo, which would have to be more flexible in its mission design and more economical. The plan was to develop an Earth-to-orbit shuttle system capable of being reused after each mission, and which would deploy, repair or retrieve satellites, or re-supply an Earth-orbiting space station. This was part of a long-range plan to develop a lunar-orbital space station (both stations based on Saturn workshop designs) and a moonbase, extending the Apollo experience and developing techniques and hardware for manned flights to Mars in the 1980s.

This all looked good on paper and in the media, but the reality was far more restrictive. Once the Apollo goal had been achieved, public interest (and support) for the programme dwindled rapidly, as did political support at a time of social unrest across America, an escalating war in Southeast Asia and more pressing needs both nationally and internationally. Though the prospects for participating on a space flight looked good in 1968 and 1969, by 1970 and the near-tragedy of Apollo 13, it was a completely different story, with only four Apollo landing missions, one Skylab space station with three manned missions, and the possibility of a joint mission with the Russians on the manifest. It would be another two years before the Space Shuttle (the only element left of the grand plan) was authorised, so instead of dozens of flight seats over the next ten years, there were now only twenty-four over the next five years. There were just over forty active astronauts, so there were more astronauts than flight seats. Clearly some would retire after flying their next missions, and while others might stick it out, it would be a long wait with no guarantee of flying.

Apollo or Skylab

With Skylab coming closer to flying by 1972 or 1973, and the hope of flying a second station at least on paper, the scientist-astronauts remaining at NASA were divided into two groups in 1969 – those who mainly worked supporting Skylab and those who worked on supporting the final Apollo missions. The division was not written in stone, however, and several crossed over on more than one of their assignments as required by the CB. As Apollo wound down, most moved to support Skylab or Space Shuttle:

Apollo	Skylab
Schmitt	Kerwin
Allen	Garriott
England	Gibson
Chapman	Musgrave
Parker	Lenoir
Henize	Thornton
	Holmquest

Supporting the landings

Support crews were assigned to Apollo early in the programme to help lighten the training load on the prime and back-up crews due to the complexity of the lunar missions. These support crews would attend meetings, visit contractors, keep the crew up to date on a variety of issues related to their specific flight and fill in for some of the more mundane simulations and chores.

Ed Gibson was assigned to the support crew for Apollo 12 (with Jerry Carr and Paul Weitz), while Karl Henize, Robert Parker and Joe Allen formed the support crew for Apollo 15. Phil Chapman and Tony England were assigned to the support crew of Apollo 16 (with Hank Hartsfield) and Bob Parker served on the support crew for Apollo 17.

In the Capcom role, an astronaut served on a shift at Mission Control in Houston and communicated to the crew in space. He was essentially a ground-based crew representative and liaised with the Flight Control Director and team. The scientist-astronauts who filled some of these roles during Apollo 11–13 and 15–17 were:[4]

Apollo 11:	Garriott, Schmitt
Apollo 12:	Gibson
Apollo 13:	Kerwin
Apollo 15:	Allen, Henize, Parker, Schmitt
Apollo 16:	England
Apollo 17:	Parker, Allen

Owen Garriott found the Apollo 11 assignment useful in getting him closer to an actual flight crew and operational participation in a "real mission". That experience helped the Skylab astronauts, especially the rookie ones, enter the programme with some experience of real space flight operations and activities. Soon after Apollo 11 landed on the Moon and while the crew rested prior to performing their EVA, Garriott served as Capcom. There was not much to do, but he had to be ready for any potential emergency.

As part of their Apollo training, astronauts were given at least one day's helicopter flight instruction, which closely simulated some of the characteristics of flying a Lunar Module. During one flight, Ed Gibson made a landing attempt on a dry looking area of the Houston ship canal. But the surface gave way and he crashed the helicopter, although he walked away from it. He thought his career as an astronaut was over –

Astronauts and flight controllers crowd around a console at MCC-Houston during the April 1970 Apollo 13 crisis. Standing at the left rear is Tony England.

and his helicopter flying certainly was – but despite being subjected to a lot of heat over the incident, he remained in the programme. However, his chances of receiving an Apollo assignment that would take him to the Moon diminished, like the other scientists, as the programme wound down. He did receive an assignment to the Apollo 12 support crew, working on procedures and timelines for the lunar EVAs and assisting in the development of the checklists to help ease the mental burden on the crew. They tested placing the checklists on the LM and on equipment, but Bob Roberts of flight crew support suggested a cuff-mounted checklist that was finally approved. It is still in use on today's ISS EVAs.

Kerwin could never work out why he was selected to Capcom on Apollo 13, but it may have been because of his work with pressure suits and environmental systems and his involvement in the thermal vacuum chamber tests. He was also leading Capcom, the one working on launch, the EVAs and entry, while Brand and Lousma worked on other phases of the mission. Specialising in areas of the mission was easier to train for than trying to cover the whole mission. Kerwin was on duty when the centre engine on stage two failed, but procedures worked and the crew made it to orbit. He was in bed when the call came that Apollo 13 had a serious problem. He had supported the work from Mission Control during the crisis and had trained as entry Capcom, but lost the lunar surface Capcom duties when the landing was abandoned. His memories of the mission are of long hours with the headset on listening to white noise (static) and picking up crew comments with the high gain antenna powered down. He was also the

Kerwin (right) and Ken Mattingly monitor communications with Apollo 13 from MCC-Houston, April 1970.

Capcom who communicated instructions for the crew on how to construct the carbon dioxide scrubber.

Karl Henize learned of his assignment to Apollo 15 from Deke Slayton at the Apollo 12 post-flight party in late 1969. He had already realised that he would not get an Apollo flight, but was not overly concerned about missing out on the Moon. He always wanted a Skylab flight, which at that time was still a strong possibility.[5] Henize worked on CSM issues, spending months at Downey in California participating in the Rockwell checkout of the CM scheduled to fly on Apollo 15. As the test programme progressed, Henize was inside the CM following procedures and throwing switches. Being the crew representative, he was then able to keep Dave Scott and his crew (and the back-ups) appraised of the myriad of small problems he encountered. "I worked extensively on the emergency procedures to ensure they were up to date for our CM. I got roughly eighty hours of training in the CM simulator, both as general experience when Al (Worden) was having a session and also on my own to confirm procedures." Henize was in the water tank at Marshal at least twice with Worden as he trained for the film cassette recovery EVA, and carried out chamber tests of the CM at the Cape along with Jack Schmitt, spending several hours in pressure suits in the vacuum chamber to confirm that critical systems actually worked in a vacuum. "I was the 'back-up CM Pilot' (a misnomer) during launch, the astronaut in the six-person crew that got the crew tucked into the CM on launch morning."[5]

Mission scientist for the Moon

The role of the mission scientist for Apollo was to interface between the crew and the scientific community. While other members of the support crew handled the hardware and flight operations, the mission scientist would coordinate all the science on the missions, primarily the surface geological surveys and EVA traverses, and would accompany the landing crew during their field trips.

Apollo 11 had the basic objective of achieving the landing and getting the crew home and science was not a priority. Apollo 12 was the back-up to Apollo 11, should the earlier mission not achieve its goal, and though it included two EVAs and more science experiments in the surface package, its primary objective was to consolidate the achievement of reaching the lunar surface. Ed Gibson worked with the Apollo 12 crew on landing choreography, but he was not formally known as the mission scientist. That designation was first used by Tony England for Apollo 13, although he was not called upon for the science on that flight due to the explosion that aborted the lunar mission en route to the Moon. Phil Chapman filled the role for Apollo 14, Joe Allen for Apollo 15, Tony England (again) for Apollo 16 and Robert Parker for Apollo 17.

For Apollo 14, Phil Chapman's role was to bring together the science of the mission, representing the crew and liaising with the scientists and PIs on what the crew could or could not do on the Moon. He also worked as Capcom during the surface EVAs.

On Apollo 15, Joe Allen was the mission scientist on the first of the J-series, super scientific missions to the Moon. It was his responsibility to pull all the science together: "I became very involved in getting scientists to train the crew members in science-related things. I was a sort of integrator, a facilitator for all of those scientific meetings and so on, and it was a great assignment. I found myself being good at translation. In this case I was translating what the scientists wanted to the test pilot astronauts, and what the test pilot astronauts needed to the scientists, who didn't understand that. I think I was quite good at that, and I added a lot to that mission." The assignment culminated in being the Capcom during the mission. This Allen found bizarre, because he came from a frugal household in a small town in Indiana, and had never even made a phone call across the Atlantic due to the expense. When he lived in Europe, he wrote letters because it was cheaper. Allen actually talked to astronauts on the Moon before he ever made a phone call across the Atlantic.[6]

Allen helped to develop Dave Scott's "hammer and feather gravity in a vacuum" experiment for the flight, based on Chapman's initial idea. He thought it would be fun to try. "I was a teacher, and I love clever little things and got to think what is it about the Moon that might lead to something rather intriguing. One [thing] is gravity, the other is that there is no air." Allen also had the responsibility of naming small local craters around the landing and exploration area which would be used as markers during the EVAs. He drew upon his hobbies and interests for the names, as well as characters from fiction and music compositions.

Allen also worked on Apollo 17 and immediately afterwards became an "astronaut without portfolio." He worked with a physics group at MSC (with considerable funding) on cosmic ray physics for a while, but although the scientists he worked with

"Mr. Galileo was right." Dave Scott drops a geological hammer and a falcon feather to demonstrate objects falling in vacuum to TV audiences during Apollo 15's lunar surface activities in August 1971. This practical demonstration was suggested by Phil Chapman and supported by Joe Allen.

were his peers and very good research physicists of similar age, Allen realised that he was not as able a physicist as he had been several years earlier, because he had been absent on astronaut training and support roles, and his interests had also developed in other areas. He then became assigned to the Outlook for Space study group, bypassing Skylab and ASTP and moving to Washington before returning to work on Shuttle issues.[6]

For the first year after returning from flight school, Bob Parker had been involved in Skylab issues, attending meetings and committees and "basically, learning by doing," as he recalled in his 2002 oral history. When the Apollo 15 support crew was formed in late 1969, Allen was assigned to follow the science and Karl Henize the CM, so Parker took to learning all he could about the Lunar Module. This included trips to the Grumman plant at Bethpage, New York. Initially, this training encompassed the H-series LM with its limited landing resources, but when the missions changed, he had to learn all about the J-series improved LM with extended duration capability and expanded resources. He recalled that during this time, he went from "planning what was going to happen somewhere out there in the future, to working on something that was going to happen in 1971. We were working with real vehicles, on real procedures and with real crew. It was a whole different world from Apollo Applications. It was great."

During geological field trips for Apollo 15, Parker worked with the back-up crew (Dick Gordon and Jack Schmitt) and Allen with the prime crew (Scott and Irwin). By the time Parker worked as mission scientist for Apollo 17, there was really no need to train the back-up crew (John Young and Charlie Duke) since they had just come off Apollo 16 and were not going to recycle to a new lunar mission. Instead, he worked with Cernan and Schmitt during their Apollo 17 geology field trips.

A stroll or a ride?

Tony England had always wanted a mission to the Moon. As a geophysicist, and a co-investigator on one of the Apollo lunar experiments that flew on Apollo 17, he knew it was still "a bit of a reach," but not totally impossible until the programme was cut. He would have been assigned as back-up Lunar Module pilot for Apollo 16, but when 18, 19 and 20 were cancelled, NASA decided to save money in training new astronauts for dead-end assignments and began recycling flown Apollo veterans to fill in back-up roles. Ed Mitchell was told before Apollo 14 flew that he would have to back up Apollo 16 before he was allowed to retire, bumping England to the support crew as mission scientist. In one of his many technical assignments prior to the cancellations, England was tasked with evaluating the landing sites in the Marius Hills region (for either Apollo 16 or 17) where, due to the proximity of primary geological sampling sites, it would be better to have walking traverses as opposed to lunar rover traverses. There were other things to consider as well, such as extra science equipment, difficulties with the landing site consumables, or even a flying vehicle, but it was not so much a defined decision as a chance to allow others in the planning groups to understand the trade-offs in using the LRV. Due to the limitations imposed by being on the Moon for only three days, England's role on Apollo 13 and 16 (as with all the mission scientists) was to optimise the science, working out the best options to maximise the returns with what they had.[7]

LOST MISSIONS AND A CREW CHANGE

While Jack Schmitt had been nominated under Deke Slayton's crew rotation system to fly a landing mission as a Lunar Module pilot, savage budget cuts would impact heavily on the Apollo programme, causing a vast upheaval in the Astronaut Office with the loss of the final three missions, Apollos 18, 19 and 20. As a consequence, one of the crews that had lost their opportunity to fly to the Moon was the one that would have been officially assigned to Apollo 18 – Schmitt's crew.

Back in 1970, Alan Shepard had called Schmitt into his office, informing the surprised scientist that he would be joining the Apollo 15 back-up crew, with Dick Gordon as mission commander and Vance Brand as Command Module pilot. Their task was to serve as back-ups for the prime crew of Dave Scott, Jim Irwin and Al

Worden. "We pushed them as hard as we could in the simulations and everything," Schmitt recalled in 1999, "and Dick Gordon and I became a pretty good spacecraft crew, too. Dick would never let me try to land it from the right side; but other than that, we worked together very closely and I think probably flew those simulators as well as the prime crew." He then added, "Of course, that's what any astronaut will tell you."[8]

Under the rotation system then in place, Dick Gordon's back-up crew would have later stepped up to become the prime crew for Apollo 18, but as they carried out their back-up duties they began to hear disconcerting rumours about Apollo 19 and 20 facing the axe, with Apollo 18 also under scrutiny and in jeopardy.

An uncertain future

In October 1969, a revised launch schedule had set the Apollo 19 landing back to November 1972 and Apollo 20 to May 1973. Between the Apollo 18 mission (then planned for February 1972) and Apollo 19, three orbital Apollo Applications flights were also tentatively on the manifest – training flights preparatory to the establishment of America's first space station in 1975. At the time, Flight Operations Director Chris Kraft expressed some concern over the busy schedule. "It's going to be difficult to handle both Apollo and Apollo Applications from an operational point of view, as well as a people point of view in 1972."[10]

A further reason for delaying the later Apollo landings was the possibility that they would be made in remote lunar areas such as the large, bright-rayed crater Copernicus just south of Mare Imbrium, and the rim of Tycho, also a relatively young crater (probably of impact origin) in the southern lunar highlands, and rated by scientists as a top-priority landing site.

By 1970, however, NASA was struggling. President Richard Nixon's administration was pouring billions of dollars into the yawning maw of the Vietnam conflict, and "soft" programmes such as Apollo were rapidly plummeting down the preferred funding list. It was a quandary for NASA Administrator Tom Paine, who that month told a press conference about the need to align space operations with the Financial Year (FY) 1971 budget.

"We recognise the many important needs and urgent problems we face here on Earth," he stated. "America's space achievements in the 1960s have rightly raised hopes that this country and all mankind can do more to overcome pressing problems of society. The space programme should inspire bolder solutions and suggest new approaches. It has already provided many direct and indirect benefits and is creating new wealth and capabilities.

"However, we recognise that under current fiscal restraints, NASA must find new ways to stretch out current programs and reduce our present operational base. NASA will press forward in 1971 at a reduced level, but in the right direction, with the basic ingredients we need for major achievements in the 1970s and beyond. We will not dissipate the strong teams that sent men to explore the Moon and automated spacecraft to observe the planets."[9] In April that year, the life-or-death flight of Apollo 13 dramatically evinced the colossal risks inherent in manned space flight. But with the crew safely back on Earth, public apathy set in once again.

Juggling the rockets

Following the first three Apollo lunar missions, it soon became abundantly clear to everyone that NASA was going to have to pare Apollo 20 from the flight schedule due to budgetary woes. President Kennedy had been right; landing men on another world was a very costly exercise. Enough Saturn Vs had already been budgeted to see out the lunar landing programme with Apollo 20, but Apollo Applications also needed a Saturn V to launch the massive station into orbit. Shortly after the Apollo 11 launch, and before the landing, Tom Paine had signed off on the decision to switch Apollo Applications to a "dry lab" concept that would require a Saturn V launch. His announcement did not automatically cancel Apollo 20, as there was still a prevailing hope that there might yet be another Saturn V production batch.

However, it was becoming increasingly clear to Paine that Saturn V production would not recommence, and Saturn 1B production would also be halted. He ultimately had no other option; Apollo 20 was officially cancelled on 4 January 1970. The Saturn rocket now freed up would be allocated instead to the Earth-orbiting programme. Even Apollo 18 and 19 now began to look a little doubtful.

The heat was truly on for NASA, and one factor which weighed heavily was the mounting push to have a geologist on whichever might be the last lunar mission. Schmitt recalls being aware of "various press and informal activities, trying to convince NASA that I should fly on a mission, either 18 or, when that wasn't available, then 17."[8]

Then, in the summer of 1970, the axe dropped. On 2 September, Apollo 15 and 19 were officially cancelled. Apollo 16 retained its original lunar destination of Descartes Plain, but was renamed and became the 'new' Apollo 15, together with the original Apollo 15 crew. Apollo 17 now became Apollo 16, while Apollo 18 was renamed Apollo 17 – and was announced as the final manned lunar flight.

After the enforced gutting of the space agency's manned and unmanned programmes by the White House, Administrator Tom Paine resigned from NASA. Unfortunately for the astronauts, there were now too many candidates for too few lunar missions, and most would miss out. The scientific community was also up in arms. Why appoint a group of scientist-astronauts back in 1965, when not even one of them would fly to the Moon?

When Dick Gordon got wind that NASA was considering sending Schmitt to the Moon, he began jockeying to have his entire crew replace Cernan's. There were no holds barred in this last-gasp effort to secure the final lunar flight, and it was left to Chief Astronaut Deke Slayton to make a decision, albeit with extreme pressure being exerted by NASA HQ to get a scientist on the flight.[10]

A difficult decision is made

Meanwhile both crews trained hard, seeking to impress on everyone that they were the best available for the job. Although it did not affect their back-up training for Apollo 15, the three men knew they were now in an undeclared competition with Cernan's crew for the last lunar landing mission.

One option was to replace Cernan's crew entirely with Gordon's coherent crew. Another option being mooted was to replace Joe Engle with Schmitt on Cernan's crew. Cernan certainly didn't help matters for his crew when he crashed a helicopter into the Banana River during training, but he didn't suffer any serious injuries or after-effects, and seemed to escape any form of official censure, so the intense rivalry continued anew. "I didn't have any mixed feelings at all," Schmitt says of that anxious time. "We were competing for a slot. I understood how Joe would feel, but I wasn't about to give up the position because he might have felt disappointed."[8]

By August 1971 it was time to announce the crew for Apollo 17, and Cernan's crew got the nod – with one exception. Lunar module pilot Joe Engle had been dropped to make way for Jack Schmitt. It later transpired that Slayton had originally submitted the Cernan–Evans–Engle crew to NASA headquarters as his recommendation, but it had been rejected. He was told in no uncertain terms to get Schmitt on the Moon, and his hands were officially tied, but he still chose Cernan and Evans over Gordon and Brand.[10]

Joe Engle was visibly upset for some time, and not a little bitter about the decision, but he finally came to accept the fact that he had been dropped from Apollo 17 for political reasons. When asked by a reporter how he felt about missing out on the flight, he said the toughest thing of all was telling his kids that daddy wasn't going to the Moon.

A month after the successful Apollo 15 mission, Schmitt had been sitting in his Nassau Bay apartment when friends started phoning and calling in to tell him that he had been selected as the Lunar Module pilot for Apollo 17, replacing Engle. "I had already, some years before, made the decision that if I wasn't assigned to a crew (and I didn't think the probability was very high at that time) then, like everybody else in Apollo, I still thought it was well worth dedicating that part of my life to it without any question.

"And so, in a sense, getting assigned to a mission was frosting on the cake. I had already done more than I had ever expected to have a chance to do and having a chance to actually now fly a mission was something that was far more than I had originally expected. Always hoped for, but never expected."[8]

SELECTING THE LAST LANDING SITE

The last manned lunar mission to date, Apollo 17, left Earth on 7 December and landed near the south-eastern edge of Mare Serenitatis, in the valley of Taurus-Littrow – a veritable geologist's paradise. Because this was to be the final manned expedition in the Apollo series, serious consideration had been given to the landing site. All of the high priority regions were revisited and examined by scientists, geologists and other lunar experts, and many were excluded for either scientific or operational reasons. This left three potential sites under final consideration: the seventy-five-mile-diameter Alphonsus crater east of Mare Nubium, the central peaks of the slightly smaller Gassendi crater to the north of Mare Humorum, and the Taurus-Littrow valley.

In order of priority, there were several primary objectives to consider in assessing these three sites. The principal priority was to obtain clean samples of old highlands material in an area far removed from the Imbrium basin, where three landings had already taken place. The elimination process had ensured that all three of the final candidate sites were at least 500 miles from this area. The second goal was to investigate the possibility (and associated detritus) of lunar volcanic activity in the previous three billion years, which was considered crucial in understanding the thermal evolution of Earth's nearest neighbour. Another, lesser, consideration was that of orbital science. Planners wanted orbital ground tracks that had minimal overlap with those of the two preceding manned missions, so that the gathering of new information could be maximised.

A place called Taurus-Littrow

At the time of selecting a suitable landing site for Apollo 16, Alphonsus crater had been regarded as the primary objective for Apollo 17, as it had the dual advantages of allowing both the sampling of highland material in the crater wall, and that of volcanic material in several dark-halo craters on the floor of the crater. However, concerns were later expressed that highlands material in the crater wall might be substantially covered by later deposits, and would therefore be difficult for the crew to retrieve. Subsequently, Alphonsus lost its place at the top of the candidate list.

Gassendi was a potentially interesting site, which also offered the opportunity to sample ancient highland rocks in the crater's central peak. Unfortunately, unlike Alphonsus, it was not an area where relatively "young" volcanism had occurred, and the terrain beyond the planned landing site was exceptionally rugged, which might create immense problems for the crew in their Lunar Rover, and even preclude them from reaching Gassendi's central peaks.

On the other hand, Taurus-Littrow, which takes its name from the Taurus mountain and Littrow crater, was a narrow valley surrounded by three steep, mile-high massifs, part of similar mountains that make up the rim of the Serenitatis impact basin. The site was thought to contain some comparatively recent examples of volcanic vents, or cylinder cones, following some activity within the Moon. Mission planners could see no problems in guiding the Lunar Module to a safe landing in this geographically rich area and the crew, using their Lunar Rover, could then obtain samples of ancient highland material from both the north and south walls of the valley. Planners knew there had been a landslide on the southern face of the valley, which might also provide material of interest to the crew and geologists back on Earth. The Apollo 15 crew had reported seeing several craters in this region, surrounded by deposits of darker material – the so-called dark-halo craters. One such crater of

Schmitt and Cernan, the last two Apollo moonwalkers, review traverse maps while training near Boulder City, Nevada.

interest, known by the less-than-ostentatious name of Shorty, was located near the proposed landing site.

With these and many other considerations taken into account, Taurus-Littrow became the landing site of choice for Apollo 17 at the beginning of February 1972.

A crew is formed

Mission commander Gene Cernan wanted this flight to be the best in the series, and his Command Module pilot, Ron Evans, worked extremely hard at perfecting routines for orbital science. However, this was a crew with a difference; the Lunar Module pilot was a professional scientist, the only one to travel to the Moon in the twentieth century. Political pressure to include a geologist on the final lunar mission had been intense, and Jack Schmitt knew a lot hinged on his performance. Placing him on the crew meant that Joe Engle was bumped from the last scheduled Apollo mission, and the former X-15 pilot was a deeply disappointed man. Attempts were made to reinstate Engle, but ultimately a decision was laid at Cernan's feet. He could either

Schmitt and Cernan ride the Rover test vehicle during geological and simulation training in the Pancake Range of south-central Nevada.

accept Schmitt into the crew, or all three members of the back-up crew, including Schmitt, would take their place. Cernan reluctantly gave in at this point, but to his credit and that of Evans, they had shown their loyalty to a crewmember, and when this hadn't worked out they showed no visible rancour to the scientist-astronaut. It was not his fault, they knew, and so they knuckled down to their training.

On the evening of 7 December 1972, NASA officials cleared the way for Apollo 17 to become the first manned night time launch by the United States. This was despite a cold front moving rapidly eastward that could have had an influence on the launch time, then set for 9:53 p.m. EST. However, launch chief Walter Kapryan was openly confident, and told reporters: "The probability is that we will have good weather at the time of the launch. We have had fewer problems than we've ever had at this point in the past."

The night before, on Cocoa Beach, many revellers, including astronauts, industrialists and socialites, had gathered and were drinking toasts to Apollo's last stand, wearing buttons that read, in hope, "Only the beginning: Apollo 17." But for most of the area's residents and workers, this was more like a wake than a celebration, and more an ending than a beginning. The budget cuts that caused the cancellation of the final three Apollo missions had consequently trimmed the Apollo workforce by half, with many more jobs to go, and the knock-on effects of the dwindling Cape population had also hit local business people in their wallets. The last Apollo mission was certainly cause for mixed emotions as the time drew near.

Prior to the final Apollo lunar mission, Schmitt briefs members of the press on the Taurus-Littrow landing site.

One moving photo from the launch site at Cape Canaveral shows two elderly guests seated in the VIP stands. The older of the two is neatly dressed in a white cowboy Stetson and bow tie, and he is sporting an Apollo 17 mission patch on his suit jacket. He was African American Charlie Smith, said to have been a former slave, who claimed to have been born thirteen years before the assassination of Abraham Lincoln, which would have made him 130 years old. Sitting beside him with a thick cigar jammed horizontally into his clay pipe like a miniature Saturn rocket is his 70-year-old son, Chester. It wasn't until after Charlie Smith's death in 1979 that a meticulous check of the records revealed the fact that he was actually 98 at the time of the launch of Apollo 17, but it made for a nice story at the time.

SETTING OFF FOR THE FINAL TIME

At 12:33 a.m., two hours and forty minutes behind schedule, Apollo 17 finally soared aloft in a radiant display of sight and sound, the combined plume of fire from the Saturn V's five F1 engines turning night into day around Cape Kennedy. "You're

The view from the launch tower as Apollo 17 lifts off for the Moon.

right down the 'pike', 17," Capcom Gordon Fullerton informed the crew. The astronauts had been strapped into their Command Module America for five hours, not knowing if they would fly that day due to a succession of technical delays. The first launch attempt had ended dramatically with only thirty seconds remaining in the countdown, when the terminal countdown sequencer detected that the liquid oxygen tank on the Saturn's third stage had not automatically pressurised. At first, it was feared that the crew might have to evacuate the Command Module and even use the emergency slide wires to get away from a potentially dangerous situation, but the cause was quickly discovered and the crew notified to stay put. The tank was eventually pressurised manually, and the countdown resumed at T-22 minutes.

Now, as the three exultant astronauts finally headed into space, people for hundreds of miles around cheered the mighty Saturn V as it blazed a brilliant path into the dark skies, lighting up the Florida east coast.

Because of the delay, controllers decided to burn the third stage of the Saturn V six seconds longer than planned when setting Apollo 17 on a course for the Moon after two Earth orbits. Later, following the firing of their third-stage engine, Command Module America was smoothly docked with Lunar Module Challenger in preparation for NASA's final and most difficult chapter in the exploration of the Moon. Four-and-a-half hours after lift-off, the crew was 15,000 miles from Earth and travelling at 10,873 miles per hour on their eighty-six-hour coast to the Moon. Fourteen hours after they had been strapped into America back at the Cape, the crew finally settled down to grab a few hours' sleep.

The Moon looms larger

The journey to Earth's celestial satellite was punctuated by shrill, faulty master alarm signals that occasionally rang out, waking and annoying the crew. That aside, the new trajectory provided the astronauts with a superb view of their destination. As the Moon drew ever closer, it was proving an unforgettable experience for geologist Schmitt.

"For three days, the fascinating hanging scenes of an ever-smaller planet dominated our thoughts until, outside, a dark looming presence increasingly made itself felt as much as seen. The disk of the black, lightless Moon grew in aspect, blocking more and more of the universal star field, and then the mountains of the Moon crossed the Earth itself."[8]

As they circled the Moon prior to undocking from the Lunar Module, the crew had a little sightseeing time to themselves. Each time they flew around to the far side of the Moon they lost radio contact with Earth, which meant that, apart from a few small procedures, they were able to look down on the lunar surface below and take photographs for later examination. On the Moon's near side, Schmitt recalls being impressed by how much light the Earth cast on the Moon. Features were very clearly discerned in the spectral blue light, "and really quite spectacular." Then he saw something that really took his breath away.

"At one point, I was looking down at the surface – it would have been way west of Copernicus, and probably even getting close to the big basin called Orientale – and

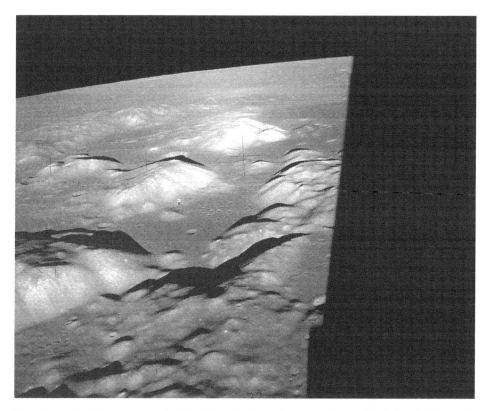

The Apollo 17 Command/Service Module, photographed from Lunar Module Challenger, is just visible against the Taurus-Littrow landing site.

I saw a little tiny pinprick of light on the surface. It was almost certainly a meteor hitting the surface of the Moon, and they will give off a little bit of visible light. So I had a chance to see what was effectively a shooting star hit the Moon."[11]

On their tenth lunar revolution, Cernan and Schmitt powered up Challenger and checked that all the systems were functioning properly. On the following revolution they deployed the landing gear, checked the guidance platform, and once both spacecraft were fully sealed, separated. "Okay, Houston, we're floating free out here. The Challenger looked real pretty," Evans reported.

Prior to beginning their thirteenth tandem orbit, Evans fired America's SPS for a total of four seconds in order to circularise his orbit. Cernan and Schmitt both reported that it had been a good burn. Less than five minutes later, Cernan fired the DPS engine on Challenger, which would bring the Lunar Module ever closer to the Moon's surface. Schmitt was already pondering the tasks ahead.

"We would enter the valley from the east with the sun behind and below us for good shadow definition of rocks and craters. Like many other major lunar valleys, Taurus-Littrow extends radially from a large circular basin; in this case the 300-mile diameter basin called Mare Serenitatis. This huge basin formed about 3.9 billion years ago as a result of a large comet or asteroid impact. Subsequently, the basin and valleys partially filled with dark volcanic lavas.

"Anticipating exciting events to come, and after several orbits and some preliminary manoeuvres, we ignited the descent rocket engine for the twelve-minute braking burn that placed us on the valley floor. Initially, we flew with our backs to the Moon, looking out into space. Beginning at about 5,300 feet per second and ten nautical miles altitude, the descent engine slowed our velocity and bent our flight path toward an intersection with the lunar surface. When the landing radar measured 6,000 feet in altitude, we pitched forward far enough that Gene could see the landing area. We could both see the nearby north and south sides of the valley. I took one quick glance at the sights out the window and went back to work, giving Gene the velocity and altitude read-outs he needed to adjust the Challenger's flight toward a touchdown away from large, potentially dangerous boulders and craters, but as close to the planned spot as possible."[8]

A "go" for landing

Five thousand feet above the Moon's dusty surface, Cernan could see that Challenger was actually dipping below the level of the massif mountain ranges to the right and left of the valley. Houston chipped in with confirmation to proceed: "Challenger, you're go for landing!" At one hundred feet altitude, Cernan took over manual control of Challenger and guided the craft down. As Schmitt recalls, the exhaust blast from the descent engine soon began to flay at the lunar surface below.

"Dust began to stream away from the rocket as we went through about 100 feet above the surface. After Gene slowed a briefly-too-rapid descent, the six-foot-long probes beneath the landing pads touched, a blue light flashed on in the cabin, we shut off the rocket engine, and fell the last few feet to the surface. Slightly less than a second later, news of our landing arrived at Earth."[8]

Almost three and a half years after Armstrong and Aldrin had set Eagle down on the Sea of Tranquillity, Cernan announced their safe arrival at Taurus-Littrow, almost echoing Armstrong's unforgettable words: "Okay, Houston, the Challenger has landed." The two pilots were busy for some time as they shut down and secured Challenger's systems, and exchanged technical data with Capcom Gordon Fullerton. Then, for Schmitt, it was time for a quick peep out of the windows, and there was cause for excitement. They were about 330 feet from a little crater they called Poppy, indicating their touchdown was right on target.

"It hit me what a magnificent place we had landed in. My first view out of the right-hand side, looking northwest across the valley at mountains 6,000 feet high, encompassed only part of a truly breathtaking vista and geologist's paradise. Indeed, one of the most majestic panoramas within the view and experience of humankind confines the Valley of Taurus-Littrow. The roll of dark hills across the valley floor blends with bright slopes that sweep evenly upwards, tracked like snow, to the rocky tops of the massifs 6,000 feet above. The valley does not have the jagged, youthful exuberance of the Himalayas, the layered canyons of the Colorado, the glacially symmetrical fjords of the north countries, or even the now-so-intriguing rifts of

Mars. Rather, it has a subdued and ancient majesty. And we were there and part of it."[8]

A GEOLOGIST WALKS ON THE MOON

Just four hours after touchdown, the crew began the first of their three planned lunar EVAs. Wearing his bulky spacesuit and life support system backpack, Cernan had eased himself backward through Challenger's tiny hatch, descended the fragile ladder and then stood on the footpad for a few moments of contemplation. As he placed his left foot firmly into the thin lunar crust he said, "As I step off at the surface of Taurus-Littrow, I'd like to dedicate the first steps of Apollo 17 to all those who made it possible."

Schmitt would follow a few minutes later, but as he set foot on the Moon his left boot slipped on the side of a rock encrusted with small beads of glass and he began to topple. He grasped the ladder tightly, tried again, and was soon standing safely beside Cernan. The two astronauts spent the next few moments looking around, describing the rugged landing site and expressing surprise at how bright it was in the sun. The scientist in Schmitt could wait a few more moments; what he saw was breathtaking.

"Only later, when I could walk a few tens of metres from the Challenger, did the full and still unexpected impact of the awe-inspiring setting hit me: a brilliant sun, brighter than any desert sun, fully illuminated valley walls outlined against a blacker-than-black sky, with our beautiful, blue-and-white marbled Earth, about a two-thirds Earth in terms of its phase, hanging over the south-western mountains."[8]

Preparing for the task

The two astronauts would eventually spend a total of seventy-five hours on the Moon, and were engaged in lunar extravehicular activity (LEVA) outside Challenger for twenty-two of those hours. At first, Cernan expressed surprise at the glittering appearance of the lunar soil, saying it looked like millions of tiny diamonds, but Schmitt was a little more pragmatic, and stated that it was in fact microscopic beads of glass reflecting sunlight. "[The soil looks] like a vesicular, very light-coloured porphyry of some kind. It's about ten or fifteen per cent vesicles [porous volcanic rock]."

The crew's first scheduled task was to unload their Lunar Rover, which was stowed outside the Lunar Module, "like a piano tied to a truck," as Cernan later said when describing the job ahead of them. Using a series of lanyards, cables and hinges, they succeeded in lowering the ungainly machine to the Moon's surface and carefully assembled it, ready for the first of their three LEVAs. Cernan bounded into the driver's seat, switched on the batteries, and noted to his relief that all seemed to be in order. Electric motors on each wheel gave them driving power, so Cernan took the vehicle on a short test run, and reported that it worked perfectly in forward and reverse.

Once the rover was loaded up for their first expedition, Cernan teased his companion a little, feeling he was too busy being a geologist to truly appreciate where they were. "Hey Jack, just stop. You owe yourself thirty seconds to look up over the South Massif at the Earth."

"What? The Earth?" Schmitt replied. "Just look up there," Cernan insisted.

"You seen one Earth, you've seen them all," Schmitt replied, somewhat flippantly. Chided for this seemingly blasé reaction in Cernan's later autobiography, Schmitt revealed there was actually a reason behind his comment. His commander may not have thought he was looking up, but "he didn't know I was!

"I always had a plan to do a lot of Earth observation on the way to the Moon, and that keeps you occupied, until you fully adapt. And so I had been looking at the Earth; I filled pages of air-to-ground transcript. Gene and Ron may not have been aware that I was doing as much as I was, and so when Gene made that comment on the surface of the Moon [and] I said something like 'you've seen one Earth you've seen them all' – [I was] being facetious about it. But it was mainly because I had already gotten through that particular thing and, being a geologist ... you are used to thinking of the Earth as a body in space, as a whole body. If you haven't had that kind of intellectual experience before, it's a surprise to see it."[12]

The proudest moment

Prior to leaving their Challenger base, the two astronauts hammered a thin metal shaft into the ground and inserted an American flag. This flag had actually been carried to the Moon and back by the crew of Apollo 11, and it had been hanging in Mission Control in Houston since that time. Now it was back on the Moon forever. "This is one of the proudest moments of my life," Cernan remarked, as he and Schmitt unfurled the flag.

The two men then drove to an area where they were scheduled to deploy and activate a science station known as ALSEP (Apollo Lunar Surface Experiment Package), powered by a small nuclear reactor. This station included experiments to investigate gravity wave-induced oscillations, active and passive seismic activity, the composition of the thin lunar atmosphere, micrometeorite impact rates, heat flow from the lunar interior, and neutron and cosmic ray fluxes. As they were unloading their gear, the handle of a rock hammer in Cernan's suit pocket snagged on one of the rover's rear fenders, dragging it off. He quickly repaired it with duct tape and then began the laborious task of boring through rock with a special cordless drill in order to set up the ALSEP. He had to drill two 2.5 m holes, about 8 m apart, for the heat-flow probe, and then a slightly deeper one in which he would insert a probe for measuring the rate at which cosmic rays produce neutrons at various depths in the regolith. It also provided a core sample for later analysis back on Earth. Schmitt, meanwhile, was fully engaged in erecting the gravity-wave detector. It was arduous work for both men, and when it was completed they were forty minutes behind their LEVA schedule.

Re-boarding the Lunar Rover, the two men heard that, because of the delay, plans for a mile-and-a-half trip to another crater had been curtailed, and they were directed to a nearby boulder field near the crater Stenno. Following their examination of the area, and gathering up some rock samples, they pressed on, but as they passed over a flat area the broken fender fell off, and the two astronauts were showered with a thick plume of dark soil. Apart from getting covered with grime, the astronauts knew they had to make repairs to the rover, as the dirt would make its way into some of the sensitive instruments they carried and cause them to fail. Filthy and exhausted, they arrived back at Challenger and cleaned each other off with a large brush they'd hung by the ladder for this very purpose.

The following day, Cernan and Schmitt rigged up a makeshift fender for the rover, manufactured from their stiff geologic maps. The repair job was effective, and they set off to do some more work on their second LEVA. This time the rover took them nearly five miles from the Challenger to the South Massif, where they sampled boulders at the base of the south wall of the valley and explored an avalanche deposit that had moved out away from the Massif. Cernan used a rock hammer and a set of tongs for most of his sampling, while Schmitt preferred a long-handled scoop with an adjustable head. They both had chest-mounted, 60 mm electric drive Hasselblad cameras to record their findings and any items or scenery of interest. Cernan's camera was loaded with colour film, while Schmitt's used black-and-white film for subsequent photometric measurements of materials in place. As Schmitt recalls, they worked well as a team.

"At most sampling sites we generally worked together, using a specific sampling and documentation routine that provided significantly greater efficiency than working alone. While Gene dusted off the equipment on the rover, I would look over the sampling area, giving a general description of the geology and what we would try to do. Then, I would take a down-sun photo while Gene took a cross-sun stereo pair of photos of the area to be sampled. One of us would pick up rock or soil samples while the other held a numbered Teflon sample bag open to receive the sample. After stowing the bagged samples in larger bags mounted on our backpacks, one of us would take both a post-sampling photo to show which samples had been collected, and a circular panorama that included the site. A running dialogue of each step in the operation as well as other geological observations accompanied this process. The rover's colour television camera, operated remotely from Earth, followed most of our activities, providing both additional scientific documentation and many humorous clips of our pratfalls to delight future audiences."[8]

Finding orange soil

As the two weary astronauts returned from the South Massif, Schmitt was preparing to take some more photographs. All of a sudden he began to cry out with surprised delight. He was about to announce finding something that certainly seemed to justify sending a scientist to the Moon.

"Hey!" he exclaimed, as he kicked over an interesting patch of dirt. "It's orange. I found orange soil!" Everyone back at Mission Control sat up with the excitement evident in Schmitt's voice.

Cernan moved over to join Schmitt. "Well, don't move till I see it," he said, and Schmitt repeated, "It's all over orange!" Once again Cernan told Schmitt to wait until he could also see the orange soil. With Cernan by his side, Schmitt remarked, "I stirred

A geologist walks on the Moon 227

It was in this area at Station 4 (Shorty Crater) that Schmitt discovered traces of orange soil during the second lunar EVA. The tripod-like object is a gnomon and photometric chart assembly used as a photographic reference to establish local vertical Sun angle, scale and lunar colour. Although it cannot be seen in this monochrome photo, the orange soil is situated midway between the gnomon and the small boulder at the bottom right.

it up with my feet!" By scuffing around in the surface with his boot, Schmitt and Cernan could see soil ranging in colour from orange to almost a ruby red. Cernan felt the excitement of discovery as well. "Hey, it is! I can see it from here." "It's orange!" Schmitt said once again, almost in disbelief. Cernan began fumbling with the gold visor on his helmet. "Wait a minute, let me put my visor up." There was a slight pause, and then he affirmed the discovery. "It's still orange."

228 A Geologist on the Moon

Schmitt collects a soil sample on the south side of the rim boulders at Station 5.

Finding this orange soil caused tremendous excitement at the time, and Schmitt was hoping it might provide proof of relatively recent volcanic activity in the area. To their moderate disappointment, the soil would later prove to be a series of microscopic glass spheres and fragments tinted by titanium and intermixed with black or black-speckled grains. It was about the same age as other rock material in the vicinity, and very similar to samples taken from the Sea of Tranquillity, several hundred miles to the south-west, by the crew of Apollo 11. Unlike the Apollo 11 samples, these spheres were found to be curiously rich in zinc but, as Schmitt said in 1996: "This chemically unusual material from 3.5 billion-year-old volcanic fire fountains has given new insights into the origin of the Moon and the nature of its interior."

A geologist walks on the Moon 229

A panoramic shot at Station 7 shows Schmitt next to the Lunar Rover.

The crew's third LEVA provided them with an opportunity to study the large boulders that had rolled and bounced down the north wall of the valley. According to Schmitt, they "hoped to learn more about what happens when large objects from space hit, break, and partially melt planetary crusts. During the detailed examination of one large boulder, the unexpected discovery of a subtle contact between two types of impact-generated debris units, one intrusive into the other, again proved the worth of the trained human eye in exploration."

In their three expeditions into the valley, the two astronauts drove their Lunar Rover about nineteen miles. For around twenty-two hours of the seventy-five hours they spent on the Moon, the two men were engaged in LEVA, during which they collected a record cargo of more than 100 kg of priceless lunar rock and soil samples, and took over 2,400 photographs.

LAST STEPS ON THE LUNAR SURFACE

The final departure of humans from the Moon was marked by a meditative ceremony, acknowledging the end to the greatest scientific undertaking and exploration ever attempted by people from Earth. In a melancholy statement read to the astronauts from Mission Control, President Nixon paid homage to the significance and majesty of the Apollo programme. He was sadly prophetic when he stated that it might be the last time any humans walked on the Moon in the twentieth century, but he did say that the exploration of space would continue. "Those are beautiful words by a great American president," Cernan responded. "We are very honoured to serve our country in a way we believe in."

With the brief ceremony at an end, Cernan and Schmitt began transferring their final haul of lunar rocks and soil into Challenger. Back on Earth, the most eagerly anticipated trophy for waiting scientists and geologists was the orange soil that Schmitt had found almost by accident during their second excursion aboard the Lunar Rover. For the final time, the two men dusted each other off and then, as Schmitt prepared to ascend the nine steps back into Challenger for the last time, he reflected on the accomplishments that had brought him to this amazing place. "This valley of history has seen mankind complete its first evolutionary steps into the universe. I can think of no more significant contribution that Apollo has made to history."

Once inside Challenger, Schmitt began to clean up his suit and the Lunar Module, while Cernan drove the rover about a mile away and left it in such a position that the television camera could record their lift-off the next day. He hopped and skipped his way back to Challenger, took one last, long look around, and then prepared to climb the ladder. He had made some farewell notes on the cuff of his sleeve, but he felt he had some more profound things to say, and spoke spontaneously, as the last man to leave his footprints in on the Moon.

"As we leave the Moon and Taurus, we leave as we came, and God willing, as we shall return, with peace and hope for all mankind." As he lifted his boot from the lunar soil, he added, "As I take these last steps from the surface for some time to come, I'd just like to record that America's challenge of today has forged man's destiny of tomorrow. Godspeed the crew of Apollo 17."

Heading home

Once Cernan was back on board, the two men shed their space suits and did some more domestic chores, cleaning up and jettisoning some valuable gear – now just unwanted dead weight – out of Challenger onto the Moon. This included several tools, cameras and their backpacks. With an additional 100 kg of lunar samples on board, every excess kilogramme they could shed was critical. As Cernan later remarked, "We threw out nearly everything that wasn't nailed down." The two exhausted astronauts then settled down for a last sleep on the Moon.

The following day, 14 December, Cernan and Schmitt donned their suits once again after running through a lengthy checklist. Then, when all was in readiness,

In what is probably the last photograph ever taken of an Apollo astronaut on the Moon, Schmitt stands next to the American flag.

Cernan flicked a yellow ignition switch and the ascent engine fired. While his last words on leaving the last footprints on the Moon the previous day may have been profound and inspirational, the last words spoken by Cernan as Challenger lifted of were hardly likely to be engraved on any plaques: "Okay, now let's get off." Something of a modern-day myth has evolved which suggests that Cernan's last words on the lunar surface were actually the more profane 'Let's get this mother out of here!' However, while this fanciful misinterpretation has never been supported by the actual voice transmission tapes and subsequent transcripts, it sadly seems destined for immortality in space quotes and trivia books. In fact, the very last human words spoken from the surface of the Moon were Jack Schmitt's, as he counted down. "Three … two … one … ignition!"

The American flag and lunar surface photographed through an LM window prior to lift-off.

A mile away, the television camera mounted on the rover recorded the lift-off for posterity. Although controlled from the ground, it required precise timing and tracking, as the commands had to be relayed six seconds ahead of the event, but it went exactly as planned, and the camera gave the world the best lunar lift-off captured on film. Although poor in resolution, it shows a huge, sudden cloud of dust and shredded gold foil from the exterior of Challenger billowing out as the upper stage of the lander rapidly soared upward, the camera slowly panning up in a predetermined arc. "We're on our way, Houston!" Cernan confirmed.

Schmitt later noted, "We barely noticed the half-G acceleration of lift-off and the slight oscillation during ascent, partly because at the instant of ignition, the uplink communications turned into raw static. Later, we found out that a mix-up on a transfer between Earth transmitting stations caused the problem. As we flew back into orbit on a direct rendezvous path towards Ron in the America, I spent the first few minutes trying to restore Challenger's communications while Gene monitored the guidance system and yelled at me to 'get the comm. back!' It turns out that nothing I could do would have helped. Mission Control finally restored communications on their own."

Remaining behind on the floor of the valley of Taurus-Littrow as an eternal monument to the Apollo programme, the dust-covered, four-legged descent section of Challenger will stand in stillness, waiting for the next human voyagers. Affixed to its side is a small, black-rimmed plaque that reads, "Here man completed his first exploration of the Moon, December 1972, A.D. May the spirit of peace in which we came be reflected in the lives of all mankind."

Two hours after blasting off from the Moon, Challenger and America successfully docked while circling seventy-two miles above its ancient craters. Once they were linked, Cernan and Schmitt floated through a narrow transfer tunnel into America, bringing with them their precious cargo of carefully catalogued rocks and soil. Later they would jettison Challenger, which silently moved away for its last mission, an eventual crash-landing in the Valley of Taurus-Littrow that would register on the seismometer they had set up. As Schmitt would later say, "Having lived for many years with Challenger (or LM-12 as it is designated in the books) through design reviews, crew function checks, vacuum tests, flight configuration checks, and final tests, I had become attached to this inanimate object as if it were a member of the family. [It was] Tough to let her go."

While in lunar orbit over the next three days, the crew continued their work programme by searching for any traces of gas, water and precious metals in the lifeless world below them. One instrument they would use was a radar sounder, which could probe up to a mile under the Moon's crust and detect hidden ores and water-bearing strata. An ultraviolet spectrometer would also look for rare belches of gas emanating from three large craters near the Taurus-Littrow Valley, which they had explored on foot and by Lunar Rover.

Several hours later, right on schedule, the astronauts fired America's SPS engine, sending the spacecraft on a return trajectory to Earth. As the Moon began to slowly recede, Cernan was once again moved to words. "We're looking back at some place I think we will use as a stepping stone to go beyond some day. We will all see it, in our lifetime, not just as a nation, but as a world. This was the beginning. This is a beginning." More than thirty years later, Cernan can only shake his head and rue the fact that this great initiative in space has never been duplicated, and will not be for many years, even decades, to come.

Deep-space EVA

When America was 184,000 miles from Earth, it was time for Command Module pilot Ron Evans to take his turn in the limelight. As Apollo 17 sped back to Earth, he made a one-hour spacewalk to recover three reels of film from the rear of the spacecraft. Linked to a long, white tether, which carried an oxygen line and communication cables, Evans made his way to the rear of America, crawling hand-over-hand for 4.5 m

Command Module pilot Ron Evans is photographed retrieving film canisters during his trans-Earth-coast EVA to the rear of the spacecraft.

along the side of the spacecraft to retrieve the film, using a series of handrails. Schmitt floated in the open hatch, keeping Evans's umbilical line from tangling. After the EVA, Schmitt guided Evans back into the Command Module feet first. By the time the hatch was sealed and the cabin re-pressurised nine minutes later, America had moved 2,000 miles closer to Earth. As Capcom Bob Overmyer had reminded the crew on waking them that day, Evans's spacewalk had taken place on the sixty-ninth anniversary of the Wright brothers' first powered flight.

America was on such a true return trajectory that when it finally crossed into the dominant influence of Earth's gravity, a midcourse correction manoeuvre was cancelled. After nearly two weeks, the final Apollo mission was nearing its end, and Schmitt says the tempo began to pick up.

An exhausted but exhilarated Gene Cernan and Jack Schmitt were photographed by Ron Evans during the lunar return journey.

"As entry into the atmosphere approached, our speed increased to about 35,000 miles per hour. We first separated from America's Service Module, relying on the Command Module's batteries to see us through to splashdown. Then, with the blunt end forward, we pointed the lift vector of the conical Command Module toward Earth ... This ensured capture by the atmosphere and avoided any danger of skipping out into space. As a consequence of this manoeuvre, we experienced a peak deceleration of seven G.

"Once captured, the spacecraft guidance computer held about four G for several minutes ... the computer rolled its lift vector back and forth, correcting our flight path toward where it believed the recovery aircraft carrier waited in the South Pacific. Actually, the computers demonstrated such accurate knowledge of the planned point of splashdown that the Navy decided to cruise several miles away in order to avoid any possibility of a hard and embarrassing landing on the carrier's deck!"[8]

Journey's end

At 25,000 feet above the ocean on 19 December, the drogue chutes deployed, slowing the fall of Apollo 17. At 10,000 feet, the three main chutes deployed and the spacecraft began to drift slowly down to the waiting ocean.

The heat-seared Command Module America returned to a successful pinpoint splashdown in the gentle waters of the South Pacific, 400 miles southeast of Samoa.

Photographed from a helicopter hovering overhead, America splashes down in the South Pacific, ending the Apollo lunar programme.

The splashdown was captured by television cameras aboard the recovery ship, the USS *Ticonderoga*, positioned just 3.1 miles away. The crew of Apollo 17 was in good health and spirits, and an exultant Cernan soon reported: "The crew is *go*! This is America, and the crew is doing fine. We've all got our sea legs." It was a flawless ending to an incredible mission, and to the Apollo lunar programme.

Fifty-three minutes after splashing down, Cernan, Schmitt and Evans had assembled on the deck of the *Ticonderoga* for a brief welcome prior to beginning extensive medical checks and debriefings. Cernan thanked the crew of the ship and recovery helicopters, and in a short speech said, "We think we accomplished something and, by golly, we're proud of it. Nothing is impossible in this world when dedicated people are involved. You must grow or die, whether that be an ideal of a man, a flower, or a country. I thank God our country has chosen to grow."

Following the safe recovery of the crew, President Richard Nixon gave an impressive speech in which he spoke of future endeavours in space exploration, even though he had been firmly at the helm in curtailing Apollo and other manned space programmes through savage budget cuts. However, his words are worth recording as an epilogue to the last human flight to the Moon in the twentieth century:

"The safe return of the [Apollo 17 crew] marks the end of one of the most significant chapters in the history of human endeavour. Since the beginning of Apollo, nine manned missions have been made to the Moon ... We have barely begun to evaluate the vast treasure store of extraterrestrial data and material from these voyages, but we have already learned much, and we know that we are probing our very origins. We are taking another long step in man's ancient search for his own beginnings, pressing beyond knowledge of the means of human existence to find, perhaps, the meaning of human existence.

"We ... pay homage to those ... whose hopes, skill and courage enabled the first man to reach the Moon ... But the more we look back, the more we are reminded that our thrust has been forward, and that our place is among the heavens where our dreams precede us, and where, in time, we shall surely follow."

As Lunar Module pilot on Apollo 17, Jack Schmitt logged 301 hours and 51 minutes in space; 22 hours and 4 minutes of which was spent outside Challenger on the lunar surface. More than three decades after the ascent stage of Challenger lifted off from the lunar surface, taking with it the last two men to have walked on another world, Schmitt is still aggrieved that there are no absolute commitments for a return to the Moon, although President George W. Bush's Vision for Space Exploration gives hope of a renewed impetus for human exploration of the Moon and Mars. Initial congressional funding has been approved for a programme aimed at sending the first manned exploration vehicles to the lunar surface in 2015, but many remain cynical. As Schmitt states, it is a vision that has stalled so many times in recent decades, mostly as a result of financial meltdowns.

"I didn't think it would ever be this long. I can understand why it has taken so long, and I can understand why it is probably going to be another ten years, but if you look; take a broader reach of history – and certainly American and British and European history shows this among all other histories – thirty or forty years isn't a very long time. There has been that long a hiatus in exploration before.

A Geologist on the Moon

Following their flight, the three Apollo 17 astronauts addressed a joint session of Congress. Here, Schmitt talks into the microphone as Gene Cernan and Ron Evans look on.

"So you can't start to feel frustrated about that – you have to try to understand why it's happened, and what has to be done to make it more permanent next time. And I think we can do that; I think the presence of energy sources on the Moon is going to be a very, very critical factor to get human beings back into space. And once they are back there, and permanently on the Moon, then we can go to Mars. We have the technology base to do that again, so I am very optimistic. I think it will happen."

WHAT THE FUTURE MAY HOLD

As an adjunct professor at the University of Wisconsin, Jack Schmitt and his dedicated team of scientists have spent nearly two decades looking at the subject of energy sources on the Moon. They are developing ways to harness the energy from Helium 3 (He3), which he believes will be found in abundance in the lunar soil, having accumulated over billions of years. It will, he says, be an ideal and plentiful fuel that could be used in the production of fusion power – a fuel with few or no adverse

A recent photo of Harrison Schmitt, the last man to set foot on the Moon. [Credit: Francis French, Reuben H. Fleet Science Center, San Diego]

environmental effects. Schmitt says a single ton of He3 would provide the energy equivalent of billions of dollars of coal, and though it would be an expensive process, it is financially viable.

"A lunar space station, industrial base, or network of bases may have a critical role to play in both the provision of alternatives to fossil fuels on Earth and the supply of resources for use by space-faring peoples. For people on Earth worried about satisfying the long-term needs of this planet with fossil fuels or other energy sources, He3 offers an alternative which is environmentally benign and highly efficient."

Uniquely, the energy of He3 fusion could theoretically be converted directly into electricity at twice the efficiency of existing thermal power plants, or even other types of proposed fusion plants. Power plants fuelled by He3 from the Moon could supply the environmentally acceptable energy that civilisation will require in succeeding generations.

Furthermore, Schmitt says that the technology subsequently developed would play a key role in sending humans to Mars, and it would also address the little-feared but potentially catastrophic problem of rogue asteroids impacting with Earth and wiping out all but a few species of life. "We're the first generation of human beings who can do something about that. As an indirect consequence of returning to the Moon, and doing that repetitively, you [will] have the rockets to divert these asteroids."

Schmitt is adamant that while we have learned much about living and working in space, humans need a much more credible scientific understanding of human physiology in space than we have today. The International Space Station, he says, is a good start, but it has a finite mission life. "The next station is already up there. It's the Moon." He may be looking to the future, but Jack Schmitt can also afford to spend time reflecting on past achievements in our unquenchable need to explore.

"Humankind sought and attained galactic stature with the first explorations of the Moon between 1969 and 1972. During these momentous years, our species took its first clear steps of evolution into the solar system and eventually into the galaxy. Now, as the Pueblo Indians of America relate the lesson of their ancestors, 'We walk on the Earth, but we live in the skies.'

"Early explorers took their eyes and minds into space and became the eyes and minds of billions of other explorers on the starship Earth. They began the long process of transplanting human civilisation into space. This fundamental change in the course of history occurred as humans also gained new insight about themselves and about their first planetary home."

THE END OF THE BEGINNING

With the conclusion of the Apollo 17 mission and the Apollo programme in December 1972, humankind had reached "the end of the beginning" of its movement away from home.

Owen Garriott never made it to the Moon, but he is pleased that his colleague Jack Schmitt did an exemplary job in representing NASA, the scientist-astronaut group, and the entire scientific community on Apollo 17.

These days, Garriott is very clear when it comes to his own memories of his days in the Astronaut Office. "I want to stress that there was relatively little friction between scientist-astronauts and senior management, such as Slayton or Shepard. Nearly none with the other pilot-astronauts. There may be a few who feel differently, but it could be largely because they felt overlooked for crew selection."

He did emphasise, however, that there was one crewing decision that impacted greatly on one of his colleagues. "The person who lost the most in the Astronaut Office was Joe Engle, my next-door neighbour for twenty years." Engle had lost his place on Apollo 17 in favour of Schmitt, but Garriott is adamant that it was the right decision. "Schmitt absolutely needed to go to the Moon and he did, albeit against Slayton's desires." Engle was disheartened by this decision, but Garriott recalled a man who swallowed his disappointment for the overall good of the programme. "I never once heard him complain about his misfortune."

On the subject of crew selections, Garriott is also comfortably philosophical. "I've often thought about how I might have placed the scientist-astronauts on the crews if I were to do it objectively. In fact, it would be almost exactly as Slayton did it in 1972–75. Schmitt, as I've said, needed to go to the Moon. Kerwin, Gibson and myself were not geologists and many of the pilots were as good (or better) observers as we, and were highly motivated to do the research as well. The three of us did have the experience and motivation to do long duration orbital missions, perhaps better than most.

"There was a case for flying two scientist-astronauts on each Skylab mission, but this was not done. The biggest calamity of all was not to fly Skylab-B for even longer missions, and including some of the second group [of scientist-astronauts], but this was decided in Washington."

Today, Skylab is almost the forgotten element in the public's knowledge of America's space programme, but it forged an entirely different path in space exploration. As the excitement of the Apollo programme began to subside, the first US space station was poised to involve NASA's remaining scientist-astronauts in a whole new and abundantly worthwhile realm of space exploration and scientific discovery.

REFERENCES

1. Joseph Kerwin, JSC Oral History project transcript, 12 May 2000.
2. *Apollo Vacuum Chamber Tests*, Part 1: Apollo 2TV-1, Ed Hengeveld, Spaceflight **42** No. 3, March 2000, pp. 127–130, BIS; also the photo essay Spaceflight **45** No. 12, December 2003, pp. 506–507.
3. *Lunar Receiving Laboratory Project History*, Susan Mangus and William Larsen, NASA CR-2004-208938, June 2004.
4. *Where No Man Has Gone Before: A History of Apollo Lunar Exploration Missions*, William David Compton, NASA Special Publication SP 4214, Washington, DC, 1989.
5. Private correspondence from Karl Henize to Dave Shayler 25 July 1983, 4 March 1988.
6. Joe Allen NASA Oral History transcripts, part one 28 January 2003, part two 16 March 2004.
7. AIS telephone interview with Tony England, 14 July 2005; Memo from A.W. England, Mission Scientist, CB, re comparison of science time available on LRV versus walking mission, filed in the NASA Apollo collection, University of Clear Lake, Houston, copy on file AIS archives.
8. NASA JSC Oral History, Dr. Harrison H. Schmitt interviewed by Carol Butler, Houston, 14 July 1999.
9. *The Partnership: A History of the Apollo-Soyuz Test Project*, Edward C. and Linda N. Ezell, NASA Special Publication SP 4209, Washington, DC, 1978.
10. *Deke! US Manned Space: From Mercury to the Shuttle*, previously cited.
11. *A Field Trip to the Moon* by H.H. Schmitt, NEEP602 Fusion Technology Institute Course Notes, 1996.
12. Harrison Schmitt's response to an audience question at the "Apollo 17 Event" in Hollywood, California, 7 December 2002.

8

Laboratories in the Sky

While Jack Schmitt was in training for his flight to the Moon, most of the other scientist-astronauts were busily engaged in preparing for the Skylab programme. Skylab was the made-over Apollo Applications Program, having evolved from plans to use converted Apollo lunar hardware in order to create America's first space station. It would become a platform for research in solar science, stellar astronomy, space physics, Earth resources, life and material sciences. Skylab would be where Americans first learned the skills and realities of truly living and working in space.

The first US space station would be by far the largest spacecraft ever launched into long-term orbit, ready for human occupancy. It would be described euphemistically by the media as "a house in space," and that was pretty close to the mark, for three teams of astronauts would occupy Skylab on science missions of twenty-eight, fifty-nine and eighty-four days respectively. Skylab was not merely a temporary work platform for American astronauts, it was the forerunner to NASA's vision for the future; a massive solar-driven stepping-stone on the path to what was called "the next logical step in space" – a permanent human presence in Earth orbit.

The prospect of extended duration meant that flight opportunities for the scientists improved dramatically, but despite a call for two scientist-astronauts on each Skylab flight, only the three remaining from the first selection group flew to Skylab, while members of the second group of scientist-astronauts only served as back-ups, or in support roles.

MICHEL RESIGNS

Meanwhile, another of the scientist-astronauts had decided enough was enough. Curt Michel, from the first group elected in 1965, saw little prospect for a flight after four years with NASA. They had once been the "Scientific Six," then the "Incredible Five," but now only four remained: Garriott, Gibson, Kerwin and Schmitt. Michel, however, had seen the writing on the wall, and it had been coming for quite some time.[1]

"The Astronaut Office, being a paramilitary organisation, is extremely sensitive to the tradition of seniority. However flights might be parcelled out with a given group (a group being those selected at a given time), in every case no-one got a flight from the next group until all those qualified had flown in the previous group. Since we were the fourth group chosen, we naturally expected that the fifth group of nineteen would not fly until we either got a flight or became disqualified (usually through medical problems, the ultimate disqualifier being death, which occurred far too often). The fifth group evidently thought the same, for there was considerable discouragement over the prospects for flights in the foundering AAP."

A dissatisfied customer

Michel would also convey his dissatisfaction in an interview with author David Shayler in 1992. "There was a presumption," he said, "that you would fly in order of groups, which of course didn't happen since they skipped over our group to fly the next group – although that was never official. There never was anything official about where you stood, or how the decisions were made."[2]

As Michel attested in his contemporary journal, a lunar landing required proficiency on the Lunar Module; and of conventional vehicles, only the helicopter had control properties that simulated those of the LM. Therefore advanced helicopter training remained an essential requirement for any astronaut who might want to fly to the Moon. But there was a shortage of these vehicles at the time and as a result, it became impossible for the Astronaut Office to conceal the identities of those who might be going to fly any forthcoming missions, and there were some nasty surprises in store. Out of Group 4, only geologist Jack Schmitt would receive this advanced helicopter training. Michel was not impressed.

"It was disappointing that only one of the five of us was to have any expectation of going to the Moon, and by implication, that the remaining four were consigned to the uncertainties of AAP. Such perhaps might have been the status of the program, but no. The stunning fact was that almost three-quarters of the fifth group received helicopter training assignments. I was appalled. Not only did the assignments suggest that the fifth group might not have to wait for AAP, but it meant that those of us untrained could not hope for anything else despite whatever future developments might take place.

"I immediately wrote a memo to Shepard urging that more from our group be trained, even if only to provide back-ups to Schmitt. To appreciate the degree of exclusion of scientists from the possibility of lunar missions, I should mention that two of the nineteen (from Group 5) either had an advanced degree (Don Lind, a physicist) or the expectation of obtaining such a degree (Bruce McCandless). They were informally lumped in with us as scientist-astronauts and neither would fly helicopters. It was therefore pilots eighty-two per cent versus scientists eighteen percent and even the most innocent of us thought that might be significant.

"Shepard asked Owen (Garriott) to prepare a priority list for his consideration.

Garriott complied with the recommendation that, in order of preference, Lind, I, he, Gibson and McCandless should be considered. Shepard responded by adding Lind and Garriott to the list. This decision was interesting in accepting the point I wished to make while at the same time rebuking me for making it. A hollow victory that."

Turning to Apollo Applications

Knowing he now had very little, if any, chance of flying to the Moon, Michel's interest turned to AAP, and in particular the Apollo Telescope Mount (ATM). This advanced solar observation facility was designed to carry a suite of instruments to study and photograph the sun and its spectra, and it would later play a prominent role in the outstanding success of the Skylab series of missions.

In 1969, according to the timetable then in place, the first ATM flight was tentatively scheduled for 1972, but like so many others in the AAP, it remained unfunded. Most of the scientist-astronauts decided to bide their time, although the future of such programmes was clouded by uncertainty. Some, like Joe Allen, remained optimistic about their chances. "I feel that we are playing a vital role," he told *Scientific Research* magazine. "A great deal of support work is required of this office. I feel that if I can serve as a mediator and help on some of the scientific experiments, I have had a big part and done my job. If it should happen that I can get to go along on a mission, that will be just perfect. But I don't feel that I'm wasting my time if I don't go into orbit."[3]

Karl Henize agreed with Allen. "We are very anxious to see at least one scientist go. I know that my personal chances are slim, since I was in the last nine chosen and there are over fifty astronauts in the program. I would like to fly one mission, but I'm not eating my heart out about it."[3]

Michel was a little more forthright in his opinions, and was well known for his outspoken comments on the many dissatisfactions associated with being a scientist in the astronaut programme. "When we joined the program in 1965," he said, "we were led to believe that some of us would be chosen for flights by 1968. Now the soonest it seems that any of us will fly will be 1972. That means we will have been in training and away from our disciplines for seven years. That's a long time."[3]

On the last day of June 1969, the day of his return from a year's break at Rice University, Michel had a meeting with Alan Shepard. Although there was no need for him to see Shepard in order to resume his astronaut duties, he was anxious to resolve several important questions and used his return as a pretext for the interview.

Possibilities fade

"After the usual innocuous preliminaries I asked what my duties might be and explained that I did not look forward enthusiastically to an extension of my previous duties. I wasn't quite sure what would happen, but the captain slightly surprised me by taking a very positive view, suggesting that I should explore things with Walt Cunningham, who was now heading up the AAP effort, and with 'William' Hess (Dr. Wilmot Hess, who held the directorship of the Science and Applications Directorate, is known as 'Bill' – the mistake was natural, but calibrates the degree of interaction between the two branches). For some reason, Walt had inherited AAP 'for the duration' and I expected a hard-charging, effective, and opinionated job with him. Hess, on the other hand, might be expected to provide some sensible accommodation between the divergent demands of science versus hardware technology. It was really quite a sensible suggestion on Shepard's part."

Next, the two men discussed the possibility of operational Apollo experience, and Shepard revealed that crews, including support crews, had been designated up to Apollo 15. Michel then estimated the next available slot, on Apollo 16, could be some two years distant. However, Shepard said there was a possibility that Michel might be offered a role as a part-time Capcom prior to this, which was hardly exciting news. He later admitted he had probably "lost some brownie points" by his determination to take the twelve-month leave of absence, and had effectively gone to the bottom of the selection heap for the very few slots available. "I could count the number of empty seats and that would put me off the whole thing, so things weren't going all that well."[2]

He then asked if the main utilisation of the scientist-astronauts would be concentrated on the space station programme, and Shepard thoughtfully conceded this was very likely. "I returned home," Michel recorded in his journal, "and wrote my resignation."

By 8 July 1969, with the Apollo 11 mission imminent, Curt Michel decided he had mulled over his decision long enough. He could no longer play the protracted waiting game with NASA, and took a draft notice of his resignation to Shepard's office. Shepard was not in, so Michel left a note along with his resignation, which said: "I think I should have your comments before sending this in to the Mill – I know you are busy with 11 but when you get a chance ..."

The next day, Shepard got in touch with his office and asked that Michel call him at the Cape. Michel called twice, but was told each time that Shepard was involved "in the simulator." The following morning, at 11:15 a.m., Michel finally reached Shepard and apologised for raising the issue of retirement at such an overwhelmingly busy time. Despite this, Shepard was surprisingly understanding, and said he couldn't argue with Michel's reasoning. He added that he thought the current projection of lunar flights would be cut from ten to around six or seven, that Apollo Applications would probably slip by a year, and might even evolve into the proposed space station. Michel decided to pursue the biggest question in his mind: would any of his group get a chance to fly an Apollo mission? Shepard would only say that Schmitt "had a reasonable opportunity," and he was also expecting other astronauts to begin dropping out with the dearth of flight prospects.

Michel had been anticipating problems when talking with Shepard – perhaps an argument, at least an annoyed objection, but there was none. He now found most of what he was going to say to the chief astronaut was irrelevant, and ended by stating that in the light of their conversation his resignation would stand. Shepard said he would send the letter on to Deke Slayton. "I thanked him for responding," Michel later jotted down in his diary notes, "and that was that."

In August, and following the successful flight of Apollo 11, Michel had a late night meeting with Deke Slayton, who also said he didn't have any argument with anything in the draft letter of resignation. The two men also discussed the possible directions of the Apollo programme, and Slayton said he expected each succeeding flight to become more difficult, and that the missions would never become "operational." Then, to Michel's amazement, Slayton began complaining about how he had been coerced into selecting the second scientist-astronaut group, and more recently the seven Group 7 astronauts from the cancelled US Air Force Manned Orbiting Laboratory (MOL) programme, when they were clearly superfluous to projected astronaut requirements. At least, he said, he had managed to trim the latter group by imposing a mandatory age limit of thirty-five, which effectively halved the group. It was a sad pronouncement of the way things were headed at NASA, and for Michel it provided the final incentive to resign.

"I signed and dated the draft and thereby made the letter official. The meeting ended cordially with some question over when to make an official announcement of resignation. He said (somewhat glumly) that I could say whatever I pleased. I said that the letter contained the substance of any statement.

"The next day I sent a letter advising Slayton that I wished to make it official on 18 August, and reiterated that I had nothing further to add beyond the letter. Slayton and Jack Riley (from NASA's Public Affairs) phoned to advise me of the release date and read me the press release which excerpted my letter. That night I called the remaining four members of my group. I discussed my decision. The general attitude was one of understanding, yet hope (but not optimism) over possible future developments."

Looking back

Thirty-six years have now passed since Curt Michel resigned from NASA, but several issues can still ignite deep-seated feelings of disappointment and even resentment. For one, he is grossly unhappy with several assertions made about him and his work in Apollo 7 astronaut Walt Cunningham's autobiography, especially the implication that he was given, according to Cunningham, "the astronaut equivalent of firing."

"Absolutely not true!" was Michel's retort, referring to these comments as "Walt's repetition of office gossip." He also spoke about his wife Beverley's dislike for his involvement in the space programme. "She figured I'd just get myself killed and she'd end up getting the shabby treatment from NASA that the actual widows have experienced. It didn't help that things just dragged along and she wasn't an officer's wife used to being part of a second tier support group. I heard about this constantly, but after [Duane] Graveline's experience, no one was inclined to discuss doings at home."

Years later, having returned to Rice full time, Michel was prepared to pass up a much higher paying job in order to stay in academia. But when he joined NASA in 1965, he had found that there was an immediate and ongoing financial imposition, particularly for the scientist-astronauts. "When I joined, NASA assigned me to a GS grade that paid almost exactly what Rice did. After about two years of tiny automatic rises I asked what was needed to get a real raise and I was told I needed to fly in space! That was the pattern with the military guys: whenever they returned from a flight the president gave them a raise in rank."

Michel is also convinced that his resignation had a lot to do with a scientist-astronaut finally being selected to an Apollo lunar mission. "Joe Engle was originally assigned to the seat that Jack Schmitt ended up with. This was pretty ironic because Joe was a pretty good friend of mine and Jack might never have gotten the flight had I not resigned. The National Academy was supposed to be looking after us scientists, but those big shots weren't going to spend much time worrying about day-to-day working conditions at NASA. However they had to pay attention when you slam the door behind you. It got their attention before NASA managed to launch every last Apollo mission without a scientist on board. I'm not sure Jack ever realised this."

On the question of his relationship with NASA these days, he is openly succinct. "Did you know I've never been invited to any NASA celebration of the space program? Some flunky probably wrote a note in my file claiming I was an 'enemy' of the space program because I had resigned. Engineers can come and go, but not astronauts."

SKYLAB – A SPACE STATION FOR AMERICA

On 14 May 1973, the final Saturn V left the launch pad at the Kennedy Space Center in Florida. This two-stage variant of the launch vehicle that took American astronauts to the Moon carried the unmanned Skylab Orbital Workshop into orbit to begin a new phase of American space exploration. Among the crowd of 25,000 onlookers were three astronauts who were looking at more than just another lift-off, as they were hoping to follow it just twenty-four hours later. It may have been a spectacular rocket launch to everyone else, but Pete Conrad, Paul Weitz and scientist-astronaut Joe Kerwin were watching what they thought would be their orbital home for the next month.

Within minutes of entering orbit, the success of finally getting the vehicle off the ground was tinged with disappointment and uncertainty as telemetry indicated that the station was in serious trouble. The proposed manned missions looked in doubt. Just sixty-three seconds after lift-off the micrometeoroid shield, designed to protect the habitability section from impacts during the mission, prematurely deployed a few centimetres. It was all but ripped off by the aerodynamic forces encountered during the powered ascent. Further, one of the two solar array wings had been lost, while the other appeared, from data received on the ground, to have been only partially deployed. With Skylab crippled in orbit, plans were devised to attempt to overcome the setback and see whether anything could be salvaged from the mission. Three teams of three astronauts were scheduled to visit the station for missions of twenty-eight, fifty-six and fifty-six days, but in mid-May 1973 the Skylab mission seemed to be over before it had started.[4]

APPLYING APOLLO TO OTHER GOALS

The Skylab programme has its origins in the early years of NASA during the late 1950s and in the development of a manned space programme to follow the original one-man Project Mercury series. Studies of a station in space had existed for decades but with the dawn of the space age and the creation of the American civilian space agency, these plans moved from pages of science fiction to the drawing boards of leading spacecraft contractors.

A three-person spacecraft, called Apollo from July 1960, was planned as the next stage in America's manned quest for space. Apollo had the capability of flying both in Earth orbit and also to the Moon and back. By 1970, NASA was also looking towards the creation of a temporary scientific research station in Earth orbit, which in turn would lead to the first interplanetary trips to Mars in the 1980s and 1990s. Originally, the Apollo Command and Service Module was to have docked to specialised modules designated "space laboratories", or to carry scientific instruments in the spare equipment bay of the Service Module (or even a converted Lunar Module). But gradually, studies turned towards converting the Saturn S-IVB stage into a laboratory for missions of between four and six weeks. This concept featured in what became the Apollo Applications Program from 1965. The original idea was to use a spent (formerly fuelled, or "wet") S-IVB stage launched on an unmanned Saturn 1B, but the design was eventually simplified to utilise an unfuelled (or "dry") stage launched by a two-stage Saturn V variant.

Mercury–Gemini–Apollo–the Moon

In May 1961, US President John F. Kennedy set NASA its Moon landing challenge. To meet it, Apollo was amended to include lunar landing among its long-term goals, and the creation of a scientific research platform or space station slipped down the list of priorities. Later that year, a new programme called Gemini was devised to test the techniques that Apollo would require to achieve its goal, but in Earth orbit. A programme of scientific experiments would also be carried, however, taking advantage of the added volume on the Gemini, the extended duration of the missions and the addition of a second crew member. Though Gemini, like Mercury, was essentially an engineering development programme on the path to the Apollo Moon landings, science would be included on each mission at an increasing rate.

In order the meet the crewing requirements for Gemini and early Apollo missions, new pilot-astronauts were selected in 1962 and 1963. In the latter selection, although the focus remained on piloting skills, the selection criteria was expanded to include academic skills and qualifications. The first few flights in the Apollo programme were test-flight orientated, with mission safety and system qualification of primary importance. Science was more relevant to plans to use Apollo hardware for other missions following the successful landing on the Moon, although some of the Earth-orbital Apollo test flights could include some science to be conducted once the primary test objectives had been met. While the flight was in progress, it made sense to take advantage of the fact that a crew (of three) was in space, as well as the length of the missions (up to fourteen days) and the increased volume of the Apollo. However, although science experiments could be assigned to a mission, if such experiments did not support the lunar landing goal or could be deferred to a later mission, they would usually be reassigned.

Applying skills to AAP

In planning for future missions beyond those devoted to reaching the Moon, NASA had instigated a series of studies to utilise Apollo-derived hardware to support expanded surface explorations of the Moon, scientific Earth-orbital missions and the development of space stations. Between 1960 and 1965, this programme was known in turn as Apollo "A"; Apollo X (for "experimental", then "extended"); XMAS (Extended Mission Apollo System); and the Apollo Extension System (AES).[5] By 6 August 1965, the project had progressed sufficiently to warrant the creation of a dedicated directorate as part of the Office of Manned Spaceflight. It was designated the Saturn/Apollo Applications Directorate or, more simply, Apollo Applications. That same month, an Apollo Applications Office was established at the Manned Spacecraft Center in Houston, just a few weeks after NASA had named its first group of scientist-astronauts.

An Astronaut Office (CB) memo from Chief Astronaut Alan Shepard, dated 3 February 1966, detailed the creation of a new branch office within the Astronaut Office. The Advanced Program Office would be headed by former Mercury astronaut Scott Carpenter. Assigned under him were Joe Kerwin (with technical responsibilities for pressure suits and EVA development) and Curt Michel (assigned to work on experiment issues). From September 1966, Owen Garriott, Ed Gibson and Jack Schmitt would be assigned to this office once they returned from flight school. During most of 1966, Kerwin also worked as CB point of contact for the developing idea of using a spent S-IVB stage as the basis of an orbital workshop. During his work on the development of this concept, Kerwin expressed concerns over both the apparent lack of experiment planning and the operational safety provisions for hardware to be installed in the spent stage. By August 1966, work on the AAP programme had evolved sufficiently for a more defined CB Branch Office to be established. Its first Chief was pilot-astronaut Al Bean.

Shepard detailed further assignments in the expanded AAP CB Branch Office in a subsequent memo dated 3 October 1966. Initially, pilot-astronauts were assigned under Bean (Bill Anders, Joe Engle, Jack Lousma, Bill Pogue and Paul Weitz), while Garriott headed up the "Experiments Branch" with Gibson, Don Lind, Bruce McCandless, Michel and Schmitt assigned under him. Kerwin was assigned to the Suit-PLSS-Recovery Branch with Ed Givens, under the leadership of John Young. In December, when Young was assigned to the Apollo 2 back-up crew, both Kerwin and Givens joined the Experiments Branch. The two branches exchanged members several times over the next few months, until 4 April 1967 when they were merged into the Apollo Applications Program (AAP) Branch Office under the leadership of Bean.

SUPPORTING AAP

As Apollo lunar efforts gathered pace, activities under AAP remained very much in its shadow. While assignment to the first Apollo "test" missions was widely seen as a step toward a higher profile lunar landing crew, an assignment to the equally important

AAP development work to ensure mission success and crew safety was often perceived as a slower progression in an astronaut's career. The other factor involved in preparing crews for missions was the availability of simulators and crew training teams. With the priority to get Apollo to the Moon at least once by the end of 1969, it was necessary to assign Apollo astronauts to training cycles to give them as much simulator time as necessary. By December 1966, there were twenty-seven astronauts (all from the pilot groups of 1959, 1962, 1963 or 1966) named in nine "crews" for the first Block I and II Apollo missions. The system developed for crew assignment under the Gemini programme was modified to work for Apollo as well. A back-up crew on one mission would skip the next two missions in order to train to fly the fourth. Despite a number of necessary individual crew member changes, this was essentially how the Apollo crewing system worked from 1966 to 1972. The Apollo 1 pad fire of 27 January 1967 was a tragic blow that set the maiden launch of a manned Apollo back over eighteen months, but the crews continued generic training until a new manifest was devised.

In May 1967, twelve astronauts assigned to AAP completed a walkthrough of the proposed Saturn S-IVB workshop mock-up, in order to evaluate habitability, stowage and simulated EVA operations. This came as part of their programme of visits to AAP contractors, NASA field centres and proposed experiments across the country. An astronaut's diary during such technical assignments was usually filled with attendance at countless meetings, reviews and briefings, before returning to report to other members of the Branch Office, or to the CB in general if it related to the wider Apollo programme. The technical support areas given to the four AAP-assigned scientist-astronauts at this time were:

Garriott:	Communications
Gibson:	Crew quarters layout and controls
Kerwin:	Food, water and IVA
Michel:	Hand holds, tethers and foot rails

As a professional geologist and scientist-astronaut, Jack Schmitt was assigned to Apollo development issues rather than to AAP, utilising his unique talents in support of the early landing missions while hoping for an assignment to the Moon himself. He did not participate in the S-IVB workshop review, focusing instead on AAP lunar plans.

In addition to the three astronauts lost in the Apollo 1 pad fire, there were two further losses to the corps in 1967. Ed Givens died in a car crash in June and C.C. Williams in a plane crash that October. The loss of five of the assigned Apollo crew members meant others had to be reassigned in their places. Al Bean moved. Pete Conrad's Apollo crew and was replaced as Head of AAP by Gordon Cooper. He in turn was moved to a back-up position on Apollo in early 1968. Owen Garriott filled the AAP role until November 1968, then Walt Cunningham took the reins. By August 1967, the second group of eleven scientist-astronauts had completed their academic training and were about to undergo a year at flight school. By 1969, only nine of the group were available for astronaut assignment, with Tony Llewellyn and Brian O'Leary having left the astronaut programme during 1968.

In April 1969, the new astronauts received their technical assignments, which divided the group between Apollo and Apollo Applications. Five of them (Joe Allen, Phil Chapman, Tony England, Karl Henize and Bob Parker) were assigned initially to Apollo support issues, while the other four (Don Holmquest, Story Musgrave, Bill Lenoir and Bill Thornton), together with Group 5 pilot (and space physicist) Don Lind, were assigned to the AAP group. Some fifteen years later, Karl Henize stated that although he was assigned initially to support an Apollo mission (Apollo 15), his real goal lay elsewhere. "By that time [1970], there was no hope of my getting a lunar flight, but that had never been my ambition anyway. My aim when I entered the programme in 1967 was to fly on a Skylab mission, of which there were several more planned at that time than were actually flown."[6]

Holmquest's AAP assignment was on habitability and medical experiments, which he worked on for the next eighteen months. According to Walt Cunningham,[7] Holmquest thought that working with the staff assigned to develop Skylab medical experiments would be an advantage in securing an early seat on the Skylab mission, while still retaining links to his medical career. This was all to no avail, however, and when it became evident that he would not fly on Skylab or receive a back-up assignment, he opted for a leave of absence. This became a second leave and led to him resigning from NASA in September 1973, around the time of the second manned Skylab flight. Lenoir, Musgrave and Thornton all received initial technical assignments on AAP/Skylab (in hardware and experiment development) pending assignment to back-up or support roles.

When Chapman came back from flight school, he was assigned to Frank Borman's team evaluating future space station designs *after* the first OWS. Part of these studies related to the potential for using the Saturn V to launch larger "Skylab"-type space stations pending the availability of more permanent structures in orbit. In December 1969, the CB became involved in evaluating crew stations for such a workshop. The design of what was termed the Saturn V Workshop (SVWS) featured elements that were being developed for Skylab A/B, such as a Multiple Docking Adapter to accommodate both Apollo CSMs and Shuttle orbiters, and an upgraded Apollo Telescope Mount facility. There were also discussions to consider whether gravitational experiments could be included on the research programme for the second "Skylab". During most of 1969, the Measurement Systems Laboratory at MIT had been investigating such experiments and their potential for future programmes. The team at MIT had hoped that the experiment could be formally proposed as an SWS-II experiment by March 1970, but it soon became clear that a second workshop was not a realistic option given the existing budget restrictions. Though the idea of "Skylab B" rumbled along for some months, the programme never formally received funding. Had the gravitational experiment actually been assigned, Chapman (with his former association with MIT) would have been a likely candidate to act as the principal investigator, or at least as a co-investigator.[8]

This experiment would be proposed for a circum-solar probe after first evaluating the technology aboard Skylab B. As Chapman pointed out in his memo, "There are good reasons for considering flying a version of the proposed apparatus on a manned orbital workshop. This would provide a zero-G checkout of the advanced inertial instrumentation required and make good use of the capabilities of man as a developmental test engineer." The design of the experiment featured a large sphere (the size of which depended upon the final configuration) mounted on three gimbals in a cube frame measuring a metre on each side, and as close to the centre of mass of the station as possible to reduce external forces. There would also have been a separate electronics box of about 30 cm square, with approximately 100 kg of mass and a power supply of up to fifty watts. It was envisaged that twenty hours operation would be sufficient during the mission, with onboard recording available, although it would have taken up to ten man-hours to set up and calibrate the experiment. Unfortunately this experiment, like the follow-on "Skylab" OWS, never proceeded to flight status.

In August 1969, seven former USAF Manned Orbiting Laboratory astronauts transferred to NASA upon the cancellation of that programme. Their previous training on Gemini systems under the MOL programme meant they would be helpful in both AAP and Apollo. From this group, Bob Crippen, Bob Overmyer and Dick Truly were assigned to the AAP Branch Office, with the other four (Karol Bobko, Gordon Fullerton, Hank Hartsfield and Don Peterson) assigned initially to the Apollo Branch Office.

With Apollo on track to achieve its primary goal, developments under AAP picked up pace as the pressure to prove the Apollo hardware eased. Now, with the prospect of launching at least one orbital workshop with three manned missions in the early 1970s, and the possibility of a second laboratory in the middle of the decade, NASA began to confirm plans that it had been debating for some years. In May 1969, the space agency decided to follow the "dry" workshop profile for AAP instead of the more complicated "wet" workshop. Then, on 17 February 1970, the name AAP was dropped in favour of a more publicly appealing name – Skylab (laboratory in the sky). In August 1970, Pete Conrad became the new Chief of the CB Skylab Branch, a position he held until the end of the flight programme in 1974.

SCIENCE PILOTS FOR SKYLAB

In 1961, the crew designations for a lunar mission had been identified as Commander, Navigator Co-pilot and Engineer-Scientist. By 1965, these were refined to become the more pilot-orientated Commander, Senior Pilot and Pilot. In most of the early press releases concerning Apollo X or AAP missions, there were no such individual designations. On 29 November 1966, Deke Slayton issued a memo[9] revising the designations for Block II (lunar-orbit capable) Apollo missions into Commander, Command Module Pilot (CMP, formerly Senior Pilot) and Lunar Module Pilot (LMP, formerly Pilot). In the same memo, Slayton indicated that for AAP-related missions, the LMP would be identified as the Mission Module Pilot (MMP), whose role and responsibilities would be defined as mission plans evolved and could therefore vary from mission to mission. If a mission required a medical specialist then the MMP would be a doctor; if astronomical observations were a primary requirement, it could be an astronomer; and if space science featured in the mission, the MMP could be a physicist or engineer. This is the position that evolved into the Science Pilot as AAP gave way to Skylab, but though the name changed, the idea of role speciality remained for some time.

In 1969, after the Wet Workshop became the Dry Workshop and the LM was dropped for an ATM that would be launched on the unmanned Workshop, the designations for the three crew members also needed updating. By 1971, these crew positions became clearer. Heading the crew would be a pilot-astronaut Commander, ideally with at least one previous mission to his credit and preferably one with rendezvous and docking experience. The Pilot would be second in command, a rookie, and also from the pilot-astronaut cadre. The third place would be filled from the scientist-astronaut group and even here, seniority counted, with the prime positions falling to the 1965 selection rather than the more recent group. Two reasons given for this were the length of astronaut service (which was about six years for the 1965 group compared to four for the 1967 group) and the amount of time the 1965 group had worked on the AAP/Skylab programme (which in Kerwin's case was for virtually the entire six years he had been part of the astronaut programme).

On the missions, the commander would hold overall responsibility for mission success and crew safety and would also be the expert on the Apollo Command and Service Module systems and procedures. He would handle the docking, undocking and fly around, as well as positioning and initiating systems for entry at the end of the mission. The pilot would be the expert on Workshop systems and electrical subsystems, while the science pilot would be the expert in onboard medical equipment, the Apollo Telescope Mount and its associated hardware. All other experiment objectives would be equally divided between the three crew members depending upon availability and choice.[10]

The plan to fly just one scientist-astronaut on each mission did not sit well with the scientific community as a whole or the scientist-astronauts themselves. They believed two would be better and in theory the crews could have consisted of three scientist-astronauts, since they were all qualified jet pilots, had all trained on the Apollo CSM and had undergone years of astronaut training and support work. Indeed, in 1970, a series of papers were presented at the thirteenth International Science School,[11] an annual event organised by the Science Foundation of Physics within the University of Sydney, Australia. One of these papers (by L.B. James, the Director of Lunar Operations at the Marshall Space Flight Center in Huntsville, Alabama) looked forward to the Skylab programme and its plans for science and investigation. Interestingly, the paper indicated the general plan to fly three missions of three-person crews but added: "The three-man crew of scientist-astronauts will ... conduct more than fifty experiments." This could have been a misinterpretation and referred to the inclusion of a scientist-astronaut on each crew, but further reading suggests this was not so. It indicated that a second scientist-astronaut crew would be launched for a fifty-six-day mission: "This crew may consist of one astronaut who is an astronomer and one who is a medical doctor, as well as a third astronaut who may have some other scientific speciality." A similar pattern was suggested for the third and final mission, postulating that nine scientist-astronauts would fly, rather than the three that actually flew. Had this been a true interpretation, then perhaps the scientist-astronauts who left due to the lack of flight opportunities (Michel, O'Leary, and even Chapman and England) may not have done so, while those who ended up awaiting the Shuttle (most of the 1967 selection) would have probably flown a decade earlier.

In late 1970, following the aborted flight of Apollo 13, the budget restrictions that forced the cancellation of the last three Apollo lunar landing missions and the termination of the programme after Apollo 17, there was a shortage of flight seats available to *any* astronaut, let alone the scientist-astronauts. With most of the crewing for Apollo resolved, there remained only the three Skylab missions and the prospect of a joint mission with the Russians on the horizon before 1980. Skylab would provide the first, and for some time, the only long-term scientific space research platform for America.

This lack of flight opportunities was bound to leave some disappointed and possibly even resentful astronauts among those who were not selected for Skylab. The prospects of a second Skylab continued to diminish and it soon became clear that, barring the loss of the first OWS, the second would not fly. Suddenly, instead of about eighteen flight seats, Skylab was reduced to just nine and with Deke Slayton leaning towards experience in his selection of crew commanders, three of those seats were effectively ruled out as far as the scientist-astronauts were concerned.

The scientist-astronaut group had already voiced their disappointment at the emphasis on operational criteria over science objectives on Apollo after the first lunar landing had been achieved, particularly as the next three crews (Apollo 12–14) were all filled from the pilot-astronaut group. The fact that none of their group had yet been selected to a lunar flight, coupled with the cancellation of three of the landing missions, meant that their chances of a flight assignment on any mission were slim. Support for the inclusion of scientist-astronauts on Skylab strengthened in late 1970 when the Space Science Board tried to obtain assurances from NASA that two scientist-astronauts would be assigned to each of the three Skylab missions. Their argument was that for each mission, the actual "flight operations" that required piloting skills would take up a small fraction of the overall mission time, whereas science and other objectives would be paramount. The safety-over-science argument that applied to the early lunar missions did not wash as far as Skylab was concerned.[12]

Several months of arguments ensued between Homer Newell, the Chief Scientist at NASA Headquarters who had listened to the scientist-astronauts' arguments, Deke Slayton, the Director of Flight Crew operations at MSC, and Bob Gilruth, the Director of MSC. Though Skylab would use proven technology, it was in a new configuration, and since the onboard systems could not be modified after launch, the direction of training at Houston leaned towards system management and malfunction procedures in order to deal with any hardware problems. That demanded a high level of systems expertise, calling upon a test pilot's natural skills. Additionally, Skylab training was designed to include cross-training in which all of the crew members would be trained to the same level of proficiency on all the major experiments. According to the 1983 official NASA Skylab history,[12] "an astronaut's specific academic background was relatively unimportant." There was an attempt to get MSC to agree to fly two scientist-astronauts on at least one of the three missions, but when three cosmonauts died during re-entry after a twenty-three-day mission on the Soviet Salyut space station in June 1971, operational and safety requirements quickly gained the higher priority over science once again.

On 6 July 1971, NASA Headquarters finally approved the MSC plans for a two pilot/one scientist astronaut crew for each manned mission, with the first crew to include a physician-astronaut. The final selection of the crews would be left to MSC and announced the following year.

Skylab assignments

During 1971, at the time of discussions over who would crew the Skylab missions, the following scientist-astronauts were still listed as active in the CB:

Astronaut	Selection	Academic attainment	CB Technical role
Allen	1967	PhD Physics	Support Apollo J missions
Chapman	1967	PhD Physics	Support Apollo J missions
England	1967	PhD Geophysics	Support Apollo J missions
Garriott	1965	PhD Electrical engineering	Skylab Branch Office
Gibson	1965	PhD Engineering	Skylab Branch Office
Henize	1967	PhD Astronomy	Support Apollo J missions
Holmquest	1967	MD; PhD Physiology	Leave of absence (from Skylab Branch Office)
Kerwin	1965	MD	Skylab Branch Office
Lenoir	1967	PhD electrical Engineering	Skylab Branch Office
Musgrave	1967	MD	Skylab Branch Office
Parker	1967	PhD Astronomy	Support Apollo J missions
Schmitt	1965	PhD Geology	Back-up LMP Apollo 15; LMP Apollo 17
Thornton	1967	MD	Skylab Branch Office

Of the four remaining 1965 members, geologist Schmitt had been working on Apollo issues for years and was in line to fly the final Apollo landing mission after backing up Apollo 15, so he was never really under consideration for an AAP/Skylab mission. The other three, however, had been working on the programme for some years, as well as fulfilling some Apollo support roles, so they were prime candidates for the three planned missions. The 1967 members had completed candidate astronaut training in 1969 and were immediately assigned to either Apollo lunar mission support or Skylab. They would provide support and back-up roles for the first station missions, with the slim hope that a second workshop might be funded and launched in late 1975 to support manned missions in 1976–7 (see below).

Given the physician requirement for the role of science pilot on the first mission, Kerwin was the natural choice, with Garriott and Gibson likely to fly the other two missions. In early 1971 at a CB pilots' meeting, Deke Slayton named the fifteen astronauts who would be assigned to the first Skylab's three-mission programme. There was a possibility of a fourth visit and, if circumstances and resources allowed, a fifth. Should the back-ups not get a chance to fly to the first station, a second station mission programme was still, at that time, a possibility for 1975–7. The fifteen were:

Crew	Prime	Back-up
1	Conrad–Kerwin–Weitz	Cunningham–Musgrave–McCandless
2	Bean–Garriott–Lousma	Schweickart–Lenoir–Lind
3	Carr–Gibson–Pogue	Schweickart–Lenoir–Lind
4	Cunningham–Musgrave–McCandless	From the crews of 1, 2 and 3
5	Schweickart–Lenoir–Lind	From the crews of 1, 2 and 3

The exact date of this announcement has been contested by some of the scientist-astronauts. Kerwin has stated it was a 1970 meeting and no firm crewing had been decided upon, nor had the system of back-up rotation to prime crew been defined. Garriott indicates that Slayton and Al Shepard had been evaluating crewing (at least for scientists) for the first mission for five years before they flew in 1973 – which meant 1968 – so Kerwin, Garriott and Gibson had been earmarked for the three missions for some time.

The first mission would be the earliest opportunity the Americans had to fly for longer than fourteen days. This raised a number of medical issues that needed to be understood, and the best way to do that was to fly a physician-astronaut to monitor the health and well-being of the crew. As he was the most senior such astronaut, the position fell to Kerwin. The second mission was aimed primarily at ATM solar studies. Garriott's background in electrical engineering and his work during the development of the experiment package (plus his administrative roles in the AAP office) made him the natural choice for science pilot on the second mission. That left Gibson, who had studied atmospheric physics and solar physics and had worked on ATM issues (including extensive EVA simulations for ATM film retrieval and replacement) since coming to the Astronaut Office from flight school. With Earth resources objectives being a key part of the third mission, Gibson's selection as science pilot on that crew was a logical one.

Several of the pilot-astronauts who did not receive a Skylab assignment were not happy, even though the order of seniority still governed crew selections and despite the possibility, however slim, of a fourth or fifth mission or a second workshop. Musgrave was chosen as back-up to Kerwin, with Lenoir serving as back-up to both Garriott and Gibson. Cunningham left the programme before the formal announcement of the crews on 16 January 1972 and was replaced by Schweickart because an experienced astronaut was needed to support the first mission. Vance Brand replaced Schweickart as back-up commander for the second and third mission, but no Mission 4 or 5 crews were announced. By now, the prospects and the budget for a second Skylab had all but disappeared as the Space Shuttle began to be heavily promoted as the next step in space for NASA. Additional assignments announced in January 1972 were those of the support crew and Capcom for all three missions: Bill Thornton and Karl Henize would join Group 7 former MOL astronauts Bob Crippen, Hank Hartsfield and Dick Truly (and members of the two back-up crews) as required. On 19 February 1973, having just finished supporting Apollo 17 as Mission Scientist, Bob Parker was named as Programme Scientist for all three Skylab manned missions.

SUPPORTING SKYLAB

With Kerwin, Garriott and Gibson undergoing specific mission training, those members of the 1967 group who were assigned to Skylab would be required to fill a number of supporting roles, an important step towards their own eventual flights into space. Following flight school, the 1967 group worked on Skylab issues for between nine and twelve months, as Bob Parker recalled in his official NASA Oral History interview: "We spent [time] working on Skylab – a little of this, a little of that, and a lot of what I would call apprentice stuff, just going to meetings [and] gradually learning. It wasn't a training program, by and large, with specific objectives. Somebody may have sketched something out like that, but then [we would find] somebody was on a committee working out this, somebody else was on a working group doing that, and somebody was worrying about this particular vehicle or those particular experiments on Skylab. So it was basically learning by doing."[13]

Skylab support roles

In addition to the support assignments already mentioned, some of the scientist-astronauts were involved with specific experiments flown on the workshop, as either principal or co-investigators. Each of these support roles had their own story to tell.

Programme scientist: Parker's role as programme scientist was to ensure that the science requirements for the manned missions were compatible with programme requirements and safety issues. During Skylab flight operations, Parker would liaise between the programme director and managers to ensure the in-flight science requirements were being met by the flight crews, utilising his background as both a scientist and an astronaut, and drawing upon his recent experience as mission scientist for Apollo 17.

Though pleased with the initial data, Parker found that the investigators felt starved of sufficient time in the flight plan to accomplish their particular experiment. Parker had difficulty securing their trust and reflected that, although he was the programme scientist and was an astronomer himself, in their eyes he was now an astronaut and no longer "one of them". Parker recalled that most of the investigators thought that their experiment was the one that had been short-changed in the planning, and it was a constant battle to convince them otherwise. The general feeling on the first mission seemed to be "that they had put an awful lot of their time and NASA money into getting very little data, and they had jolly well better get more next time." For the second and third missions, Parker worked hard to placate his scientific colleagues, introducing a programme of periodic planning sessions that helped each

PI gain an understanding of the others' problems. Even more helpful was the ever-increasing flow of data from Skylab as the flights progressed.

Back-up crewman: Story Musgrave gave an insight into his preparation as a back-up to the Skylab 2 mission in a lecture he gave in October 1973.[14] He recalled that shortly after returning to Houston from flight school, they began more detailed briefings and schematic reviews on the Skylab space station, with a part-task approach to sub-systems and mechanisms. According to Musgrave, after several years of briefings, getting simulator time on various spacecraft systems, working on the hardware design and mission development (potentially for some considerable time) and supporting other flights in specific areas or as Capcom, gave a crew member "a good start prior to commencing with formal mission training." The general approach to learning spacecraft systems and operations, according to Musgrave's experience as an assigned back-up science pilot was:

- Instruction on the mission requirements and objectives.
- Briefings on the rationale, science and technology behind the spacecraft design.
- Briefings on spacecraft systems and schematic reviews.
- Part-task systems training in simulators.
- Nominal systems operations in simulators.
- Systems malfunction procedures in simulators.
- Normal operations of the entire spacecraft.
- Normal operations and malfunctions of the total spacecraft through specific mission events, such as launch, rendezvous, on-orbit operations and re-entry.
- Integrated simulations where the spacecraft simulator is patched into the mission control centre and both the real-time mission computers and the spacecraft simulator operations are supported by those flight controllers who are assigned to the mission.

Musgrave stated that training on mission experiments followed a similar timetable, except that far more effort was put into acquiring both the fundamental scientific principles and the special objectives peculiar to each experiment. On Skylab, there were approximately 100 experiments in medicine, physiology, biology, solar physics, remote sensing and Earth resources, astronomy, manufacturing in space, habitation, navigation, particles and radiation, crew manoeuvring units, technology, and student experiments. The key to the development of adequate learning processes, according to Musgrave, was early training on each experiment, conducted by the principal investigator.

Capcom: Musgrave also gave an insight into the role of Capcom during the three missions. The Capsule Communicator, or Capcom (derived from the early days of the space programme, where a spacecraft was referred to as a "Capsule" and those who talked to the crew from the ground were "Communicators" – traditionally an astronaut support role), at least in the case of Skylab, was trained for specific tasks that were being performed in flight. The Capcom knew the science hardware and procedures involved and was a test bed for any voice call or teleprinter message. If he could not understand the call or carry off the procedure, it was unlikely that the flight crew could either. The Capcom usually worked and trained with the flight crew for years, coming to know both the crew themselves and the way they would approach the mission or a specific task. This approach was modified during Shuttle missions, because the Capcom would work on several missions and with different crews during a "tour" in the role, but it was still, in a sense, an additional member of the flight crew who remained on the ground.

Support crewman: The support crew roles during the Apollo era were filled by a third tier of astronauts, normally unflown, who would use their technical assignment skills in either the Apollo CSM or LM to help alleviate the workload on the prime and back-up crews. This assignment was seen as a stepping stone towards a potential future flight and included the role of shift Capcom. This role developed in 1966 and lasted until the end of Apollo-type missions in 1975. It evolved partly in response to CB concerns that, with Apollo contractors spread across the United States and the added complexity of each lunar mission, crew representation was being lost on important decisions and developments. This role was even more relevant to Skylab, where the mission duration would increase dramatically over the short Apollo flights. Additionally, there was a larger scientific programme to contend with on Skylab than there had been on Apollo. The support crew would also relieve the prime and back-up crews of some of the more mundane activities. The scientist-astronauts assigned to the Skylab support crew (alongside Bob Crippen and Hank Hartsfield) were Henize and Thornton. Crippen and Thornton had worked together on the SMEAT test chamber exercise in 1972.

Principal investigator: In September 1967, Karl Henize left Northwestern University to enter astronaut training, but continued as a professor (on leave) until 1972. He then transferred his affiliation, as well as his Skylab experiment (S019 – ultraviolet stellar astronomy) to the University of Texas at Austin, where he became an adjunct professor. During the Gemini programme in the mid 1960s, Henize provided a small hand-held camera that could be used during stand-up EVAs to capture UV spectrometer images of hot stars. Flown on the last three Gemini missions in 1966, the camera obtained some good results, and Henize continued this research with a follow-up camera for Skylab. He developed a six-inch aperture telescope with an objective prism that could penetrate deeper into the far UV region. In assigning his experiment to a manned spacecraft, the idea was to obtain "quick-look" data that was not readily available from larger unmanned astronomical observatories. These were not trivial experiments, but a development towards a better understanding of astronomy observations from space using limited resources and hardware.

Henize looked at the feasibility of deploying and operating the experiment during an early Apollo Earth-orbital mission (Block I Earth-orbital or early AAP mission). However, after the Apollo 1 pad fire, the redesign of the side hatch and the renewed commitment to getting Apollo to the Moon by 1969, few "science experiments" were carried on the early Apollo missions (7–11). Henize was told that the experiment would be reassigned to Skylab, with a three year delay to 1972 or 1973. During those three years, some of the more interesting and easier observations were accomplished via sounding rockets. However, they could only record data of a few dozen stars.

Henize's experiment covered a much broader scale and thus gave him data with which he could classify stars based on their UV spectrum.[15]

The S019 UV Stellar Astronomy experiment was designed to take UV photographs of large areas of the Milky Way in which young, hot stars are abundant. Located in the Skylab Air Lock (SAL) on the side of the Orbital Workshop, the manually-operated equipment featured a six-inch reflecting telescope and moveable mirror. The objective was to image fifty areas of five degrees by four degrees two or three times each, with exposures of thirty, ninety and 270 seconds. Each film cassette held 164 frames of special UV-sensitive film. A fourth exposure capability was available if the stability of the workshop permitted. The experiment was operated by all three crews, resulting in 1,600 photos of 188 star fields. Though some images were smeared due to the movement of the station, these were subsequently processed through computers to recover as much data as possible.

Henize was appreciative of the frustrations of the other principal investigators, but was equally well aware of the difficulties facing both the flight crew and Mission Control. He felt that the PIs' frustration was bordering on paranoia: "You never quite knew what the other man's problems were, and you would put in your requirements and get them back all mangled. Everybody was mad at each other."[12]

Henize clearly demonstrated (as did Thornton and Lind) that on Skylab, a balance of astronaut duties and research science was achievable given suitable attention to detail and priorities at the right time. No one said it would be easy, but no one said it could not be done either. The choice was down to the individual.

Simulation crewmember: Bill Thornton was selected for a ground-based fifty-six-day Skylab simulation in June 1971 and began training the following month. The "mission" was known as the Skylab Medical Experiment Attitude Test (SMEAT) and was conducted in the 6 m altitude chamber located in Building 7 at the Manned Spacecraft Center in Houston. Thornton would serve as the science pilot on the "crew", alongside Robert Crippen (commander) and Karol Bobko (pilot). The simulation lasted from 26 July to 20 September 1972 and not only established a baseline medical data bank for the experiments, but also provided a wealth of information on crew habitability, hardware design and procedural issues. The test helped to iron out several problems that would otherwise have needed to be dealt with on orbit during the actual missions.[4]

Dr. Bill and SMEAT

As a physician with a background in exercise and an investigator of one of the experiments to be flown on the station, Bill Thornton could draw upon years of personal research and his development of techniques, procedures and hardware for exercise in space, as well as assignments in the USAF Manned Orbiting Laboratory Program.

In the planning for SMEAT, a total of 412 hours of training were allocated to each man. In reality, all three exceeded 500 hours, with Thornton recording 509.25 hours of training. Thornton has always been emphatic that without SMEAT, Skylab would have failed. The problems the SMEAT crew encountered and overcame during

The Skylab Medical Experiment Altitude Test crew (left to right): Commander Bob "Crip" Crippen, Science Pilot "Dr. Bill" Thornton and Karol "Bo" Bobko, who completed a 56-day ground simulation of a Skylab mission to gather baseline medical data.

the ground test saved so much time and alleviated so many difficulties that it is hard to see how the three live missions would have coped had this test not taken place. One item that received Thornton's full evaluation – to the point of collapse – was the bicycle ergometer. For several reasons, the one used in the ground test was not as structurally sound as the one planned to fly the Skylab. Thornton has always underlined the need for anything that is to be flown in space to be adequately tested on the ground beforehand, and with as accurate a duplicate training model as is possible. So, if SMEAT was to establish operating procedures and data for the space missions, then the equipment in the chamber should achieve this standard. Clearly the ergometer was one item of ground equipment that had to be used to demonstrate that second best

The Skylab SMEAT crew prepares to enter the altitude chamber at MSC-Houston for the 56-day test.

simply would not do. Sure enough, the ergometer used by the SMEAT crew failed during the test and had to be removed from the chamber for unscheduled repairs. When it was returned to the chamber, Thornton's frustration at the inadequacy of the machine and the waste of time and effort in not providing a suitable machine in the first place was taken out on the ergometer. This time, he tested it to destruction.

After SMEAT, Thornton continued his work on developing crew health regimes. He was especially concerned about the condition of the leg muscles of the first two Skylab crews upon their return. In a move to alleviate the problem for the final crew on the longest mission yet, Thornton designed a simple (and economical) device to help maintain the condition of their leg muscles. A sheet of shiny Teflon was attached to the floor over which a sock-wearing crew member could slide their feet, while being held in place by elastic bungee cords. By adjusting these cords, varying loads could be applied to the exercise. This quickly prompted the crew to call the device "Thornton's Revenge". However, the positive results gained from this device served as the basis for more sophisticated equipment that has been employed on the Space Shuttle, and later on the International Space Station. Although it was a successful test, the kind of problems he experienced in SMEAT (with inadequate government-furnished equipment and bureaucratic hurdles) continually plagued Thornton for the next twenty years as he tried to provide effective and economical exercise and hygiene equipment for the Shuttle, Spacelab and Space Station.

SCIENCE PILOT TRAINING

The Skylab flight crew training programme tended to follow a generic plan. In summary, each crew's training featured the following topics:

Activity	Average hrs per crew
Briefings	450
System training	257
EVA & IVA	167
Medical	98
Simulators	695
Experiments	424
Rescue	16
Total	2107

When the Skylab crew pool came together under Conrad in the Skylab CB Branch Office in 1970, there was very little actual hardware to train upon. While the pilot group focused on developing the training programme (using the first crew as pathfinders) and concentrated on systems and procedures, the scientists in the group focused on the experiments, specialising in research fields that were closely associated with their own areas of expertise.

Kerwin estimated they (the Skylab group) were perhaps more trained for a flight than any previous crew, due to the complexity of the science programme and the length of preparation. Kerwin had joined AAP/Skylab almost as soon as he arrived at NASA and had worked on the development of medical experiments and the protocol behind them. He left the work on ATM development to Garriott and Gibson, but because there was only one scientist-astronaut on each crew he would have to look after the ATM on his mission, so he spent a lot of time training to operate it. In becoming the lead crew member on ATM and the medical experiments, he had little time to devote to the Earth Resource Experiment Package (EREP), apart from the S-190 five-inch camera he was responsible for, so Pete Conrad and Paul Weitz handled the EREP payload on Skylab 2. In addition to ATM operations, Kerwin also had to learn the malfunction procedures in case of failure. Further, he was the crew's navigator and handled the telescope, sextant and computer, and was back-up to Conrad and Weitz on other mission objectives and procedures.[16] This was a pattern that was followed for all three missions.

In December 1971, a Skylab training review revealed that the crews would group under the following headings:

- CSM operations: systems, mission, stowage walkthroughs.
- Saturn workshop operations: systems and missions, stowage and walkthrough,

EVA/IVA, maintenance and housekeeping, photo and TV, In-flight Medical Support System (IMSS).
- Experiments: Apollo Telescope Mount (ATM), Earth Resources Experiment Package (EREP), medical, others.
- Miscellaneous: egress and fire; rescue; SMEAT.

Full-time use of the Houston Command Module Simulator for Skylab crew training began after Apollo 16 (April 1972), when the Apollo 17 crew transferred their training to the Cape. Following Apollo 17 (December 1972), the CMS at the Cape was also committed to Skylab crew training, with the prime crews using the Houston simulator before transferring to the one at the Cape as their mission approached. Orbital workshop systems and mission training ran from February 1972, as did ATM experiment operations training. EREP simulator integrated training began in June 1972.[17]

Reviewing the Skylab training programme

The commander and pilot on each crew conducted most of the training with regard to ascent to orbit, rendezvous, docking, and entry operations. Stowage training involved the whole crew and amounted to two exercises of two hours each for the CSM, the MDA and ATM and four exercises of four hours for the OWS. This included a walkthrough to familiarise the crew with loose equipment stowage arrangements and to perform crew equipment transfers which were not included in activation or deactivation procedures. It also gave them the opportunity to review the in-flight stowage provisions and decal identification configurations, as well as tracking the status of consumable crew equipment. There were also approximately twenty hours of briefings for each crew on Skylab photographic hardware and systems. For biomedical training, the science pilot would serve as an observer for commander and pilot operations, and the commander would observe when the science pilot participated in an experiment.

None of the science pilots were trained in any of the manoeuvring experiments prior to flight. They also did not participate in all of the Skylab Airlock Experiment training. For EREP training, the whole crew trained on the equipment, but the commander and pilot received the majority of the training. The whole crew also participated in ATM training, with the commander or pilot taking responsibility for the other technological experiments. Nine of the experiments aboard the station required very little training by the crew, and ten others required no crew training at all.

For the EVA programme, preparation and post-EVA procedures and operations required ten 1-G walkthroughs for the SL-2 crew, fifteen for the SL-3 crew and fourteen for the SL-4 crew. In addition, each man completed two altitude chamber runs for EMU familiarisation. Part-task training was required prior to the use of the Neutral Buoyancy Simulator (NBS) at Marshall Space Flight Center. During a week at MSFC, groups of two or three EVA simulations were completed, with approximately forty per cent unsuited and sixty per cent suited, in addition to fifteen to twenty NBS runs for each crew. The crews also participated in three of four flights in a KC-135 aircraft, for umbilical management and clothesline deployment training purposes.

For in-flight maintenance, there were over 100 individual training tasks, including ATM (four tasks), Communications (eleven tasks), EPS (ten tasks) and Environmental Control System/Housekeeping (over 100 tasks by themselves). Egress training covered pre-flight, post-flight nominal and emergency egress from the Command Module, the ATM and the MDA during vacuum conditions. Closed-hatch spacecraft tests, launch pad and special facilities evaluations and end-of-mission recovery operations added a further thirty to thirty-five hours over eight or nine exercises, while fire training (detection, prevention and suppression) contributed up to twelve hours to the workload in three to five further exercises.

There were also eighty-two hours of In-flight Medical Support Systems training involving illness and injury, dental, microbiological (for microbial prevention and identification) and laboratory training (blood sampling/count techniques, urine analysis procedures). Physician-astronauts Kerwin, Musgrave and Thornton participated in the analysis of this phase of the training.

Musgrave summarised his Skylab training as a back-up crewmember as:

- Crew reviews of the hardware while still in the manufacturing process.
- Crew participation as operators during spacecraft tests, both at the manufacturer and at the Kennedy Space Center.
- Bench reviews of equipment prior to final stowage into the vehicle.
- Participation in all systems tests in which the interfaces between spacecraft components, equipment, and experiments were checked out.
- Crew participation in electromechanical closeout, where the spacecraft is put into the launch configuration and left that way until it launches.

For the assigned crews, physical examinations and biomedical studies (in addition to the baseline data gathered periodically for sixteen medical experiments) were performed at set intervals: launch minus twelve months; minus six months; minus thirty days; minus twenty-one days; minus fourteen days; minus seven days; minus three days; minus one day; and finally on the day of launch. In October 1973 Musgrave wrote, "There is little doubt that Skylab crews are the most extensively studied biological creatures on the planet." Their total exposure to formal Skylab mission training averaged about 2,800 hours for each mission.

SKYLAB – HUMAN EXPERIENCE

For in-depth historical and operational details of the Skylab programme, see *Skylab: America's Space Station*, David J. Shayler, Springer-Praxis, 2001. What follows are additional observations and comments from the three science pilots who flew the missions.

The first manned mission (Skylab 2 – 25 May–22 Jun 1973)

After a delayed launch and the Skylab repair, the science programme of Skylab 2 crammed as much science and investigations into the final two weeks of the mission as possible. Problems with the ergometer surfaced during the first run of the unit, and because of the increased heat, Kerwin recommended shortening the experiment run. It was also difficult to restrain themselves to the device, so trying to ride it as they had done on the ground was a mistake. On top of that, the flight day schedule was too tight, which reduced their exercise period even further. Kerwin commented, "It's been scheduled strictly on paper, as far as we were concerned, because the other scheduled

The Skylab 2 crew, who would fly the first manned mission to America's space station, stand in front of the Skylab 1 Orbital Workshop atop the last Saturn V, May 1973. Left to right: Pilot Paul Weitz, Commander Pete Conrad and Science Pilot Joe Kerwin.

The 7 June 1973 EVA to deploy the jammed solar array and rescue the Skylab space station was performed by Conrad and Kerwin.

tasks have taken so much time that they have completely absorbed and wiped out the physical training." Kerwin thought that scheduling physical exercise before or after a major activity with a defined time schedule was a serious mistake and expressed concerns that Mission Control would give the exercise period more priority over other objectives. The harness had been designed by Story Musgrave and looked fine on the ground, but it proved impractical on orbit, restricting circulation in the thighs and having a tendency to ride up when pedalling the bike. It was found that maintaining posture with no restraint worked best, so the Musgrave harness was "carefully folded up and shoved in the trash airlock."[16]

Due to the extra workload caused by the EVAs and trying to catch up, recorded pulse rates were higher than expected, but the doctors only reviewed the data several days after it had been taken. The crew were surprised when the doctors asked for further information on their health before allowing a further EVA. Pete Conrad was mostly upset by the fact that the ground had not asked the in-flight doctor for his opinion. One of the reasons for flying a doctor in space was to study the workload and adaptability of the crew in real time, and Kerwin was in a much better position to observe the condition of the crew and the relevance of the data than the doctors on the ground. It seemed that even with a doctor in space, the decisions would be made by the team on the ground – without consultation. An apology was received by the crew, but it was still early days for flying scientists trained as astronauts in space and allowing

The Skylab 2 crew (in Navy Whites) present mementos to President Richard M. Nixon and Soviet Premier Leonid Brezhnev.

them to make independent decisions, particularly when they affected operational issues.

As the flight progressed, the workload lightened to allow the crew to prepare for the return to Earth. Sometimes the work seemed tedious, to such an extent that Kerwin noted that it seemed like they had remained on Flight Day 18 for a week. Kerwin was responsible for the ATM and medical issues on the mission, and was the CM navigator handling the telescope, sextant and computer. He also acknowledged that the crew never really got into a suitable ATM routine on Skylab, at least not for more than a few days at a time. This was basically due to a number of interruptions and frustrations with the ATM solar flare detector, which could also be triggered by the South Atlantic anomaly. In one case, Kerwin began the procedure for recording a solar flare, but fortunately realised what had happened before he used too much film. He was never enthusiastic about the alarm system, and in the post-flight report he labelled it "absolutely worthless."

Kerwin recalled in his 2000 oral interview that the lessons learned on Skylab had direct application to ISS, in terms of habitability, diet and exercise, and the structure of the working day. That the gap between Skylab and ISS was over twenty-five years disappointed him, but the fact that the construction had started was a move in the right direction. To Kerwin, a permanent presence in space was important for establishing a way station for planetary exploration and as a "very useful national laboratory in weightlessness." In fact, Kerwin commented that the one good thing about the substantial gap between Skylab and ISS was that it kept him in useful employment, as he was constantly being asked how things were done on Skylab and his opinion about how things should be done on ISS.

The second manned mission (Skylab 3 – 28 Jul–25 Sep 1973)

Science was given a higher priority on the longer second Skylab mission, as Garriott recalled in 2000. "Ninety per cent of your time is spent doing useful science research on board the Skylab, and everybody was highly trained and highly motivated to get that done. I could tell no significant difference between Alan Bean or Jack Lousma's motivation from my own."[19] One of Garriott's primary tasks, in addition to shifts at the ATM console, was operating the S190B Earth terrain camera during Earth resources passes while his colleagues operated the view finder tracking system for the S191 spectrometer or manned controls at the main display console. Garriott also found time to conduct several science demonstrations concerning the effects of weightlessness on drops of water, magnets, and spinning objects. These demonstrations were filmed and though less important than the primary science investigations,

The Skylab 3 crew, assigned to the second manned mission to the Skylab station. Left to right: Science Pilot Owen Garriott, Pilot Jack Lousma and Commander Al Bean.

The Skylab 3 crew in the 1-G Skylab trainer review their procedures. Left to right: Garriott, Bean, Lousma.

provided the layman with an understanding of flying in space. Along with Jack Lousma's "space fan" broadcasts on life in space, this gave Skylab inexpensive publicity and a lighter approach than the shorter, but more dramatic flights to the Moon. Skylab may have been perceived as a "routine" programme in space, but in some respects it was far more important to the long-term exploration of the solar system than the Apollo missions, though this was not necessarily apparent to the watching public.

Garriott took the opportunity to fly the Experiment 509 prototype of the Manned Manoeuvring Unit inside the orbital workshop after Bean and Lousma had completed the assigned test flights. Garriott had not trained to use the device, but drawing upon his modest amount of training to fly a spacecraft with translation and attitude control thrusters, he soon mastered it. "It was very pleasant to see how easy it was to fly, and perhaps it was useful to the designers as well to see that a person with essentially no training could learn to fly it so quickly."

Garriott also recalled the sheer fun of "running" around the ring of lockers, first demonstrated by the Skylab 2 crew. Al Bean was a competent gymnast, but it was not long before all three members of the second crew were flipping and twisting across the domed upper workshop. "I'm sure you could win all the gold medals on Earth if you were just allowed to compete from space," was Garriott's opinion.

In reflecting on the greatest contribution of his Skylab flight, Garriott chose the gathering of solar measurements of the sun. This was achieved because they were able to see the solar disc for longer periods at UV and X-ray wavelengths, and could

272 Laboratories in the Sky

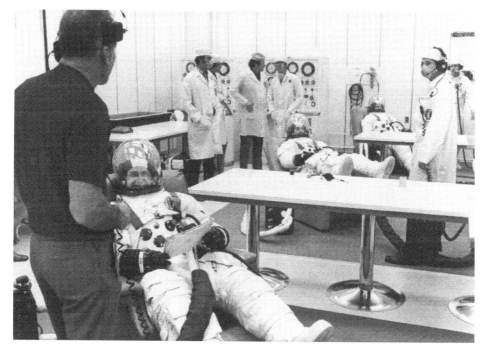

Director of Flight Crew Operations Deke Slayton talks to Skylab 3 commander Al Bean during suiting operations prior to launch. Lousma and Garriott can be seen at rear.

examine the corona and mass ejections, data that was still being used for comparison with unmanned spacecraft data twenty-five years later. The other highlight was the focus on adaptation to long-duration space flight, proving that it is not a handicap if an emphasis is placed on the importance of exercise. This had direct application to extended duration flights on the Shuttle and ISS and will apply particularly on prolonged flights out to Mars.

The third manned mission (Skylab 4 – 16 Nov 1973–18 Feb 1974)

Ed Gibson likes to describe his only space flight, of eighty-four days, as over 2,000 hours of high performance rocket time – although he slept though a third of it.[20] Having worked on AAP/Skylab issues since returning from flight school, Gibson became the point of contact in the Astronaut Office for the development of the ATM on Skylab, along with Garriott.

In 1975, Gibson was interviewed about his work on Skylab and way the third crew developed a working pattern based on the experiences of the first two crews. Apart from a much published disagreement over the initial workload and the illness of Bill Pogue, the results of the third mission were as good if not better than those of the first

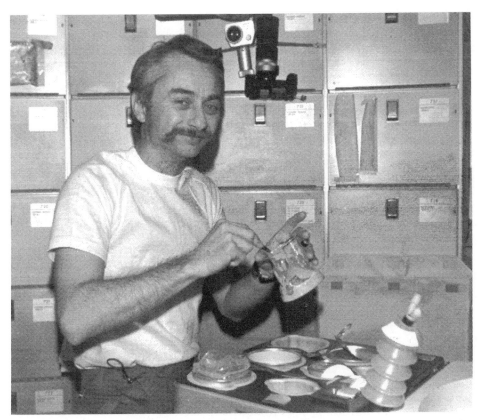

SL-3 Science pilot Owen Garriott enjoys a meal in the wardroom of Skylab during his 59-day mission.

two missions. Gibson recalled that the third crew worked all their days off and stayed up late as much as the second crew had. Eventually, though they struggled early in the mission, they would return with a quantity and quality of work comparable to that of the other two crews, due in part to learning from their colleagues' experiences. Gibson speculated that if the second and third crews had flown in reverse order, then Bean's crew would have come out with the higher quality data, simply due to the learning curve of the "system" based on the previous missions.

Gibson wrote the textbook *The Quiet Sun* (NASA SP-303) in the early 1970s, when the crew training programme for Skylab was being evolved. As most of the astronauts in the Skylab office had piloting or engineering backgrounds, there was little available material to educate them on solar physics to enable them to know what there were doing when using the ATM. Gibson worked with Al Holt of Flight Crew Support to remedy the situation, providing the CB with a handbook of solar physics that explained things to a lesser degree than professional solar scientists might expect, but in a way that the astronauts could understand. Holt worked on the "active sun" whilst Gibson worked on "the quiet sun" and developed it into the textbook published

Owen Garriott on EVA from Skylab during the second manned mission.

in 1973. "It turned out to be a very good textbook, and I've been surprised since then. People come up and tell me they've used it as a textbook in their training and thought it was very good. That's probably one of the more positive feedbacks I've ever gotten in my life about something I've done." But did it help him to secure a seat on Skylab? "Well I don't know. I could have written anything and maybe still got the seat because there were three scientists and three seats. It was that simple. I didn't know at the time."

Early in the mission, a major problem occurred when Bill Pogue fell ill. There was considerable pressure to complete the mission and to prove that astronauts could sustain a long flight in space, so that the Shuttle could fly a week or more in space with no problems. If an astronaut became sick in the first couple of days, then trying to land the Shuttle might present a problem, and this worried NASA into stating to Congress that they would solve the space sickness problem. Aware of this pressure, Gibson suggested bagging the vomit for the mineral balance experiment and telling the ground later, not wishing to "stir up a hornet's nest." But, unbeknown to the crew, the onboard tape was recording all of their comments. These were later dumped to Earth and played back, causing some consternation back on the ground and earning the crew a reprimand from Al Shepard. Then the crew fell behind schedule while trying to

The Skylab 4 crew arrive at KSC in their T-38s ahead of final preparations for the last mission to the Skylab space station. Left to right: Gibson, Carr and Pogue.

acclimatise to space flight and find where everything was stored. There were no open communications to help them and they fell even further behind, until things reached a point when they voiced their opinions to the ground, and the ground responded by expressing their concerns to the crew. Following this frank exchange, the mission finally began to gather pace. By the end of the flight, the crew was surpassing many of the records set by the previous occupants.

Gibson reflected on the lessons learned on his mission. "I still think NASA ought to allow private communication between the ground and the crew, and if the press wants to know what it is, tell them to pound sand. For the efficiency of running that space station (ISS), you need private communications."

Gibson considers Skylab 4 to have been a mission at the forefront of something which was going to grow and become much larger in the future. "The enjoyment was in looking back at Earth, the EVAs and some of the good science that was done and being able to use human ingenuity to accomplish it. Not to be just a button pusher, but to exert some human judgement into how you operate the experiments and then improve the quality of what was brought back. I realise how lucky I was to be in the right spot at the right time." One thing he would readily like to forget was the huge amount of food bars stuffed into the Command Module to supplement the food supplies on the space station. After three months of eating them, he has never been able stomach one since.

Ed Gibson at the ATM console in the Multiple Docking Adapter of the Skylab OWS during Skylab 4.

Skylab Rescue – a fifth mission?

When a leaking RCS thruster problem surfaced early in the Skylab 3 mission, the prospect of a two-man rescue flight to return the three astronauts was seriously considered, to the point of preparing the launch vehicle and training the crew in the profiles required to complete the mission. In the event, the rescue flight was not needed, but had it flown then it could have seen the flight of the first of the Group 6 scientist-astronauts several years before his colleagues. Bill Lenoir was training officially as the back-up science pilot for the second and third missions in a crew that included commander Vance Brand and pilot Don Lind. Brand and Lind were also assigned as the two-man Skylab rescue crew for all three missions.

Don Lind was officially selected as a NASA pilot-astronaut in April 1966, but he was also a space physicist, having gained a PhD in high energy nuclear physics in 1964 from the University of California at Berkeley before entering the astronaut programme. Lind had served in the US Navy from 1954 to 1957, and from 1957 to 1964 he was employed at the Lawrence Radiation Laboratory in Berkeley. For the two years prior to entering the astronaut programme, he worked at NASA's Goddard Space Flight Center in Maryland.

A selection of Skylab onboard shots featuring scientist-astronaut Kerwin (top right) and the pilot-astronauts of SL3 (top left), SL2 (bottom left) and SL4 (bottom right).

Qualified to fly an Apollo spacecraft, and eventually to command a mission, Lind was assigned as a back-up pilot for the Skylab missions, but with his interest in science, and being a co-investigator of an experiment on Skylab, he suggested to Lenoir at one point that they should swap roles. If they happened to fly a replacement mission, he would operate as science pilot and Lenoir as pilot, and their cross-training would cover any responsibilities they would not have time to retrain for.

At the time, there was a suggestion that a fourth manned mission to Skylab might be launched (Skylab 5), lasting about twenty-one days. This would give the new crew the chance to finish off any experiments and close out the station, preparing it for orbital storage until the Shuttle was operational or a re-visit could be planned. The back-up crew for Skylab 3 and 4 (Brand–Lind–Lenoir) was under consideration for this flight, having already trained extensively for such a contingency. However, with the success of Skylab 2 and 3 (the latter increased from fifty-six to fifty-nine days) and indications that Skylab might not support another mission after the third crew, it was decided to extend the Skylab 4 mission by a month from fifty-six to eighty-four days in order to gain the most from the last planned mission, to observe Comet Kohoutek and to alleviate the need for the short Skylab 5 mission. With this decision, Lenoir lost his chance to fly into space aboard Skylab. During the course of their training, when it became obvious that any rescue mission would require only a commander and pilot to bring home a three-person Skylab resident crew, Lind reasoned that it was more sensible to remain as the back-up pilot and be available for a rescue mission.

The assignment of two astronauts who were both pilot trained and scientifically trained to a Skylab crew made a lot of sense and was, in part, a response to the early call to fly two scientist-astronauts on each Skylab crew instead of one. Lind was categorised as a pilot-astronaut, having three years of active pilot assignments in the USN but having also spent nine years as a physicist. Lenoir had qualified from USAF pilot school as part of his astronaut training and had worked as an instructor while studying for his doctorate at MIT. Later, he spent two years at the institute as an associate professor of electrical engineering, developing scientific experiments for satellites before entering the astronaut programme.

Lind had tried to enter the space programme in 1963 (pilot class) and 1965 (scientist-astronaut class) without success, but was finally selected in 1966. He could equally have been part of the 1967 selection and avoided the need to attend the jet pilot training course (as Kerwin and Michel had in 1965). In 1963, Lind had only 850 flying hours, 250 short of the required 1,000 hours, but he also had a PhD (which only Buzz Aldrin had the equivalent of in the group of fourteen finally selected). By 1964, when the call came for scientist-astronauts, he had made up his 250 flying hours deficit in the Naval Reserve and was an almost perfect candidate for the scientist-astronaut selection. But he was a mere seventy-six days over the age limit and was turned down, despite reasoning with NASA that the seventy-six days could effectively be recovered by not having to attend flying school (as four of the six candidates selected would have to do). Over three decades later, he recalled with amusement that his selection in 1966 was probably due in part to NASA being fed up with his persistent applications.[24]

Skylab B

When the Group 6 astronauts arrived at MSC, they were under the impression that they would be trained and assigned to missions under the Apollo Applications Program. For many of them, this meant space stations and not lunar trips. However, the early comments from Slayton that they were in fact surplus to requirements and that their chance of *any* space flight was by no means certain, meant that the prospect of flying on an AAP workshop (Skylab) was, at best, remote.

When Saturn production lines closed in the late 1960s, it had a significant impact on the opportunities for the scientist-astronauts to enter space. In 1971, unassigned hardware from Apollo and Skylab A raised the possibility of "low Earth-orbital manned missions", which potentially included joint missions with the Russians, Earth resources survey missions using the Apollo CSM, or even a second Skylab workshop mission. Though the docking mission with the Russians progressed and was flown as the Apollo-Soyuz Test Project in 1975, the other studies remained on the drawing boards, including the second workshop – Skylab B.

The possibility of a second orbital workshop persisted for the next two years, but was never really funded to a point where its launch was a realistic possibility. This was partly due to NASA's desire not to dilute funds and divert focus away from the new Space Shuttle programme and its plans for a far larger space station. Had the Saturn production lines continued, along with funding, political and public support, then a permanent American presence in space could have been established in the 1970s, long before ISS and as a direct rival to the Soviet Salyut and Mir programmes.

The loss of the second Skylab laboratory still upsets those who could have visited it. Instead, the hardware was relocated to the National Air and Space Museum in Washington, as a representation of the flown Skylab A. Even thirty years on, some of the astronauts who were likely to have visited the second Skylab are still reluctant to go inside the museum display. The thought that it could instead have been their home in orbit is still difficult for them to come to terms with.

For Owen Garriott, the biggest annoyance about some of the "foolish decisions, largely political, partly financial," that were made at the end of the Skylab period was that NASA failed to make use of the flight hardware that had already been purchased and was available. According to Garriott, what NASA should have done was to launch the second Skylab and fly crews to gain six to twelve months of experience, effectively doubling and tripling the advances and results from the first Skylab. Of course, the re-supply issue was a limiting factor (but this may have been overcome) and the hardware was not available to fly more than a few missions, but flying such durations on Skylab B would have given NASA good quality biomedical and operational data on long-duration space flight operations years before cooperation with the Russians on Mir and before ISS was launched. Indeed, Skylab B could have been flown in the late 1970s, with the option of a Skylab C in the early 1980s that would perhaps have been supported by early Shuttle flights.

Countless variations could be suggested for possible crewing. However, it is clear that several of the scientist-astronauts were told that they would be assigned only *if* a second Skylab was authorised. With the loss of two science seats on Skylab A, senior scientist-astronauts Kerwin, Garriott and Gibson were always in line for the three remaining seats, having worked on the programme for almost eight years. Skylab B, however, was the station that would have seen the first flights of the 1967 scientist-astronauts.

Based on the existing crew rotation system used by Deke Slayton during Gemini and Apollo, there is nothing to suggest that the back-up and support crews of Skylab A would not have rotated to crew Skylab B, utilising their training and experiences to the full. Of course, some of the Skylab A crewmen may have been rotated to command the new missions, including some of the science pilots, but those scientists-astronauts who were assigned to Skylab A and who did not fly would have been ideal choices to complete the long-duration crews. Depending on the flight manifest, it could have been possible to fly a pilot commander and two scientist-astronauts aboard each crew. As Karl Henize recalled in 1986, "One of the things that really upset us [the scientist-astronauts] was the crew selection for the Skylab flights. We consciously agitated to get two scientist-astronauts and one pilot on them – it was a scientific mission you know! But Deke Slayton, who was the big boss, didn't see it that way. He condescended to include at least one scientist on each mission and fortunately the Skylab scientist-astronauts and Jack Schmitt proved that scientists could be very capable operational people. So we finally won some grudging acceptance, but it is still not as fair a system as we would like."[25]

As all the scientist-astronauts were pilot trained and capable of flying the CSM, there was an argument in favour of the idea of two or three long-duration flights with a pilot commander and two science astronauts/science pilots, with smaller visiting crews of two pilot-astronauts launched to re-supply Skylab B. They could also have exchanged the docked Apollo CSM for a new one to allow continuous occupation of the second Skylab for added safety and increasing the mission returns, something the Soviets accomplished during Salyut 6/7 operations and continued to operate for Mir and ISS.

The Skylab B scientist-astronaut candidates may therefore have been:

Henize (astronomy): former Skylab A support crew, Capcom and principle investigator of a Skylab A experiment, who joined the programme to fly on a mission such as Skylab B.

Lenoir (electrical engineering): former Skylab A back-up crewman for the second and third missions, whose training would have helped in gaining assignment to an early Skylab B mission.

Musgrave (physician): former back-up Skylab 2 science pilot, who would have been a leading contender for the first long-duration mission to Skylab B.

Parker (astronomer): former Skylab A mission scientist and another astronomer who could have made it to space via a Skylab B assignment.

Thornton (physician): former SMEAT crewman and Skylab support, who would have probably flown the last extended duration mission on Skylab B.

Lind: although he was a Group 5 pilot-astronaut, his back-up experience with the second and third missions, his PhD and his co-investigator status on a Skylab experiment could well have seen him fly as either a pilot- or scientist-astronaut, taking a seat from one of the Group 6 scientist-astronauts.

Had the programme been more defined and funded for adequate hardware, perhaps Allen (physicist), England (geologist) and Chapman (physicist) would have been assigned in back-up roles after completing their Apollo duties. They may well have then rotated to the rumoured Skylab C station in the late 1970s or early 1980s, supported by the Space Shuttle. If that had been the case, the story of NASA's scientist-astronauts would clearly have been very different.

With the cancellation of Skylab B and the abandonment of Skylab A in orbit, any prospect of flying in space for the scientist-astronauts now rested with the Space Shuttle. For some, the wait to fly was too long, but others accepted the fact that their feet would remain on the ground for many years, and that getting involved in the Shuttle programme at an early opportunity might secure that coveted first flight into space once the proving flights had been completed. Their chances of flying were still there, but not for some time to come.

REFERENCES

1 Quotes taken from the Michel Papers held at the Rice University Archives, Houston, except where subsequently noted.
2 Taped interview between Curt Michel and David Shayler, 24 June 1992.
3 Scientific Research magazine, 4 August 1969, article written by Marvin Reid and Nancy De Sanders.
4 *Skylab: America's Space Station*, David J. Shayler, Springer-Praxis, 2001.
5 *Apollo: The Lost and Forgotten Missions*, David J. Shayler, Springer-Praxis, 2002, pp. 1–45
6 Private correspondence from Karl Henize to David Shayler, 1984.
7 *All American Boys*, Walter Cunningham, Macmillan Books, 1977, pp. 248–249.
8 A Gravitational Experiment for SWS-II, CB Memo from P.K. Chapman to KW (G.O. Smith) and KF (T.R. Kloves), 23 December 1969; located in the Skylab Archives, JSC history collection, University of Houston at Clear Lake; copy in AIS Archives.
9 Block II Apollo Flight Crew Designations. Memo from Director of Flight Crew Operations (Deke Slayton), 29 November 1966. From the Apollo History Archives, University of Houston at Clear Lake, copy on file AIS Archives.
10 *Skylab: A Chronology*, NASA SP-4011, 1977, pp. 256–257.
11 *Pioneering in Outer Space*, Hermann Bond et al., Heinemann Educational Books, England, 1971.
12 *Living and Working in Space: A History of Skylab*, NASA SP-4208, 1983.
13 Robert A.R. Parker, JSC Oral History Project transcript, 23 October 2002.

14 *The Selection, Training and Activities of a Surgeon-Astronaut*, Story Musgrave MS, JSC. I.S. Ravdin Lecture Series, American College of Surgeons, Chicago, Illinois, 17 October 1973, supplied by Story Musgrave, copy on file AIS Archives.
15 Karl Henize interview by Andy Salmon, NASA JSC Houston, 22 July 1991.
16 Joseph P. Kerwin JSC Oral History Project transcript, 12 May 2000.
17 Skylab Training Review, briefing notes, 8 December 1971, presented at Building 4 Room 261, MSC; Skylab Collection, JSC History Archives, University of Houston at Clear Lake, copy in AIS archives.
18 Skylab 1/2 Technical Crew Debriefing, 30 June 1973.
19 Owen Garriott JSC Oral History Project transcript, 6 November 2000.
20 Ed Gibson presentation notes dated November 2001, copy in AIS archives.
21 Skylab interview with Edward G. Gibson, Department of Health, Education and Welfare, 25 August 1975, copy in AIS archives.
22 Edward G. Gibson, NASA JSC Oral History Project transcript, 1 December 2000.
23 Private discussion with Ed Gibson, Dave Shayler, May 2001, Nottingham, England.
24 AIS interview with Don Lind, 23 May 2000, London, England.
25 Conversations with an Astronomer-Astronaut, *Sky & Telescope* November 1986, p. 446.

9

Shuttling into Space

Completing a manned lunar landing and safe return to Earth on the first attempt was a huge achievement for NASA. But accomplishing the goal set by President John F. Kennedy in 1961 so successfully also contributed to the demise of the Apollo programme and the era of single-mission spacecraft. Political and public support for the huge investment the space programme required in the 1960s declined as fast as it had risen. Indeed, by 1969, several of the astronauts involved in America's pioneering space programmes had also departed for new career goals. The aborted flight of Apollo 13 in 1970, further budget restrictions and other domestic issues, all contributed to the end of the manned lunar programme in 1972, with the final flight of an Apollo-type spacecraft in 1975.

Most of the veteran astronauts between 1961 and 1975 were from the pilot groups and only four scientist-astronauts, all from the 1965 selection, flew in space in the first fifteen years following the Mercury Seven selection in 1959. With only Jack Schmitt assigned to an Apollo mission and three others (Joe Kerwin, Owen Garriott and Ed Gibson) to Skylab missions, the chances of a flight into space for the class of 1967 seemed no closer eight years later than it had when they had joined. The only opportunity they had was to tough out the long wait for the Space Shuttle that was planned to follow Apollo by the end of the 1970s. But, as the early test flights would probably be crewed by test pilots and not scientist-astronauts, the wait might even be as long as a further eight years. In fact, the first members of the 1967 group finally made it into space in 1982, some fifteen years after selection. Others had to wait until 1985, and some never achieved their dream at all.

SPACE SHUTTLE – A RELIABLE ACCESS TO SPACE?

When America achieved the first manned lunar landing in July 1969, there were plans for at least a further nine landings using the basic Apollo hardware before the expansion of the programme to support longer surface stays and the creation of a rudimentary lunar base. In addition, a new space infrastructure was being planned to operate in conjunction with Apollo hardware, at least for the 1970s and 1980s. This included a new manned spacecraft called Space Shuttle. When first envisaged in 1968, the Shuttle was just one element in a whole space infrastructure that included a large Earth-orbiting space station, a space tug, a lunar base, an Earth–Moon–Earth supply system, and the first manned flights to Mars. Severe budget cuts between 1968 and 1972 and design changes in the Shuttle (with the decision to proceed not with a manned booster but with only a manned orbiter using solid rocket boosters and an external tank) resulted in all but the Shuttle being shelved.

"An entirely new type of space transportation system"

After more than a decade of development and debate, authorisation to proceed with the new programme to replace the Apollo–Saturn series of vehicles was given by President Richard M. Nixon on 5 January 1972. In his press conference statement, Nixon said: "I have decided today that the United States should proceed at once with the development of an entirely new type of space transportation system designed to help transform the space frontier of the 1970s into familiar territory, easily accessible for human endeavour in the 1980s and 1990s. This system will centre on a space vehicle that can shuttle repeatedly from Earth into orbit and back. It will revolutionise transportation into near space, by routinising it. It will take the astronomical costs out of astronautics. In short, it will go a long way towards delivering the rich benefits from practical space utilisation and the valuable spin-offs from space efforts into the daily lives of Americans and all people."[1]

This statement helped to create the impression that the Shuttle would be the answer to all of America's launch problems, with routine, cheap and reliable access to space. In a joint statement, NASA administrator, Dr. James Fletcher, called the decision "a most historic step in the nation's space program. It will change the nature of what man can do in space. By the end of this decade the nation will have the means of getting men and equipment to and from space routinely, on a moment's notice if necessary, and at a small fraction of today's cost." Fletcher described the new vehicle as an aircraft-like orbiter, about the size of a commercial DC-9 airliner, with the capacity to take payloads up to 4.6 metres in diameter and 18 metres long (with a mass of up to 29,500 kg) into orbit and back again. It would be launched by unmanned boosters on missions of one week, during which the crew would launch, service or recover unmanned spacecraft, perform experiments and, in time, re-supply and re-staff space modules brought into orbit by other Shuttles to create a space station. All this would be "in a modest budget to make space operations less complex and, more importantly, less costly."

This view from 1972 was, at best, optimistic and at worst, seriously flawed. With the benefit of hindsight in 2006, after twenty-five years of Shuttle operations, over 114 launches and two major fatal accidents, the reality of what the Shuttle has achieved is a very different picture to the plans proposed thirty-four years ago. With the loss of a space platform in the early 1970s, the Shuttle also lost its assembly and logistics support role; one for which it was well designed, as evidenced by Shuttle operations during Shuttle–Mir and ISS construction.

Instead, the Shuttle's role became more commercial, providing launch and payload services to recoup some of the investment in the programme. After a series of orbital test flights, the system would be designated "operational", with a programme of commercial satellite deployments and payload capacity sold to those who wished to gain access to space but did not have the means to do so on their own. This "selling of the Shuttle" was a misinterpretation in the press, and consequently in the public eye. In reality, the Shuttle could never achieve a "commercially" operational status. In retrospect, it *has* achieved a role in space operations by deploying, servicing and retrieving satellites, space probes and great observatories. It has also proven its worth in supporting various payloads of scientific, engineering and military usefulness, and as a short-term space platform for scientific research.

On 28 March 1972, the *New York Times* published a response by University of Michigan astronomer, James A. Louden, to a 16 January letter from former scientist-astronaut Brian O'Leary, who had questioned the compatibility of the Shuttle with national space goals. Louden re-emphasised that the most important aspect of the programme was the ability of the Shuttle to carry passengers: "For the first time, scientists will be able to perform experiments in space without spending years in irrelevant pilot training first." Of course, this is exactly what the two scientist-astronaut groups had had to undertake before any assignment as a NASA astronaut, but now the Shuttle offered them the chance to make at least one space flight. Four months later, on 29 July, Administrator Fletcher indicated that NASA had plans to expand astronaut opportunities by selecting minority and women candidates. The following year, discussions with European space organisations would investigate the possibility of European assistance in post-Apollo activities in the Shuttle programme, which could include flying foreign astronauts on the Shuttle.

The first documented use of the "Sortie Can", the prelude to Spacelab, came in September 1971, as part of an in-house NASA design study for a research laboratory that could be carried in the payload bay of the Shuttle for short duration missions. After more than two years of negotiations between the European space organisations and the United States, an agreement was reached in August 1973 to jointly develop a "space laboratory module" for the Shuttle. In October 1973, the NASA HQ Sortie Lab Task Force was renamed the Spacelab Program Office.

With the introduction of the Shuttle, offering as many as seven flight seats, the NASA scientist-astronauts could be forgiven for thinking that their chances of a flight would greatly increase. But if new astronauts with scientific backgrounds were being brought into the programme, and with the added prospect of foreign crew members, the incumbent scientist-astronauts felt that they might still have no flight prospects at all. Some of their peers had already left NASA due to the lack of space flight opportunities and the lack of scientific research on the flights that had already taken place. With the Shuttle, seat numbers increased and science would – hopefully – follow, but it was still not certain that astronaut seniority would play any part in crew selection. The method of crew selection was devised by Director of Crew Operations, Deke Slayton, in which assignment to a back-up crew on one flight meant (normally) missing the next two missions and flying as the prime crew on the fourth. How one was assigned to the back-up crew in the first place, however, remained a mystery.

As many of the scientist-astronauts had not expected a lunar flight, their work on Apollo Applications and Skylab helped focus their experience into working with space science investigations, objectives and hardware. This would also be useful for early support assignments on the Shuttle programme, and these technical assignments were directed to the astronauts who were not already deep in training for Skylab.

Reorganising the scientist-astronaut office

By 1974, several astronauts had already begun working on development and support issues in the Space Shuttle Branch Office of the Astronaut Office (CB) at the Johnson Space Center (JSC) in Houston. As part of a re-focus of activities from the Apollo/Skylab era to the new programme, the departments in the Astronaut Office were restructured to allow for retirements from the astronaut team and to better utilise the talents, skills and experiences of the astronauts (especially the scientist-astronauts) in supporting the developing Shuttle programme.

A major reorganisation of JSC was planned for February 1974 and was announced on 7 December 1973. As part of this restructuring, scientist-astronauts still active in the programme were assigned to CB offices in either the Science and Applications or the Life Sciences directorates. Owen Garriott became Deputy to the Director of the Science and Applications Directorate, Jack Schmitt was named Chief of the Science and Applications Astronaut Office, and Joe Kerwin would be the Chief of the Life Science Astronaut Office.[2] The group was assigned thus:

Science and Applications CB
Schmitt (Branch Chief and Chief of the Scientist-Astronaut group)
Garriott (also Deputy to the Director, Science and Applications Directorate)
Gibson (at that time in orbit, during the Skylab 4 mission)
Allen
England
Henize
Lenoir
Parker

Life Science CB
Kerwin (Branch Chief)
Musgrave
Thornton

In May 1974, Garriott replaced Schmitt as Chief of the Scientist-Astronaut group, due to the latter being named (effective 13 May) as the new NASA Administrator for Energy Programs in Washington DC. In fact, Schmitt had been on temporary assignment at Headquarters as the Special Assistant to the Administrator for Energy Research and Development since 29 January. The former moonwalker remained in this position until he left NASA on 30 August 1975. Garriott continued as Chief Scientist-Astronaut until September 1974 (replaced by Bob Parker), when he became the Acting Director for Science and Applications at NASA JSC in Houston. He held that post until taking a sabbatical at Stanford University in 1976, before returning to the Astronaut Office as Director of Science and Applications and Assistant Director for Space and Life Sciences.

Gibson was disappointed at the lack of future space flight opportunities and was not looking forward to a wait of several years for a second space flight. On 30 November 1974, he resigned from NASA to become a Senior Staff Scientist at The Aerospace Corporation in Los Angeles, California. Here, he conducted research on the solar physics data collected during the three Skylab missions, concentrating his research on solar activity, its development, the underlying causes and its effects on the Earth. The other eight scientist-astronauts were on assignments that they hoped would lead to selection to Shuttle missions – probably not on the early test flights, but certainly on the first "operational missions".

Initial CB technical assignments for these astronauts included supporting the design and development of workstations and aides in the crew module of the Shuttle, and crew support equipment. In many cases these assignments lasted several years, from 1973/4 until assignment to a crew training cycle:[3]

Allen worked on payload support issues, and crew stations, displays and controls for the physical sciences.

England was assigned to the Operations Mission Development Group.

Henize, a professional astronomer, was assigned to the development of Shuttle-borne astronomical payloads, payload support and handling, and crew stations, displays and controls for the physical sciences.

Lenoir continued his personal interest in the role of humans in the remote sensing of Earth. His CB assignments included work on payloads; crew stations, displays and controls for the physical sciences; payload deployment and retrieval systems and procedures; the development of the Shuttle Extravehicular Mobility Unit (EMU) and Portable Life Support Systems (PLSS); supporting preparations for the Orbital Flight Test (OFT) missions; the development of the Remote Manipulator System (RMS); and studies of Power Satellites.

Parker worked in the Payload Operations Working Group with NASA and ESA on payload support and crew stations, displays and controls for the physical sciences. In September 1974, he replaced Garriott as Chief of the Scientist-Astronauts, until the position was redefined in 1978.

Kerwin worked in crew station design, controls and, drawing on his professional experience, medical monitoring of crew members.

Musgrave was assigned to work on developing payload crew stations, and displays and controls for the life sciences. He was also assigned as CB point of contact for all Shuttle EVA-related equipment, include pressure suits, PLSS, the airlock, EVA toolkit, and Manned Manoeuvring Unit issues, and in the development of the Shuttle Avionics Integration Laboratory (SAIL).

Thornton was assigned payload support and crew station issues, controls and displays for the life sciences, and developmental studies on deployable payloads, although his primary field of responsibility and interest was in evolving crew procedures and techniques for monitoring crew health on orbit.

SIMULATING SPACELAB

Spacelab was Europe's contribution to the Space Shuttle programme, borne out of negotiations held during 1969 between NASA and the European Space Research Organisation (ESRO – from 1975, the European Space Agency, ESA), which decided to develop a manned orbital laboratory to be carried to and from orbit by the Shuttle in its cargo bay. The idea of a "Sortie Module" evolved in 1971, with a first concept appearing in 1972 and a final design for the Space Laboratory (Spacelab) emerging in 1973. Contracts were awarded in 1974, with a design and development phase held between 1974 and 1978. Construction commenced early in 1977 and was completed with delivery of the engineering mock-up to KSC in 1980, the first flight unit (consisting of a long pressurised module and five pallets) in 1981 and the second in 1982.[4]

Shuttle's laboratory

Spacelab was a purpose-built laboratory, providing a pressurised shell for outfitting with science racks that could be exchanged and adapted for a specific flight. In addition to power and thermal control subsystems, there was also a communications system to and from the experiments and provision for airlocks and pointing systems where required. An array of computers and associated software handled the management of the subsystems, experiments and data flow. The crew could oversee and intervene in experiment operations, monitor subsystems and, where required, repair faulty equipment or amend the collection of data to reflect real-time situations during an experiment process.

Experiments aboard Mercury, Gemini and Apollo spacecraft were limited by the size of the spacecraft, the design of the mission and the availability of the crew. With up to seven astronauts, a Shuttle flying a Spacelab mission would allow the division of labour between those who looked after the vehicle (normally the commander, pilot and mission specialist 2/flight engineer) and those who managed and operated the science package (NASA mission specialists and non-NASA payload specialists). Spacelab missions in the payload bay of the Shuttle could last up to two weeks

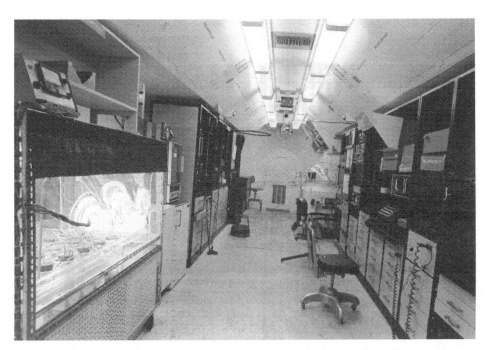

The interior of the October 1974 SMS module showing the laboratory-like racks to simulate Spacelab science racks.

Musgrave injects nutrients into the root system of pine seedlings during the October 1974 Spacelab life-sciences simulation.

(this was later extended up to seventeen days), offering greater science capability than the pioneering US space programmes, but far less than the Skylab missions.

Though restricted by the design parameters of the Shuttle in terms of launch weight, mission duration, consumables and crew time, missions could repeat flights of the same hardware, or with amendments to the hardware on later flights based upon the experiences of previous missions. A dedicated science crew could operate independently of an orbiter crew and work over two twelve-hour shift cycles offering, on a ten-day mission, twenty man-days of operations. For the NASA scientist-astronaut team, it provided the opportunity to ply their trade in space and combine their astronaut and pilot training with their scientific and engineering background. Experience gathered from Skylab also helped identify effective working practices that were applied by several scientist-astronauts in their work on Shuttle/Spacelab development issues.

But it was clear that Spacelab was no Skylab, and what NASA and America really needed was a Shuttle programme that included Spacelab for short-term research and a space station for long-term research. In the 1970s that prospect, though often promoted, looked increasingly remote, as it was becoming a challenge just to get the first Shuttle off the launch pad, let alone fly a dedicated science mission.

The term "Spacelab" is generally applied to the missions carrying the pressurised laboratory, but in fact it also applied to the missions carrying a pallet-only science research package, or a combination of both. The size and combination of the "Spacelab payload" depended on the specific mission being flown inside the Shuttle payload bay. Access to the laboratory was via a 1-m-diameter transfer tunnel from the mid-deck aft hatch to the Spacelab module. The EVA hatch was incorporated into the forward end of this transfer tunnel, giving a "shirtsleeve" environment from the Shuttle flight and mid-deck, via the tunnel, to the laboratory. In addition to the pressurised module, there was a limited space available on the mid-deck as well as facilities for a small amount of Spacelab support work on the aft flight deck of the orbiter.

The *Short Module (SM)* consisted of one, 4-m-diameter pressurised cylindrical module, 2.7 m long, enclosed between two end cones. This was the core segment which contained the Spacelab subsystems supporting the laboratory facilities. There were also eight cubic metres of space available for experiment hardware. This configuration was never flown on a Shuttle mission.

The *Long Module (LM)* comprised two short module elements, with the core segment augmented by an experiment segment giving a further 14.6 cubic metres for experiments. Inside the laboratory sections, experimental research hardware was located in 48-cm laboratory equipment trays, which could also be mounted centrally on the "floor" of the module.

The *Spacelab Pallets (SP)* were available to support unpressurised equipment, experiments and associated hardware. Each U-shaped pallet measured 3 m long and 4 m wide. These could be mounted singly, or in a series of up to five called a "pallet train". In the case of a pallet-only configuration, all the essential subsystems were located in a 1.1-m-diameter cylinder, termed an "igloo", located at the front of the first pallet. Experiments on the pallet-only payload could be controlled from the aft flight deck, the module or the ground, or in some cases adjusted by contingency EVA.

With a combination of short or long modules, modules and pallets, or pallets only, the system was extremely versatile and could support a range of missions in various fields. Laboratory modules usually supported life sciences, material and microgravity studies, while pallet-only missions supported astronomy, Earth studies or space technical development. Not all Spacelab missions were termed "Spacelab" and between 1981 and 1998, over thirty missions carried or featured Spacelab components. Members of the scientist-astronaut group were instrumental in the development of, and operations from, the early Spacelab missions, but before these could be flown, a significant amount of work had to be completed on the ground in developing the hardware, experimental protocols, procedures and operations. Between 1973 and 1978, when the first crews for Spacelab were named, several members of the scientist-astronaut group participated in ground-breaking and important simulations geared towards flight-qualifying the hardware.

Ground and airborne simulations

In order to define flight requirements, a series of ground and airborne simulations was completed by NASA with the support of several field centres. At the Marshall Space Flight Center (MSFC) in Huntsville, Alabama, a Concept Verification Test (CVT) programme was assembled on 25 July 1973, with the aim of simulating high data-rate experiments on the ground to test data compression techniques. These also included data interaction and onboard processing.[5] In order to support ESA's development work on Spacelab (and in part to reduce costs in developing experiments that could be flown in space), a general-purpose laboratory resembling the Spacelab configuration was added to the CVT at Marshall in January 1974. After reviewing the original facility, it was suggested that the CVT version should be upgraded to resemble the design of Spacelab more closely. A Preliminary Requirements Review for the new simulator was held on 29–30 May 1974. The CVT Spacelab test programme was established to provide participation by principal investigators who were developing potential experiments for Spacelab missions. This would allow them to test hardware, techniques and interfaces in order to verify their operation prior to assignment to specific missions.

The CVT programme was established at Marshall in 1971 to support that centre's bid for space station concepts that were under development and consideration across the agency, as part of post-Apollo studies. The CVT simulated environmental control and life support systems that would be applicable to future manned spacecraft – ideally a space station, but also with applications to Shuttle and to Spacelab. This desire at Marshall to enter the realm of manned space flight conflicted with the Houston field centre's management and mandate. This often led to contention with

Houston, of the type that surfaced over the previous decade in both the Apollo and Skylab programmes, and which would not be resolved for some time. With Marshall developing the CVT for space station, and Houston arguing for total payload control for the Shuttle, including Spacelab, there were disagreements over how the CVT could help with the development of the "Sortie Can" that became Spacelab.[6]

Several CVT simulations were completed from 1972, related to developing the facility and future life support systems, but those specifically Spacelab-related were:

CVT Test No. 1 (Jan 1974) with experiments conducted in ionospheric disturbances, atmospheric cloud physics, metal alloy preparation, high-energy astronomy and super fluid helium.

CVT Test No. 2 (Spring 1974). Little information is available about this test, other than the date it was conducted.

CVT Test No. 3 (15–19 Jul 1974) with an integrated life sciences mission planned and conducted by scientists from the Ames Research Center. The objectives were to demonstrate the protocols of each candidate experiment, module housing units, the rack-mounted equipment and other techniques.

CTV Test No. 4 (16–21 Dec 1974) was a five-day simulated Spacelab mission with eleven materials sciences experiments that the four female crew members had developed. These four women, all employees of Marshall Space Flight Center, were Ann F. Whitaker, Carolyn S. Griner, Mary Helen Johnston and Doris Chandler. This simulation provided a wealth of data on the benefits of having a scientifically trained crew aboard the facility to identify and repair minor malfunctions. In the 1976 report of this test, it was stated: "The value of the well trained scientist crew was emphasised during the test. Had it not been for the extremely knowledgeable science crew, two experiments at least would have been lost early in the simulation. [They] were saved by their knowledge of both the hardware and the science that was to be obtained."[7]

CVT Test No. 5 (10 Aug 1975) was a multi-discipline, multi-centre simulation, which also provided valuable information on the standardisation of hardware, crew training, communication requirements and procedures for dealing with equipment failures.

At this time (1975), funding for the CVT programme was a serious issue. Experiments were under development from a variety of NASA centres, but funding for the test missions to evaluate them came from the Office of Manned Spaceflight. The Deputy Associate Administrator, John E. Naugle, issued a memo asking for additional funding (and further simulations) to the Directorates of Science, Applications and

(From left) Dr. Mary Helen Johnston, Carolyn Griner and Dr. Ann Whitaker complete a Scuba diving training exercise at the Marshall Space Flight Center's Spacelab Neutral Buoyancy Simulator in preparation for CTV Test No. 4 in December 1974.

Manned Spaceflight, but their replies were very negative. Other facilities and simulations had superseded the CVT programme and this response signalled its demise, although two other planned simulations were completed first.

CVT Test No. 6 (17–21 Nov 1975) was a materials sciences simulation to determine whether a team of scientists, with moderate experiment operations training, could operate a package of experiments "in orbit". During the test, they were monitored via downlink TV and two-way voice communications by a team of principal investigators.

CVT Test no 7 (15 Jul 1976) was the final CVT simulation and used a high-energy cosmic ray balloon experiment to gather scientific data.

In the NASA Spacelab history, author Douglas Lord stated that the CVT "had provided very useful information and operational experience, [but] the program fell victim to the vagaries of organisational and budgetary life."[8] At the same time as the CVT programme was running, another series of Spacelab simulations was being conducted, with more participation by the scientist-astronauts.

AIRBORNE SCIENCE/SPACELAB EXPERIMENT SYSTEM SIMULATION (ASSESS)

The Ames Airborne Science Program had been established in 1971 using the Convair 990 (CV-990) aircraft. At the beginning of the Spacelab programme, discussions were held about whether these flights could add significantly to the planning for Spacelab missions. Representatives of NASA and ESRO (later ESA) were to participate in CV-990 missions to evaluate the facility for more dedicated Spacelab simulations. The first scientist-astronauts assigned to support these simulations were Joe Allen, Bob Parker and Karl Henize. Story Musgrave and Bill Thornton would work on ground-based Spacelab simulations, gathering baseline medical data.

Learjet simulation programme 1972–4

In order to evaluate the Shuttle/Spacelab concept in more detail, a series of simpler simulations was planned, flying an instrument-laden Learjet aircraft based at Ames Research Center to replicate the constraints of a Shuttle mission using a team of PIs associated with each experiment. This series of four flights ran between 1972 and 1974. In addition, a flight on the CV-990 was used to evaluate experiment operation procedures, using a limited number of specially selected experiment operators, designated "Payload Specialists". During the week of 30 September to 4 October 1974, the fourth Spacelab Learjet Simulation was completed at Ames. Assigned as Experiment Operators (EO) were astronaut Karl Henize from JSC and Lee Weaver of MSFC. This was the first time that experiment operators were trained as "crew members" and the first direct CB involvement in a Spacelab development mission. It was also the first time that specially trained EOs had been used, rather than a PI and an assistant, as on the earlier flights in the series.

Learjet 4 simulation mission

A goal of two flights per night was established for maximum experiment operation, and the Experiment Operators would be confined to the accommodation trailer or the aircraft for the duration of the mission. The one instrument provided (due, in part, to space limitations in the Learjet) was a twelve-inch broadband infrared photometer that operated in the region between 40 and 200 micrometres. The purpose of this flight was to measure the concentration of IR radiation from various astronomical objectives. During the flight, Henize directed the telescope, while Weaver operated the electronics rack containing the equipment for signal amplification, data recording and in-flight data processing.

In the draft copy of the preliminary report on the simulation mission,[9] Karl Henize added his own comments and amendments to the text. Henize noted that total training time for this mission was about 140 hours but stated that, although training provided a significant part of this mission, they were conscious of their own shortcomings with regard to knowledge of the equipment, especially in malfunctions and remedies. The report noted that through the experiences of the Learjet 4 mission, the results would "be helpful in planning subsequent missions."

Training for the flight began at the end of May 1974 with an introduction to the equipment involved. After three days spent at the PI's home institution at the end of August, the programme was completed by actual flight training with the PI at Ames during September. For the mission, ten flights were planned over five successive nights. In fact, nine flights were flown over six nights, due to aircraft or signal problems. The EOs were occupied with activities directly related to the mission for about twelve hours out of every twenty-four. This time was divided, roughly, into five hours of pre-flight preparations, four hours of flying time and three hours of post-flight discussion of data and future planning. To simulate future Spacelab operations more accurately, the workbench area was reduced to the equivalent of that planned for actual Spacelab Shuttle missions. Storage was also limited, but was found to be satisfactory for this mission.

Henize noted that although extra equipment was added after the "launch date" in order to keep the mission operational, this did not violate any constraints, as they used back-up equipment already acquired that would probably be in the Spacelab anyway. In respect to the inadequacy of EO training for repair and maintenance of equipment without the detailed and direct assistance of the PI, Henize noted that such direct involvement of the PI would have been more efficient, with prompt troubleshooting and repair. But it was not the purpose of the flight to prove that an EO should be trained to the efficiency of a PI, although this might be advantageous in the future should the PI not be selected for flight.

The report noted that minor mechanical servicing was all that was required to restore some failed equipment, but the EO training did not cover this possibility. Henize noted, however, that training should not include such minor repairs. "It is the opinion of the EOs that the extensive training required to become efficient in all phases of equipment troubleshooting and repair would not be sensible. This mission clearly indicates that reliance on replacement units and detailed help from the ground is an efficient way to handle repairs." Essentially, Henize was stating that Experiment Operators were just that – operators of experiments, not system repairmen – a factor that should be noted for future in-flight failures during Spacelab missions.

Training activities would now move on to a series of airborne simulations in a larger aircraft, and ground simulations in a mock-up Spacelab module that would be more detailed than the CVT or Learjet simulations had been able to achieve.

ORIGINS OF ASSESS

On 19 March 1974, at the ninth meeting of the Joint Spacelab Working Group (JSWG), Dr. Ortner of ESRO proposed a joint ESRO/NASA programme, to be called the Airborne Science/Spacelab Experiment System Simulation. Leading NASA's participation in this programme was Joe Allen. A mission planning group (and requests for flight experiments) was already working towards a proposed 1975 mission that would be used to determine Spacelab design parameters, study operational concepts, and perform a variety of scientific experiments. On 12 September, the ASSESS Mission Planning Group (MPG) identified potential Experiment Operators for the initial ASSESS mission as Dr. John Beckman from Queen Mary College (UK) of ESTEC and Nick Wells from the University of Sussex (UK), both representing ESRO, and Americans Lee Weaver of MSFC (assigned as back-up and who was about to participate in the Learjet 4 simulation), D. Harper of the University of Chicago (later replaced by Kenneth Dick of the University of Maryland) and Joe Allen of JSC. No European back-up operator was assigned. Bob Parker was subsequently requested for support duties in planning the mission.[10]

On 5 November 1974, Philip E. Culbertson, Director of Mission and Payload Integration at NASA HQ, wrote to Bob Parker with reference to an ASSESS planning meeting scheduled for 19–21 November at ESRO HQ in Paris, France. This was the only joint meeting of US and European participants prior to the installation of experiments and the actual flights. Culbertson urged either Parker or Allen to attend in their role of NASA Experimental Operator: "Your attendance is of critical importance due to the nature of the subjects to be discussed, the knowledge you both can provide to the deliberations of the group as scientists trained in space operations, and the impact the resulting decision will have on NASA's role in the ASSESS mission."[11] The meeting was to identify primary operations for the experiments, operator training definition and scheduling and the allocation of observation times. Culbertson stressed that the education and skills of both scientist-astronauts, were invaluable in themselves but moreover, they were not obtainable from "any other attendee". The role of ASSESS as a precursor to Spacelab missions was a key to providing a model of successful and meaningful cooperation in later years. In addition, it was clear from these memos that the specific skills of the scientist-astronauts were being recognised and used to plan and prepare the scientific Shuttle missions that would follow the engineering Orbital Flight Tests.

Also in November, Owen Garriott, then Acting Director of Science and Applications at JSC, informed the Spacelab Program Office in Washington DC that either Allen or Parker would be put forward for participation in the ASSESS mission. Primary candidate Joe Allen had previously been committed to the "Outlook for Space" Study Group and if this assignment conflicted with ASSESS, he would be replaced by Parker for the planned May 1975 mission.[12] This proved to be the case, as Parker became more involved in ASSESS-I and replaced Allen as primary NASA Experiment Operator for the mission. By the end of 1974, planning for the first in the series of ASSESS simulation flights was underway.

Scientist-astronauts' role on Space Shuttle missions

On 18 November 1974, at the request of NASA, the Space Program Advisory Council (SPAC) undertook a study of the role of the scientist-astronauts within the Shuttle programme. This report sought constructive suggestions from a broad segment of the community of scientists, scientist-astronauts and programme managers associated with past manned space science and application programmes, or those who were likely to be associated with the forthcoming Shuttle era. In conclusion, the committee emphasised the high degree of usefulness a scientist-astronaut could have on experiment and payload integration issues between the principal investigators (PIs) and the CB office for Spacelab-type missions, and as the primary experiment operator on dedicated Spacelab missions.[13] At the time, the terminology for non-pilot crew members on Shuttle flights was being defined. In Spacelab simulations, the term "Experiment Operator (EO)" was being used, but this would evolve into 'Mission Specialist (MS)' for career astronauts and "Payload Specialist (PS)" for non-career crew members.

Responding to the request for suggestions on the utilisation of scientist-astronauts in the Shuttle programme, Louis C. Haughney, Geophysics Program Manager of the Airborne Science Office at Ames, wrote to the SPAC suggesting that a scientist-astronaut would be best suited to flights involving Spacelab: "A scientist-astronaut's years of study and training has allowed him to gain comprehension, experience and skills in both aspects of the Spacelab program; the scientific and engineering investigations on the one hand, and the spacecraft systems and operations on the other." In addition to serving as an overall coordinator of a specific Spacelab flight, Haughney suggested, "it may be worthwhile to consider SAs (scientist-astronauts) as mission managers for Spacelab flights."[14]

Haughney proceeded to suggest an example of how such a concept might work. When a Spacelab flight is first proposed or considered, a scientist-astronaut could be assigned based on a particular scientific or engineering background relating to that specific flight. Using his experience and training, he could then provide input into the NASA decision to fly the mission or not, develop the flight plan, and review proposed experiments, being the point of contact with PIs and becoming familiar with both the science objectives and the equipment. When the flight is approved, between six and twelve months prior to launch, the SA could be assigned as "the full-time manger of the mission and also the Mission Specialist for the flight." A second SA with similar background could be assigned as Assistant Mission Manager and back-up MS. With the right support, between them they could organise the team of specialists, including the training of payload specialists in both Spacelab systems and experiments. On the flight, Haughney suggested, the SA/Mission Manager could serve as MS and, in effect, the on-orbit flight director. He would be responsible for all Spacelab systems relating to experimental payloads, including electrical power supply and data processing. The back-up SA/Assistant Mission Manager would remain on the ground coordinating ground-based activities, working with the on-orbit SA to amend the flight plan and coordinating with the PIs over the day-to-day evaluation of the flight.

In concluding his suggestion, Haughney stated, "The assignment of scientist-astronauts as the managers of Spacelab missions utilises their dual backgrounds. Their understanding and appreciation of the scientific and engineering objectives are combined with their intimate knowledge of the Spacelab's configuration. Thus they are in a unique position to direct the most effective and efficient use of the Spacelab and the Space Shuttle to carry out the desired investigations." This is not the way the role evolved, but it gave rise to the concept of a science team working on the experiments on research flights in the 1980s, with an orbiter crew handling the Shuttle. This led in the 1990s to a leading MS being designated Payload Commander and, more recently, developed into the position of NASA Science Officer on board ISS. (See Chapter 12.)

ASSESS-I

The first mission was planned as a series of five flights aboard the CV-990 Galileo II aircraft on consecutive days, to resemble the useful experiment time aboard the proposed seven-day Spacelab mission. Since the experiments were "real", a further two weeks of airborne flights were scheduled for the PIs following the "Spacelab mission" to ensure the achievement of scientific objectives for each experiment. The programme of experiments included research into infrared astronomy and upper atmosphere physics. The chosen group of Experiment Operators from Europe and America reflected the proposed research fields for future Spacelab missions; in this case, airborne astronomers, a doctoral graduate student, an engineer and (with Parker) a science-trained astronaut with a background in astronomy. The mission began on 7 June 1975 with the international crew flying a six-day mission and living in specially designed quarters adjacent to the aircraft parking area between flights. During the mission, the crew completed infrared observations of Earth's upper atmosphere, Venus and various stars (including NGC 7000, Rho-Ophluchus and other significant celestial features), as well as UV measurements of planetary atmospheres.

This series of flights not only provided useful information from the flown experiments, but also evaluated key aspects of the proposed Spacelab mission operations. The crew worked in teams of three, with the fourth retiring to another part of the flying laboratory, simulating a crew member moving into the mid-deck or flight deck area of the Shuttle. This was planned for the actual space missions, as the Spacelab life support system (under development) could only support up to three crew members working in the module simultaneously for six hours, with a maximum one-hour overlap by a fourth person. The CV-990 flew a night-time triangular ground track (at about 11,300 m for astronomical observations), typically from Moffett Field over Santa Barbara in California, to Payette in Idaho, continuing on to El Paso in Texas, and finally returning to Santa Barbara.

Using standard racks provided by the Ames Research Center's Airborne Science Office and locked into the aircraft's seat tracks, the experimenters evaluated their mission as EOs. They conducted several experiments each at the same time, developed "space"-to-ground communications with the mission operations centre (located at the same hanger at Ames in which the vehicle was parked between flights and where the crew rested), undertook real-time maintenance and repair of experiment hardware, automated certain functions and evaluated the benefits of both dedicated and general purpose data systems. Isolated for the duration of the mission, the airborne crew moved from the special living quarters to the aircraft by means of a lift van, simulating transfer from the mid-deck to the Spacelab module. The decision to isolate the crew reflected potential decision-making processes on Spacelab missions, although there was no physiological or psychological monitoring of the crew. The flight crew of the aircraft, observers and data system operators were not confined during the experiment. A briefing was held about four hours before each flight to confirm the day's flight plan and experiment objectives, and each six- to seven-hour research flight was followed by an experiment debriefing.[15]

The flights provided valuable results, not only from the science experiments, but also from the hardware and the crew. Crew training was found to be an important factor in the success of the missions, as was the fact that real-time communications between the PI and the EO needed to be controlled to avoid overloading the work schedule and diverting attention from any task the EO was trying to perform.

ASSESS-II

Despite the success of ASSESS-I, NASA did not initially schedule a second ASSESS mission, due in part to experiences with the CVT programme, securing further funding and scheduling the aircraft. However, by late 1975, ASSESS-II was approved. This was confirmed in March 1976, with the "launch" occurring fourteen months later in the spring of 1977. During planning for the second ASSESS flight, there were discussions within NASA about what the role of the mission specialist should be, both in simulations and on actual Shuttle flights. With the selection of the first official "mission specialist" astronauts underway, this discussion focused on the existing members of the scientist-astronaut group, as well as the prospect of accepting new astronauts for assignment on Shuttle flights.

Defining the role of mission specialist

By April 1976, scientist-astronaut William Lenoir had been assigned to participate in the ASSESS-II mission, although he expressed concerns over his ability to fulfil both his CB assignments and those with ASSESS-II at the same time. Writing to the Director of JSC, the Acting Director of Flight Crew Operations, the Acting Director of Science and Applications and the Chief of CB Science and Applications, Bob Parker requested that Lenoir be relieved from some of his other assignments in order to participate as MS on ASSESS-II.[16] Parker felt it was important for the Science and Applications Branch (Code TE) of the Astronaut Office to participate in the forthcoming airborne mission, which was expected to define the function and role of the mission specialist. The TE had always maintained that the MS should be "professionally knowledgeable in the prime discipline of the mission to which he was assigned." The MS also had responsibility for the "in-flight integration of experimental objectives and, overall, for the successful completion of the experiment mission objectives and to reduce training loads." As the primary objective of ASSESS-II was in atmospheric physics, with a synthetic aperture radar and microwave limb sounder as leading experiments, Lenoir was judged the most knowledgeable member of TE in these areas. As Parker noted, "I have no doubt [that] if this were a real mission, we would be proposing him as a mission specialist."

On 20 May 1976, following an ASSESS and Shuttle planning meeting with NASA HQ officials, Eugene Kranz, the Deputy Director of Flight Crew Operations, forwarded a memo to a number of directorates at JSC, including the Astronaut Office.[17]

In his summary, Kranz noted, "The Headquarters ASSESS meeting was long, involved and frequently emotional ... I believe the JSC presentation went as well as could be expected since there were so many differences of opinion prior to the meeting." After dealing with payload management concepts, interaction with STS, and defining direct interfaces between user (PI) and operator (NASA), the discussions concerning the function of a payload operations working group caused much controversy. Kranz noted that the intent at JSC was to establish this group to work on technical problems with the Orbital Flight Test series of Shuttle flights, along with "early operational concepts, requirements, and vehicle operational interfaces, as well as subjects associated with training, scheduling, and flight planning integration."

Several participating directorates and field centres at the Washington meeting strongly objected to flight operations (JSC) leading this activity (science), and though Kranz was told he could indeed call a meeting if he wanted, "no user representation would be provided by the science organisation." Apparently, agency-wide science working groups were already in place for such purposes. Kranz commented in his memo: "Again, history is repeating itself. We had science working groups for all previous programs; however, they never satisfied the needs involved in preparation for a real mission." Kranz predicted some flak on this issue when the memorandum of the first meeting was released. The long-standing battle between field centres dealing with science and flight operations was still deep-rooted in NASA, and would remain so for some time.

Several smaller issues also surfaced at the meeting, which generated responses from the CB. Apparently, the mission manager concept for Spacelab would be pursued by HQ for OFT missions as well, where there would be little "science". Kranz also noted that "the mission specialist job description and functions are not recognised by some branches of NASA, and even some of the personnel at the Office of Space Flight are unclear as to the intent in this area." Kranz stated that further definition should be pursued and a standard guideline for this position agreed throughout the agency. In addition, it was suggested that a Skylab flight-experienced astronaut should be assigned to act as MS in support of ASSESS, with the objective of using the availability of ASSESS to help define the role of MS. Kranz asked the CB if a Skylab astronaut assignment was necessary. Jerry Carr, commander of the third Skylab flight added the note, "Heck, no!" Carr did not see any need for such an assignment and did not wish to deprive an unflown scientist-astronaut of the opportunity to participate on a "mission", even if it were only airborne, not space-borne.[18]

In summary, Kranz noted, "This Headquarters meeting was quite disturbing. Flight Operations has been frequently accused of a 'business as usual' approach. My feeling is that there is such a disagreement between the Office of Space Flight and the Office of Space Science (with JSC and Flight Operations Directorate) in matters pertaining to science operations, that it is impossible to develop a plan that will work. Somehow, we have to convince these groups of the need for an integrated plan for the Shuttle program." It had to be easy for a user to visualise, and effective in utilisation, but remain flexible to accommodate the large amount of payload users and Shuttle mission types. "The basic concept of using a coordinator outside of Flight Operations to 'protect' a scientific user from the vagaries of flight operations cannot be tolerated," Kranz stated.

In reply, John Young, Chief of the Astronaut Office at JSC, noted, "The whole thing [the memo from Kranz] is the same old story; a power struggle that won't produce anything except hard feelings (and no scientific data)." Astronaut Ken Mattingly was working on Shuttle development issues at this time, and his technical responsibilities included working towards the Orbital Flight Test missions. He was also part of the CB group assigned to Spacelab development issues. He added his own notes to Young's comments: "John isn't going to volunteer anyone [to liaise with Ames] yet. But I [Mattingly] suspect the inevitable is coming. In order to relax and enjoy it, I'd appreciate comments on a] ASSESS, b] what should be the role of the MS and c] should we get into the fray on how to fly payloads?" One of the responses attached to this series of memos and notes came from Joe Kerwin, who recognised the need to refine the role of MS, "if Ames thinks one person can be mission specialist and mission manager." Kerwin offered to take the lead in this effort, working with key personnel at Ames. He also suggested using a copy of the Shuttle Flight Operations Planning document, in conjunction with discussions with Ames, to review the role of the MS and devise a recommendation about providing one for ASSESS.

At a staff meeting of Flight Operations Directorate (FOD) members on 25 May at JSC,[19] at which support was given for the ASSESS programme, it was noted that the FOD would be required to participate, with the inclusion of "a flight-experienced astronaut", and that further discussion with Chris Kraft, the Director of JSC, and John Yardley, the Associate Administrator for Space Flight at Washington DC, would be forthcoming on this issue. Again, John Young responded: "We need to look into this to see how *little* time can reasonably be devoted to it (ASSESS). We should *not* send a crewman to Ames for 50 per cent of his time between now and ASSESS and 100 per cent of his time for the ASSESS program, which is what Bill Lenoir says it will take. It is all or nothing. I vote zero." Kerwin again added his thoughts in a memo to Mattingly, in which he suggested that ASSESS was "the engineering equivalent of baseline testing on astronauts – its good for them (Ames) but not for us (JSC). What can a pilot do on ASSESS except fly the plane? [I] recommend sending a scientist *only* if the payload is proportionally rewarding to him, as determined by him."

In July 1976, Robert Parker addressed both the role of the MS on Shuttle and participation in ASSESS-II in a memo to fellow astronauts.[20] Parker tackled the question of how seriously the role of MS on ASSESS-II would be viewed by the "outside world" (Goddard Space Flight Center, Marshall Space Flight Center, Headquarters and others). How other centres viewed the role was not how JSC would view it, nor how they would *like* the other centres to look upon the role. Parker suggested that JSC should "go along with the others and play like it [ASSESS] is real," unless they had the power to completely terminate the participation of an MS in the simulation. It was also not clear, even at JSC, that such participation would be used to revise the role of an MS on a Shuttle flight. With the re-wording of the documentation defining the new role, and with the first selection process that would include "mission specialists" about to begin, Parker made it clear that the MS role in ASSESS-II would be important in the years to come and that JSC and the FOD should actively participate in order to highlight their point of view in the wider discussion. Parker also indicated that MS definition documentation could result in "a bloody fight over a piece of paper which we may or may not win."

With respect to ASSESS-II, Parker reaffirmed the plan to assign Bill Lenoir to the flight due to his experience and qualifications. "If we are going to treat ASSESS-II seriously, he [Lenoir] should be our participant. As a Skylab back-up crewman, he has the operational background to be able to crucially evaluate the operations of ASSESS." Parker also noted that a real space mission would not be prepared using several mission specialists on a rotating basis to avoid wasting time in bringing other crew members up to speed. Though it would alleviate the requirement to replace a crewman on other significant activities, this would not support the JSC argument that any MS should be in a position to take charge of the payload in flight. "He has got to be around enough on a continuing basis to carve out this responsibility," Parker wrote. There was great confidence that Lenoir could do this, regardless of the paper positions of MSFC and others. The expected time allocation was given as 30–50 per cent after September 1976 and over 50 per cent from February 1977, increasing to 100 per cent in April for the planned May 1977 flight.

As for MS system training, Parker wrote, "An issue as far as training is concerned is the question of how much knowledge of the 990 [Convair aircraft] is required and how much training will be required to get it; after all, we expect the MS to be knowledgeable of orbiter systems and limitations." But the point of the discussions taking place was not the systems role on the Shuttle, but the science input during a mission. By using ASSESS, the experience would help define the overall role further: "In summary, if we are going to participate [in ASSESS-II] ... and it would seem that unless we can completely squelch it then we should participate ... it certainly behoves us to do our usual super job."

Later in July 1976, Ken Mattingly issued a CB Discussion Item to all current astronauts on the subject of Spacelab design and integration.[21] Opening the discussion, Mattingly stated: "I have been concerned for some time that we (CB) were not sufficiently active in the Spacelab design evolution. During the past few weeks, my concern has intensified. I now have the feeling that not only CB, but also FOD, JSC, STS and NASA may be missing the boat. There is even some indication that ESA may also be internally less than coordinated."

According to the former Apollo astronaut, the programme fell into a variety of groupings. His memo clarified the amount of participation that the astronauts had, or wished to have, in developing procedures and participation in new manned space programmes in the 1970s. Mattingly's memo also observed that:

- The role of the MS was not agreed within NASA.
- There was apparently a major lack of agreement between JSC, MSFC, and perhaps Headquarters, concerning the FOD's role in Spacelab operations.
- There seemed to be a general attitude that crew interfaces would come later and would reflect the design rather than drive it.
- There did not appear to be anyone who was universally accepted as being responsible for the operational aspect of the design. There did not appear to be an

operations-oriented plan for Spacelab testing, checkout or integration into the STS.
- There were numerous loose ends in the hardware/software interface area.

Mattingly was convinced that on the day of flight, NASA would "bear the lion's share of public responsibility" and that the major responsibility for "pulling off one of these flights will ultimately come to roost with FOD. I believe our resources will be severely taxed when this day comes; in fact, we're probably already over our heads. Since we have not been asked to participate, it seems reasonable to ask why that might be. I suspect our internal disagreements have kept us from becoming more aggressive, even more than our reluctance to embarrass our European counterparts. Therefore, I propose that a detailed straw-man outlining the MS role and responsibilities must be our primary activity. We should define our approach and then, through FOD, sell our concept formally up the line to Headquarters. Once this is accomplished, I believe many of our other problems will work themselves out."

To help solve these problems, Mattingly requested that ideas addressing these issues should be in the form of "a data dump on Joe Kerwin ASAP."

Over the next few months, Kerwin worked on these issues with regard to ASSESS-II and in September 1976 issued a memo on MS participation in the flight.[22] The plan emphasised that the nominated MS should operate as per the revised crew functions definition and that he should not be said to serve as payload specialist or perform the functions of a payload specialist, as it was important to clearly establish the MS function in experiment operation as much as the PS function. It was also important to include the assigned MS in summaries and Crew Activity Plans, post-test reports, and responsibility for the submission of final crew reports. "Training coordination should be provided to schedule payload and integration training for the MS, and provision should be made for a European training visit to familiarise him with European experiments, although he will not be prime in any of these on the ASSESS mission." It was clear that the development of the MS role would progress beyond ASSESS to the selection of the first Shuttle-era astronauts, planned for later in 1977.

ASSESS-II crew assignments

Aside from the discussions about defining the MS role, there was also the need to assign crewmen to support the flights. In November 1976, Chris Kraft wrote to John Yardley, informing him of the decision to formally assign an astronaut to ASSESS-II by 15 December 1976.[23] "In order to establish proper selection criteria, the FOD is assigning Dr. Robert A. Parker and Dr. Karl G. Henize to work jointly in establishing formal relations with ASSESS Program elements, and to identify those requirements that should apply in selecting the ASSESS crewman." On the question of MS responsibilities, Kraft also assured Yardley that JSC personnel "will constructively approach and support the resolution of this issue within the agency."

Based on a general MS responsibilities memo, issued on 5 November 1976 by Bernard T. Nolan, the Program Manager of ASSESS-II, and on further discussions between Karl Henize and Lee Weaver in December 1976 to refine the role, the detailed responsibilities for the MS assigned to ASSESS-II were defined.[24]

Pre-mission, the MS would participate in both the crew training plan and the crew activity plan, contributing to payload crew safety and assisting the mission manager as required.

During the mission, the responsibilities of the MS would be divided into three phases. During pre-flight, he would schedule pre-flight activities and coordinate pre-flight planning. During the mission itself, he would coordinate PS activities with aircraft manoeuvres, resolve experiment conflicts caused by short-term changes to the flight plans (as a result of unexpected winds, changes in altitude and other factors), and coordinate long-term flight operation changes with the Payload Operations Control Center (POCC). He would also coordinate POCC-PS communications, act as a single interface between the aircraft flight crew and the payload crew, provide status reports to POCC and be responsible to the commander of the flight for in-flight payload and payload crew safety. Further, he would operate designated experiments and equipment, monitor power distribution and assist the PS as time permitted. During post-flight, he would schedule post-flight activities, complete the MS report, and participate in briefings.

Post-mission, the MS would write his own individual report and participate in the overall mission report.

The 5 November memo and additional notes by Henize and Weaver constituted the core document for a more general, NASA-wide agreement on the role of the MS, although this would not be reached for several months. With the role having been initially discussed in general terms, it was not until July 1976 that it was given more serious consideration. As a result, a stand-off developed between JSC and MSFC mission management, and even referring the dispute to NASA HQ failed to resolve the problem until November 1976, when the ASSESS Program Manager finally proposed MS role criteria that were accepted by the Office of Space Flight, JSC and MSC. All that remained now was to select the MS.

Initially Bill Lenoir was nominated, but due to his prior workload, he did not feel he could devote 100 per cent of his time to the role. Therefore, Dr. Karl Henize was nominated as prime MS for ASSESS-II early in December 1976, based upon his academic experience (astronomy), past astronaut assignments (support for Apollo and Skylab) and previous participation. In the first interim progress report on the ASSESS-II mission, it was noted that Henize was "working into the role very smoothly. His functions are developing and include management of several mission systems, experiments and ground communications. It appears this very difficult problem [that of defining the role of the MS] has reached solution in ASSESS and may set a proper pattern for Spacelab," That final comment raised some issues back at JSC, where (unidentified) handwritten notes questioned the statement: "Whose views does this comment represent ...with what authority?" Clearly, the subject of exactly what a mission specialist would be responsible for on Spacelab missions was still under debate.[25]

Training for ASSESS-II

Payload specialist training had commenced in September 1976. Henize joined the programme in time to become involved in the mid-phase review and revision of flight plans, and in the early definitions of the crew activity plan. He was also able to participate in meetings of the Investigators Working Group (IWG) and the Mission Steering Group (MSG). In February 1977, he participated in the first of two four-day simulations at the SPICE facility in Porz-Wahn, Germany, taking the role of MS.

In reporting on this event, Henize wrote a memo to other members of the CB on his experiences during the visit to ESA,[26] indicating that "Many lessons were learned and many interesting issues came up." The definition of the role of MS was becoming clearer in the minds of both ESA and MSFC personnel, and included:

- Chief coordinator of in-flight activities, where it is useful to have one person monitoring timelines and issuing warnings of expected task changes while the PS focuses on the matter in hand.
- Chief coordinator of in-flight communications, filtering ground communications away from "a strong tendency to bug the PS with useless questions." On orbit, an MS would screen the questions to allow the PS to continue with critical tasks.
- Representative of the payload crew at high-level meetings, where the payload specialists are neither experienced in anticipating how management decisions may affect them, nor particularly bold in defending themselves. Henize frequently informed the management about what they could and could not expect the crew to do and defended them from well-intentioned but ill-conceived planning for sleeping, eating and briefing periods, and in how POCC policies could affect the crew. Being involved with the IWG and the MSG – though without official sanction – Henize was "initially tolerated as a guest" and it became evident to him that crew input in these meetings would be very useful.
- Back-up operator of several, if not all the experiments, prepared to step in and take over the operation of the remaining experiments if a problem occurred that required the intervention of the PS to resolve.
- Primary operation of a few, but not too many, experiments. Henize made it clear that the MS could and should share the load in operating experiments that did not fit the expertise of any individual PS. However, such a load should not occupy more than fifty per cent of their work time, leaving the remainder free for their coordination or planning roles.

Henize recognised that the above roles were clearly valid for ASSESS, in which up to four payload specialists were working simultaneously and the pilots were fully occupied with flying the plane. But he also suggested that some of these roles could be shared with the Shuttle commander or pilot on a real mission, and that some might not be required at all. Henize's work with the Europeans in defining roles on ASSESS-II was also becoming valid for future Shuttle missions with Spacelab/science payloads.

Henize also supported the need for longer simulations, a recommendation made by Musgrave after completing the first Spacelab Medical Simulations at JSC (see below) and for CB involvement with the screening of payload specialists as soon as possible in future assignments. The subject of the overall relationship between MS and PS was very sensitive at the time in the US, but early involvement would help blend the PS into the Astronaut Office training philosophy more smoothly. Indeed, MS cross-training would be an advantage in supporting and understanding the requirements of the PS on each flight. One unresolved issue at the time of Henize's memo was how to back-up a PS: "If the candidates are PS-type scientists and engineers, they have little to gain and much to lose (with time out from their professional activities) by playing the back-up role. It might be expected that they may not wish to. Who, then, should play the back-up role? No one? The mission specialist? Some other scientist-astronaut? Some other NASA person from the payload centre?" Henize realised this was a thorny issue, and one that would continue to be debated within the FOD for some years to come.

For Henize, his pre-mission responsibilities were of greatest importance, yet these were the most poorly defined in existing documentation and were not addressed at all in the Nolan memo. Even with the small cost and effort invested in ASSESS, it soon became clear that a back-up for the MS role was required, and this had implications for Spacelab missions already planned. According to Henize's post-mission assessment,[27] "It is inconceivable to me that we will be able to fly early Spacelab missions without some provision for back-up to the MS. This is perhaps the chief lesson learned in this area [from ASSESS-II] ... how the MS should be backed-up is still an open question."

For ASSESS, a shortage of manpower made it impossible for the CB to provide a full-time back-up MS, so scientist-astronaut Bob Parker (a professional astronomer with experience in the physical sciences, Apollo and Skylab support experience, and operational experience on ASSESS-I) was assigned to the role, with the stipulation of a bare minimum of training time that would not exceed three weeks.[28] Parker's three weeks included a week for CV-990 command and data systems, a week for observing integration activities and participation in cross-familiarisation experiment briefings, and a week devoted to participation in a mission sequence test and observation of an integral mission simulation. In his post-mission report, Henize noted that although Parker only had minimal training in the role, the experience lent itself to considering part-time MS back-up assignments on Spacelab missions, "should manpower limitations continue to be severe in the mission specialist group." As Henize envisioned the MS role in future Spacelab missions, the concept of a part-time back-up might be valid for later more "mature" Spacelab missions. "The importance of ensuring success in early Spacelab missions may be such that a full-time back-up MS will be required."

In the event, this did not mature as suggested. All NASA back-up roles were suspended from STS-4 in 1982, when NASA felt it had a sufficient pool of experienced astronauts available to replace any flight crew member should the need arise. Back-up crew assignments in the NASA astronaut corps did not become the norm again until mid-1994 and the Shuttle–Mir programme of long duration residence on the Russian space station, when the Russians required such an assignment in their crewing policy for resident crew members.

Henize's training was "almost entirely scoped and scheduled by myself," but was arranged in consultation with the Deputy Mission Manager and the Training Coordinator. His training for ASSESS-II began on 6 December 1976 with familiarisation briefings on the management of the programme, the US PS, objectives of the experiments, and summaries of mission plans. In January 1977, he received training on the CV-990 command and data system and experiment briefings, followed by a three-week orientation visit to ESA in February/March. Also in March, he flew on three flights aboard the CV-990 to observe and operate experiment support systems as part of his Mission Director training development, and received instruction on payload crew safety procedures. During April, Henize completed further experiment equipment familiarisation training, and a Mission Sequence Test with all four PSs. The month was rounded out by experiment operation briefings. During the first two weeks of May 1977, CV-990 safety equipment training was followed by a Mission Sequence Test, an Integrated Mission Simulation and final experiment briefings. Four days prior to the start of the mission, Henize received final flight track and experiment priority reviews.[27]

ASSESS-II in flight

The actual "mission" began on the afternoon of 16 May 1977 and continued for the next nine days until the evening of 25 May. There were nine, six-hour flights (seven night flights and two day flights) of the CV-990. Between each flight, the payload crew was confined to the CV-990 and the living quarters van, which was connected to the aft door of the aircraft between flights. A five-man payload crew (four PSs – two from the US and two from Europe – and one NASA MS – Henize) conducted single-shift operations of ten experiments over six hours for each flight.[29] The most significant feature of the series of flights was that it involved personnel from both NASA and ESA, who were (or would be) directly involved with payloads associated with the early Spacelab series of missions. (The crew included Robert Menzies, who would later become a US Spacelab 1 PS candidate, and Claude Nicollier, who would later become an ESA astronaut.) According to the official post-mission reports, this allowed them to test and evaluate management policies that were being considered for Spacelab. In addition, valid scientific data was obtained in IR astronomy and atmospheric physics.

Though the data gathered was limited to a six-hour shift, this was assumed to be the profile for later Spacelab flights, and it also gave good baseline data for MS training for Spacelab 1, where more involvement in experiment operations was foreseen. Henize found that the ESA integrated simulation held in Germany three months prior to the mission gave him a good insight into the total twenty-four-hour daily crew schedule, the scope of air-to-ground communications, and "in keeping the POCC PIs happy without allowing them to interfere unduly with PS operations." Henize also strongly suggested that the US side give more consideration to integrated mission simulations, or integrated crew training for Spacelab missions, of at least

The crew for the second Spacelab medical simulation. Left to right: Dr. Charles Sawin, Story Musgrave, and Dr Robert Clark.

three to four days of simulated on-orbit operations involving the entire crew, the PIs and POCC. This process was adapted in part for later Spacelab missions, depending on the discipline.

The assignment of an MS to the responsibility for onboard payload operations management and onboard coordination of STS/Payload integrated activities was demonstrated as both workable and beneficial. In the final report, it was re-emphasised that there was strong evidence in favour of adapting this approach in real Spacelab missions because of the increased complexity of the Shuttle system and Spacelab payload activities, and of how critical in-flight team work would be. Other benefits from the ASSESS-II mission in general were payload selection and funding, management relations, pre-flight planning and integration of payloads, documentation, European payload integration, and processing of the "launch site" safety issues. The selection, make up, training, scheduling and care of the flight crew, ground and flight operations, real-time rescheduling and the handling of data also benefited from the test. All of these parameters would aid future planning of actual Shuttle and Spacelab science missions, with advantages and disadvantages discovered by both the US and European partners, although not all of the lessons learned would be carried over to future operations.

ASSESS-II represented the final major airborne simulation in the development of Spacelab missions. Though much work still needed to be completed, there had also

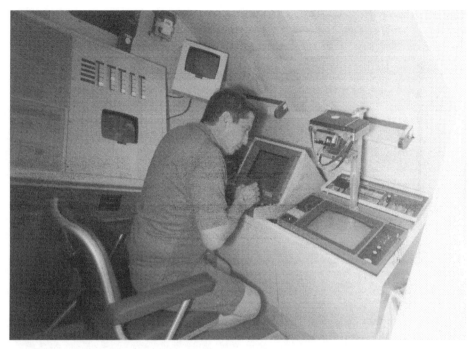

The mock-up of the aft flight deck on the Shuttle during the second simulation – evaluating the potential location of a payload specialist station that would be used on a real mission.

been a series of ground simulations aimed at defining the role of the MS in Spacelab-type mock-ups which, combined with the airborne flights, provided a baseline for training refinements and mission preparations leading to the first Spacelab orbital operations.

SPACELAB MEDICAL SIMULATIONS

Running parallel to the airborne simulations was a series of ground-based Spacelab simulations held at NASA JSC in Houston. The mock-up facility was constructed by members of the Bioengineering Systems Division and Technical Services Division at JSC. It was located in the former Lunar Receiving Laboratory (Building 37) and outfitted just as the Spacelabs planned for the 1980s would be. Measuring 6.8 m long by 4.06 m in diameter, the facility would be used in a series of life science simulations that would extend from 1974 until the Shuttle and Spacelab were ready for flight operations (which was planned for the late 1970s). Initially termed the Life Sciences Payload Facility (LSPF), this mock-up would support only three Spacelab Mission Development tests (SMD), between October 1974 and May 1977, prior to being terminated four years before the Shuttle flew into orbit and six years before the first Spacelab mission.

The SMD mock-up configured for the second test, with the cosmic ray experiment located at right rear simulating a Shuttle pallet payload.

Construction of the first mock-up began ahead of the selection of the final Spacelab contractor, so it was not an exact replica of baseline data. Additionally, the payload and orbiter interfaces were not precisely representative, and the ground support equipment and control and monitoring of experiments did not reflect the planned Spacelab Command and Data Management System. But these limitations did not seriously affect the contributions of the crew to the success of the simulation.

Spacelab Medical Development Test I

The first SMD, also termed Life Science Payload Mission Simulation (LSPMS) 1, was conducted between 1 and 8 October 1974 by physician-astronaut Story Musgrave (MS) and Dr. Dennis Morrison (PS) of the Bioscience Payloads Office at JSC. They would complete "a seven day shakedown test of operational procedures and experiment demonstrations."[30] Though the pair occupied the Spacelab mock-up for the "working day" period, they lived in a mobile home adjacent to the laboratory during off-duty hours. Located near to the mock-up was an array of control consoles, manned by test directors, science managers and data managers. In addition to twelve biomedical demonstrations (which were selected to be representative of the experiments likely to be carried on real Spacelab orbital missions), this first test was also a rehearsal for future tests in the series and served as a baseline for perfecting operating procedures, techniques for handling data gathering, and the integration of man-machine interfaces.

In the ensuing post-flight reviews, Story Musgrave's Mission Specialist Report identified 127 recommendations, of which twenty were specifically related to the life sciences experiments, while the remaining 107 were applicable to future Shuttle/Spacelab operations.[31] None were to be considered as directions or policy, merely as ideas and points of discussion to be supported, monitored, or even rejected. They were a result of lessons learned directly from experiences during SMD-I which were, as far as was possible at that time and in that simulator, identical to future Spacelab operations.

The pair began their training on 20 June 1974, about three-and-a-half months prior to commencing the simulation. Both men trained to operate all experiments and on all tasks, as there was no commander or pilot assigned. Musgrave supported the use of integrated simulations in the preparation for the "mission", stating, "Traditionally, the flight crew have been the operators in the integrated testing of experiment flight hardware and the actual spacecraft, and this may be the only opportunity for the crew to operate actual flight hardware in the environment of actual spacecraft.

A selection of scenes (this and the following two pages) showing some of the experiments conducted by Story Musgrave during the Spacelab medical demonstration in 1974.

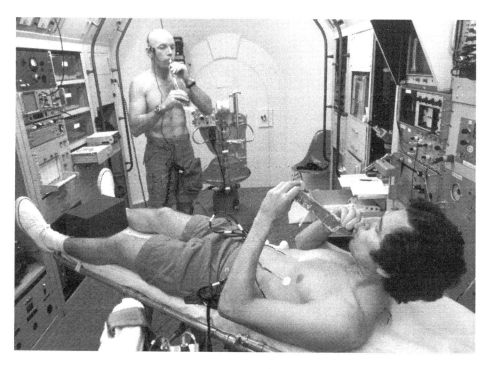

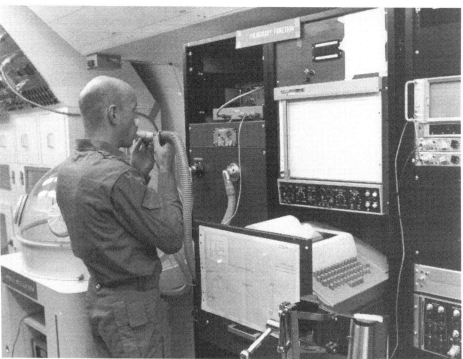

Spacelab medical simulations 313

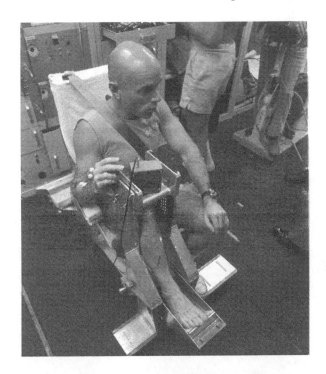

SMD hardware used during the 1977 simulation, with support consoles, showing the attached Shuttle mock-up mid-deck used by the crew for rest and relaxation.

This approach was found to be very useful for the SMD simulation, and should also be integrated into further Spacelab/Shuttle mission training." Musgrave also noted that integrating complete experiments into Spacelab racks as early into the training process as possible assisted in the proficiency and confidence of the crew in developing their operational skills.

Musgrave also wrote that, "While the life science simulations will probably never achieve the ultimate fidelity ... they do produce many of the essential elements of mission training and integration not provided by the operation of individual experiments." Musgrave further stated that one of the primary objectives of this simulation was to evaluate the processes of moving an experiment from the laboratory to flight readiness and to develop concepts for accomplishing this. Many of the lessons learned from this were, in fact, confirmations of the Apollo and Skylab programmes, where the ease with which their experiments progressed through the laboratory-to-simulation process on SMD-I was proportional to their participation in previous programmes. As in Apollo or Skylab, the early assignment of a scientist-astronaut to flight experiments and early communication between them and the experiment group allowed the experiment to be integrated much more smoothly.

In this simulation, the flight crew consisted only of an MS and a PS, and though the paper timeline accounted for commander and pilot positions, these were not included in the simulation. On real missions, the commander and pilot in all Shuttle/Spacelab missions would participate in payload operations. Musgrave observed that "the ability of pilot- and scientist-astronauts to conduct space flight experiments has been established ... particularly in the Apollo and Skylab programs ..." Timeline analyses in 1974 revealed that a commander and pilot would have about sixteen hours between them each flight day to devote to experiment operations, taking into account their participation in orbiter and Spacelab system requirements. Any division of labour would depend on the education and experiences of the crew member, their personality, motivation and training, and other factors. Musgrave used the analogy: "Football coaches should not assign the team responsibilities until they have met the players."

Musgrave also believed that, even for a fourteen- to thirty-day life science mission, a seven-day simulation would suffice, but less than five days was not recommended. Even though they were allotted an hour for physical training, Musgrave preferred to devote this time to running experiments and working on malfunctions. For seven-day missions, Musgrave suggested that exercise hardware should be provided but not timelined. For up to fourteen days, exercise should be left to the discretion of the crew, but for over fourteen days, formal exercise timelining should be included. The results from the initial Shuttle missions and an increase in Space Adaptation Syndrome saw the inclusion of exercise devices on subsequent Shuttle missions, regardless of duration. On SMD-I, the sleep pattern started at seven hours for the first night but reduced to four hours by the final night. This was expected to be reflected on actual flight missions, but on this simulation, this was a result of trying to get the most out of a seven-day mission. This philosophy *was* reflected in many actual Shuttle missions, and had to be readdressed in the longer missions on Mir and ISS.

Spacelab Medical Development Test II

The second simulation ran from 26 to 31 January 1976. Once again Musgrave was involved, but this time with Robert S.C. Clark, a nuclear chemist from the Planetary and Earth Sciences Division, and Dr. Charles F. Sawin, a cardiopulmonary physiologist with the Biomedical Research Division. Again, the Spacelab mock-up was used for the working day, but this time they spent their rest and recreational time in a full-size mock-up of the Shuttle orbiter crew compartment (mid-deck and aft flight deck), where the three men would eat, sleep and conduct a variety of related duties. For this second mission, twenty biomedical experiments (fourteen primary, six alternate) were assigned and there was an additional space physics experiment – a cosmic ray laboratory. There were fourteen operational test requirements, which were designed to evaluate personal hygiene, general housekeeping, and special-purpose cleaning and maintenance concepts for the orbiter. They also evaluated the capability of the aft flight deck to support Spacelab experiment monitoring and performance (essentially, the cosmic ray experiment that was set up behind the laboratory mock-up, representing a future pallet-mounted experiment).[32]

During the post-simulation press conference, the crew gave their impressions of the week of experiments and evaluations.[33] Musgrave stated that the reason for such simulations was to develop operating procedures for the Shuttle/Spacelab system in the future, "helping to design the Shuttle vehicle, to design a Spacelab, and to develop ways of integrating off-the-shelf laboratory experiments and astronomy experiments into that system." They also emphasised the evaluation of handling payloads from the aft flight deck, the habitability of the crew quarters and further interaction with the command section just outside the mock-up. Generally, the crew were awake for eighteen hours each day and spent most of that working on the experiments. They completed over forty per cent more experiment runs than originally planned.

Musgrave was asked if he felt enclosed in the mock-up during the simulation: "I never did. We had too much to do, and like space flight, it isn't sensory deprivation. There's communication with the Mission Control Center and communication with many experiments and so on. Even though we weren't flying in space, we were enclosed inside a can, you might say, for a week. As opposed to sensory deprivation, there's a multiplicity of sensory input, so it really isn't [like] being closed in at all, and you're relating with the science." In response to a comment about being separated from his family, Musgrave added: "It's total dedication to a mission; identical to a space flight. You don't feel like you're hemmed in; you've got an awful lot to do. So it's really like running the mission and trying to get as much done as you can."

Spacelab Medical Development Test III

During the press conference for the second SMD, there were comments about a third, even longer simulation – perhaps as long as thirty days – planned for later that year, as well as future simulations that could include female crew members or representatives from ESA. A meeting about following up the second SMD was held on 25 March 1976, at which new areas of involvement were suggested, including "support from the foreign community, as well as any potential commercial user involvement."[34]

There was a request to expand the crew complement to include pilot personnel and to identify training requirements. It was also suggested that pallet modes should be evaluated as well as laboratory simulations, and that the programme should be continued, but with a less confusing acronym. The concern for the FOD was that, by going agency wide, they would lose control of the tests – with integrated training ending up with Marshall or Kennedy Space Center – and that by offering other centres participation, it might remind too many people of the CVT series. Broadening the range of PIs beyond JSC was thought to be a good move, however, especially if there was more than one additional run planned. Whatever the plan, Bob Parker recognised that any proposals had to be "very well prepared, and we need to proceed carefully." As a result of these conversations, Ames Research Center became a major contributor to the third SMD, which was finally agreed to be of sufficient interest and importance to schedule for the summer of 1977.

This time, physician-astronaut Bill Thornton was assigned as prime MS. A leading expert on space adaptation, nutrition and exercise requirements by space crews, his experiences in MOL and Skylab would be of great help in the simulation

Support scientists and controllers monitor the progress of a Spacelab simulation.

itself and his physics background, coupled with medical training and astronaut experiences, provided a valuable asset in arguing the case for the CB in developing the simulation.

By 18 December 1976, Thornton had noted a number of issues concerning the planned simulation date, now set for May 1977.[35] In his memo, Thornton indicated the increasing complexity of these simulations as the series more closely imitated actual space flight mission procedures and plans. On SMD III, there were plans for fifteen animal (non-human), seven human and four mixed experiments, plus twelve other investigations. The animal investigations included over 100 rats, four monkeys, more than forty mice, two frogs and thousands of flies. Thornton reported some concerns over the role of the MS on this simulation. One problem that he felt might not be alleviated was that the MS (himself) might become a "puppet on an electronic

318 Shuttling into Space

Emblem for SMD-III showing a primate riding the back of the Shuttle holding an astronaut in his hand. To the crew this emphasised the difficulties in handling the animals in the holding facility and in particular one awkward monkey.

string manipulated by a 'cookbook schedule' from controllers and investigators following his every move on the ground." With the complexity of these experiments, Thornton emphasised that it was imperative that the MS and the PS should aim their training to be an extension of the principal investigators. Thornton was assuming the developing CB role that would see the mission specialist on a Shuttle/Spacelab flight being responsible for the scientific success of the mission and the lead scientist on board, as well as assuming other vehicle and mission responsibilities.

Thornton also expressed concerns about the way the selection of payload specialists might be amended at short notice, adding to the programme costs, diluting joint efforts and increasing the load on the MS. Apparently, the division of experiments between the MS and PS was agreed to ensure they shared the load in the simulation. A tentative selection of PS was made between NASA JSC and NASA Ames and unofficial joint "training" was being accomplished to an acceptable degree. But news that the PS selection might not be ratified and that Ames was looking to train its own selection (plus a second, less qualified person in parallel, to make the final selection shortly before "flight") did not go down well at JSC. There was even a rumour that the JSC PS selected might be replaced by a second person of unknown technical experience.

If such a move were made, it would be very difficult for the MS on the simulation to ensure mission success. A long training programme with a familiar crew had already produced good results on previous space missions. The CB would wish to continue this proven system and would seek considerable input into selection of the SMD crew. Even though they would not fly in space, it was important to establish clear guidelines that were applicable to further flight crews. Making a politically attractive selection just prior to final selection of the SMD crew would not help smooth out the differences between the field centres as the first flights of Shuttle and Spacelab approached.

In preparation for the simulation, Thornton held discussions with former Skylab astronaut Alan Bean and other "physical sciences astronauts" in order to determine the most appropriate training process. It was decided that the two PSs would share responsibilities as both primary or back-up on the range of experiments, to provide dual coverage and redundancy where required. The prime experimenter would receive training to a level of knowledge that allowed routine performance of the experiment, and for being prime contact for the investigators and PIs, as well as being the leading troubleshooter for that experiment. The back-up experimenter would only receive a level of training sufficient to be able to substitute if required. As MS, Thornton would receive enough training on all the experiments to allow him to "cookbook" all bar the most complex. He would become familiar with all the goals and limits of the experiments, be responsible for the overall troubleshooting, and serve as prime experimenter on appropriate experiments.

Reflecting some experiment difficulties encountered during Skylab, Thornton was aware that, although some experiments looked good on paper, the research behind them was poor and they would require much more involvement than was indicated. One such experiment the astronaut cited called for the simple throwing of a switch, but in reality, it featured tens of data channels, changing of film, manipulation of boiling water and monitoring and tweaking of several channels. It was suggested that the crew should have full indication of the experiment and intended input as early in the training flow as possible, with equipment as close to flight standard as possible, in order to both attain the necessary skills and techniques to operate the experiment and to understand its objectives and the specific individual requirements of the crew member.

Thornton also expressed concern about receiving appropriate documentation early enough in the preparation for the simulation to allow for any effective troubleshooting. He suggested in his memo that the CB "should have an agreement that, when something fails, we cut it off and forget it, rather than spending hours flailing

320 **Shuttling into Space**

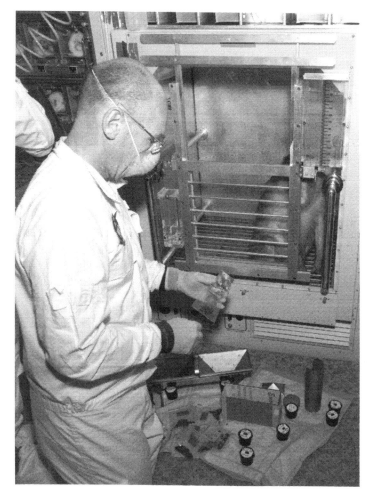

Dr. Bill with his monkeys ... one of the more challenging assignments for the scientist-astronaut, in more ways than one.

around with makeshift apparatus and procedures thrown together by a crash team on the ground." The medical experiments were also being revised to a more manageable package by the new head of the Life Science Directorate, Sam L. Pool. The "friendly doctors" wanted the simulation crew to perform seven days of isolation and a physical every day prior to commencing the simulation, record all food eaten and all waste products and undergo a nightly medical report. In addition, it was important to test elements of crew equipment prior to installing them in the SMD programme. A case in point was the planned Shuttle Waste Management System, which had failed many times in simulations. Thornton expressed the hope that someone was working on the system to solve the problem prior to its inclusion on the Shuttle, so that the Skylab WMS difficulties were not repeated.

By 1 December 1976, the prime and back-up PS were undergoing medical tests as part of the evaluation to gather baseline data for future PSs on Shuttle flights. The PS group consisted of W. Carter Alexander, PhD, of JSC, and Bill A. Williams, PhD, Patricia S. Cowing, PhD, and Richard E. Grindeland, PhD, all from NASA Ames Research Center at Moffett Field in California. Thornton was accompanied on the simulation by Alexander and Williams, using the high-fidelity laboratory and crew module mock-ups between 17 and 23 May 1977. Twenty life sciences experiments from Ames and six from JSC were operated, with medical monitoring and health service provided by Thornton.

SMD-III – an overview

The three SMD simulations were preparations for future medical Spacelab missions to be flown on the Shuttle. They were based upon the strong precedent set by SMEAT and Skylab, in that any mission of this complexity required a ground-based simulation to iron out any developmental problems and to enable operations on orbit to run more smoothly.

The crew for SMD-III. Left to right: Carter Alexander, Bill Thornton and Bill Williams inside the simulator during preparations for the test.

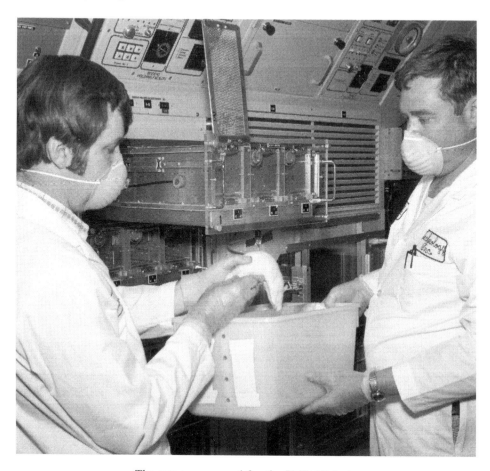

The rats are prepared for the SMD-III test.

The development of the series, and especially the third simulation, caused some friction between the NASA field centres involved, principally JSC and Ames. The Ames Research Center initiated the programme because it was developing animal research facilities for Spacelab life science missions. They acted as managers of the SMD program. Ames intended to have its own mission control, with its own scientist controlling the simulation (which they did to a degree for SMD-III). This caused many heated management meetings and discussions between JSC and Ames. There was also the issue of the varied depth of experience of the principal investigators of the range of experiments that would be included, as well as hardware issues – in particular with the animal holding facility that eventually flew on Spacelab 3 in 1985.

Crew selection for SMD-III resulted in some of the strongest political in-fighting, because each of the scientific factions had their own candidates for inclusion in the final simulation. It almost reached the point of a team of payload specialists completing the simulation, rather than including NASA astronauts in the crew.

The intent was to simulate the procedures and hardware of all the experiments, many of which had undergone shakedown tests on the first two simulations. SMD-III provided an opportunity to try all the experiments, hardware and procedures in one go, but many were nowhere near ready to proceed to flight assignment. This was also one of the stages in the history of the CB where the influence of the NASA astronauts was being diluted by politicians and bureaucracy. The Astronaut Office philosophy at that time (mid-1970s) was that "any job would be executed as perfectly as it could be by the people on the flight and those in flight were primarily responsible for production of the results." Therefore, the most logical course was for the simulations to be prepared as though they were an actual space mission (with checklists, timelines and a formal way of approaching the task and objectives). The philosophy of astronaut training at that time was for repetitive development of procedures, using hardware that was as identical as possible to the actual flight hardware, allowing the flight crew to become so familiar with the experiment or objective that when it was performed in space, it had become second nature to them. On any space flight there are enough unexpected events that occur without having to overcome unfamiliar procedures or situations.

The science community involved in developing experiments for life sciences missions on Spacelab at that time were totally unfamiliar with the operational constraints. There was a serious lack of understanding not only of space flight operations but also of conditions relevant to weightlessness – to such an extent that one proposal included the use of ethyl in an open dish on board the SMD-III simulation, something totally unacceptable for an experiment on orbit. Another absurd proposal was for the astronauts to monitor the mating habits of fruit flies in weightlessness and then describe them, rather than setting up an automatic camera to film the event. Clearly, no astronaut would be prepared (or would have enough time) to watch fruit flies copulating in space – or anywhere else – for hours.

Once he was assigned to SMD-III, Bill Thornton participated in numerous meetings in which he tried to establish procedures for operating each of the experiments on the simulation. This included potential failure modes and how crewmen might be able to recover from them successfully.

There was a huge lobby at the time to perform experiments on animals in space, just as there had been with animals on Earth. The argument was that it was too dangerous to attempt such experiments on humans in space first. However, most of the experiments proposed were out of date and enough of the information had already been discovered to make many of the research fields irrelevant to human adaptation to space flight. The proposals went ahead, but there would be other considerations, such as the handling and care of the animals (including feeding and waste management). There were also numerous problems in developing the cages; too many to deal with in the simulations, but problems which would come back to haunt the crews who worked with these cages on Shuttle flights. There was also a problem with some of the test rats attacking each other, which made it difficult to collect viable data.

One of the monkeys involved in the experiment became "such a brute" that the simulation crew featured him on their emblem as an ape sitting on the back of the Shuttle holding an astronaut in its hand, reflecting both the pain of the experiment for

Bill Thornton inside the Spacelab mock-up for the third SMD test. The development of hardware to simulate what would fly on a real Spacelab mission is evident in a comparison of this photo to that of the 1974 simulation.

the crew and the pressure being put upon them by scientists to work with inadequate experiments and equipment.

On SMEAT the astronauts could at least become involved in the development of hardware at JSC, and could inspect and repair the equipment where necessary. But since SMD was not driven by JSC, it was more difficult to instigate changes or amendments to the tests.

The human experiments were driven to some extent by Thornton's determination, but were restricted by the inadequacy of the hardware through lack of design forethought. Thornton's dilemma was that, whereas on SMEAT he could see and participate in the development of the hardware before its inclusion in the simulation,

Bill Thornton at one of the work stations during the SMD-III test.

on SMD-III, he could not do this. Hardware for SMEAT had been designed by people already involved in space flight – and more specifically, Skylab – hardware development at Marshall Space Flight Center. They had never constructed medical equipment before, but understood the problems and were willing to listen and amend the experiments to "get it fixed." The CB was very much in control of the test operation – not in management decisions as such, but certainly in having a voice that would be heard.

The "voice" that was listened to for SMD-III was that of the scientists, rather than the Astronaut Office, and the equipment came from a group unfamiliar with developing space hardware or operations. On nominal space flights, volumes of material are written or filmed (or passed on from personal experience) and available if troubleshooting is required to resolve a problem. The idea was not to wait for the crew to get into difficulty and have to solve the problem through communications with the ground. It was the crew that were the front line for resolving problems. On SMD-III, however, there were no troubleshooting efforts, nor wiring diagrams for the experiments.

When the Astronaut Office assigned Thornton to the "crew", the hardware was still being built, as were the experiments, and there were myriad examples of totally inappropriate procedures that hindered the training programme's development.

Eventually, after much wrangling, Thornton developed troubleshooting procedures and a suitable timeline for an operational "day" during the simulations. Still, there were delays in setting up the operation centre at Ames, and the difficulties encountered in placing an operational mission control at JSC and a scientific mission control at Ames showed that such a set up would not work, so the scientific mission control for SMD-III was located at JSC.

Thornton's perception was to do what the astronauts always aimed to do – to complete any job to the best of their ability, with sufficient training and adequate tools to ensure their tasks were accomplished. If something was seen to be going wrong, controlled communications would assist in troubleshooting and resolving the problem, or terminating the task if necessary. Proper calls would be adhered to and no inappropriate "chatting on line" would be allowed. But this methodology ran contrary to the scientists' concepts of what should be done on the simulation. SMD-III, for them, was an extension of their own labs, in which they would make real-time decisions and tell the "lab crew" what to do. As trained scientists, the crew would do what the ground scientists wanted but how they actually accomplished this was down to them. For Thornton, this meant by-the-book in space flight fashion, using procedures and checklists. Unless the crew needed to talk to the ground scientists, he felt they should stay off the line until requested to discuss something by the "flight crew". This system worked well on past missions, with everything going through the astronaut Capcom apart from any direct intervention by the Flight Director. The conflict between the two views caused friction until Thornton was advised to relax a bit on the communications. His frustration, and fear, was over uncontrolled communications between several scientists and "crew members" without a central control filter (a Capcom) blocking communications, thus wasting crew time on irrelevant discussions and probably bypassing checklist procedures.

Thornton's argument was that NASA should have flown scientists who worked as part of an investigating team as "servants" of that team. There were so many problems that there was no room for time off. The crew stayed on top of the situation but it was "messy" and frustrating to the astronaut. Additionally, little actual Shuttle equipment could be tried out as it was too early in the programme, so Skylab equipment was used. They did try to use a prototype of the Waste Collection Systems, however, and in a perfect prelude to the problems encountered later on Shuttle missions, it failed completely and was replaced by a Port-A-Loo device.

The simulation was certainly far advanced from where they started, but there was a lot to learn from the situation. It illustrated that, apart from very few exceptions, the future payload specialist/scientist on orbit would not so much be a scientist as a servant of the investigators.

Was SMD-III worth it with all the problems? Thornton thinks so. "It identified

The crew celebrates the end of a successful but challenging simulation, thus ending the series of simulations begun in 1974.

the problems, but by golly, I shudder to think what might have been if we had not done it. There were some first rate trainers, but the procedures in getting some of the experiments defined for the simulation were hard enough, without trying to evaluate them for space flight conditions." It was a hard learning curve for all, and one that was not totally learned by the time Spacelab 3 flew almost a decade later in 1985!

SMD-III was the final Spacelab-based simulation prior to specific mission training and preparations for orbital operations.

THE VALUE OF PARTICIPATION

From the early 1970s, while the Shuttle was being developed, members of the scientist-astronaut group were assigned to support roles which resulted in their early involvement in the design of workstations, crew equipment (including space suits and EVA tools), and Shuttle payloads or experiments. After Skylab and ASTP had flown, the whole of the Astronaut Office was reorganised to support the Shuttle effort during

1974–5. With the selection process for the first Shuttle era intake underway between 1976 and 1977, the Approach and Landing Tests of the Enterprise Shuttle, and the Orbital Flight Test programme fast approaching, crews for the first Spacelab missions needed to be assigned to begin work on integrating the payload with the ever-changing equipment of both the Shuttle and the CB.

From 1978, with the arrival of the thirty-five new astronauts, the scientist-astronaut group became known as Senior Scientist-Astronauts for a short time, before being designated as mission specialists available for assignment on forthcoming Shuttle crews. Several scientist-astronauts who had left the CB following the demise of the Apollo programme returned to resume training for a possible flight on Shuttle and Spacelab missions.

The early participation of the scientist-astronauts in the development of scientific payload operations from the Shuttle (especially the Spacelab series) clearly helped define the role of the mission specialist and focused both the criteria for selecting payload specialists and the crew responsibilities for each category. They were also influential in determining the operational constraints for experiments, timelines and hardware on Spacelab-type missions by conducting the Learjet, ASSESS and SMD simulations. The next phase was to assign astronauts to the first Spacelab mission, allowing further evaluations of CB participation in the preparations for that mission, its payload and flight plan. This would help to create a foundation upon which to build further Spacelab crew training and participation. It would also further define the role of the NASA mission specialists and non-NASA payload specialists on other missions.

Mission specialists for the Shuttle

In 1976, NASA announced a call for the first group of astronauts to train specifically for the Space Shuttle. In addition to the pilot category, a new type of astronaut would be accepted under the mission specialist category. These astronauts could include scientists and engineers (with or without piloting skill), who would perform duties such as EVA, satellite deployment, repair and retrieval, and the operation of scientific equipment. Because of the reduced g-forces that would be experienced during a Shuttle mission in comparison to the earlier Mercury–Gemini–Apollo programmes, the physical requirements for this new group were less stringent. There would also be no age limit and no requirement to attend jet pilot school.

Candidates had to have attained at least a Bachelor's degree in engineering, biological or physical sciences or mathematics, however, although a more advanced degree or equivalent experience was preferred. They also had to pass a Class II physical examination (which had a broader acceptable range for vision and hearing) and stand between 150 cm and 193 cm (5 feet and 6 feet 4 inches). The height parameters were now dictated by the size of the EVA suit rather than the spacecraft. Targeted recruitment drives urged minority groups, women and non-pilots to apply for the positions available.

In January 1978, the names of thirty-five new candidates were announced by NASA. They included twenty mission specialists (six of whom were women, the first selected for NASA astronaut training) with a range of academic skills. Fourteen of the mission specialists selected could have easily qualified as a third group of 'scientist-astronauts'; eleven of them held PhDs, while three others were qualified physicians: Guion Bluford – PhD in aerospace engineering; John Fabian – PhD aeronautics/astronautics; Steven Hawley, Jeff Hoffman and George Nelson – PhD astronomy; Shannon Lucid – PhD biochemistry; Ronald McNair and Sally Ride – PhD physics; Judy Resnik – PhD electrical engineering; Kathryn Sullivan – PhD geology; James Van Hoften – PhD fluid mechanics; and Anna Fisher, Margaret Seddon and Norman Thagard – MDs.

Other early Spacelab assignments

In addition to participation in the various Spacelab simulation programmes, a number of astronauts took part in the development of the Spacelab module. Between 22 and 23 November 1976, pilot-astronaut Paul Weitz accompanied scientist-astronauts Ed Gibson, Bill Lenoir and Joe Kerwin (all former Skylab assigned astronauts) to MBB-ERNO, the German prime contractor for Spacelab payload integration. They conducted a walk-through of the Spacelab module, where they simulated various airlock operations and noted a number of improvements that needed to be included prior to the first mission. The following year, between 25 and 29 April 1977, a formal Crew Station Review was held at ERNO, which involved astronauts Paul Weitz, Bob Parker and Ed Gibson working with specialists from NASA, ESA and ERNO to review crew habitability of the Spacelab design.[36]

Selecting the first Spacelab crew

In January 1978, the year after the ASSESS-II mission, NASA named the first astronaut candidates to train for Space Shuttle missions as pilots and mission specialists. After a period of candidate training, they would be assigned to technical areas of support pending their first flight assignments. In March, NASA named the crews for the series of Orbital Flight Tests and in August, as Europe named their first astronauts for Spacelab 1, NASA released details of who would fill the mission specialist role on that first pioneering Spacelab mission – Owen Garriott and Robert Parker. The focus now shifted to qualifying the Shuttle system in the first four to six missions and verifying its operational capabilities by deploying communication satellites, prior to flying the first of a planned long series of Spacelab missions. For the team of scientist-astronauts who remained active in the CB, the prospect of either a second flight, or perhaps their only chance of their first mission, was getting closer – providing, of course, that the Shuttle could perform as it was designed to.

REFERENCES

1   *Astronautics and Aeronautics 1972*, NASA SP 4017, 1974, pp. 2–5.
2   JSC *Roundup*, 7 December 1973, p. 1.

3 *The Shuttlenauts 1981–1992: The First 50 Missions*, Volume 2, Shuttle Flight Crew Assignments STS-1 through STS-47, compiled by David J. Shayler, AIS publications, December 1992, pp. 6–10.
4 *Spacelab: Research in Earth Orbit*, David Shapland and Michael Rycroft, Cambridge University Press, 1984, pp. 25–48.
5 *Chronology of Spacelab Development*, NASA Historical Data Book Volume V, NASA SP-4012, pp. 462–479.
6 *Power to Explore, A History of the Marshall Space Flight Center 1960–1990*, Andrew J. Dunar and Stephen P. Waring, NASA SP 4313, NASA 1999, pp. 431–432 & 541–543.
7 *Women in Space: Following Valentina*, David J. Shayler, Ian Moule, Springer-Praxis, 2005 pp. 156–160.
8 *Spacelab: An International Success Story*, Douglas R. Lord, NASA SP-487, 1986, p. 116.
9 *Preliminary Report on Learjet Shuttle Simulation Mission No. 4*, John F. Reeves and Bruce R. Vernier, ASSESS Observer Team, Northrop Service Inc., Moffett Field, CA 94035, undated draft copy with hand-written notes by Karl Henize, on file AIS archive.
10 Notes of MPG meeting on 12 September 1974, in a memo dated 1 October 1974 from Douglas R. Lord, Director of the Spacelab Program, and William O. Armstrong, Director of Mission Analysis and Systems Requirements, to a distribution list of NASA and ESRO personnel. From the files of R.A. Parker, NASA JSC history archive, University of Clear Lake (U of CL), Houston; copy on file AIS archive.
11 Memo from Philip E. Culbertson, Director of Mission and Payload Integration, NASA HQ, to Robert Parker, NASA JSC, 5 November 1974; copy on file AIS archive.
12 Memo dated 8 November 1974 from Owen Garriott to Director of Spacelab Program, NASA HQ., original in the R.A. Parker collection, JSC history archive, U of CL, Houston.
13 Memo from Homer E. Newell, Chairman, SPAC Ad Hoc Subcommittee on Scientist-Astronauts, Washington DC, dated 18 November 1974.
14 Memo to Chairman of SPAC Ad Hoc Subcommittee on Scientist-Astronauts from Louis C. Haughney, NASA Ames, undated – but probably late 1974.
15 Spacelab Simulation Test Procedures, Benjamin M. Elson, *Aviation Week and Space Technology* 14 July 1975, pp. 40–45.
16 Memo from R.A. Parker to Director of JSC, Acting Directors of Science and Application and Flight Operations Directorate, and Chief of CB Science and Applications, 14 April 1976. From the R.A. Parker Files, NASA JSC history archive, U of CL, Houston.
17 Memo CA-EFK-76-82 from E.F. Kranz to Distribution (inc CB) 25 May 1976, subject: ASSESS Meeting with NASA Headquarters on 20 May 1976, with attached comments by John Young, Ken Mattingly, Joe Kerwin and Jerry Carr; copy on file AIS archive.
18 Private communication from Jerry Carr to David Shayler, 23 July 2005.
19 FOD Staff Meeting Notes/Actions, 25 May 1976, with additional notes from J. Young and J. Kerwin; copy on file AIS archive.
20 Participation in ASSESS-II for the purpose defining the responsibilities of the Mission Specialist, R.A Parker, 2 July 1976. Original filed in the R.A Parker collection, ASSESS Box, JSC history archive, U of CL Houston; copy on file in AIS archive.
21 Spacelab Design and Integration CB Discussion Item, from T.K. Mattingly to all CB astronauts, dated 26 July 1976; copy on file in AIS archive.
22 Comments on ASSESS-II Support Plan, Joe Kerwin, 3 September 1976. Original in the R.A. Parker Files, JSC history archive, U of CL Houston; copy in AIS archive.
23 Memo from C. Kraft to J. Yardley, dated 29 November 1976, CA-EFK-76-177; copy on file AIS archive.

24 Mission Responsibilities for the ASSESS-II Mission, from the R.A. Parker files, ASSESS box, JSC history archive, U of CL, Houston; copy on file AIS archive.
25 Progress Report – ASSESS-II Mission, from Donald L. Anderson, Assistant Chief, Medium Altitude Missions Branch, NASA Ames Research Center (SEM:211-12), dated 10 March 1977. Original in the R.A. Parker collection, ASSESS Box, JSC history archive, U of CL Houston; copy on file AIS archive.
26 Insights gained from participation in the ESA confirmed simulation for ASSESS 2, Karl G. Henize, CB/KGHenize:lmc:3/21/77:2311, dated 21 March 1977. Original in the R.A. Parker collection, ASSESS Box, JSC history archive, U of CL Houston; copy on file AIS archive.
27 ASSESS-II Mission Specialist Post-Mission Report, Karl G. Henize, 28 July 1977. Original in the R.A. Parker collection, ASSESS Box, JSC history archive, U of CL Houston; copy on file AIS archive.
28 Memo Re Back-up Mission Specialist for ASSESS-II, from George W. Abbey to Manger of Applications MSFC, dated 15 March 1977. Original in the R.A. Parker collection, ASSESS Box, JSC history archive, U of CL Houston; copy on file AIS archive.
29 ASSESS-II STS Flight Operations Post-Mission Report, October 1977, FFOD ASSESS-II Flight Operations Support Team, NASA JSC; and ASSESS-II MS Post-Mission Report, Karl Henize, 28 July 1977. Both originals filed in the R.A. Parker collection, ASSESS Box, JSC history archive, U of CL Houston; copy on file AIS archive.
30 Spacelab simulation underway at JSC, NASA News Release 74-255, 1 October 1974; *JSC Roundup* 13, No. 23, 18 October, and No. 24, 25 October 1974.
31 Life Sciences Payload Mission Simulation I, Story Musgrave MD, Scientist-Astronaut, Life Science Astronaut Office, NASA JSC, March 1975. Filed in the R.A. Parker collection, ASSESS Box, JSC history archive, U of CL Houston; copy on file AIS archive.
32 Spacelab simulation, NASA News 76-04 23, January 1976; *JSC Roundup* 15, No. 2, 30 Jan 1976.
33 Spacelab Sim II Press conference, JSC, Houston, Texas, 6 February 1976. Filed in the R.A. Parker colllection, ASSESS Box, JSC history archive, U of CL Houston; copy on file AIS archive.
34 SMD II Follow-on memo from E.F. Kranz (CA-EFK-76-59) and attached notes from R.A. Parker. Filed in the R.A. Parker collection, ASSESS Box, JSC history archive, U of CL Houston; copy on file AIS archive.
35 CB Memo from William Thornton to Joe Kerwin, George Abbey, John Young and all Scientist- Astronauts, 18 October 1976 (CB/WEThornton:lmc:10/18/76:2421). Filed in the R.A. Parker collection, ASSESS Box, JSC history archive, U of CL Houston; copy on file AIS archive.
36 NASA Historical Data Book Volume V, NASA Launch Systems, Space Transportation, Human Spaceflight, and Space Sciences 1979-1988, by Judy A. Rumerman, NASA SP-4012, 1999, specifically Table 4-44, Chronology of Spacelab Development, pp. 462–479.

Significant assistance in the preparation of this chapter was also received from Karl Henize and Bill Thornton.

10

The Long Wait

The departure of Curt Michel from the astronaut programme in 1969 left four scientists from the 1965 selection (Garriott, Gibson, Kerwin and Schmitt) and nine from the 1967 selection (Allen, Chapman, England, Henize, Holmquest, Lenoir, Musgrave, Parker and Thornton) still at NASA. For the four remaining members of the 1965 scientist-astronaut group, the time between selection and first flight had been seven or eight years, with Schmitt flying to the Moon on Apollo 17 in 1972 and Kerwin, Garriott and Gibson flying to the Skylab space station during 1973. For the members of the 1967 selection, however, the wait would be considerably longer. With the Apollo-type spacecraft and Saturn launch system now defunct following the joint flight with the Russians in 1975 (the Apollo-Soyuz Test Project), the 1967 astronauts would have to await the introduction of the Space Shuttle for their chance of a space flight. For some, this wait would be too long and they would leave the agency without making a flight. Others decided to persevere, keeping themselves fit and available for an early Shuttle flight.

Originally intended to make orbital flights in the late 1970s, it was not until the early 1980s that the Shuttle was finally sent into orbit. It was quickly certified to progress from orbital test flights to full operational mission status. The long wait for the remaining 1967 scientist-astronaut members at last seemed worth it, as they finally saw their chance to fly into space. But with the first of their group flying in 1982 and the last active member making his first flight in 1985, it was still a long wait of fifteen to eighteen years between selection and launch, with the added frustration of seeing members of the 1969 (former MOL astronaut transfers) and 1978 (first Shuttle era) groups flying before them.

SUPPORTING THE SHUTTLE

While Garriott and Parker were training for the first flight of Spacelab from 1978, the other scientist-astronauts (now designated senior mission specialists) at last began receiving assignments supporting the first flights of the Shuttle. A series of six (later amended to four) initial flights by OV-102 Columbia would be flown under the Orbital Flight Test Program, designed to qualify the Shuttle hardware, systems, procedures, infrastructure and support areas for full-scale operational activities. Other scientist-astronauts were also assigned to future Spacelab missions (following the development of the payload) while still awaiting official assignment to a flight crew.

In March 1977, former Skylab astronaut Ed Gibson returned to NASA as a senior scientist-astronaut (MS), accepting the role of Chief of Selection and Training for the new mission specialist group (Group 8), as well as assignment to support duties (Capcom) for the first manned Shuttle mission. Gibson was invited to return to NASA by JSC Director Chris Kraft to "help develop a space station," but after a year or so, it became apparent that the Administration and those on Capitol Hill would not yet support such commitment. This was a discouraging time in the Astronaut Office, particularly for Gibson.[1]

In late 1977, Karl Henize began work on the Spacelab 2 payload (mostly solar physics) with a view to flying the mission as a mission specialist.[2]

Thirty-five new guys

Following their selection on 16 January 1978,[3] the thirty-five members of Group 8 underwent a two-year (later amended to one year) Astronaut Candidate training programme beginning in July, which would qualify them for technical assignments leading to selection to a Shuttle flight crew. The pool of astronauts from which a Shuttle crew (between five and seven persons) could be selected had now increased, and the scientist-astronauts still awaiting their first missions were now hoping that NASA's projected flight rate would be met. They wanted to be assigned to early seats before the queue became too long!

On 26 July 1978, Joe Allen returned to the CB as a senior scientist-astronaut from his position as Director of Legislative Affairs at NASA HQ in Washington.[4] He arrived back at the CB one day ahead of the new astronauts, as he did not wish to be considered "junior" to them by arriving later, even though he had been an astronaut for almost eleven years by then. Allen gave a little talk to the new group, joking that he had had some problems being an astronaut candidate and had been banished to NASA HQ for three years. He warned them to ensure this did not happen to them. But no one in the group of new astronauts laughed at all, assuming he was being serious and that if they messed up, they too would be banished. They seemed worried about exactly what Allen had done to warrant such a sentence.[5] Allen had informed the friends he had made in Washington that "It had been a very interesting and quite enjoyable three years," not letting on that he considered his NASA assignment to have been the "worst headache in the world." He also told his friends and colleagues on Capitol Hill that he wanted a less stressful line of work, and was returning to the astronaut office to prepare to fly into space. After all, the potential for high blood pressure and heart attacks would surely be more worrying than sitting on a Shuttle and being blasted off the launch pad!

The following month, Garriott and Parker were officially named to the Spacelab 1

Ed Gibson photographed at the UACC (Universal Autograph Collectors Club) show in Washington DC, May 2003 (Credit Bruce Rogalska).

mission as mission specialists (MS), and would work with the European and American payload specialist (PS) candidates on the Spacelab 1 science programme and crew activity plan. Six days later, on 9 August 1978, the four Spacelab 2 PS candidates were named. Some years later, Chris Kraft mentioned that he had suggested saving money by halting the PS programme for Spacelab missions and instead utilising mission specialists, who could draw upon their career astronaut training and could perform many of the tasks assigned to the payload specialists.[6] This argument would continue for several years, especially when the PS programme was opened up to non-scientist candidates for missions other than Spacelab science research missions. The following February (1979), USAF Under Secretary Dr. Hans Mark initiated the Manned Spaceflight Engineer (MSE) Program, in which military payload specialists would be trained to accompany Department of Defense (DoD) satellites and payloads on classified Shuttle missions. Contrary to Kraft's suggestion, the pool of payload specialists was clearly about to expand, squeezing the number of seats available to NASA's astronauts on the Shuttle each year.

On 3 June 1979, Tony England returned to the CB as a senior scientist-astronaut and was assigned to technical duties (with Henize) for Spacelab 2 payload development issues. For the two former scientist-astronauts (England and Ed Gibson) returning to NASA with the imminent launch of the Shuttle flight programme, the prospect of finally reaching orbit must have seemed brighter than it had only a few years before. In October 1979, Don Holmquest wrote a letter to George Abbey also asking to return to the CB as a senior scientist-astronaut, recognising that mistakes had been made on his first astronaut tour. He did not return, however, and it was clear that not all the former scientist-astronauts would be required for the Shuttle. The only other astronaut to return to the agency after retirement was Group 5 pilot-astronaut Bill Pogue (Skylab 4). Pogue had retired in 1975, but returned in 1976 as a consultant on programmes to study the Earth from space. He departed the space agency a second time in 1977.

Given a choice between flying on the Shuttle or on another space station, Ed Gibson has said he would have chosen the latter. But with hindsight, an assignment to a Hubble Space Telescope service mission may well also have persuaded him to remain at NASA. His EVA experience on Skylab would have been a factor in his wanting to fly that type of mission, but of course he did not know that such a mission would even be manifested until much later. A place in the back seats of the Shuttle seemed a step backwards to the former Skylab astronaut, so he decided that it was time to forget flying in space again and move on. On 31 October, Gibson resigned from NASA a second time to take a position with TRW (formerly Thompson Ramo Wooldridge Inc.). Had he stayed, he may have received an assignment to a Shuttle flight launched between 1982 and 1985. For Gibson, that wait was too long, and he estimated that the new space station was at least another fifteen years away at the time he left. That proved to be an optimistic estimate.[1]

By 1980, the Shuttle had still not orbited the Earth and it seemed that the inaugural flights would still be months away, pushing the operational flights further into the mid-1980s. NASA planning documents of the time included crew planning charts, evaluating the manpower requirement over the coming few years. When Shuttle flights exceeded twenty launches per year, crews would be recycled intact within a year of flying a mission, to both shorten training time and to take advantage of their experience of working together as a unit. These manifests, showing the proposed twenty-six missions a year (a flight every two weeks using both the Kennedy Space Center in Florida and Vandenberg AFB in California), would require at least twenty-six crews of between five and seven NASA personnel, depending on the mission. That would mean having an astronaut corps of between 130 and 190, all trained and ready to fly at least once a year. At the time, NASA had twenty-seven astronauts from the 1959–69 selections, and thirty-five in training from the recent 1978 selection – sixty-two in all – only a third of what was indicated in the planning charts.

"America's greatest flying machine"

On 12 April 1981, the twentieth anniversary of Yuri Gagarin's first flight into space, the Space Shuttle system lifted off on its maiden launch. Pilot-astronauts John Young and Bob Crippen took Columbia on a two-day test flight around the Earth. Shortly after coming back to the Astronaut Office in 1978, Joe Allen was assigned to the

In February 1980, Joe Kerwin and Group 8 astronaut Anna Fisher undertook weightless training for an axial scientific instrument change out on a mock-up modular section of the Hubble Space Telescope in the Marshall Space Flight Center's Neutral Buoyancy Simulator (NBS).

support crew for this first Shuttle mission, becoming involved in early flight techniques meetings and flying some simulations to familiarise himself with the way the Shuttle would return to Earth, to aid in his role as Capcom. These simulations lasted from two or three hours up to a couple of days. For this first flight, there was a great deal of nervousness about the re-entry phase of the mission. The launch had occurred without incident and the Shuttle had gone through its paces in orbit, so the most dangerous part of the mission now was the return to Earth. Images revealed some damaged tiles, but not in critical areas, and it seemed that the thermal protection system would be fine for entry and landing.

On Apollo 15, Allen's support role was based on science and he had little to do with the actual spacecraft, but on STS-1 there was little science on board. The whole mission was an engineering experiment to confirm that the Shuttle system worked. In his 2004 oral history, Allen compared the STS-1 mission flight techniques meeting to those of Apollo 11 and 12, the flights assigned to evaluate and prove the complete Apollo mission profile. Unlike Apollo, there had been no unmanned orbital test flights prior to STS-1, nor any manned missions to evaluate parts of the mission profile. For STS-1, it was all or nothing. Due to the short duration of the mission, Allen only served in Mission Control three times, with the final stint coming during re-entry. No data was received during the blackout period, as the (Shuttle-launched) Tracking and

338 The Long Wait

Kerwin and Anna Fisher take part in underwater evaluation of a Hubble Service Mission at the Marshall Space Flight Center in February 1980, ten years before the telescope was launched and nearly fourteen years before the first service mission was flown.

Data Relay Satellite (TDRS) system was not yet deployed. After exiting the blackout, data received on the ground indicated Columbia was performing superbly, as Allen described: "We were thrilled. Columbia was still hypersonic and still with lots of speed and altitude to lose before we were safely on the ground ... but this new invention, the orbiter, seemed to be performing absolutely perfectly and it remained perfect all the way to the ground." For the early missions returning from space, NASA had decided to use the dry lake bed runway at Edwards Air Base in California (where there was plenty of room for any overshoot) before committing to the specially constructed Shuttle Landing Facility at the Kennedy Space Center in Florida. On 14 April 1981, the main landing gear of orbiter Columbia hit the ground at 215 knots. The nose wheel touched down seconds later and STS-1 rolled 2,743 metres (9000 feet) along the runway, kicking up dust as it went, before coming to a stop 2 days 6 hours 20 minutes and 32 seconds – and 37 orbits – after lift-off. The relief and excitement about what had been accomplished was evident in the voices of the crew and Capcom Joe Allen as the vehicle rolled along the runway:[7]

> *Allen:* "Welcome home, Columbia. Beautiful!"
> *Young:* "Do I have to take it up to the hanger Joe?"
> *Allen:* "We're going to dust it off first."
> *Young:* "This is the world's greatest flying machine, I'll tell you that. It worked super."

Following the tradition of the Apollo programme, there was a huge post-flight party, celebrating the success that had taken so long to achieve. However, with three more test flights to fly before declaring the STS system operational, there was still plenty of work to do. These flights would be flown by two-man teams of astronauts from the pilot cadre, but for the scientist-astronauts, a successful Orbital Flight Test Program would bring their own chances of flying on subsequent missions that much closer.

The second flight of the Shuttle occurred in November 1981 and was intended to be a five-day mission. Problems with a fuel cell saw the flight reduced to a fifty-four-hour "minimum mission", but the crew still managed to complete most of its assigned tasks, including testing the Canadian-built Remote Manipulator System (RMS), or robot arm. For this mission, Joe Allen was assigned to media support, working with the representatives of the ABC network. STS-2 also highlighted a potential problem with Space Adaptation Syndrome (SAS), namely the possibility of having to bring home the Shuttle when the pilot had not fully adapted to space flight. Joe Engle suffered from SAS upon entering orbit, but would have had time to recover on the full five-day mission (as illustrated by his second flight in 1985). However, due to the minimum mission requirement, he had to bring Columbia home early while he was still dehydrated, causing some concerns during entry and landing. Medical officials decided that a situation could arise during an abort mode, an early return from space or due to crew illness, in which SAS could affect the performance of one or both of the pilots flying the vehicle. They urged that more studies needed to be done early in the Shuttle programme to look specifically at SAS on the ascent into space and the trip home. It was quickly decided to assign physician-astronauts to an early Shuttle flight to determine the symptoms, causes and countermeasures for such an illness. At that time (late 1981), crews for STS-3, 4, 5 and 6 were already in various stages of training, and while it would be possible to place some medical investigations into these flight manifests, it would not be practical to assign a doctor to a crew until STS-7 or STS-8.

For STS-4, and again for STS-5, physician Bill Thornton was assigned to medical support out at Edwards Air Base, standing by in case of emergency situations. This was one of several early assignments for the few available physician-astronauts, who completed familiarisation and procedures training on helicopters and for evacuation of the crew in the event of a emergency landing and crew retrieval situation. For STS-4 Thornton worked with Jim Bagian, and on STS-5 with Norman Thagard.

By late 1981, STS planning schedules included projections for the attrition of "veteran astronauts" – those chosen prior to the 1978 initial Shuttle-era selection. No names were assigned to these documents which were merely intended as planning documents to manage the large number of flight personnel. It was estimated that by the late 1980s, most of the pre-1980 astronauts (including the 1978 group) would have left the programme. These planning documents also revealed that at the time, NASA was considering teaming up mission specialists in pairs and flying them together on between five and eight missions, to capitalise on their mission experience and to alleviate repetitive training from what would have become a stretched training syllabus.

340 The Long Wait

STS-5: WE DELIVER

With two OFT missions remaining, it seemed that the first flight of mission specialists would indeed occur on STS-5. Although NASA was not giving anything away, *Rockwell News* reported on 23 December 1981: "Although no crews have yet been named officially for the fourth through sixth flights, NASA sources have said that eight astronauts have started Shuttle mission training in addition to the STS-3 prime (Jack Lousma and Gordon Fullerton) and back-up (Ken Mattingly and Hank Hartsfield) crews." These were astronauts Vance Brand, Bob Overmyer, Joe Allen, Bill Lenoir, Paul Weitz, Karol Bobko, Story Musgrave and Don Peterson. Clearly, seniority played its part in these assignments, as the recently selected 1978 group of astronauts had only completed Ascan training in 1979 and were still in their early technical assignments pending review and selection to a flight crew.

Assigning the first mission specialists

On 2 March 1982, NASA named the crews for STS-4, 5 and 6. Mattingly and Hartsfield took the flight seats for the fourth and final OFT, with Allen (designated

The Ace Moving Company proudly advertises the fact that their delivery service is fast and courteous. The crew of STS-5 poses for the first "star-burst" crew photo, now familiar on most Shuttle missions. Commander Brand holds the sign, with Lenoir and Allen in flight suits and Overmyer in a T-shirt.

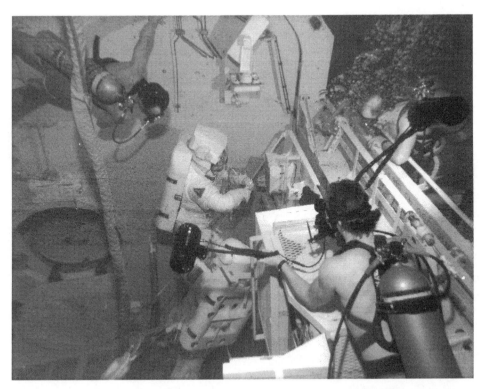

Allen and Lenoir, supported by Scuba divers, practice their planned EVA activities in the Weightless Training Facility (WETF) at JSC. Both suits are pressurised and weighted down to simulate the microgravity conditions of space.

mission specialist, or MS 1) and Lenoir (as MS 2) joining Vance Brand and Bob Overmyer on STS-5. Story Musgrave and Don Peterson were in training for STS-6, alongside Paul Weitz and Karol Bobko.[8] Back-up crews would no longer be assigned to STS missions, as there was now a pool of experienced Shuttle pilots from which to draw and any crew member could be replaced with minimal impact to crew training and scheduling. Crews would still continue to provide support roles on like-for-like missions for some time, however.

In March 1982, STS-3 flew a near-perfect mission, further testing the robotic arm (by lifting a diagnostics package out of the payload bay for the first time) and carrying an expanded payload of science and engineering experiments. Three months later, STS-4 completed the OFT programme with a textbook mission, which was partially classified due to the DoD payload it carried. When it landed on 4 July, America celebrated the inauguration of its Shuttle programme to operational flight status. The Shuttle had proven it could fly, operate in orbit, return and fly again. Now it was time to see what work it could really do.

When Joe Allen was told he would fly on STS-5, he was absolutely thrilled. It was too early for the 1978 selection to be assigned to a flight (although they were being considered for STS-7 and STS-8), so the choice of mission specialists for the first operational flight came from the 1967 selection. The remaining two 1965 scientist-astronauts were Garriott (already assigned to Spacelab 1) and Kerwin. At one time, Kerwin had been considered for assignment to STS-13, but he had become the NASA representative in Australia in April and was out of consideration for crew assignment. Allen thought the selection would go alphabetically because he and Lenoir were assigned to STS-5 (Parker was working on Spacelab 1, Henize and England were already working on Spacelab 2 and Thornton, according to some reports, on Spacelab 3 issues).

Lenoir, however, thought his selection to the crew was based on the back-up work he had performed on Skylab (along with that of Musgrave who was assigned to STS-6).[9] In addition, he assumed that his experiences in simulators for Skylab (as he put it, "tested under fire") revealed how he would probably respond and behave on an actual space flight. But, as with Gemini and Apollo, the mysteries of crew assignment remained just that – a mystery. Once Lenoir came off Skylab, he had been assigned by Jack Schmitt, then Assistant Administrator for Energy Programmes at NASA, to lead a NASA team of eight people studying the potential and economics of space orbiting power systems, an assignment which lasted from 1974 to 1976. In 1976, Deke Slayton announced the crews for the Approach and Landing Tests (flown in 1977) and in 1978, the names of the four Orbital Flight Test crews for the initial test flights of the Shuttle programme. Flights one to four would be two-person crews, but flights five and six were expected to be four-person crews, including the first mission specialists. It was the first opportunity for Lenoir (and the other scientist-astronauts not assigned to Spacelab development) to be considered for a flight seat.

The challenge and the responsibility

Preparing for a space flight can be a challenge, especially your first, and after a fifteen year wait, one could imagine a sense of anticipation, excitement and thrill to know you were finally going to make the ride into space. Lenoir never found the preparation daunting, although the added assignments of deploying the first satellites from the Shuttle and participating in the first EVA of the programme (and the first US EVA since Skylab 4 eight years before) were perhaps the most challenging aspects of the preparation. No one had actually deployed a satellite payload from the Shuttle before, so Allen and Lenoir worked with the engineering department in developing the procedures they would use on orbit. As there was no Tracking and Data Relay Satellite network in place, air-to-ground communications were only available for about twenty-five per cent of the time. The other seventy-five per cent of their on-orbit time would be spent in radio blackout, which placed a lot of responsibility on the crew since it was during this air-to-ground down-time that both satellite deployments would take place.

The first operational Shuttle mission

The STS-5 mission launched on 11 November 1982 from Pad 39A at the Kennedy Space Center and landed 5 days 2 hours 14 minutes 26 seconds and 82 orbits later on

Lenoir (foreground) and Allen (rear) are assisted in their translation across the mock-up Shuttle payload bay in the WETF at JSC during training for their planned, but subsequently cancelled EVA on STS-5.

16 November, on the concrete runway at Edwards Air Base in California. The SBS-C business communication satellite was successfully deployed during the sixth orbit some eight hours into the mission, while the second Comsat, Canada's Anik C-3, was deployed on orbit twenty-two. A planned EVA demonstration was at first postponed due to crew illness, then cancelled due to faulty equipment. The nausea episode suffered by both Overmyer and Lenoir was published and endlessly discussed by the media, much to the annoyance of the crew. The openness of private medical matters was becoming an irritation to the astronauts as a group and plans were being devised to amend the reporting of such incidents from STS-6 onwards, unless it threatened the mission. Proud of their satellite deployment success, the crew posed for an in-flight crew portrait with a card advertising the fast and courteous satellite deployment service provided by the "Ace Moving Co", whose motto was "We Deliver".

344   The Long Wait

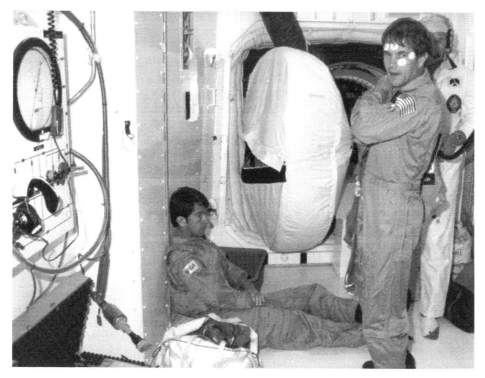

Lenoir (seated) and Allen, who is wearing electrodes for monitoring his reflexes during launch as part of Bill Thornton's biomedical tests. They are waiting to enter Columbia via the side hatch on 11 November 1982.

Upgrading the Columbia

For the flight of the first two mission specialists, certain changes had to be made to Columbia to accommodate the additional astronauts. A flight seat was installed at the aft flight deck station between and behind the two pilot seats, whose ejection seat pyrotechnics were deactivated much to the relief of the mission specialists – their seats were not fitted with such a system! The two overhead ejection panels remained and would be available for emergency ground egress should the need arise. A second new seat was installed in the mid-deck and both had quick disconnect features, allowing them to be removed and stowed for on-orbit operations and easily reinstalled for entry and landing. For this flight, the RMS flown on STS-2, 3 and 4 was removed to accommodate the additional weight of the two satellites and their deployment support hardware.[10] Because the large ejection seats remained in place on Columbia, there was only room for one additional seat on the flight deck, so Lenoir would sit on the flight deck for launch with Allen on the mid-deck, and they would swap positions for the ride home. Allen was not nervous about the lonely ride to orbit, as he would not have much to do down on the mid-deck or have much of a view of the events unfolding as they ascended, although he joked about having a security blanket for company.

Lenoir would fly as the first third-seat crew member, a position that evolved into the flight engineer role, assisting the commander and pilot on aspects of ascent, entry and landing by reading off checklists and monitoring the flight deck operations, ready to support any abort situation that might occur. One of Lenoir's chief responsibilities for this mission was to evaluate and develop this role of flight engineer. Both Allen and Lenoir worked together on deciding who would fly in which seat. With Lenoir's experiences in developing Shuttle controls and display systems and Allen's experience as entry and landing Capcom, it was logical that Lenoir's should occupy the flight deck seat for ascent and Allen for the re-entry. The techniques developed by Lenoir and Allen for STS-5 have continued in the MS 2/flight engineer role throughout the Shuttle programme.

A laid-back approach to launch

The crew climbed aboard Columbia about three hours before launch and strapped into their seats after the initial checks were completed. There were a couple of hours with practically nothing to do, so Lenoir, being a laid-back type of person, decided to unstrap himself and catch up on some sleep in his seat. He would be woken by Brand

Lenoir is photographed by Allen in the White Room just prior to entering Columbia. Brand and Overmyer are already in their seats on the flight deck.

when he was required for the pre-launch activities leading up to lift-off. With delays and holds experienced on the first four missions, Lenoir never really thought they would go on the first attempt, so when "zero" was reached without delay, he was a little surprised. The two minutes spent "riding the solids" (the Solid Rocket Boosters, or SRBs) seemed to Lenoir to pass in two seconds, although he remembered the ride being very rough. As soon as they separated, the ride to orbit became smoother, with more time to ensure that everything was working correctly. At Main Engine Cut-Off (MECO), Lenoir's first reaction was that something had suddenly gone wrong and he grabbed the malfunction checklist to be prepared for any contingency they might be faced with. When he let go of the board and it just stayed there, he realised that nothing had happened. He had been strapped in so tightly that when the engines cut off he had momentarily registered their absence as a problem. Once he let go of a pen and it also floated in before his eyes, he realised that everything was fine.

Down on the mid-deck, Joe Allen was all alone with little to do other than to enjoy the experience of finally leaving Earth. Unlike his colleagues on the flight deck he did not have a window to look out of, but he was able to fully experience the dynamics of the event. In the pre-flight press conference, Allen had been asked what his duties would be in comparison to Lenoir on the flight deck: "Well I'll actually be a passenger. I'll be in the mid-deck, but I've requested my shipmates that they not send any radio transmission or ask any questions on the intercom that end in the word 'that', such as, 'What was that?' They will be very specific in what they say."[11] In 1983 he wrote of the experience: "I told friends I'd believe I was riding Columbia when I heard the solid rockets light for lift-off [having already felt the power of the three Shuttle main engines ignite seconds before]. Your whole soul knows when the solids light. Your body shakes, and you know you're on the front end of the world's most powerful afterburner, going straight up. You accelerate rapidly, and the noise suddenly stops when the solids burn themselves out. The Shuttle is now running on the three main liquid hydrogen and oxygen engines, the smoothest running machinery I've ever been around. The solids are a rocket man's dream, making loud crackles and pops, but the liquid engines are so quiet you don't even hear them when they shut down."[12]

Welcome to space

Joe Allen has described the first day in space as being, "like a baby deer on ice. Your feet go out from under you, you bang into everything, and you move around with your arms held out in front of you. Space is like an undersea world, with a dream-like quality in which things seems to move more slowly. By pushing off from the wall of the spacecraft you keep going until you reach the other side, holding out a hand to stop your momentum. Early on you awkwardly bump into things, but slowly you learn to adapt and use gentle pushes to move around."

The downside of experiencing microgravity for the first time is the onset of what has become known as Space Adaptation Syndrome (SAS), more commonly referred to as space sickness. On average, about half of all space explorers have some kind of dehydration upon entering space for the first couple of days of the mission. In the early programmes, movement inside the spacecraft was severely restricted and it was only

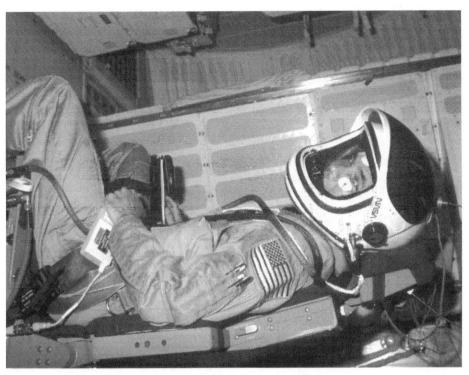

Allen wears the Shuttle constant-wear garment in his seat on the mid-deck of Columbia for his first ride into space.

when crews could translate between spacecraft that the first indications of motion sickness became apparent. First appearing during the Vostok 2 flight for the Soviets, and on Apollo 8 and 9 for the Americans, it was during the Skylab programme that the increased volume of the spacecraft (and quick movements of the crew) really brought the problem to the fore. On STS-5, both Bob Overmyer and Bill Lenoir suffered from SAS, while Brand, the only veteran on the crew, also felt a little discomfort. Joe Allen found no problem at all in adjusting to weightlessness. As Overmyer recovered, Lenoir's sickness worsened and on the third day, he threw up – something he described as more like a "wet belch" – prompting the mission management team and flight surgeons to postpone the planned EVA for twenty-four hours from Flight Day (FD) 4 to FD 5. By the fourth day, Lenoir felt much better.

Lenoir felt generally fine for the first two days prior to his sickness, but he ate a large supper at the end of FD 2 and woke early the next morning with a vomiting episode, general malaise and stomach problems. During FD 3 he ate nothing, but he drank fluids and took medication and after a night's rest, he improved considerably.[13]

Fellow scientist-astronaut Bill Thornton, a specialist in human adaptation to space flight, told Lenoir after the mission that he thought he would get sick, because his vestibular sensors were so acute that he could sense minute movements that no one else could. Lenoir had never become sea sick or air sick, even during jet pilot training,

348 The Long Wait

The first two mission specialists in space. Lenoir and Allen are shown on the aft flight deck of Columbia during the STS-5 mission.

so he had no reason to expect that this would happen. Thornton suggested that Lenoir's mental adrenalin during the build-up to flight, the launch, and activities during the first day – and after waiting so long to fly – was the reason he did not become ill until after the second satellite deployment. This had completed the primary objectives of the mission, allowing him to relax. Lenoir has described the symptoms as similar to "a low grade hangover."[9]

With Columbia safely on orbit, it was time to get on with those primary objectives – the deployment of the two communication satellites and the EVA demonstration. However, one of the more unusual activities in securing the vehicle for orbital flight occurred when Vance Brand taped up the side hatch handle, to remind everyone that it would not be advisable to try to use the double lock handle to steady themselves in the mid-deck since it opened outwards, away from the pressurisation compartment!

We deliver!

On the ground during simulations, the crew could hear the mechanics of the communication satellite's rotation, but in space, the stabilisation of the seven tons of satellite was silent. Allen later commented that the only way to know the satellites were spinning was to see them doing so out of the aft flight deck windows and via computer confirmation that they were rotating at 49.9 revs per minute – something that still amazes him. The firing of the pyrotechnics and the mechanical deployment could still be felt inside the vehicle, however.

Although it was a two-person job, all four of the astronauts were involved in the deployment of both satellites. For the first deployment (SBS-C on FD 1), Lenoir acted as "flight director" from the command seat on the flight deck, monitoring computer displays. Allen looked out of the aft flight deck windows, reporting what he saw to Lenoir, who operated the deployment procedures. On the second deployment (Anik C-3 on FD 2), these roles were reversed. According to Lenoir, Brand was overseeing the operation, "making sure we did not screw up since he was the commander of the whole thing." Overmyer took photos when Allen (the prime mission photographer) was occupied with the deployments.

No EVA this time

The objective of the STS-5 EVA was to have Lenoir (EV-1) and Allen (EV-2) verify the operational status of the end-to-end Shuttle EVA system. This included preparing and using the Extravehicular Mobility Unit (EMU), the airlock and payload bay provisions, and demonstrating procedures, timelines and training. The EVA was scheduled for three-and-a-half hours and included a range of major tasks; specifically, evaluating communications, translation rates, Shuttle EVA tools, restraints and torque measurements at a work station. The latter involved working at a mock-up Main Electronic Box (which was to be replaced on the Solar Max satellite during STS-13, or 41-C), during which thermal blankets would be removed, fasteners mounted, connectors uncoupled, a ground strap severed and the connectors re-coupled (all demonstrations of manipulative tasks). They would also be operating the winch to simulate contingency modes for closing the payload bay doors and conducting translations down the payload bay door hinge line with a bag of EVA tools (with an approximate mass of 27 kg) to demonstrate techniques for moving mass across the payload bay.[14]

Lenoir's illness led to the EVA being delayed, to avoid the risk of the astronaut vomiting inside the helmet and potentially choking himself. The following day, after waking up to music from *The Stroll*, the two astronauts prepared their equipment and began their pre-breathing period. After troubleshooting some minor problems as the spacecraft came back into communications range of Houston, the first signs of a more serious problem were reported by the crew as they prepared the suits. The fan on Allen's suit developed a fault and, despite some troubleshooting efforts, refused to work correctly.

It soon became apparent that Allen's suit would not support an EVA so, while repair activities continued in an attempt to fix the problem, the crew suggested to Mission Control that a solo EVA demonstration should be undertaken to try to salvage something from the EVA objective. Joe Allen put forward his thoughts on the idea to the ground: "I would sure like to make a strong suggestion that ... Bill proceeds with this [solo EVA]. He is well trained and ought to go to it." Lenoir also tried to influence the decision himself: "In looking at our flight plan here for the EVA, I would suggest that ... we just do a bare bones [demonstration] to verify the

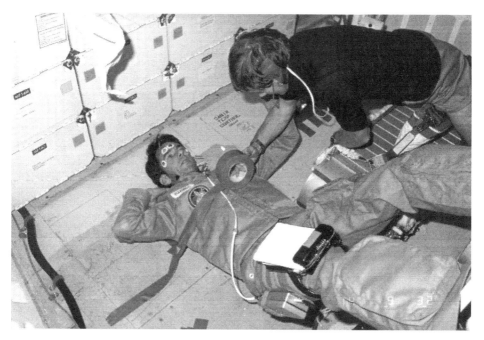

Lenoir is taped to the mid-deck floor by Allen as part of the biomedical tests conceived by Bill Thornton. The treadmill designed by Thornton is beneath Allen.

EVA, which in essence would be the aft translation and some of the little tasks. I would not be very inclined to want to put my foot into any foot restraints, being the only guy out there."

Mission Control came back with an amended plan to have Lenoir to check out the EMU while remaining in the airlock. The crew, though disappointed, was not about to knock the idea – an "almost EVA" was better than no EVA at all. For Lenoir, this would have been a frustrating activity, being so close to the exit but not being able to even put his head outside. However, the crew would conform to the rules from Mission Control and prepare for their experiment, the rationale being that their work would help later crews with their EVA preparations and post-EVA activities. But while Lenoir was beginning to pressurise his suit and the crew and ground were discussing the problem with Allen's suit, a new glitch surfaced. Lenoir was unable to raise the pressure in his suit from 3.8 psi to the required 4.3 psi. After receiving the "go" from Houston to reattempt the pressurisation, the same problem occurred. It was decided to perform a leak check and press on with the pre-breathing stage before Lenoir pressurised the suit and let it stabilise for three minutes. This was completed successfully, but still not at the mandatory 4.3 psi.

Mission Control broke the bad news to the crew that the EVA had to be terminated as Lenoir's suit was regulating too low to continue. However, they did request that Allen's suit (without the astronaut inside as it would have been too hot to wear comfortably) should be pressurised to see if its regulator operated

Allen, taped to the middeck floor, participates in experiments to monitor his responses to pre-designated activities. These would be evaluated and compared with the results of similar experiments conducted in 1 G.

correctly. During the re-pressurisation of Allen's suit, further problems were encountered, with the regulator logging just 4.1 psi. It was expected that full pressurisation would take longer as there was no astronaut inside the suit, which meant that there was a greater internal volume to pressurise, but after installing a new LiHO canister the regulator recorded 4.3 psi, underlining the fact that the problem was in Lenoir's suit alone. There would be no EVA for the astronauts this time, but with most of their objectives accomplished, the crew observed that only a safe entry and landing remained. As Allen commented, "Out of these two [the EVA and a safe landing], if we had to make a choice, let's choose the safe landing."[15]

In hindsight, Lenoir thought that – from an engineering standpoint – far more was learned from the failures and abandoning the EVA than by completing the task, as those failures would likely have occurred at some point in the future. It was far better to have experienced them when an EVA was not necessary than when it was essential. This perspective was not picked up by the media to the same degree, the failures highlighted as a setback rather than simply part of the learning curve in the development of the systems and procedures.

Flying for work, not comfort

Flying in the Shuttle, at least on the earlier, short missions, was not exactly "comfortable", but it was far better than the early capsules. Knowing the mission would only last five days, the crew were not overly concerned about habitability. Each of them had "roughed it" on camping trips lasting far longer than a flight on the Shuttle. As Lenoir put it, "We didn't think we were going up to be comfortable. We were going up to do some work."[9] Habitability engineers thought that the crew would each use a sleeping bag down in the mid-deck, but it did not work out that way in reality. Brand did not like floating around and used a small bungee cord from a clipboard to attach himself to the most suitable location, still floating but tethered while asleep. Lenoir squeezed into a place in the aft port side of the flight deck, which was small enough that if he relaxed he would not float out and inadvertently flip any switches. Joe Allen was the "freewheeler", going to sleep in the mid-deck but waking on the flight deck. Overmyer ended up as the evaluator of the sleeping bag for the engineers, attached by Velcro to one of the mid-deck bulkheads.

Although they were "up there to work," this was the first flight for both Lenoir and Allen, and neither could resist taking in the view out of the windows. What struck Lenoir was the way in which the view changed so quickly in such a short time, bringing home the realisation that he was orbiting a planet at five miles per second. After a while, a mere glance out of the window was enough to pinpoint which continent they were flying over and he even found it possible, eventually, to recognise the oceans, because the meteorology over each was so different. Bill Lenoir had long been interested in the remote sensing of the Earth and its resources, with particular emphasis on the role of man. Flying on STS-5 finally gave him the opportunity to gain first-hand experience of looking at Earth from space to help determine the potential of humans to support remote sensing of the planet.

Allen was awestruck by the view of Earth from space: "You know the Earth is round because you see its roundness, and then you realise there is another dimension, because you see layers as you look down. You see clouds towering up, you see their shadows on sunlit plains, a ship's wake in the Indian Ocean, bush fires in Africa and the reds and pinks of the Australian desert. In space the sun truly comes up like

Lenoir participates in a vision test during the STS-5 mission.

thunder, and sets just as fast, but in that time at least eight different bands of colour come and go, from a brilliant red to the brightest and deepest blue. No sunset or sunrise is ever the same [and] night falls with breathtaking abruptness. One moment you see the Earth, the next you don't." Allen assumed he would know where the Earth would be even in darkness, due to city lights or light from the sunrise and sunset, but he encountered the darkest black he had ever seen. To try to find the Earth, he would track stars until they disappeared behind the planet.[12]

Experiments and hardware

As with most space flights, there were opportunities to add smaller objectives to the primary goals set out for STS-5. The deployment of the two Comsats and the EVA (and the first flight of a four-person crew) were the primary objectives, but there was a range of small mid-deck experiments for the crew to conduct during the mission. These included a number of life sciences Detailed Supplementary Objectives (DSO). For Lenoir, those devised by fellow astronaut Bill Thornton were the most interesting, and were directly applicable to human adaptation to space flight. Some were conducted during the ascent and others during re-entry. One test required Lenoir to be restrained in a prone position in order to take data readings. The problem of trying to do this in weightlessness was resolved quite simply by taping the astronaut to the bulkhead with grey electrical ducting tape. Such improvisation was not confined to the medical experiments.

354 **The Long Wait**

Joe Allen and the sheer joy of space flight – especially after waiting fifteen years to get there.

As chief photographer, Joe Allen was very pleased to be given flight cameras to take training shots for practice, evaluating the rolls of film to hone his skills prior to the actual flight. Allen wanted to keep the cameras until the very last minute prior to flight, which to him meant when he was strapped into his seat on the mid-deck for launch. He took images of the trip out to the pad and of the ingress and gave the camera to a technician prior to closing the hatch. Some of these photos were so good that they appeared in *Time* magazine.

Allen was given a NASA training book for his Nikon camera, but could not find out how to do a delayed shutter release. He went to the local photo shop and asked for an owner's manual for the same model, which revealed where the delayed shutter release mechanism should be. However, the NASA space flight version had been modified, at a cost of tens of thousands of dollars, "in order to make it astronaut proof, such that astronauts didn't, by mistake, put the camera on delayed timing and thus mess up a picture." Allen investigated further and found that costly and totally unnecessary modifications had been made, despite the fact that the cameras had been successfully used in wars, violent storms, at the top of mountains and at the bottom of oceans without such changes. Still, as this was the first flight of four astronauts on one spacecraft, Allen wanted to record the event for posterity. Ever mindful of strict NASA regulations about taking personal items into space (since the Apollo 15 stamp incident from 1971, a mission he had supported as Capcom and mission scientist), he was careful to bend but not break the rules for STS-5. From a camera store, he purchased a cheap shutter release mechanism that attached to the shutter release directly, not the camera body facility that had been blocked. The device allowed him

A scientist-astronaut demonstrating the physics of liquids in space. Famous for his experiments on liquid dynamics, Joe Allen watches a globule of orange juice float in front of him.

to take the first four-crew photos, including some of the first "crew starburst" images, something that has become a standard feature on subsequent flights. Despite the camera not having a delayed shutter release, no one at NASA ever said a word to Allen about the images, but he suspected that staff in the photo lab must have wondered how the shots were achieved.

For the return leg of the STS-5 mission, Lenoir and Allen exchanged places and Allen found himself on the flight deck – but with very little to do. He decided to take photos of the entry, and stood up to take what turned out to be spectacular views from inside the flight deck during entry, including views through the overhead windows of the plasma coming back together above and behind the orbiter. These were of particular interest to the engineers, as they had no such photographic records previously. Allen was also particularly pleased with a shot of one of the OMS burns (or, more correctly, an enormous flash of light at the back of the orbiter, which has since been reproduced in several books). The flash apparently lasted for only a fifth of a second and the camera exposure was a sixtieth of a second, "So I had to shoot a sixtieth of second during that fifth of a second, which is virtually impossible to do. But I got very lucky and was quite pleased by that result."[16]

During the re-entry, Lenoir continued Thornton's SAS investigations. He also noted that about half way though entry, the inside of the mid-deck seemed to get warmer. On landing, however, Brand had to tell him that they were down as it was so smooth. The two scientist-astronauts were now the very first flight-experienced mission specialists and had finally made their trip into space. They were looking to return

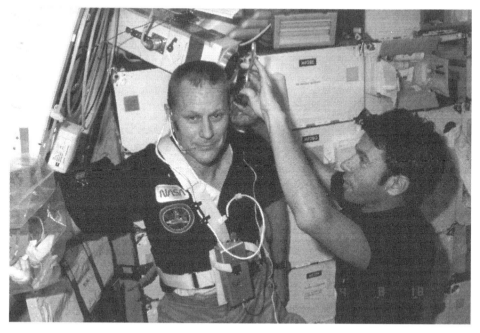

"Hair cut sir?" Lenoir trims the hair of Bob Overmyer to Marine standards.

to orbit soon, and both would begin training for new missions into space from 1983. Only one of them would return to orbit.

The late oceanographer Bob Stevenson, who trained many NASA astronauts and crews in observation and photography, once described Bill Lenoir as "a real happy chappy," and mentioned his particular fondness for fresh jalapeño peppers, which he grew in his back yard. When they were still green, Lenoir would often bring a bagful to work and happily nibble away on them during the day. Sometimes, he would offer them to younger astronauts and take a little impish pleasure at their reaction when they bit down and found out the true nature of what they were eating.

Lenoir somehow received clearance to take a paper bag full of his beloved jalapeño peppers on the STS-5 mission. He planned on just popping one into his mouth every so often, and that whenever there was a TV broadcast to be made he would be seen just chewing away on them – just for fun. Unfortunately he became ill with a bout of space sickness early in the mission, and by the time he'd recovered sufficiently toward the end of his flight the rest of the crew had naturally finished off his lovely fresh peppers.

STS-6: THE CHALLENGE OF EVA

The crew for STS-6 had been announced on 2 March 1982 (NASA News 82-012) as Paul Weitz (commander) Karol Bobko (pilot) Story Musgrave (MS 1) and Don

STS-6: the challenge of EVA

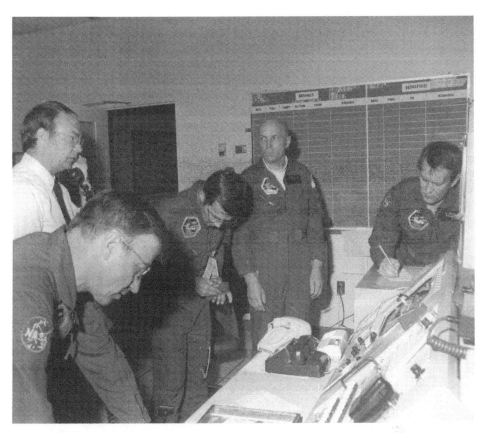

The STS-6 crew receives preliminary flight instruction from Shuttle Landing Facility Manager Roger Gould prior to flying formation manoeuvres in their T-38s the day prior to their launch into space. Left to right: Gould, Commander Paul Weitz, MS Don Peterson, MS Musgrave and Pilot Karol Bobko.

Peterson (MS 2). This crew included one veteran (Weitz) and three rookie astronauts, but all had been at NASA since the 1960s, like the STS-5 crew. The combined age of the STS-6 crew (191 years) was humorously highlighted by the astronauts in TV transmissions, with all four posing for the cameras wearing vintage spectacles. The objectives for STS-6 were to deploy the first Tracking and Data Relay Satellite (TDRS) by means of the IUS upper stage, and to inaugurate the Challenger orbiter OV-099 to operational flight. The cancelled EVA demonstration on STS-5 was reassigned to STS-6, again with the primary objective of verifying the operational status of the basic EVA system. From a Shuttle programme standpoint, this EVA would be the first step in verifying the elements necessary to support the Solar Max satellite repair mission scheduled for STS-13 (later renamed STS 41-C). The EVA activities would begin with functional equipment checks on FD 3, with the EVA itself conducted on FD 4.[17]

358 The Long Wait

Musgrave's STS-6 training load

Though formally identified in March 1982, the crew commenced their training in October 1981 with ascent, orbit and entry training lessons in the Shuttle Mission Simulator (SMS). Their first ascent and entry integrated simulation occurred in September 1982. The STS-6 crew became the first to have a dedicated SMS team from the start of their training programme through to launch. Training for the crew included ascent, orbit and entry flight operations, payload (IUS) flight operations, crew subsystems training (including pre-launch ingress/egress), EVA operations, and integrated simulations. The IUS training course was completely redesigned and sequenced between the publication of the STS-6 training plan and actual training. The only crew member who had any formal IUS training from the training flow was Musgrave, while the rest of the crew received informal briefings not listed in the training flow.

Table 3. MS 1 Story Musgrave planned vs. actual training hours STS-6

Course	Planned hours	Actual hours
Ascent integrated simulations	0	0
Orbit integrated simulations	92	167.5
Entry integrated simulations	0	29
Ascent flight operations	0	0
Orbit flight operations	83	69
Entry flight operations	1	40.5
Crew systems	59	58
EVA operations	65	120
IUS training	63.5	37.5
Orbiter support systems	0	0
CFES	24	16
TDRS	8	8
Secondary experiments	8	10
Total	*403.5*	*555.5*

Challenger flies

Following the decision not to use the first orbiter (OV-101 Enterprise) for orbital operations due to its excess weight, Structural Test Article OV-099 was upgraded to become the second orbiter to enter space. Originally planned for a 27 January 1983 launch, the discovery of serious problems with the SSMEs resulted in an exchange of engines and a launch delay until 4 April 1983. Had the engine problem not been detected, a disastrous launch pad explosion could have resulted. The primary payload of this flight was the first Tracking and Data Relay Satellite (TDRS), which would be launched from the payload bay using the Inertial Upper Stage (IUS). The TDRS would be the first of a series of communications satellites designed to expand the on-orbit coverage for direct contact with a Shuttle crew including, eventually, during re-

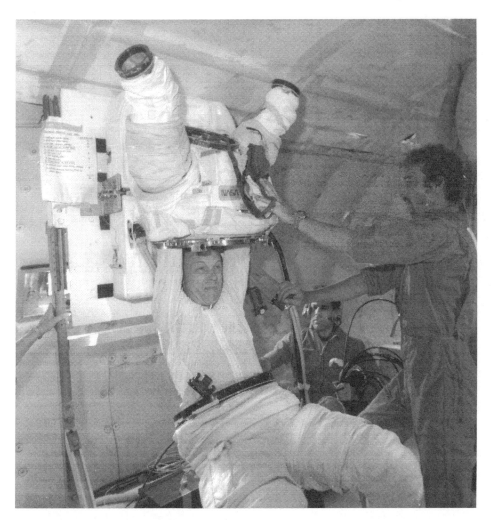

Musgrave dons the Hard Upper Torso of the Shuttle EVA suit during KC-135 training in November 1982. This "zero-gravity" exercise was to practice donning and doffing the suit under simulated weightless conditions.

entry. The TDRS network would gradually replace the existing network of fifteen ground stations that gave only thirty per cent direct communication time with the Shuttle in any ninety-minute orbit. Three operational TDRS satellites placed 120 degrees apart in synchronous orbit would provide ninety-eight per cent real-time coverage each orbit. TDRS was also a vital element in the US/European Spacelab mission manifested for later in 1983 aboard STS-9 (Spacelab 1). This was a planned round-the-clock science mission over nine days. The second TDRS was manifested for launch on STS-8 and the third on STS-12.

The launch of STS-6 occurred on time and without major incident, and the

STS-6 Pilot Karol Bobko assists Musgrave during preparations for an underwater EVA simulation at the WETF at JSC. Musgrave stands on the platform that would take him and Don Peterson (behind Musgrave) into the pool to begin their simulation.

deployment of the IUS carrying the TDRS-A satellite was achieved successfully ten hours into the mission. However, although the crew's operations in the deployment of the payload went without a hitch, the pre-planning sequence to place the satellite into operational orbit did not. The IUS first stage burned for 2 minutes 31 seconds to place the combination into its transfer orbit. The second stage should then have burned for 1 minute 43 seconds to place the TDRS in a synchronous orbit at 56 degrees W. However, an oil-filled seal deflated, causing the steering mechanism to malfunction. Only prompt action by the ground controllers saved the satellite as they quickly separated it from the erroneous IUS stage, far short of its planned orbit. Over the next fifty-eight days, delicate manoeuvres eventually placed the satellite on station at 41 degrees W, but with 370 kg of the 598 kg of propellant used, its operational mission was seriously compromised. The STS-9/Spacelab 1 mission was secure, but future IUS launches were put on hold as the problem was investigated. The IUS from STS-8 and STS-10 (a planned military satellite deployment) were taken off the manifest until the problem could be rectified.

Aboard Challenger, after the TDRS deployment, the crew and vehicle performed flawlessly, including during the first EVA from the Shuttle in the payload bay of the orbiter. For over three hours, Musgrave and Peterson worked outside, clearly enjoying the experience. The crew also performed a number of mid-deck and secondary experiments and completed their mission with a 9 April landing on Runway 22 at Edwards AFB in California, after a flight of 5 days 0 hours 23 minutes 42 seconds and 81 orbits.

Story's story

Ever since the configuration of the Shuttle (with its solid rocket boosters) was developed in the early 1970s, Story Musgrave was more than a little apprehensive about the design. When it came to flying the missions, though, he was not as scared on his first flight as he would later admit to being on his second. During his first flight into space, he was more involved with the EVA and orbital operations than he was with activities during ascent. Peterson, the other mission specialist on STS-6, was a test pilot, so he helped Weitz and Bobko on the ascent and entry phases and looked after the Shuttle on orbit. Musgrave essentially "handled everything else." With less focus on operational activities during ascent and descent and more on the orbital phase of the mission, Musgrave's "fear factor" was less on this first mission, as he was "basically along for the ride."[18]

With a nominal TDRS deployment (which Musgrave has mentioned was basically down to repetitive crew training, learning a sequence of switch throwing and working as a team to get the job done without too much variance to the pre-flight plan), he and Peterson began preparing to conduct their EVA. During the EVA, Musgrave translated to the aft bulkhead of the payload bay and looked back over the three main engine bells, marking the difference between this real EVA and his training in the WETF water tank back in Houston with the comment, "This is a little deeper pool than I'm used to working in." Working through the timeline, Musgrave and Peterson completed a translation to the aft bulkhead and performed safety tether dynamics, a mobility evaluation of the EMU, and a range of operations at the tool box, as well as simulating contingency operations by lowering the IUS tilt table and closing the payload bay doors using the forward winch system. For Musgrave, there were no surprises in the EVA and he later commented that the use of the water tank was excellent preparation for it.

Story enjoyed the food and took a sleeping pill to help him sleep on the mission, but he suffered no symptoms of SAS. He never found adapting to space a problem (he ended up by accomplishing the transition from Earth gravity to microgravity six times in his career) and could float right out of the seat and complete a somersault with no effect at all. It was coming home that he found more challenging.

362 The Long Wait

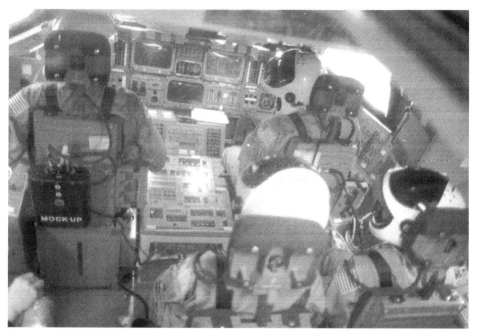

The four STS-6 crew members participate in a December 1982 simulated launch during training in the Shuttle engineering mock-up at JSC. These are the positions the astronauts would occupy during their actual launch four months later.

During re-entry, Musgrave stood up to take photos over the shoulders of the pilots. He found it very difficult to hold the camera, as the force of gravity made him feel very, very heavy. Once the crew disembarked from the orbiter after landing, Musgrave held on to the hand rail of the steps "for dear life," to stop himself falling over. The plan to stand up on the flight deck was not a standard procedure but the crew wanted to determine how reasonable it was, in the event of an emergency or off-nominal situation, to send a crew member down to the mid-deck to, for example, close a circuit breaker or throw a switch, and this was the first insight into evaluating that capability. Of course, following the Challenger accident, the slide-pole method of escape would have required a crew member to leave the flight deck and descend to the mid-deck to evacuate the vehicle, although this procedure was not available at the time of the 1983 STS-6 mission. Musgrave's activities helped formulate such plans and capabilities years later.[19]

In the post-flight press conference, the crew was asked if the flight had changed their lives in any way. Musgrave replied that the flight had not changed his life too significantly, although it had "brought to fruition something that I've been working hard on for at least sixteen years and I'm just looking forward to going again as soon as I can."

Following STS-6, Musgrave put the experience of EVA from the Shuttle and years of support work developing EVA systems and techniques to use, reviewing

Story Musgrave on the 7 April 1983 EVA during STS-6. Note the slide wire tether system that connects him to the Shuttle at all times.

the plans for refuelling the Landsat 4 satellite at the General Electric facilities with astronaut Jerry Ross in October 1983, just prior to the announcement of his second flight crew assignment. The next former scientist-astronaut to reach orbit, however, would be Dr. William Thornton aboard STS-8.

STS-7 flew in June 1983, under the command of STS-1 veteran Bob Crippen. Challenger was also on its second mission and carried the first representatives from the first (1978) Shuttle-era selection: pilot Rick Hauck, and mission specialists, John Fabian, Sally Ride (the first US female to fly into space) and Dr. Norman Thagard, who studied Space Adaptation Syndrome issues in support of Bill Thornton's programme.

Medicine takes precedence over Earth science

During the Apollo programme, NASA decided to invite a select group of Earth-sciences people to train its astronauts in Earth surveillance from orbit. Oceanographer Bob Stevenson from the Scripps Institution of Oceanography had already been training the Gemini astronauts, and he would be joined in giving these crew lectures by Australian-born Dr. Paul Scully-Power during the latter's first work assignment in the

A close-up of Musgrave during the EVA, revealing the chest-mounted control panel, cuff-mounted procedures checklist, restraint devices and helmet-mounted lights of the Shuttle EVA suit design he helped develop and evaluate as a technical assignment in the Astronaut Office.

United States for the Naval Underwater Systems Center in Connecticut. In 1978, Scully-Power was officially invited by NASA to join Stevenson in further briefing sessions at JSC, specifically designed for Shuttle crews. By the time Columbia was first launched into space in April 1981, Stevenson and Scully-Power had developed a strong rapport with the Shuttle crews, who rewarded their involvement by bringing back magnificent and specifically-requested photographs of ocean phenomena.

As Stevenson revealed, there was already a big push from within NASA to have specialised oceanographers aboard future Shuttle missions, and they were looking at STS-7 and STS-8 to launch this particular programme.

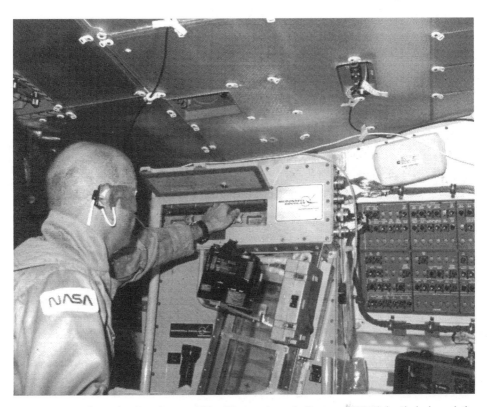

Musgrave monitors the Continuous Flow Electrophoresis System (CFES) loaded aboard the mid-deck of Challenger during STS-6

"One of the strongest advocates of this was Dick Truly, who flew on the second Shuttle mission. While the first Shuttle flight had only been a two-day proving mission, STS-2 was sent up for five days, and carried the first Synthetic Imaging Radar (SIR-A). Paul and I were the Navy oceanographers assigned to that experiment.

"It was after that flight … that the thought of flying an oceanographer was expressed by Dick Truly to George Abbey, who was the Director of the Flight Crew Operations Division – an influential guy at NASA who had the final say on crew selections. The idea did not advance beyond the few of us until mid-1982 when the Shuttle test programme had been completed with the flight of STS-4, and the crews could then be expanded to four or five members.

"Had the problem of 'space sickness' not begun to cause concern amongst NASA Headquarters people, then I would have flown on STS-7 with Bob Crippen and Sally Ride, and Paul would have flown on STS-8 with Dick Truly and his gang. The follow-on would have been for both of us to fly on 41-G, as it was to be (and was) a high-inclination orbit, and would carry SIR-B and the large format camera. It was not to be, however, as physician Norman Thagard was assigned to STS-7 and Bill Thornton to STS-8. At least they learned that space sickness could not be solved with a pill."

366 The Long Wait

Stevenson would later be asked if he wanted to join the STS 41-G crew as a payload specialist oceanographer, but he would selflessly decline due to his wife's advanced cancer treatment. Instead, Scully-Power would fly the mission. Stevenson would later be reassigned to STS 61-L in August 1986, but this flight, and his only opportunity to fly into orbit, disappeared with the loss of orbiter Challenger.

STS-8: DR. BILL FLIES

On 24 April 1982, the crew for STS-8 was named, with Dick Truly as commander, Dan Brandenstein as pilot and Guion Bluford and Dale Gardner as mission specialists. On 21 December 1982, NASA announced that Dr. Bill Thornton had been added to the crew to study SAS phenomena (Norman Thagard was assigned to the crew of STS-7 at the same time, for the same reason). The removal of the second TDRS from the manifest following the STS-6 problems had little effect on Thornton, who had his own package of experiments to perform on himself and his crew. His excitement at finally getting a flight was very evident, especially to his family. His wife, Jennifer, thinks that the happiest she ever saw him (during his years with NASA) was the Christmas before he flew for the first time.[20]

A workaholic astronaut

Bill Thornton's association with the study of human adaptation to space flight stretched back to the 1960s and his involvement in the USAF Manned Orbiting Laboratory Program. He worked on an exercise programme that would have been conducted by the MOL crews during the thirty-day missions had the programme not been cancelled in 1969. By then, Thornton had joined NASA and, drawing on his medical and physics background, became assigned to the Skylab programme and early developments in the Shuttle/Spacelab programme in the 1970s. After Skylab and a year's leave in 1976 to study at the Texas Medical School in Galveston (where he became a clinical instructor), Thornton returned to the CB to work on developing Spacelab procedures, maintenance techniques for crew health and experiments for Space Adaptation Syndrome for STS-4 through 7. In 1977 Thornton, still an active astronaut but with no sign of getting an early space flight, applied as a candidate payload specialist for Spacelab 1 but was not selected. He developed his ideas for treadmill exercises from the rudimentary device flown on Skylab as well as about a dozen other exercise devices for use on the Shuttle.

Following his work on developing the Teflon sheet slider "treadmill" on Skylab, Thornton continued this work for the Shuttle. His Shuttle treadmill flew for the first time on STS-2, using a shiny plate (not Teflon) to allow the astronauts to "jog in place, not slide in stocking feet." This was something he should have developed for Skylab but did not have time to devote to it. His basic but effective Skylab device could have been much improved with longer development time and with the knowledge that he gained about human muscle physiology in space after the missions. He received a small grant and fabricated the new treadmill for later Shuttle missions, despite strin-

Dr. Bill, a very busy mission specialist collecting a mass of biomedical experimentation data, checks a prolific roll of data on the mid-deck of Challenger during STS-8. The patches on his head reveal that his crewmates were not the only medical test subjects.

gent weight and volume restrictions and the need to be able to stow and unpack the device without too much difficulty on orbit. It flew on the Shuttle for several years from 1982 though, as Thornton readily admitted, "you did not really need it. It was useful in Shuttle but it was not essential the way it was on Skylab."

For the sixteen years he had been an astronaut Thornton, more than any other member of the CB, had developed techniques and ideas for keeping his colleagues healthy in orbit. These included a lower body negative pressure suit to keep fluids from leaving the legs and migrating to the upper torso and head in microgravity (the Russians had used a similar device – called Chibis – on their Salyut stations for years. It was far more bulky than Thornton's device). During those sixteen years, Thornton began to think that he would never fly into space, but Space Adaptation Syndrome – space sickness – was exactly what he had been studying for years. Of the sixteen members of the first six Shuttle flights of between two and five days, five astronauts had suffered a loss of appetite, four had experienced general malaise, five had suffered headaches, four had stomach complaints, three had nausea and six had vomited.

Concerns were being raised within NASA that if this pattern continued without addressing the problem, it could seriously affect the workload and health of future crews, as well as the ability of some astronauts to bring the orbiter home safely. Thornton was therefore assigned to STS-8 to investigate this problem first-hand, with similar experiments being conducted by Dr. Norman Thagard on STS-7.

His CB colleagues dubbed Thornton a workaholic, as he would arrive at the office at 7:00 a.m. and not leave until 7:00 p.m. seven days a week. But they also acknowledged that if anyone could find the causes of space sickness and determine the best preventative methods and procedures to overcome them, it would be Thornton. He was also known as the "last angry man" for the way he battled against bureaucratic decisions over what could or could not be accomplished in space. Thornton would simply tell the bureaucrats in no uncertain terms how he would achieve the objective and leave it at that. It was clear around JSC that Bill Thornton would not take "no" for an answer.

Dr. Bill's orbital clinic

With feedback from STS-2, Thornton was able to bring to the forefront his own thoughts and ideas for investigations into the SAS problem. He drew upon his own experiences from MOL, SMEAT, Skylab and ground simulations to develop a package of simple but effective investigations to try to address the problem. The first opportunity to study Space Adaptation Sickness on the Shuttle would really be on the STS-9 Spacelab 1 mission. However, Thornton was able to demonstrate that key data could be obtained relatively simply and with little expense by using basic techniques from which more intense studies could be performed, such as on longer Spacelab missions. Thornton worked with STS-4 commander Ken Mattingly on increasing the EKG amplifier gain rate (the EKG was used for ascent but then not used again) to obtain the first eye motion recording in flight, giving Thornton a powerful tool to support his case for flying his suite of investigations early. He then tried some further investigations with Allen and Lenoir on STS-5 before repeating the experiments on STS-6. Thagard was assigned to STS-7 and "did a great job" with Thornton's experiment package prior to Thornton's own flight on STS-8.[21]

Due to limited preparation time prior to the mission, and restricted crew time (other than Thornton's) and volume inside the vehicle, a precise determination of which experiments and investigations to fly had to be made. As Thornton recalled: "STS-8 was the first, and probably only, flight that an investigator was ever allowed to make his own selection of experiments and fly with it. It was a combination of what I considered could be done and the most essential things that needed doing, because if you use Méniére's disease (the inner ear ailment that had grounded Alan Shepard for years) for example as a model of space motion sickness, which it wasn't, there were all kinds of abnormal eye motions. The biggest single drive was to record eye motion during space motion sickness and that's why I concentrated on that. It was a combination of what was possible, such as recording bowel sounds (which turned out to be an excellent marker of space motion sickness especially when you recover), versus the things that needed to be done." Some of these, to Thornton's frustration, were never followed up after the flight.

Aboard STS-8, Thornton's studies encompassed seven basic disciplines:[22]

- *Audiometry* – testing of aural sensitivity thresholds.
- *Bio monitoring* – monitoring of crew health and medical status.
- *Electrical-oculography* – recording and measuring of eye movements.
- *Kinesymmetry* – study of the repeatability of physical motion.
- *Photography* – photo records of leg volume changes, if any.
- *Plethysmography* – volume of limbs measured in circumference.
- *Tonometry* – measurement of external tissue pressure.

Thornton flew as MS 3 aboard STS-8. As he was assigned just ten months prior to launch specifically to conduct SAS studies and various medical test objectives, and was deeply involved with these objectives on STS-7 as well, he received minimal orbiter system training, concentrating his efforts on the medical test objectives.[23]

In his preparation for the mission, Thornton attended only the post-insertion, orbit timeline and de-orbit lessons from the STS-8 training programme. While training for the major payload involving the mission specialists was beneficial, the crew reported that the biofeedback experiment and preparation for the CFES were highly unsatisfactory. Bill Thornton felt that technical support associated with the biofeedback experiment was poor, as the appropriate hardware was never available until just prior to the flight. On more than one occasion, he felt he had to drive both the preparation of the experiment for flight and the training for the sample preparation.[24] In summary, Thornton's training load (planned/actual) for his first mission was recorded as:

Table 4. MS 3 Bill Thornton planned vs. actual training hours STS-8

Course	Planned hours	Actual hours
Ascent integrated sims	0	0
Orbit integrated sims	110	57
Entry integrated sims	6	3
Ascent flight operations	0	0
Orbit flight operations	0	4
Entry flight operations	0	0
Shuttle training aircraft	0	0
GNC/DPS	0	0
Orbiter support systems	17	13
Crew systems	78	64
EVA operations	0	0
Rndz/Prox Ops	0	0
PDRS	0	0
PAM systems	0	0
Payloads	17	17
Total	*228*	*158*

370 The Long Wait

Comparison with Musgrave's preparation for STS-6 reveals the dedicated work Thornton was assigned to on STS-8 (almost payload specialist-level duration), mainly due to his late assignment to the mission.

First Shuttle night launch and night landing

After a wait of sixteen years, there was great excitement not only for the Thornton family, but also among the residents of his hometown of Faison in North Carolina. With a population not much larger than 600, it seemed that half the town had come out for the launch, which was still "big news" at the time. Thornton told his wife that she could at last have her own launch party. For commander Dick Truly's wife Cody, her party had occurred during STS-2, so this time the focus was on the rookie astronauts' families. Brandenstein, Gardner and Bluford joined the Thorntons at one big party at Patrick Air Base, near the Cape. The Thorntons' two sons stayed at one of the hotels near the beach and twelve of their friends stayed at a condo close by, under strict instructions from Jennifer Thornton to behave themselves.[20]

The launch of STS-8 was scheduled for 30 August 1983 but was delayed ten days due to TDRS difficulties. The inclusion of the Insat 1B Comsat dictated a spectacular night launch, which was threatened for a while by weather conditions that eventually cleared sufficiently for the launch to occur. The satellite was deployed by PAM upper

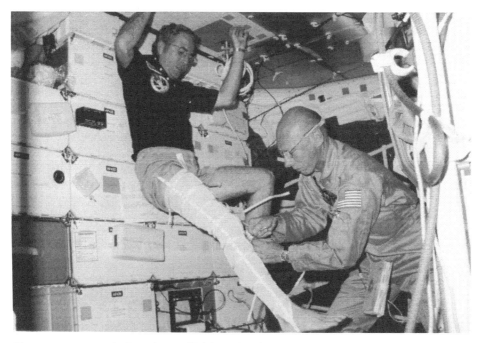

Thornton measures the leg volume of Dick Truly using a stocking plethysmograph. The test was part of the ongoing studies into in-flight fluid changes, part of Thornton's DSO programme.

stage on the second flight day at Mission Elapsed Time 23 hours 30 minutes. All five crew members were involved in the intricate operation, with Thornton operating cameras and VTR equipment throughout the entire sequence. The launch panel display used for the deployment was part of Thornton's CB technical assignment to develop crew station displays, controls and deployment procedures and plans in the mid-1970s. It fascinated him to see it in action years later.

The TDRS-B had been taken off the flight and was replaced by a 3,375-kg dumbbell-shaped Payload Flight Test Article (PFTA), designed to be grappled by the Remote Manipulator for a series of tests that involved moving the arm with a load attached. This was a simulation of satellite deployment and retrieval activities that would be conducted using the RMS on later missions. While the crew operated a series of experiments, including testing communications with Mission Control via the TDRS-A satellite, Bill Thornton performed a suite of experiments on the mid-deck aimed at collecting data on the SAS phenomenon.

Thornton's "chamber of horrors"

Shortly after the OMS burn, Thornton and MS 1 Dale Gardner left their seats and began setting up the medical measurement apparatus. The experiments had to be conducted shortly after ascent to ensure accurate measurements of the body's initial adaptation to weightlessness. Both crewmen moved slowly while adapting to the new environment but still managed to accomplish the task prior to OMS 2, with measurements being taken shortly afterwards some forty-five minutes after leaving the ground. With Gardner having to assist Thornton during the post-insertion period, this took MS 1 away from support duties with MS 2 Bluford, who performed not only his own post-insertion tasks but also most of those assigned to Gardner. The crew reported that, although the post-insertion period was normally kept as free of payload activities as possible, having the three MS aboard meant that the timeline could be more lenient, especially if this involved the movement or the manipulation and set-up of payloads.[25]

Thornton's new treadmill was available for STS-8 having been tested on earlier flights from STS-4, but Thornton was the only crew member to use the treadmill on a regular basis, evaluating his own design in orbit. His fellow crew members tried it only once during the mission. What surprised them was the amount of noise and vibration produced in the crew cabin when a user began a slow run or jog. More positively, however, they found that the treadmill gave some relief to back discomfort. After standing in the harness for a few minutes, the bungee recompressed the spine slightly, relieving the minor lower back discomfort that had been reported during previous space flights.

On the mission, both Gardner and Thornton became sick, "which was just great timing," according to Thornton. Collecting data on the other crew members was more difficult, with Thornton having to "grab them when he could," although some of the crew became interested in the physician's work. Truly, for example, came down from the flight deck during the mission and seemed almost pleased to let Thornton "have at him" with his box of medical delights. However, Thornton would not dream of asking

The STS-8 crew displays one of the official US postal covers flown as part of the mission's secondary payloads.

the rest of the crew to give blood. This was still a restricted procedure during space flight, so all the blood drawn on STS-8 was Thornton's own.

The in-flight investigations of human adaptation to space featured over fifty different studies, some of which were conducted several times on each crewman. The hardware was stored in five of the mid-deck lockers and was set up as required. Thornton also used both a removable seat and associated gear (which was fixed almost permanently in place during the mission) and the airlock to support his research. Most of the experiments were non-invasive and non-provocative, with each crew member taking part during pre-flight, in-flight and post-flight tests to provide three sets of data for comparison. Some investigations were performed only by Thornton, who spent virtually all of his time on these studies, while other crew members participated whenever their primary assignments allowed.

It was not the number of investigations that was important, but rather the significance of direct clinical observations while in orbit. The experiments helped to determine that at least one of the primary causes of SAS was an intra-vestibular conflict between the semicircular ear canals and the otolith mechanisms, though this was not the complete answer. Temporary – and painful – obstruction of the intestine (ileus) was observed to be the cause of vomiting (first observed by Thagard on STS-7) and its reversal by the use of a drug was also an important finding. Thornton determined that negative findings or results were as important as the positive ones. For example, the documentation of a rapid (one hour or less) and large fluid shift from the legs followed by slight puffiness of tissue and blood vessels in the head helped to confirm that neither the vestibular apparatus nor the central nervous system was affected by any excess of pressure or fluids. It was also found that despite constant and significant (up to 4–5 cm in some cases) increases in the length of the spinal column and changes in the body's size, posture and shape, there were no adverse affects on the central nervous system. Thornton also observed that while changes in heart rate, blood pressure and other parameters were documented during SAS, he could not find any abnormal physical functions. Both blood pressure and heart rate during re-entry and post-flight activities were normal and the anti-g bladder suit was found to be an adequate method of alleviating pressure during the return to Earth.[26]

During FD 5 Thornton, normally out of the frame during TV broadcasts, participated in a twenty-minute televised news conference to demonstrate what he had been doing on orbit, using each of his colleagues as "volunteers" to model some of his experiments as he explained the theory to the audience on the ground. More akin to an educational presentation than a PAO telecast, it was nevertheless a fascinating glimpse into the world of space medicine and space flight adaptation. Thornton began by explaining that instant changes occur to the human body when it enters orbit. As much as 1.5 litres of fluid shifts from the legs towards the head during the first day on orbit, which, as Thornton observed, "makes our faces rather broad and puffy." Thornton also explained that the shift of body fluids puts added pressure on the brain and can be a factor in the headaches some astronauts have reported. Fluid shift may also be connected to how the eye perceives motion and light while in space, which could in turn help to trigger the nausea and malaise that also affects some, but not all astronauts. Holding up Gardner's leg, Thornton showed his colleague wearing a stocking device that extended from the ankle to the hip, secured by several straps and designed to limit the movement of fluids.

While Guion Bluford demonstrated Thornton's treadmill, wearing a harness and bungee cords to create the force to work the treadmill and stop him floating off into the mid-deck, Thornton pointed out a small video recorder used to record eye and hand motions and a second device (also designed by Thornton) to record blood pressure and heart rate. Thornton then called upon his commander, Dick Truly, to model instruments that measured brain reactions to light stimuli: "We all know commanders have large brains, which is why I'm demonstrating this equipment with our commander," Thornton joked. At the end of the telecast, the joke was turned on the good doctor. Referring to Thornton's working habits in space, Truly acknowledged his hard work, but then quipped that he and his colleagues were fed up of what they called "Thornton's chamber of horrors," whereupon he picked up a hammer and floated across the mid-deck to reveal Thornton being securely restrained against the bulkhead with grey tape. His three colleagues each wielded knives, wrenches, pliers and hammers and as the screen faded, a muffled scream from the good doctor was heard to close the telecast.

After a highly successful mission, Challenger came home to the programme's first night landing on 5 September 1983, landing on Runway 22 at Edwards AFB after a flight of 6 days 1 hour 8 minutes 43 seconds and 98 orbits. Following the landing,

The crew of STS-8 participate in a post-flight news broadcast in which they discussed their mission.

Thornton commented, "I learned more in the first hour-and-a-half of the flight than I did in all the previous years that I put into this study on Earth. Space sickness is a transient problem and not the dreaded monster it's made out to be. I'm convinced the problem is solvable." To Thornton, this was a confirmation of what he had assumed and with the added "bonus" of Gardner and himself becoming sick as a result of their movements on orbit, he had immediately discovered what a huge role the otoliths played in SAS.

Reality of space flight

After two decades of developing systems and procedures for medical monitoring during space flight, developing hardware and protocols to help crews maintain condition while in orbit, and working on ground simulations of Skylab and Spacelab missions, Bill Thornton knew pretty much what to expect from his first space flight experience. He knew that pre-planning and following the timeline was absolutely necessary in order to get the job done, especially on the short missions being flown by the Shuttle. Good housekeeping and getting things stowed was critical to a successful mission, according to Thornton, despite being the self-confessed "sloppiest person on Earth, if you saw my office desk."

For sleep periods, Thornton found that wedging himself between two EVA suits in the mid-deck airlock was the most comfortable position, probably due to the isolation. After the thrill of space flight, the night landing and a flight back to Houston, Thornton tried to stay awake as long as he could to fulfil some of the post-flight medical requirements of the co-investigators. But there would be no immediate post-flight party for him, as he staggered into bed and slept soundly for his first night back on Earth.[27]

A long wait and a short wait

While completing his first space flight after serving as an astronaut for sixteen years, Thornton already knew that he had been assigned to a second mission – the third flight of the Spacelab – originally scheduled for the end of 1984. Some of his colleagues had jokingly observed that he might not make a second flight if he did not slow down his work pace a little, but then acknowledged that there was probably no one who could get him to slow down, not even his family, so he would probably make the flight anyway.

While Thornton was recovering from his first space flight experience, the next flight carried two scientist-astronauts into space; one each from the fourth and sixth astronaut groups. STS-9 was the first flight to carry the European-designed and built Spacelab science module, the largest volume of research area on a US spacecraft since Skylab, so who better to fly the inaugural mission than former Skylab 3 astronaut Owen Garriott and Skylab mission scientist Bob Parker.

STS-9/SPACELAB 1

Just before 11:00 a.m. EST on 28 November 1983, the joint NASA/ESA STS-9/ Spacelab 1 mission was boosted into orbit. It had been a problem-free countdown, although some two months later than the previously planned launch date of 30 September due to possibly faulty material in a booster thrust chamber nozzle recovered after the preceding STS-8 flight. These concerns were only raised after Columbia had already been moved out to the launch pad. NASA engineers believed that the excessive erosion of the SRB nozzle throat, causing a near burn-though, might have been related to a batch of ablative carbon fibre-cloth resin used to line the throat of the nozzle that had not been properly cured. Unless remedied, this could have led to a launch catastrophe. The orbiter was returned to the Vehicle Assembly Building (VAB) on 17 October for de-stacking, and for replacement work to be carried out using less sensitive nozzle-lining material from another manufacturer. The delay was further compounded by difficulties with a new communication network. Shuttle Columbia was on its last mission prior to a stand-down period for extensive modification. At the time of its lift-off from launch pad 39A, with the European-built Spacelab 1 secured in the payload bay, it was the heaviest Shuttle stack ever launched into space.

It would be the first and only time a scientist-astronaut from each of Groups 4 and 6 would fly into space together. Owen Garriott (MS 1) and space "rookie" Bob Parker (MS 2), along with payload specialists Byron Lichtenberg and Ulf Merbold, were ready to carry out a comprehensive work programme inside the billion-dollar orbital laboratory.

Occupying the Spacelab module

Spacelab had been commissioned in the aftermath of the 1972 decision to abandon plans for the development of a large orbiting space station, beyond Skylab. With this decision, NASA realised that it would need to come up with an interim means to conduct the space science that the station would have provided, and they began looking at the proposed Space Shuttle for solutions. With congressional pressure mounting for NASA to privatise, the Spacelab module concept provided planners with an agreeable solution that would also allow the space agency to foster international cooperation – a key mandate for the future of space science and exploration.

Spacelab had initially been developed in the 1970s under the auspices of the eleven-member European Space Agency (ESA). During early development planning for the Space Shuttle, NASA realised that it needed a working facility independent of the cramped, noisy and crowded flight and mid-deck areas. At one time NASA had entertained thoughts of building the laboratory itself, but post-Apollo retrenchments

The crew of STS-9 and the first Spacelab mission hold a press conference. Left to right: Commander John Young, Pilot Brewster Shaw, MS Bob Parker (speaking), MS Owen Garriott, PS Byron Lichtenberg and ESA PS Ulf Merbold.

caused them to forego this option. The space agency turned instead to its European counterpart for assistance. ESA representatives began productive discussions with NASA on developing a pressurised, non-deployable modular research facility that could be launched inside the proposed Shuttle's cargo bay, operating in various configurations and combinations for specific types of missions.

On a flight such as STS-9, with seventy-one experiments to be conducted, there was simply not enough room in the Shuttle mid-deck area to accommodate all the experiment controls and displays. "They all took volume and panel area to hold the hardware and controls for their operation," according to Garriott, "and all this activity would otherwise have to go in the mid-deck area, where the off-duty crew was having to sleep! There was simply no room to run this many experiments, around the clock, without a separate module."[28]

The pressurised module used on STS-9 was the "long" version, 7 m in length and 3.96 m in diameter. Experiments were carried out using floor-mounted single or double racks (all side-by-side) and a workbench. Access to the laboratory, mounted in the aft section of the payload bay (due to centre-of-gravity limitations), was achieved by entering a pressurised, cylindrical tunnel through the crew module airlock in the mid-deck. There was a 107 cm vertical offset of the mid-deck airlock to the airlock at the centreline of Spacelab, so the tunnel and its adaptor had been designed with a compensatory "joggle" section, which the astronauts had to negotiate. The "joggle" also permitted a small amount of longitudinal movement during the ascent and descent phases of the flight, helped absorb movements caused by the differential expansion rates of the orbiter and Spacelab module, and minimised any overstressing that might have occurred in a straight-tunnel/adaptor configuration.

Spacelab, like Skylab before it, represented a true amalgamation of space engineering and fundamental scientific research in the manned space effort. As Spacelab Mission Manager Harry Craft stated prior to the STS-9 launch: "We are reaching a major milestone in the space program."[29]

The Shuttle/Spacelab combination offered several advantages for space science, mainly onboard scientists and other experts who could conduct and monitor experiments, maintain equipment, serve as test subjects, evaluate data, and make crucial decisions on the spot. The facility would also provide an observatory base for a global view of Earth and an unobscured view of the universe using larger, more capable instruments, the ability to retrieve and return experiment samples and apparatus for later analysis and possible re-flight, and serve as a test-bed for new equipment and research techniques.

More doctors than pilots

The STS-9 flight was also notable in that it had a crew of six astronauts – the largest crew to that point – and it was the first to carry crew members bearing the new designation of payload specialist. These two men were Dr. Byron Lichtenberg, a former Vietnam fighter pilot and biomedical engineer from MIT, and Dr. Ulf Merbold, a West German materials science specialist from the Max-Planck Institute

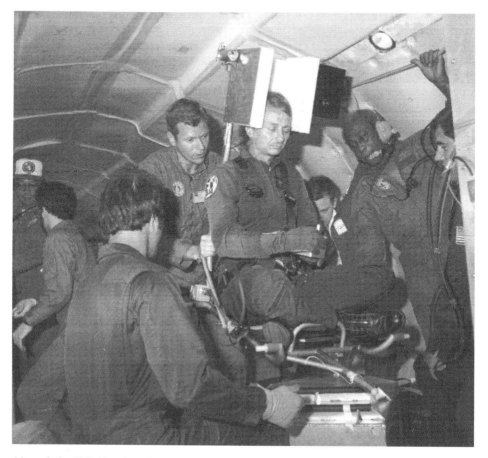

Aboard the KC-135 aircraft, Garriott participates in a simulation of a vestibular function experiment planned for Spacelab 1.

in Stuttgart, Germany. With this flight, Merbold also became the first non-American crew member aboard an American manned space mission.

The six-man crew was divided into two teams, each with a pilot, mission specialist and payload specialist. On this flight the nominated "Red" team consisted of mission commander and Chief of the Astronaut Office John Young, making his sixth and final space flight, plus Bob Parker and Ulf Merbold. The "Blue" team consisted of first-time Shuttle pilot Brewster Shaw, plus Owen Garriott and Byron Lichtenberg. They would work alternating twelve-hour shifts for the duration of the mission, with the "Red" team working from 9:00 p.m. (EST) to 9:00 a.m. and the "Blue" team from 9:00 a.m. to 9:00 p.m. All of the crew members prepared for the mission weeks in advance by adjusting their waking/eating/sleeping routine to suit the shifts they would operate on Spacelab. During the flight, they would meet twice a day at changeover time to discuss operations and share meals, but it also provided a curious scene, as one team would be eating breakfast, while the other was enjoying dinner.

"For the first time in space flight," a bemused commander Young observed before the launch, "the doctors outnumber the pilots four to two. It's going to be a very interesting and unusual mission."[30] Also worthy of note is the fact that the flight would mark the beginning of a strong and lasting friendship between veteran pilot Young and scientist Merbold.

It would be Owen Garriott's first space flight in ten years, but the fifty-three-year-old specialist in ionospheric physics had kept himself quite active in that time. Following his Skylab-3 mission in 1973, he had transferred out of the Astronaut Office to take up a new position as the Deputy Director, and later Director, of NASA's Space Science and Applications Office. In the latter post he was responsible for all research in the physical sciences at JSC. In 1975, he undertook a one-year government fellowship away from NASA at Stanford University, and in doing so renewed his academic qualifications. "I didn't specialise," he said. "I purposely followed a number

Garriott demonstrates a hand-held amateur radio transceiver of the type he used during the Spacelab 1 mission. A licensed "ham" since his teens, he conducted the first demonstration of amateur radio operation from a manned spacecraft during his off-duty hours.

of different disciplines as broadly as possible, just to get reacquainted with the academic community."[31]

When Garriott returned to his office the following year, he found the Physical and Life Sciences Directorates had been combined. "I was Deputy for Physical Sciences there for a year or so, until it became time to select crews for Spacelab, and that's when I went back to the Astronaut Office."[32]

In August 1978, Garriott and Parker were named as mission specialists for the first Spacelab mission, although Garriott has always preferred their official title of "scientist-astronaut" to the more generic designation. The actual mission was then scheduled for launch in two years, allowing them plenty of time to visit Europe and follow the construction of Spacelab 1, but it would actually be more than five years before the mission finally took place. Meanwhile, in September 1982, Lichtenberg and Merbold were named as the two payload specialists assigned to the flight.

The four scientists had actually begun their first flight experiments prior to the launch, and these would only conclude two weeks after landing. During both periods – terrestrial and orbital – various samples were taken, including blood, as part of an attempt to analyse and understand the way in which the human body adapts to weightlessness. Ground-based scientists would compare data from orbit with that gathered before and after the mission.

A busy schedule

There was a potentially serious problem on the first day of the mission. Once the payload bay doors had been opened, with their critical heat-dispensing radiators, Garriott and Merbold tried to open the hatch leading into Spacelab. Their joint efforts were in vain – the hatch was jammed tight. All six men – pilots and scientists alike – turned their immediate attention to remedying the problem and offering solutions. Fifteen minutes later, to their immense relief, the hatch finally yielded, allowing Garriott, Merbold and Lichtenberg to make their way through into the laboratory and ready it for occupancy. Reflecting on the dramas with the balky hatch, Garriott said: "My recollection is that Brewster Shaw was actually the one who pressed the hardest, and it finally opened for him. None of us wanted to use too much force in fear of doing some permanent damage and then not getting the hatch open."[28]

Spacelab was equipped in part with side-by-side standard experiment racks, each about nineteen inches wide. In the long module there would normally be sufficient room to load ten such racks on each side of the laboratory, although that number would decrease if any double-racks were carried – as on this flight, when two double-rack modules were loaded.

The seventy-one scientific experiments and investigations that would be conducted during the flight, ranged across five disciplines: atmospheric physics and Earth observations; space plasma physics; material sciences and technology; astronomy and solar physics; and life sciences. Scientists from the United States, Canada, Western Europe and Japan had been selected as principal investigators (PIs) for various experiments and each had their own team of experts for design, testing,

Garriott works at the aft flight deck station of the Shuttle 1 G simulator at JSC in preparation for STS-9.

crew training, in-flight operations and science interpretations. Among them was a French experiment, the Atmospheric Lyman-Alpha Emissions detector (ALAE), which would be used to measure the radiation produced by sunlight's action on hydrogen, and to make the first measurements of atomic deuterium, a heavy form of hydrogen, in the atmosphere. The Far Ultraviolet Space Telescope (FAUST) and the Very Wide Field Camera (VWFC) were studies conceived and put together by the University of California and the Space Astronomy Laboratory in Marseilles, France respectively, and were designed to help explain the life cycle of stars and galaxies. Targets of the FAUST ultraviolet telescope included distant quasars, hot stars and galaxies, while the VWFC would search out and take ultraviolet images of new astronomical targets and also help researchers gain a better understanding of known objects.

Two instruments, known as the West German Metric Camera and NASA's Large Format Camera, were also included. They combined techniques and equipment to provide high-resolution images intended for use in mapping parts of the Earth's surface. The Metric Camera was a modified aerial survey-mapping camera, mounted on the optical-quality window in the ceiling of the Spacelab module.

Material science experiments were included, to study processes and determine the advantages of fabricating materials such as crystals, alloys and ceramics in conditions of weightlessness. As NASA was hoping that the private sector might demonstrate an interest in manufacturing in space, experiments in materials processing aboard Spacelab 1 were of particular importance. These experiments were carried out in a double-rack module with so much equipment that it was effectively an entire materials science laboratory.

One of the materials science racks included isothermal and gradient furnaces used to conduct a variety of experiments on different metals. A mirror-heating facility furnace reduced silicon rods to a molten state through radiation from two halogen lamps, then solidified and re-crystallised them into a single rod to gauge the effects of weightlessness on the re-crystallisation process. Other processing experiments included a study of the diffusion coefficient of liquid metals, or how metals diffuse through each other. Another module incorporated into the double-rack was an Italian-designed fluid physics laboratory, allowing the first significant opportunity to observe fluid dynamics on orbit. Long, freestanding columns of different liquids were created in this module, and their reaction to zero gravity, stretching, rotating and vibrating was closely monitored. Alongside this unit was a special chamber in which ultra-high vacuum conditions could be created to allow a study of the way different metals adhered to each other.

The second double-rack experiment module was dedicated to life sciences. Sixteen vital life science tests would be conducted during the flight, including those related to physiology, the cardiovascular system, haematology and immunology, the musculoskeletal and neurovascular systems, cellular functions, circadian rhythms and biological processing.

Life sciences were an important facet of Spacelab research on this and later flights. Many experiments were carried out to help scientists understand the varied and sometimes mysterious ways in which the human body responds or adapts to space flight. In the long process of evolution, our bodies have become accustomed to the demands and foibles of gravity in many ways, but when gravity is taken away those same bodies undergo certain physiological changes. Blood, for example, is redistributed differently, affecting the circulatory, cardiovascular and endocrine systems (the latter involving hormone production and functions to regulate and control the body's

Mission Specialist Garriott (left) and ESA (German) physicist and Payload Specialist Merbold work in the first Spacelab mission long module during STS-9. The workload of the science programme is amply demonstrated by the documents they hold. Garriott has a data log book for the solar spectrum experiment, while Merbold holds a ground map for monitoring objectives of the matrix camera experiment.

metabolic activity). Muscles and bones also begin to deteriorate in conditions of weightlessness, and a number of sensory signals become confused and scrambled.

As part of their haematology and immunology studies the scientists would draw blood and then conduct laboratory tests and experiments under conditions of weightlessness. The blood samples were used to measure changes to the red cell mass and lymphocytes (white blood cells) in microgravity. Haematology studies seemed to indicate that bone marrow function is inhibited in weightlessness, resulting in the suppression of erythropoietin, a hormone that stimulates the creation of red blood cells. Lymphocytes help the human body resist infection by recognising and eliminating any harmful foreign agents. "Operation of many of the experiments required two people working together as a team, particularly in the life sciences area," Garriott reflected. "You would perform a test on one person, and then you would reverse the roles and they would perform that test on you. Experimenter and subject, and vice versa."[31]

In one experiment, known to the science team as the "hop and drop" test, they were involved before, during and after the flight in research into what is known as the otolith-spinal reflex. This is a postural reflex that normally and instinctively prepares the human body for the jolt associated with landing after a fall. To substitute for gravity in-flight, the Spacelab crew members were held to the floor by several bungee cords attached to a torso harness, and surface electrodes were placed over the calf muscles to record their neuromuscular reactions. The subject would then be suspended about a foot from the floor against the pull of these bungees Then the suspension handle would be released at an unexpected moment, and the crewman would drop to the deck. Electrical sensors on the leg measured the reflexive response as the crewman caught himself as his feet hit the deck.[28] The experiment would demonstrate that otolith-spinal reflexes progressively decrease in microgravity, indicating that the otolith organs are inhibited in weightlessness, and are gradually ignored by the body's nervous system during space flight.

One startling physiological discovery came when a Nobel Prize-winning theory on the mechanics of the inner ear was disproved. This long-held theory asserted that nystagmus, or rapid eye movement, could be triggered by thermal convection in fluid in the semicircular canals of the inner ear. But convection does not occur in microgravity, so no eye movement should have been detected. Tests were carried out on two subjects during the flight, and both responded with eye movements. This surprising discovery demonstrated that bodily mechanics other than thermal convection are involved in caloric nystagmus.

There was also the vexing problem of the physiological response to space flight that quickly came to be known as Space Adaptation Syndrome, or SAS – an affliction peculiar to more than half of all space travellers, ranging from mild discomfort to brief, episodic periods of nausea. To aid in their research the science team carried out a number of neuro-vestibular studies into the adaptation of the human brain to the environment of space. In order to record continuous measurements of brain and heart activity, and head and eye movements, the Spacelab crew members wore a physiological tape recorder – similar in appearance to a Walkman tape player – to provide data on whether head movements and visual disorientation actually provoked SAS.

As it happened, three of Columbia's science team developed symptoms of SAS. With a researcher's diligence, they kept detailed notes on the different stages and time frames of these episodes, as well as monitoring their head movements with accelerometers. While personally undesirable for those on board Columbia, this unpredictable illness would prove quite fortuitous for ground researchers, as it became the first fully documented, clinical case study of SAS and the vital role that vision played in the adaptation process in a weightless environment.

Life science aboard Spacelab was a multi-faceted programme. There was even one experiment to determine whether plants such as sunflowers might still grow in their characteristically spiral patterns, known as nutation, in the virtual absence of gravity. Plant physiologists had long wondered whether this particular movement depended on gravity or an internal growth mechanism. One component of this experiment was a plant growth unit containing two small centrifuges that could produce an artificial gravity of around 1 G – the same gravitational force encountered on Earth. The plants were allowed to grow for about three days under Earth-gravitation conditions, then were hastily removed and transferred to another chamber, this time with zero gravity influence. They were then automatically monitored and photographed by time-lapse photography to see how they responded.

Table 5. Experiments carried out on STS-9/Spacelab 1.

Life Sciences Investigations

- Advanced Biostack Experiment, H. Bücker, DFVLR, Cologne, Germany
- Circadian Rhythms during Spaceflight: Neurspora, F.M. Sulzman, NASA Headquarters, Washington, D.C.
- Effect of Weightlessness on Lymphocyte Proliferation, A. Cogoli, Swiss Federal Institute of Technology, Zurich, Switzerland
- Humoral Immune Response, E.W. Voss, University of Illinois, Urbana, Illinois
- Influence of Spaceflight on Erythrokinetics in Man, C.S. Leach, NASA Johnson Space Center, Houston, Texas
- Mass Discrimination during Weightlessness, H.E. Ross, University of Stirling, Scotland
- Measurement of Central Venous Pressure and Hormones in Blood Serum during Weightlessness, K. Kirsch, Free University of Berlin, West Germany
- Microorganisms and Biomolecules in the Space Environment, G. Horneck, DFVLR, Cologne, West Germany
- Nutation of Sunflower Seedlings in Microgravity, A.H. Brown, University of Pennsylvania, Philadelphia, Pennsylvania
- Personal Electrophysiological Tape Recorder, H. Green, Clinical Research Centre, Harrow, England
- Crystal Growth of Proteins, W. Littke, University of Freiburg, West Germany
- Radiation Environment Mapping, E.V. Benton, University of San Francisco, California
- Rectilinear Accelerations, Optokinetic and Caloric Stimulations, R. von Baumgarten, University of Mainz, West Germany
- Three-Dimensional Ballistocardiography in Weightlessness, A. Scano, University of Rome, Italy
- Vestibular Experiments, L.R. Young, Massachusetts Institute of Technology, Cambridge, Massachusetts
- Vestibulo-Spinal Reflex Mechanisms, M.F. Reschke, NASA Johnson Space Center, Houston, Texas

Material Science Investigations

Fluid Physics Module

- Capillary Forces in a Low-Gravity Environment, J.F. Padday, Kodak Research Laboratory, Harrow, England
- Coupled Motion of Liquid-Solid Systems in Near-Zero Gravity, J.P.B. Vreeburg, National Aerospace Laboratory, Amsterdam, The Netherlands
- Floating Zone Stability in Zero-Gravity, I. Da Riva, University of Madrid, Spain
- Free Convection in Low Gravity, L.G. Napolitano, University of Naples, Italy
- Interfacial Instability and Capillary Hysteresis, J.M. Haynes, University of Bristol, United Kingdom
- Kinetics of the Spreading of Liquid in Solids, J.M. Haynes, University of Bristol, United Kingdom Oscillation of Semi-Free Liquid Spheres in Space, H. Rodot, National Centre for Scientific Research, Paris, France

Gradient Heating Facility

- Lead-Telluride Crystal Growth, H. Rodot, National Centre for Scientific Research, Paris, France
- Solidification of Aluminium-Zinc Vapour Emulsion, C. Potard, Centre for Nuclear Studies, Grenoble, France

Table 5 (*cont.*)

- Solidification of Eutectic Alloys, J.J. Favier and J.P Praizey, Centre for Nuclear Studies, Grenoble, France
- Thermodiffusion in Tin Alloys, Y. Malméjac and J.P. Praizey, Center for Nuclear Studies, Grenoble, France
- Unidirectional Solidification of Eutectics, G. Müller, University of Erlangen, Germany

Isothermal Heating Facility

- Bubble-Reinforced Materials, P. Gondi, University of Bologna, Italy
- Dendrite Growth and Microsegregation of Binary Alloys, H. Fredriksson, The Royal Institute of Technology, Stockholm, Sweden
- Emulsions and Dispersion Alloys, H. Ahlborn, University of Hamburg, Germany
- Interaction Between An Advancing Solidification Front and Suspended Particles, D. Neuschütz and J. Pötschke, Krupp Research Centre, Essen, Germany
- Melting and Solidification of Metallic Composites, A. Deruyttere, University of Leuven, Belgium
- Metallic Emulsion Aluminium-Lead, P.D. Caton, Fulmer Research Institute, Stoke Poges, United Kingdom
- Nucleation of Eutectic Alloys, Y. Malméjac, Centre for Nuclear Studies, Grenoble, France
- Reaction Kinetics in Glass, G.H. Frischat, Technical University of Clausthal, Germany
- Skin Technology, H. Sprenger, MAN Advanced Technology, Munich, Germany
- Solidification of Immiscible Alloys, H. Ahlborn, University of Hamburg, Germany
- Solidification of Near-Monotectic Zinc-Lead Alloys, H.F. Fischmeister, Max Planck Institute, Stuttgart, Germany
- Undirectional Solidification of Cast Iron, T. Luyendijk, Delft University of Technology, The Netherlands
- Vacuum Brazing, W. Schönherr and E. Siegfried, Federal Institution for Material Testing, Berlin, Germany
- Vacuum Brazing, R. Stickler and K. Frieler, University of Vienna, Australia

Mirror Heating Facility

- Crystalisation of a Silicon Drop, H. Kölker, Wacker-Chemie, Munich, Germany
- Floating Zone Growth of Silicon, R. Nitsche and E. Eyer, University of Freiburg, Germany
- Growth of Cadmium Telluride by the Travelling Heater Method, R. Nitsche, R. Dian, and R. Schönholz, University of Freiburg, Germany
- Growth of Semiconductor Crystals by the Travelling Heater Method, K.W. Benz, Stuttgart University, and G. Müller, University of Erlangen, Germany

Special Equipment

- Adhesion of Metals in UHV Chamber, G. Ghersini, Information Centre of Experimental Studies, Italy
- Crystal Growth by Co-Precipitation in Liquid Phase, A. Authier, F. Le Faucheux, and M.C. Robert, University of Pierre and Marie Curie, Paris, France
- Crystal Growth of Proteins, W. Littke, University of Freiburg, Germany
- Mercury Iodide Crystal Growth, R. Cadoret, Laboratory for Crystallography and Physics, Les, Cezeaux, France
- Organic Crystal Growth, K.F. Neilsen, G. Galster, and I. Johannson, Technical University of Denmark, Lyngbyg, Denmark
- Selfdiffusion and Interdiffusion in Liquid Metals, K. Kraatz, Technical University of Berlin, Germany

Space Plasma Physics Investigations:

- Atmospheric Emission Photometric Imaging (AEPI), S.B. Mende, Lockheed Solar Observatory, Palo Alto, California
- Electron Spectrometer, K. Wilhelm, Max Planck Institute, Stuttgart, Germany
- Magnetometer, R. Schmidt, Academy of Sciences, Vienna, Austria
- Phenomena Induced by Charged Particle Beams (PICPAB), C. Beghin, National Centre for Scientific Research, Paris, France
- Space Experiments with Particle Accelorators (SEPAC), T. Obayashi, Institute of Space and Astronautical Sciences, Tokyo, Japan

Atmospheric Science Investigations

- Active Cavity Radiometer (ACR), R.C. Willson, NASA Jet Propulsion Laboratory, Pasadena, California
- Grille Spectrometer, M. Acherman, Space Aeronomy Institute, Brussel, Belgium
- Imaging Spectrometric Observatory (ISO), M.R. Torr, NASA Marshall Space Flight Center, Huntsville, Alabama
- Investigation of Atmospheric Hydrogen and Deuterium through Measurement of Lyman-Alpha Emission (ALAE), J.L. Bertaux, National Centre for Scientific Research, Paris, France
- Solar Constant (SolCon), D. Crommelynck, Royal Meteorological Institute, Brussels, Belgium
- Solar Spectrum (SolSpec), G. Thullier, National Centre for Scientific Research, Paris, France
- Waves in the OH Emissive Layer, M. Hersé, National Centre for Scientific Research, Paris, France

Earth Observation Investigations

- Metric Camera, M. Reynolds, European Space Agency, Noordwijk, The Netherlands, G. Konecny, University of Hannover, Germany
- Microwave Remote Sensing Experiment, G. Dieterie, European Space Agency, Paris, France

Astronomy and Astrophysics Investigations

- Far Ultraviolet Space Telescope (FAUST), C.S. Bowyer, University of California, Berkeley, California
- Isotope Stack, R. Beaujean, Kiel University, Germany
- Spectroscopy in X-Ray Astronomy, R.D. Andresen, European Space Research & Technology Centre, Noordwijk, The Netherlands
- Very Wide Field Camera, Go. Courtés, Space Astronomy Laboratory, Marseilles, France

Space Technology Investigations

- Bearing Lubricant Wetting, Spreading & Characteristics, C.H.T. Pan, Columbia University, New York, New York, A. Whitaker, NASA Marshall Space Flight Center, Huntsville, Alabama

Problems and progress

Early in the flight, there would be several problems for the crew to contend with and resolve if they could. Electronic equipment failed, causing the loss of vital experiment data. A computer somehow forgot what day it was, resulting in an infrared sensor being pointed at the wrong targets. Then a timer on several experiments failed until the Red Team of Parker and Merbold managed to repair it with a pair of pliers and a

388 The Long Wait

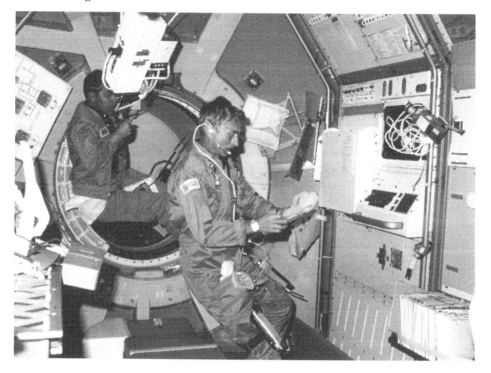

Parker (rear) and Garriott work in the long module of Spacelab 1. The scientist-astronauts had tried to fly two of their group on each Skylab mission but only one was assigned (including Garriott). Spacelab 1 offered the opportunity that Skylab did not, utilising the skills and availability of the combined science background and astronaut training to the best advantage.

cheap screwdriver. There was also an unexplained power surge in a high-speed data recorder that blacked out television transmissions from space.

This flight was also notable in that the scientists on board could communicate directly through voice- and science-data downlink with their colleagues on the ground, rather than through an astronaut Capcom. Lichtenberg would later say he was extremely grateful for this timesaving and necessary innovation. "It was the first time we've ever had direct contact between scientists on board a spacecraft and scientists on the ground. Always in the past, all of the communications had gone through an astronaut on the ground. On Spacelab 1, the concept of having practicing scientists in direct voice and video contact with their peers on the ground was a major step forward.

"I think we were able to be a lot more productive in the fact that we could talk our scientific jargon amongst ourselves. It's a shorthand, if you will, in the same way that the pilot-astronauts up on the flight deck have their own form of shorthand communication going with the Capcoms."

The Marshall Space Flight Center (MSFC), allowed the opportunity to broaden the operational experience it had acquired during Skylab, was given responsibility for

Spacelab payloads, while the Johnson Space Center was responsible for orbiter operations. A mission management team from Marshall worked out of an area known as the Payload Operations Control Center (POCC) located in JSC's Building 30, while at the same time orbiter operations were conducted at Johnson's Mission Control Center (MCC). The Huntsville Operations Support Center (HOSC) in Alabama supplied technical advice, operating much as it had during the Skylab programme.

The TDRS-1 tracking and relay satellite provided the essential downlink between the orbiter and POCC. However, there were communications problems because the TDRS satellite was faulty, proving incapable of relaying more than a fraction of its orbit-to-ground capacity. At one stage, Bob Parker got a little edgy with over-excited ground colleagues from POCC while trying to effect essential repairs with Ulf Merbold and get on with his own scheduled experiment activity. "You guys should recognise that there are two people up here trying to get all your stuff done," Parker finally snapped back. "I think you might be quiet until we get one or the other [experiment] done." Then, when further instructions began to come in, he cut them short with: "Just wait! Would you guys please tell us exactly what you want done when, and we'll forget about what else we're doing at the present time!" Taken aback a little by the outbursts, Payload Operations subsequently eased up on their requests.[33]

Despite these problems and equipment breakdowns, the Blue and Red teams worked as a highly efficient and productive crew. By using some of the most sophisticated equipment ever flown into space, their work resulted in many unprecedented and useful results. The ultraviolet telescope, for instance, provided the best sightings ever of dying stars, while the Metric Camera mapped tiny details of the Earth's surface. Unfortunately, the Very Wide Field Camera performed poorly and would have to be re-flown.

In a little zero-G alchemy, a furnace that could be fired up to 3,800 degrees Fahrenheit managed to form an amalgam of aluminium and zinc, creating a strong, lightweight metal alloy impossible to produce in the presence of gravity. The French ALAE pallet experiment detected and successfully obtained the first-ever measurements of deuterium in the upper atmosphere. This not only helped scientists study weather patterns on Earth, but if large quantities of deuterium could be detected elsewhere in our universe, it would suggest that the targeted celestial body may once have held elements of water – an essential prerequisite for life.

On this flight Owen Garriott, who had been an amateur radio operator for many years, became the first astronaut to take a small, hand-held ham radio into space. In doing so, he began a trend that created an increasingly sophisticated amateur radio space programme with other interested astronauts and contacts on the ground, and an ongoing programme known as SAREX (Space Amateur Radio Experiment). "In my spare time only, I managed to hold up an antenna to the window and talk to amateurs on Earth. When in orbit over land, I could make a CQ, which is a general call, and hear the hundreds of hams on the ground who were trying to establish communications. I used a well-designed, hand-held antenna, known as a 'cavity antenna,' which could be Velcroed to the window. It was about twenty-four inches in diameter and looked somewhat like a large, square aluminium cake pan. The transceiver then connected to

Garriott operated the amateur radio equipment during the STS-9 mission.

the antenna." Garriott's communications with fellow ham radio enthusiasts on the ground were the first outside of "official" channels.[34] He used a four-watt QRP (a ham expression for very low power) transceiver to contact fellow enthusiasts such as Senator Barry Goldwater and King Hussein of Jordan, and was able to log nearly seven hours of free operating time between 30 November and 8 December.[35]

While Garriott recalls that the science team were "a bit lethargic" for the first few days, "our efficiency and productivity improved throughout the flight. This was also a major result of published Skylab reports. Everyone recovers from early symptoms of SAS (if they have any) and feels better as time progresses, while productivity improves as everyone learns how to operate efficiently in weightlessness."

Towards the end of the mission, NASA spokesmen said they were pleased with the work carried out, and greatly impressed by the number of worthwhile results they were obtaining. Cryogenic fuel usage had been unexpectedly low, and the crew had been exceptionally careful with their electrical energy consumption, so it was decided to extend the science mission from nine to ten days.

Monkeying around with the media

On the eighth day of the mission, the international politics associated with the joint NASA/ESA mission meant that a little flag-waving would take place during a televised international press conference. Crew members Young, Merbold and Lichtenberg would be shown during a live transmission from space, planned to begin with the reading of prepared statements from President Ronald Reagan in Washington, and West German Chancellor Helmut Kohl, who was in Athens attending an economic meeting. Following this, the press would be "free" to ask questions of the three crew members. However, there was little spontaneity to this event, as the questions had been prepared in advance, and the crew were already in possession of suitable responses. Like other crews, this one did not mind a little flag-waving, but the heavy-handed politics and a decided lack of freedom of speech caused them to pull a little stunt of their own, as Garriott recalls:

"A spot in the day's timeline was picked when all six of us were awake to have the brief conversation. However, six people at a microphone for a brief conversation never works out very well, so they decided to restrict it to only three. Young, the commander, was an obvious choice as was Merbold, the German, and Lichtenberg, as a US civilian scientist. But then they wanted to set up all the communications

Taking a brief moment of respite from their work, the Spacelab 1 science crew utilise the scientific airlock (in the roof of the Spacelab module) as a "card table" and "cards" from the targets used in the Awareness of Position experiment to play what appears to be a game of space cards.

equipment and cameras and check it out before the actual link was made to the two heads of state. So they wanted the other three of us to set up all the equipment and run the checks first, then quietly exit "stage left" while the real "stars" took their places for the interview. This seemed [to be] pushing the priorities a little hard, so the three of us decided to get it all set up as requested, but then when the test video link was first established, we would be seen sitting side-by-side, positioned as 'The Three Monkeys'."

When they knew they were 'on the air' and live on the pre-interview test link, one of the crew switched transmission cameras. Suddenly, those on the ground saw Garriott, Parker and Shaw huddled together, with Garriott holding his hands over his eyes, Parker covering his ears, and Shaw doing likewise with his mouth. Their "see no evil, hear no evil, speak no evil" message was abundantly clear. After five seconds, and though they were unable to see the reaction of the ground crews, all three burst out laughing.

"We were later told on the ground that when the image came up in the control centre they also burst out laughing as they realised the implication of the pose – that we three had been to some (insignificant) degree slighted or placed in a "second class" category. We all enjoyed it a lot, and doubt if Reagan or Kohl ever knew anything about it."

The international press conference took place as scheduled, but the crew's little bit of levity ensured that they got their message across about the time-wasting absurdity of a so-called "spontaneous" interview. The image seen in the control centre had also been viewed by the press, who considered it humorous enough to place a photo of the three-man performance in the following week's *Time* magazine. "I still keep a small brass statue of the three monkeys on my memento shelf at home, and a larger pottery version is in our garden," Garriott states, still highly amused at the memory.[36]

With the four scientists busily engaged in their Spacelab experiments, there was very little time for peering out of the window at the Earth below. Garriott, however, made an interesting post-flight observation. "We were flying at 57 degrees inclination on this flight, of course, and the high latitudes made the opportunities for looking at the Earth beneath us of even greater fascination. We could see both northern and southern latitudes much better than even on Skylab at 50 degrees."[31]

Bob Parker was also quite happy that they were flying with a 57-degree inclination on his first space flight, as he explained:

"A typical Shuttle orbit comes due east out of the Cape in Florida, in a 28-and-a-half-degree inclination. It goes from a 28-and-a-half-degree north latitude to a 28-and-a-half-degree south latitude. It encompasses only the southern part of Florida and Texas and Hawaii and does not fly over the rest of the United States. In fact, it does not even fly over the Mediterranean – it just barely gets close to it. It does not quite go to the tip of South Africa, and it certainly does not get to Cape Horn, or New Zealand. You miss parts of Australia, most of Asia and Europe – a whole lot of interesting parts of the world. You see a whole lot less stuck in that orbit. It does see a whole lot of ocean! In a word, to those of us who fly, it is known as 'boring'.

"If the orbit has an inclination of 57 degrees, close to what the International Space Station has, you get to Hudson's Bay. You get below Cape Horn, above Moscow, most of Great Britain, parts of Denmark, the very bottom of Sweden, and below Australia and New Zealand. Probably ninety-five percent of the world's population is below us, there is a lot of land mass for us to see, and a lot of interesting things to photograph."[37]

A fire on landing

One unexpected bonus of the flight occurred over Kazakhstan, when the crew was able to observe and photograph the secluded and (at the time) covert Russian space launch facility at Tyuratam, situated on the bank of the Syr Darya River, which empties into the Aral Sea. In June 2002, Parker described this successful operation during a talk at San Diego's Reuben H. Fleet Science Center:

"Photographing the Russian launch site was not a public mission objective, but it was something that, as a crew, we really wanted to do. We were the first people to get that picture. We could see the Russian shuttle runway that they used for the Buran, that they flew only once, and see big assembly buildings. They don't show up very clearly because the concrete comes from the land around. You don't truck in gravel from a thousand miles away; you take it from somewhere in the area. So the land and the runway kind of look the same. It's the same with the Great Wall of China – it does not stand out.

"I was flying with John Young, and it was his sixth flight. He knew more about what was going on than I did. A month or so before we launched, he said to me, 'Bob, we are going to fly over the Russian launch site. We need to plan ahead to take pictures.' So what we did was take the map of our orbit – we knew what time we were launching, therefore we knew what day and time we were going to be where. We saw at one point we'd be passing over the southern border of the Caspian Sea, over the Aral Sea, then a river, then a bend in the river, and just after we got to that bend, we were going to see the Soviet launch site. So we marked it out, put it aside, and took the map with us.

"Fifteen minutes before we crossed the Caspian Sea, John called me up from down in the lab where I couldn't see out of the windows very much. He said, 'Okay, let's go and do this', and we started looking. We saw the landmarks, the bend in the river, looked where it ought to be, and there it was. We found it, we took our photos.

"A later mission, three or four years later, asked us about it. They told us they looked out and couldn't see it, and asked us how we'd done it. We asked them, 'Did you plan ahead?' and they said, 'No, we just looked for it.' It's important to plan ahead for low-contrast things like that."

Five hours before Columbia's scheduled landing, a dramatic situation occurred when two of the orbiter's five computers went awry while John Young was firing the nose-mounted Reaction Control System (RCS) thrusters to orient the spacecraft for de-orbit. An unexpected, jolting shock travelled through the orbiter at the first firing, and the primary General Purpose Computer (GPC) dropped out. Young later said he heard "a loud bang and the whole spacecraft shook. My stomach turned and my legs turned to jelly."[38] As programmed, a back-up GPC began to take over to compensate, but a further "hard burst" of the RCS four minutes later sent that GPC off-line.

"I think it was the up-firing jets... that made the computers fail," Young reported in a transmission to the ground. "It really hit the computers hard." He recommended that they close the forward RCS, "and not run any more of those rascals."

The thrusters left Columbia without any computer-controlled guidance and navigation for nearly a minute, before a third GPC finally kicked in. Although the Shuttle can land with only one of its quad-redundant computers operating, flight engineers were so concerned about the possible consequences of another thruster firing that Mission Control decided to call off the landing sequence for another four orbits, while engineers requested Columbia be allowed to "free drift" until they had looked at the situation. To add to their many woes, IMU-1 then malfunctioned. This was one of three inertial-measurement units designed to sense the Shuttle's acceleration and position. Fortunately, this was not crucial to the re-entry phase. Flight engineers finally gave their clearance to proceed with the re-entry burn, having already delayed the landing by around seven-and-a-half hours. On Columbia's 166th orbit, a 156-second burn on the two OMS engines took place without incident, following which the orbiter was realigned for its nose-first return passage back to Earth.

Dropping out of orbit over the Pacific on the morning of 8 December, Columbia plummeted through the dense atmosphere, soon surrounded by a white-hot halo of flame. Owen Garriott was seated in the mid-deck next to the window in the side hatch, and was able to make a comparison between his Skylab and Shuttle re-entries.

"Looking up, I could see through a bit of the overhead windows, too. You see the interesting colours – oranges and pink – as you come down through the more and more dense layers of the atmosphere. We were flying at a lower altitude on Spacelab 1 than I flew on Skylab. Although the orbital velocities are about the same, in that you're travelling a little under five miles per second in any of these lower altitudes, the angular rate at which the nadir moves by is about twice as high when you're at half the altitude. That means the ground beneath you appears to move by about twice as fast, which is certainly one noticeable difference."[31]

Eventually, Columbia crossed the California coast at nine times the speed of sound, and Young glided the 114-ton spacecraft to a textbook, unpowered landing on dry lakebed Runway 17 at Edwards AFB. Following shutdown, the recovery team moved in to vent the engine bay, and then another major problem erupted. Smoke had been seen venting out of the orbiter's aft section containing the Auxiliary Power Units (APUs). Space Shuttles have three APUs, which are used to power the orbiter's hydraulics during ascent and re-entry. They use highly toxic hydrazine propellant to drive a high-speed turbine that in turn provides hydraulic power. Later analysis found that two minutes from touchdown, leaking fuel lines had sparked a fire in wiring around two of the hot APUs. Had this occurred earlier in the re-entry process, the results might have been calamitous. Overall, however, the flight of Columbia and the Spacelab 1 module had proven to be a major success.

Following their mission, the science crew members spent the next week at Edwards AFB undergoing what Garriott termed "very extensive physiological testing" for all four mission and payload specialists. They underwent repeats of most of the experiments they had conducted in space, and their return to normal Earth-based responses were observed. Among many post-flight checks, these tests provided interesting results in vestibular rehabilitation, reflex results to the "hop and drop" experiment, and sensitivity to acceleration. Pursuant to the "hop and drop" experiment carried out before and during the flight, it was noted that normal reflexes would return a few days after regular gravitational influence and activity was resumed. According to Garriott, however, "the initial response after returning from weightlessness was quite different from that before flight. It was almost bizarre! Our legs had almost forgotten how to arrest our fall and we had to have another person help catch us from behind to avoid falling to the floor. But the normal reflexive response returned to baseline within a day or so."

The scientists also carried out cardiovascular experiments to determine the degree and rapidity of fluid shifts and blood volume loss in microgravity, and whether a decrease in heart volume caused any reduction in heart performance. They did this by such means as measuring central venous pressure in the arm veins. Prior to this mission, no direct on-orbit measurements of venous pressure had been available to check hypotheses associated with the cardiovascular process. When venous pressure was first taken twenty hours into the flight, it proved to be lower, rather than higher, than pre-flight measurements. However, one hour after the crew's return to Earth, venous pressures were higher for all four scientists, indicating fluid shifts associated with the body's re-adaptation to a 1-G environment. Investigators were subsequently able to conclude that the fluid shift is a highly dynamic process that might even begin with crew members seated in their couches for a couple of hours waiting for lift-off. These results provided a good cornerstone for further experiments on later missions.

There was good news in store for the crew of Columbia, and particularly the four-man science team: by mission's end they had not only achieved most of its goals, but they had accumulated twenty million pictures and two trillion bits of data. Results of the mission, jointly released by MSFC and ESA several months later, determined that the crew members had accomplished all systems verification objectives, with only minor anomalies, and achieved eighty percent of the overall mission objectives in all but atmospheric physics and Earth observations (which achieved sixty-five percent). Samuel Keller, deputy associate administrator for Space Science and Applications, declared Spacelab "an unqualified success," while mission scientist (and later MSFC's Director of Science) Dr. Charles (Rick) Chappell deemed the flight "a very successful merger of manned space flight and space science."

"I think the key element of this flight was its multi-disciplinary nature," Owen Garriott reflected in summing up his second and final space mission. "There was no single experiment, no single discipline, that monopolised the majority of the time.

"Spacelab 1 was intended to show how well all disciplines could make use of the Shuttle orbiter as a laboratory in space, and in that I feel our crew, our ground-based team and indeed Spacelab, achieved a remarkable success."[31]

During 1982 and 1983, of the nine remaining scientist-astronauts who still were working at NASA, six had made it to space on board the Space Shuttle. Garriott had flown a second time, but his Skylab colleague Joe Kerwin would not make that second flight, having been reassigned in 1982 to a management role in the space agency which took him out of consideration for a flight assignment. For the others, it was their first experience of space flight, and while all hoped it would not be their last, for some that

396 The Long Wait

They may be the wrong creatures, but this stone feature in the Garriotts' front garden is a whimsical reminder of the STS-9 "Three Wise Monkeys" episode. [Photo: Colin Burgess.]

long awaited first mission would indeed be their only chance to experience the thrill of launch, the excitement of orbiting our Earth and the adrenaline pumping fiery ride back home. During 1984 and 1985, five of the group would fly on the Shuttle: Allen, Musgrave and Thornton on their second missions, and Henize and England finally realising their long-awaited dream of a trip into space.

STS 51-A: WE DELIVER AND PICK UP – TWICE

Ten months after flying on STS-5, Joe Allen was formally assigned to a new crew with the announcement, on 2 September 1983, of the STS-16 crew members. As MS 1, Allen would be flying with another four astronauts, all from the 1978 group. In command was Rick Hauck (previously the pilot on STS-7), the pilot would be Dave Walker, MS 2 was Anna Fisher and MS 3 was Dale Gardner (who was still flying on STS-8 at the time of the announcement). The military payload specialists for this classified Department of Defense (DoD) mission were still to be announced. A week later (9 September), NASA announced a new Shuttle designation code for upcoming missions, based on a three digit numbering system, with the STS-16 flight being renamed as STS 41-H. Twelve days later, on 21 September, the crew were moved to STS 41-G (now manifested to deploy commercial satellites for Telstar, SBS, and Hughes Aerospace, as well as the astronomy science package free-flyer called Spartan). On 17 November 1983, the crew were again reassigned, from STS 41-G back to 41-H, this time manifested to carry either a DoD payload or TDRS-B. During 1984, the crew worked together as a unit without knowing for sure which payload they would be carrying. Then, following payload changes and slips in the launch schedule of 1984, and as a result of the 26 June 1984 pad abort of STS-12, the crew became part of a major reorganisation resulting in their fourth change, to 51-A, manifested to deploy the Telsat H and Syncom IV-1 Comsats. On top of all this, events from an earlier mission led to additional objectives being included on the flight.

Deployment and retrieval

In February 1984, during the STS 41-B mission, two Comsats were successfully deployed by the crew from the payload bay of Challenger using the Payload Assist Module (PAM) upper stage system. The Indonesian Palapa B2 and Western Union Westar VI were not able to reach their proper orbits, however, due to failures on each of the PAM Perigee Kick Motors (PKM). Each engine firing ended prematurely and stranded both satellites well below their intended geosynchronous orbits. Fortunately both PKM were successfully ejected from the satellites and each of them appeared to be undamaged, if useless, in their new orbits. Despite all crew activities being performed flawlessly (as with the TDRS-A/IUS deployment from STS-6 the previous April), the media were quick to highlight the expensive failures.

Even before STS 41-B had landed, NASA and Hughes officials (the makers of the satellites) began to study the possibility of retrieving both satellites on a future Shuttle mission. By May, both satellites had been placed in the same circular parking orbit about 1,000 km above the Earth by means of the solid fuel contained in the Apogee Kick Motor (AKM) on each satellite. This was also an essential safety procedure in the event of a rescue mission taking place, disposing of the fuel prior to either the Shuttle or any astronauts approaching the stranded satellites. Over the next few months, Hughes controllers initiated a series of manoeuvres to rendezvous both satellites in the same orbit. By August, both were stable, within 0.03 degrees of each other though 180 degrees apart and by early October, a programme of further manoeuvres was completed to both close the separation distance of the satellites to approximately 965 km and to lower their orbits to 360 km, within range of the Shuttle. Over a three-week period, more than 100 orbital adjustments were made for each satellite. By the end of October, their spin stabilisation rates had been slowed and they were ready for recovery. As both satellites were commercial, they had been insured against loss by underwriters, represented by Merrett Syndicates of London and International Technologies Underwriters of Washington DC. In August, after discussion with NASA and Hughes, a contract for $10.5 million was agreed for a dual satellite recovery.

With the satellites ready for retrieval and the cash to support a mission, NASA looked to the manifest to determine which flight and crew would be best prepared to

Joe Allen about to enter the mid-deck of Discovery prior to starting his second flight into space.

attempt the task. Despite the loss of the two satellites, STS 41-B had flawlessly demonstrated the use of the Manned Manoeuvring Unit (MMU) for the first time. The successful EVA repair of the Solar Max satellite (on STS 41-C in April) had given great confidence to the NASA team that satellite retrieval and repair was indeed possible. The mission would have to be completed as soon as possible while still giving the crew (and engineers) about six months to develop techniques and hardware to capture two satellites that were not designed to be returned to Earth. It was too late in the cycle to add the task to the crew training for the twelfth and thirteenth missions, and the fifteenth was a classified DoD mission, so the required media openness for the dramatic satellite rescue would not be possible on that flight. The flights of Spacelab 2 and 3 could not support returning a satellite payload as well as the Spacelab equipment and with the June abort of STS 41-D, the subsequent combination of 41-D and 41-F payloads, and the reassignment of the 41-F crew to a new mission (51-E in 1985), this left the Hauck crew, due to launch in the spring of 1984. The crew would now be dispatching two satellites before attempting to retrieve two more. This would be a daunting mission and a challenging one.

Flight-specific EVA training

Mission training for the flight resembled that accomplished by earlier crews for ascent, orbital operations and descent, but with an additional programme of EVA added several months into the training cycle. The EVA retrieval option was added to the

STS 51-A: we deliver and pick up – twice 399

The Crew of STS 51-A pose for their in-flight portrait on the flight deck of Discovery. Left to right (at rear) are MS Dale Gardner, Commander Rick Hauck and Pilot Dave Walker. At front are MS Anna Fisher and MS Joe Allen.

crew's training programme in April 1984, giving them just seven months to train for the operation. Official training records revealed that final procedures and high-fidelity hardware were only available for the last month or two prior to launch.[39] The EVA team consisted of Joe Allen (EV 1) and Dale Gardner (EV 2), with pilot Dave Walker acting as IV crew member and Anna Fisher as RMS operator. Commander Rick Hauck would support the EVA as required by backing up Fisher on the RMS, ensuring the orbiter was stable and taking photos through the flight deck windows.

With both Allen and Gardner having flown before, however, they were able to spend more of their limited time on developing their EVA skills, requiring only a small amount of time for basic STS refresher courses. Both had also been EVA trained (Allen for the planned EVA on STS-5 and Gardner as contingency EVA crew member on STS-8), so they could also bypass a large amount of EVA training basics. In addition, Walker had previously received a CB technical assignment in EVA/MMU training years before, in support of the Shuttle tile repair effort, and Fisher had worked on EMU design qualification for female EVA test subjects, and had performed simulated EVAs in the WETF, as well as becoming a proficient RMS operator.

Allen trained for his EVA by attending numerous flight technique meetings and hardware design reviews and undertaking hardware and procedural development exercises in the WETF and EVA retrieval equipment (stinger) design. He also undertook procedural development runs on the Air Bearing Floor (ABF); went through

Space Operations Simulator (SOS) and Space Environment Simulator (SES) runs, participated in hardware fit checks, dry runs and equipment selection and conducted hardware qualification runs in the thermal vacuum chamber. In addition, he attended management, hardware, procedures and programme reviews and numerous JSC, contractor, and customer meetings and teleconferences throughout the seven months. A scientist-astronaut with a PhD in physics and a background in nuclear physics, Allen was adding to his talents and skills as an astronaut-engineer in a very compacted time frame.

A total of fifteen water tank exercises were performed by both EVA crew members. Initially, Allen and Gardner completed individual WETF familiarisation runs, followed by fourteen two-man runs comprising one orbiter contingency EVA run, two combined orbiter contingency and retrieval development training runs, and eleven retrieval development training runs. Seven further simulations were completed at the MMU simulator at the SOS located at Martin Marietta in Denver, where an MMU familiarisation run was followed by six satellite docking simulations using the low-fidelity stinger (pole) mock-up and a simulated satellite aft end. More realistic visual scenes were provided in the SES, allowing the crew to perform MMU flights from the payload bay though to docking. Three MMU docking exercises were also completed using the ABF and three more in the RMS Training Facility, where Allen and Gardner stood on the longerons of the mock-up payload bay facility to simulate being in the bay, while the RMS was operated from the simulated aft flight deck in order to test communications and coordinate activities.

The crew also participated in the retrieval session of the Crew Equipment Interface Test (CEIT) conducted at the Kennedy Space Center several weeks prior to launch, where almost all aspects of hardware fabrication and stowage were practiced on both pallets. This proved not only to be a great training tool, but also revealed problems and differences that could be corrected prior to flight. In addition to the EV crew of Allen and Gardner, all three IV crew members (Hauck, Walker and Fisher) participated to great benefit, as their knowledge of both the pallets and the associated hardware increased to a level that made coordination and communications during the actual EVAs much simpler and more effective. Based on their training, it was decided that Allen would attempt the first retrieval. If that worked, Gardner would make the recovery attempt on the second retrieval.

In their closing comments on the EVA training programme, Allen and his colleagues noted in their flight crew report that: "It must be re-emphasised that the crew's intimate involvement in the procedures and hardware development cycle was a major contributor to the overall EVA training program and, ultimately, a prime factor in the success of a mission that had remarkably condensed preparation development, test and training time allowance."[39]

Satellites for sale – the fourteenth Shuttle mission

The fourteenth mission of the Shuttle programme and the second for orbiter Discovery (OV-103) was originally scheduled for 7 November 1984. Despite a crystal clear day in Florida and all systems seemingly working properly, the launch was scrubbed due to excessive wind speed between the 20,000- and 40,000-foot levels. All was fine the following day, 8 November, with Discovery making its return to space without incident. The two commercial satellites were deployed by the crew during FD 2 and FD 3, with Allen managing the deployment of Anik D2 and Anna Fisher deploying Leasat 2 the next day. The precise chase to reach the two stranded satellites deployed from STS 41-B passed without incident and the two resulting EVAs (6 hours on 12 November and 5 hours 42 minutes on 14 November) successfully brought both rogue satellites back into the payload bay of the orbiter for the journey home. After the completion of a number of mid-deck experiments, Discovery returned to Earth with the two satellites safely aboard on 16 November, after a flight of 7 days 23 hours 44 minutes 56 seconds, during orbit 127. Flight Director Jay Greene highlighted the huge achievement of the mission by saying: "We've deployed satellites before; we've picked up satellites before; we've rendezvoused before; and we've repaired a satellite before. But we've never before done all of these things together on one flight."

"Mighty Joe" returns to space

As MS 1 on STS 51-A, Joe Allen became the first member of the second group of scientist-astronauts to return to space and only the second (after Owen Garriott) to make two flights. He became one of only four scientist-astronauts in total that would make two Shuttle flights.

The crew were disappointed by the launch scrub on 7 November, but with hindsight they also considered themselves very lucky. In November 2004, two weeks after the crew's twentieth anniversary reunion, Joe Allen participated in his fourth session for the JSC Oral History Project. He commented that both the Challenger (1986) and Columbia (2003) accidents involved launching during very high wind sheer, "and there's some thinking now that high wind sheers and Space Shuttle do not safely go together."[40]

Allen recalled having a feeling of anxiety on the morning of 7 November, something he did not have on his first launch, perhaps because of naïveté. But when he boarded Discovery the next day, that feeling of anxiety had disappeared and he recalled feeling more confident on that second launch attempt.

With a task of deploying two satellites and picking up two others, "somewhat rude notes" from his fellow astronauts commented that neither Gardner nor Allen should be confused as to which to deploy and which to pick up: "In other words, don't bring home satellites that we'd just taken there." In fact, Allen's part in the crew's success with the satellite operations, and his "Herculean" activities during the first EVA, earned him the nickname "Mighty Joe Allen", in parody of the classic 1949 feature film *Mighty Joe Young*.

A bone-rattling lift-off

One of the new assignments for the 51-A crew was to work with the insurance industry, which was becoming more involved in the space business. Allen found it amusing to hear from an English insurance representative who, throughout his

Joe Allen and Dale Gardner each flew the MMU to capture two rogue satellites during the STS 51-A mission.

professional life, had insured things against fire or the chances of explosion. Now he was working with astronauts in the space business, who "purposely set fire to a massive amount of explosives [at launch]." He was incredulous to find that he was now betting on whether the launch "explosions" could be controlled enough to rescue the satellites and complete the mission. Allen thought that this was a graphic way of explaining what could not be controlled during the fatal Challenger launch just over a year later. He also recalled the bone-rattling lift-off during his first launch, " ... the rockets' push accelerating us beyond the edge of the Earth, the sudden silence of these same engines as they shut down, and then the eerie quiet of coasting in unending orbit about our beautiful planet."[41]

After the two satellites had been deployed without problems and several in-flight experiments were completed, the day of the first EVA arrived. Allen, with the frustration of the cancelled STS-5 EVA behind him, was eager to get on with the task. It was unlikely that STS 51-A would have had an EVA on the flight had it not been for the rogue satellite retrieval task, so for Allen it was a dream realised.

A butter cookie for good luck

"Putting on the EVA spacesuit always reminds me of the feeling I had when my mother dressed me in a very heavy snowsuit. In this case, your shipmates bundle you up. They stuff you in the spacesuit, often with a pat on the back and a butter cookie in the mouth for good luck. Then they put the helmet over your head, snap it into place at the neck ring, and from that moment on you float in the suit, your toes gently touching the boots, your head occasionally bobbing up against the helmet. You are now floating in a space-age cocoon."[42]

Fortunately, the rogue satellites were of the same design, so the retrieval and recovery hardware could work on both. As neither was designed for retrieval, new equipment had to be designed to capture and restrain the satellites. The apogee kick motors were still attached, so a capture device called a "stinger" (due to its resemblance to an insect sting) was devised that would attach to the front of the MMU and could be inserted into the kick motor engine bell. This would secure the satellites and allow them to be manoeuvred back into the Shuttle's payload bay. There, they would be locked into place using an A-frame device for the return to Earth. By closing the grapple ring on the "stinger", the astronaut flying the MMU would "dock to the satellite" to steady its rotational rate. The MMU would then be used to bring the satellite close to the RMS, which would grapple a feature on the side of the stinger. The second astronaut was then supposed to attach a bracket to the top of the satellite so that the RMS grapple could adjust its grip location and the MMU pilot could withdraw, prior to the satellite being lowered into the payload bay.

Both stingers worked as designed, but the dimensions of the common clamp of the bracket were slightly different to those of the retention rings on the satellite, necessitating a change of plan. In developing the original plan, the astronauts and EVA trainers had devised a contingency in the event of the RMS being unable to grapple the satellites properly. In this case, the RMS would remain attached to the stinger grapple, holding onto the satellite while the MMU pilot disengaged from the stinger, stowed the MMU and translated to a portable foot restraint. He would then grab hold of the still-folded main antenna to hold the satellite steady in turn while the RMS was disengaged and the second astronaut manually attached the A-frame to the floor of the payload bay. This was the method used on the actual task. Joe Allen flew the MMU to capture the first satellite while Dale Gardner flew the unit to retrieve the second. Their experiences added a significant understanding of extended EVA servicing operations and procedures that were subsequently adapted for other EVA operations from the Shuttle.[43]

Flying free

Allen emerged head first out of the airlock on the first EVA and, to his astonishment, found the Palapa satellite floating just beyond the bulkhead of the orbiter. Only a short time before, when they looked out of the window prior to suiting up, the satellite had been a bright spot 100 km away. Now they were parked barely twelve metres (forty feet) away from the satellite, whose rotation had been reduced from 40 rpm to a more sedate 2 rpm. After donning and testing the MMU, Allen flew over towards Palapa, feeling none of the jerks and jolts that the simulator on the ground had produced. As he rounded the back of the satellite to the kick motor end, he was dazzled by bright sunlight and he thought he would have trouble seeing the docking.

Allen, tethered to the foot restraint on Discovery's robot arm, holds onto the Palapa B-2 satellite with his right hand while Gardner works to the right of the frame.

He knew from Palapa's silhouette that he was heading in the right direction – and at the right rate – when suddenly he entered the shadow of the satellite, clearly viewing the nozzle and where he was to place the stinger. After achieving a temporary soft dock, he activated the device to achieve a hard dock, calling to his colleagues back in the Shuttle to stop the clock as he had captured the first satellite – much more easily than he thought he would have done.

The next task was to stabilise the satellite using the MMU. But as they were both slowly spinning, he had no reference point and no sense of motion. After the RMS grappled the satellite, Allen remained docked to it while Gardner attached the A-frame on the opposite end. As he waited, Allen took the opportunity to view the Earth below him, receiving a warning from Gardner not to fidget because his movements were shifting the satellite as well. After encountering the problem with securing the satellite, the back-up plan was adopted, with Allen stowing the MMU and moving to the foot restraints on one of the orbiter's longerons.[44]

Having your hands full

During a briefing by the crew to the Subcommittee of Science, Technology and Space in Washington DC on 28 January 1985, Joe Allen was asked by Senator Gorton, the chairman of the subcommittee, about how serious or exhausting the physical demands of holding the satellite motionless for seventy to eighty minutes were. Allen replied: "I think it is fair to say that it was considerably easier than it may have appeared to those of you on the ground. By that I mean the satellite looks very large and awkward. However, it weighs nothing. It's just a matter of babysitting it, or tending it, not holding it 'up'. This is true even though your hands may be over your head. You are not holding your hands or arms up in doing that. The assignment to work outside a spacecraft in a spacesuit for many hours, however, is not a bed of roses. The suit is bulky, you feel quite cumbersome and it is not easy to move yourself along with great control. To do that takes considerable practice on the ground, and then it takes a certain deliberate approach to do it in the zero gravity of space. Dale and I worked for about eight hours a time. At the end of that time we were pleased that the task was completed and that it was time to come in. I might say that much of the demand is mental, not physical."[45]

Allen was floating inside the suit, but due to the precise measurements it was a snug fit and required no extra effort to remain in place. What he did discover was something that Isaac Newton had realised almost 200 years earlier; "The more massive an object is, the more it tends to resist motion and thus the easier it is to handle. I was even able to hold Palapa with one hand from time to time to help relieve muscle fatigue in my arms."[46]

Having worked out how to overcome the additional problems that were encountered during the first EVA, the second would see Gardner retrieve the satellite and position it for Allen to hold, before stowing the MMU and attaching the adapter. Though Allen had flown the MMU two days before, he had not felt a sense of altitude. He was in control and at ease with the responsive MMU. But positioned atop the RMS was different: "Riding the end of the arm, high above the cargo bay, was like

A "For Sale" sign makes light reference to the successful retrieval of two errant satellites and their impending return to Earth for later re-launch.

standing on the top of the world's highest diving board ... and a moveable board at that. Because of my helmet's limited visibility I could not see my feet, nor the rail at my side. Only my knock-kneed stance kept me in the foot restraint, and my ride was as nerve-wracking as anything I had done before. Although I knew better, I had the sensation that if I fell, the payload bay would have to catch me or I would plummet to Earth. It was a relief to take hold of the satellite when Dale brought it within my reach. I felt like a man on a high wire being handed a balance bar and the round end of the cylindrical satellite provided some comfort and security."[46]

The second EVA passed smoothly and with both satellites now safely secured in the payload bay, the two proud astronauts displayed a "For Sale" sign as the rest of the crew recorded events on camera.

Fun in space

Despite the hard work of retrieving both satellites Allen, like all other astronauts, took time to enjoy his experiences. He shared this with those on the ground through fun demonstrations with liquids (using orange juice, coffee and water) and of movement in space (by performing no-handed push-ups and somersaults in the mid-deck). The most amusing sequence, however, was also one that the crew kept secret for some years.

With a few hours to themselves after retrieving the satellites, Gardner and Allen made a video that showed Allen being "extracted" from one of the small mid-deck lockers.[47] After reviewing the tape, the optical illusion seemed highly amusing and they decided to downlink it to the night shift team of ground controllers. Unbeknown to the crew, the video was shared with other NASA centres and those who saw it in Florida's KSC, where the lockers were configured for flight, knew it was impossible for an astronaut – even one as small as Joe Allen – to fit inside the locker. When the mission landed, the technicians all but ignored the astronauts and looked around to see how the trick was done. They soon realised how it could be implied that Allen had indeed come from inside a locker. On a normal mission, the lockers are arranged in groups of sixteen (four across and four down), but on 51-A, centre of gravity parameters created by the hardware required to retrieve the satellites meant that several lockers were removed, although their frames remained. By placing the camera looking across the face of occupied lockers, Gardner could open one adjacent to the two-locker gap that Allen had squeezed into and, being careful not to show any part of his body floating "below" the locker base, could make it appear on video that Allen was being extracted from the locker.

After the flight, the crew were honoured by Lloyds of London, who arranged a trip to England to visit with the British Prime Minister (Margaret Thatcher), Prince Charles, Oxford University and City Hall of London to celebrate their achievement in retrieving the satellites Lloyds had insured. It had been an exciting and fulfilling flight for Allen, one that he was unlikely to surpass.

STS 51-B: SPACELAB 3 AND THOSE MONKEYS

Spacelab 3, the first operational mission of the series, had originally been intended to fly after Spacelab 2, but delays to the instrument pointing system of that developmental mission meant the flight order was reversed. Spacelab 3 would carry a payload of experiments in the fields of materials science, space technology and life sciences. Payload specialists for the flight were identified on 8 June 1984 as Taylor Wang and Lodewijk van den Berg, with their back-ups named as Eugene Trinh and Mary Helen Johnston.

The NASA crew was announced on 18 February 1983, with STS-5 veteran Robert Overmyer as commander, rookie astronaut Fred Gregory (Group 8) as pilot, and three mission specialists; MS 1 Dr. Don Lind (a 1966 pilot-astronaut on his first mission), MS 2/FE Dr. Norman Thagard (at that time, in training for STS-7) and

408 The Long Wait

In an update to the STS-5 advert the STS 51-A crew proudly display new signs marking their achievement during the mission.

MS 3 Dr. Bill Thornton (in training for STS-8). The crew would operate a two-shift round-the-clock mission to gather the maximum possible amount of science data from the experiments during the week-long mission, effectively obtaining two-weeks worth of data. Gregory, Thagard and Van Den Berg formed the Gold Team, while Thornton was assigned to the Silver Team alongside Overmyer, Lind and Wang.

The second Spacelab mission

Launched on 29 April 1985, the seventeenth mission of the programme was the first not to make the major TV bulletins, which demonstrated that Shuttle flights were seemingly becoming routine and less interesting to the media and the general public. The flight set a new space record, as it was the first mission to include three men over 50 years old. After a brief delay, the launch went without incident and the crew were soon busy inside the Spacelab module preparing for the week of activities. Despite some small problems, the flight progressed smoothly and included extensive work by PS Wang on his fluid mechanics experiment. He even successfully rewired the device to get it working after it failed early in the mission, clearly demonstrating the benefit of having the principal investigator on board to repair his own equipment and once again proving that having a crew in space to overcome problems was worthwhile. Aboard the Spacelab, there were three experiments in material science, four in life sciences, two in fluid mechanics, four atmospheric and astronomical experiments and two small satellites, one of which failed to deploy. In addition to the seven astronauts, the "crew" included two monkeys and twenty-four rats, in holding facilities that caused problems of their own. The mission, which landed on 6 May, was deemed an outstanding success by NASA, logging 7 days 0 hours 8 minutes and 46 seconds in flight and landing during the 111th orbit. During their flight, the crew amassed an impressive 250,000 million bits of data from the suite of experiments which, according to the press releases, was enough to fill 44,000, 200-page volumes. They also gathered over three million video frames.

Thornton's return

According to Bill Thornton, one of the worst aspects of the first two days back in orbit was indiscriminate underseat stowage. On several occasions, Thornton had to ask for help from the ground, sometimes taking over an hour-and-a-half to locate items. "We had no control over where the underseat stowage was. They weren't properly listed.

The STS 51-B crew pose for photos after a successful 18 April 1985 Terminal Countdown Demonstration Test prior to launch eleven days later. Left to right: Commander Bob Overmyer, MS Don Lind, MS Thornton, MS Norman Thagard, Pilot Fred Gregory, PS Lodewijk van den Berg and PS Taylor Wang.

We had certainly never practiced with it and, unlike the initial intent which was to have a few key simple items stowed there, they were used indiscriminately for overflow stowage."[48] According to Thornton, the numbering of items in lockers was fine for accountants on the ground, but for operations in space there was still a bit of catching up to do to make activities on orbit easier and more efficient.

His experiences on STS-8 confirmed that pre-flight preparations and organisation were critical to keeping ahead in any space flight, and his problems on STS 51-B served to underline the importance of this still further. On Spacelab 3, the crew was under a heavy work schedule from the time they entered orbit. This was not helped by having one crew member taken ill and technical issues delaying the opening of the airlock door in order to access the science lab, let alone the stowage problems. Thornton called that first day on his second mission "the longest day of my life," and a mild case of SAS added to his woes. His problem this time was not in trying to get his own equipment and investigations up and running, but in helping to complete other crew member tasks.

After searching though all the stowage bags and finally locating the biological sampling devices in the last one, when he opened it, the collection devices flew in numerous different directions as they all floated out of the bag. After collecting them all up, Thornton was not in a good mood and trying to write on the very small slippery labels on each phial added to his frustration. This later drew comments from the researchers who could not read his writing, which did not surprise him. This was but one example of how not to prepare equipment for space flight. Thornton argued for months that everything should be properly assigned, with clear identification marks for stowage, proper training and clear operating instructions, and a logical timeline to the activities. An experiment that had worked well in the lab would not necessarily function in the same way on orbit.

Monkeys and men

Spacelab 3 carried the Ames Research Center-developed Research Animal Holding Facility (RAHF). The stated purpose of this life sciences investigation was "to perform engineering tests to ensure that the RAHF is a safe and adequate facility for housing and studying animals in the space environment; to observe the animals' reactions to the space environment and to evaluate the operations and procedures for in-flight animal care."[49] For the Spacelab 3 mission, two squirrel monkeys were held in identical cages in the Primate RAHF, while twenty-four rats were located in individual cages in the Rodent RAHF. According to the pre-flight literature, "Food and water will be dispensed automatically on animal demand, and waste will be directed by air flow into absorbent trays beneath the cages. Periodically, crew members will replace the food in the dispensers and remove the easily accessible waste trays, replacing them with clean ones." There were no plans for direct crew handling of the animals, although both Thornton and Norman Thagard, who were already medically qualified, were given additional veterinary training to care for them in the event of illness or injury.

Neither Thornton, Thagard, nor commander Overmyer wanted to bring home sick, injured or deceased animals, so Thornton spent a considerable amount of time caring for them on the flight. As soon as he could, Thornton reported on the status of the animal "crew members" after their launch into space: "The rodents are in good shape, and the monkeys appears to be in good shape. One of them even came and greeted me," he told Mission Control.

The Astronaut Office had expressed their concerns for years about flying a large animal payload in space, but to no real avail. The media soon picked up on the problems the astronauts were having with the RAHF leaking food stuffs and faeces, but many of these reports were initially overlooked or counteracted by officials on the ground. According to one report,[50] "sources" at JSC revealed that it was obvious before launch that the cages would not be trouble-free, but it was "too late to do anything about it." Earlier tests showed that food bars would crumble and that the airflow system would carry odours, food and faeces out of the cages when serviced. These technical problems could be fixed, but the differences of opinion between managers and astronauts would have serious consequences for future flights of such facilities. Significant improvements needed to be made to ensure that the problems on Spacelab 3 could not reoccur.

Two astronauts had to don surgical gowns, masks and gloves to service the cages, while a third held a vacuum cleaner to catch free floating debris and a fourth, Thornton, recorded the process on film. The operation was supposed to take one astronaut's time, not four, and one of the payload specialists lost valuable time for his own experiments by having to assist in tending the RAHF.

Thornton had mentioned prior to the flight that the monkeys had a reputation for being disagreeable, having bitten a trainer's finger and scuffled with the astronaut on one occasion. Troubles with the feeding system occurred on FD 3 and one of the monkeys became so sick that Thornton had to resort to hand feeding it after a two-day fast. The monkey then promptly ate forty banana pellets in one go. Problems with servicing the cages began on FD 2 when a food bar broke up into numerous crumbs when taken out of its wrapper. As well as the food crumbs, the airflow also blew out faeces from the cages. At one point, commander Overmyer voiced his concerns, which were inadvertently picked up by the air-to-ground link: "Faeces in the cockpit is not much fun. How many years did we tell them these cages would not work?" Initial reaction suggested that the crew had incorrectly identified the debris and that it was food particles, not animal waste, in the cabin. But in 1988, Thornton clearly confirmed that this was not the case: "I can absolutely tell you that, as a boy that grew up on a farm with chickens, pigs and so forth, I knew at the age of three years old [the difference between] faeces and food pellets."[27]

The problem, clearly evident to the crew, was in the design of the cages themselves. Gaps in their construction, combined with positive pressure, allowed faeces, hair, food and any other small material in the cage to build up. Then, as soon as the door was opened, the positive pressure blew the items out into the cabin. According to Thornton, "The first time I cracked the food tray an eighth of an inch, it was almost as if you had fired a gun with material that blew out."[48] The primate cage problem was solved by switching off the air flow and using a vacuum cleaner, but even this did not prevent problems with the rodent cage. A water leak in the cage also caused concerns.

An astronaut tradition – launch day breakfast with a cake decorated with the mission emblem. The crew is wearing the gold and silver shirts of the two shifts they would participate in during the mission.

The monkey's illness was not thought by Thornton to be totally due to SAS, but to a lack of interaction with humans, which it had been used to on the ground. His intervention by hand-feeding the monkey certainly helped it to recover.

As stated by the crew in their post-flight report, "A large volume could be written on the problems with this system," and serious reviews, in light of the experiences of Spacelab 3, would be required before such a device was flown again.[51] Even when Thornton tried to explain what he had seen in space, officials tried to discourage this until "official reports and investigations" had been conducted.

Problem after problem

The seventeenth flight of the Shuttle was plagued by mishaps from the first day. Though the first satellite was deployed from its small container, the second one refused to eject; the water supply faucet would not work until a bypass was added; and a urine collection system failed to work when attached to a life science experiment, in that it failed to collect urine as it was supposed to do and instead sprayed it into the cabin when the toilet was flushed. After encountering more than his share of floating yellow globules, Thornton announced that he would not continue to use the experiment. The toilet functioned perfectly well thereafter. Five of the fifteen experiments either failed or refused to start initially, and others were not working to their full potential, but by the end of the flight, the crew had managed to repair twelve of the experiments to full operation and a further two to partial operation. One of the most demanding repairs was that by PS Taylor Wang on his own experiment, which failed early in flight but was restored to full operation thanks to his detailed knowledge of its workings.

Running around the world

By 1985, Thornton's treadmill had been used for about three years and over the months, the bungee forces had become out of balance with the harness. It was also becoming very noisy and vibrations were being transmitted to the Spacelab. Thornton was developing a new treadmill for use on longer Shuttle flights, but this new device could not be used wearing just socks, as they would be caught up in the conveyor belt track and ripped off. Thornton suggested using more secure footwear on the later treadmills: "For safety's sake, we should have some sort of footwear. We simply resorted to the flight boots that we had [on 51-B], but some sort of solid footwear that would stay [in place] would be a very simple solution to this. Speaking for myself, it [the treadmill] sure did my legs good. I didn't even notice [any problems] coming back this time and I only used it that one long period. And I will say that the thing can be used for the space station."[48]

During the mission, Thornton wanted to really test himself and the treadmill by running for ninety minutes, or a complete orbit. He asked his colleagues to keep track of the time so that he knew exactly when to stop, but they decided to play a little trick on the good doctor by telling him a different orbital period for the specific orbit and adding a couple of minutes to the real flight time around the world. Thornton conceded that this was a tough run.

Back on the ground

After his second mission, Thornton adapted well to the gravity environment and his resistance to motion sickness increased. A few days after coming home, he flew in a T-38 with a colleague "to try to make one another sick," but they ran out of things to do without inducing sickness. He assumed that, as with the adjustment to space flight, getting reacquainted with life back on Earth would take a couple of days. He did not encounter this on either of this two flights, but the theory was borne out by the crews from much longer flights than his own.

During STS-8, Thornton was more occupied with what he called "good science" on his own science investigations, but this was not the case on Spacelab 3. This seriously restricted the investigations he had begun on STS-8, as he was only able to "slip in a few little things" this time around. Most of his time was spent looking after the animals and supporting other activities that were the result of bad timelining. He had no significant data from his second flight to compare with the first mission, apart from some early leg measurements. The lack of carryover of Thornton's investigation from STS-8 to 51-B meant that there was no real benefit in flying Thagard and Thornton together on 51-B as far as SAS was concerned. The timeline and in-flight

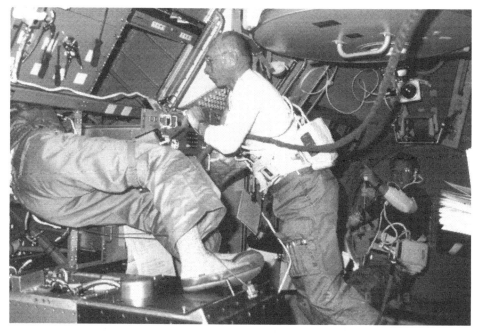

MS Bill Thornton works in the long module of Spacelab 3 as PS Taylor Wang repairs his experiment, MS Don Lind is in the background.

problems restricted any cooperative data gathering on the Spacelab mission that could have been used for comparison with the findings from their first missions.

Thornton believed that crew selection should be based on requirement, not politics, so his assignment to a third mission would have been fine, as long as it was beneficial to the programme and to his studies. A long duration mission would have suited in that respect. In 1986, however, Challenger was lost, grounding all assigned crews and cancelling planned missions, Thornton was not assigned to a crew at that time and even two years later, in August 1988 (just prior to the resumption of flights), Thornton stated that he should not be considered for flight assignment when the most important need was to re-qualify the Shuttle system. When that had been achieved and crews resumed his work on long duration space adaptability, he would have loved to have flown again.

STS 51-F: SPACELAB 2 AND THREE SCIENTIST-ASTRONAUTS

The mission specialists for the second Spacelab mission were announced at the same time as the Spacelab 3 crew, and included Karl Henize (MS 1) and Tony England (MS 3). England had recently returned from Rockwell, where he had supported avionics development since returning to NASA. At one time, he had been considered for assignment as MS to an early Shuttle flight but, for various reasons, was instead

The crew for STS 51-F takes a break in training. Left to right: Tony England, Karl Henize, Story Musgrave, Gordon Fullerton, Loren Acton, Roy Bridges and John-David Bartoe.

nominated for Spacelab 2. For about a year, both astronauts worked on mission issues, mainly at Marshall Space Flight Center, before being joined by commander Gordon Fullerton, pilot David Griggs and MS 2/FE Story Musgrave, who were announced on 17 November. Payload specialists Loren Acton and John-David Bartoe were announced on 8 June, with their back-ups being named as Diane Prinz and George Simon.

Again, a two-shift system would be utilised, with Griggs, Henize and Acton operating the Red Shift and Musgrave, England and Bartoe operating the Blue Shift. Commander Fullerton was not assigned to a shift and would work with either, as required. For the first and only time, three scientist-astronauts – Henize, England and Musgrave – would fly together on one mission, and all three would support contingency EVA operations. Musgrave was designated as EV 1, with England as EV 2 and Henize as the IV astronaut, who would have supported them had an emergency EVA been required. Delays in the mission (mainly due to qualifying the Instrument Pointing System) and changes to the Shuttle manifest resulted in pilot Griggs being reassigned from STS 51-F on 26 September 1984, due to the close proximity of his other assigned mission, STS 51-E. He was replaced on the Spacelab 2 mission by Roy Bridges.

False starts but a fine mission

The initial launch of STS 51-F on 12 July was scrubbed at just T − 3 seconds with all three main engines up and running. A coolant valve in engine No 2's hydrogen chamber failed to close and the redundant safety system triggered an abort. They finally launched on 29 July, but barely five minutes into the mission, one of the three main engines shut down due to a failed sensor. By dumping the fuel to lighten the load, the mission made it to orbit on the other two engines, but fifty miles lower than planned. As a result, significant reprogramming of the experiments was needed, by controllers, investigators and the crew, in order to get as much scientific data from the mission as possible. The mission was extended by one day to gather more data. Spacelab 2 carried fifteen experiments in seven scientific disciplines, including three solar physics experiments, one in atmospheric physics, three in plasma physics, two in high-energy astrophysics, one in infrared astronomy, one in technology research, two in life science, and four smaller experiments. The mission did not feature a pressured laboratory module, as flown on Spacelab 1 or 3, but instead carried three pallets to support the experiments, as well as the instrument pointing system and an "igloo" containing instrumentation for data collection, commands, and thermal control. At the conclusion of the mission, and with eighty per cent of the pre-flight objectives met or exceeded despite the early setback, Jesse Moore, Chief of the Shuttle programme stated, "This may be the most exciting science mission the Space Shuttle has flown, and we're delighted with it." The crew had taken thousands of photos, forty-five hours of video tape, and 370 km of data tape. The flight landed on 6 August 1985 after a mission lasting 7 days 22 hours 45 minutes 26 seconds and 127 orbits.

A long preparation

Karl Henize spent almost a decade working on a specific astronomical payload for the Space Shuttle before finally making it to orbit with that same package. A package of proposals for what became Spacelab 2 was first suggested by scientists in 1976. Henize was one of the scientists who nominated experiments for the payload, but his was not compatible and was not selected. It would have been a 0.5-m (twenty-inch) telescope, derived from his earlier work in Gemini and Apollo. Marshall Space Flight Center reviewed the proposal and selected the payload for the mission in 1977, with the first meeting of principal investigators that same year. Karl Henize was nominated as the CB representative to attend these meetings. Though he was not selected to fly the mission until 1983, he had always pretty much expected that this would be his flight ... eventually.[52]

Once assigned to the flight, Henize, in addition to countless meetings and reviews, had a programme of payload training at JSC (five days), KSC (twenty-one days, which he deemed too much) and at Marshall (fifty-six days which, at two days a week, would take him over six months to complete). In addition, he completed over ten

The science crew for Spacelab 2 visits the University of Birmingham in England in 1983 to review some of the payload hardware being prepared there. Left to right: Back-up PS Diane Prinz, PS John-David Bartoe, PS Loren Acton, back-up PS George Simon, MS Tony England, MS Karl Henize, and Professor Peter Wilmore of the University. [Photo credit University of Birmingham.]

hours a week on orbiter training for thirty-seven weeks, broken down into 110 hours for Spacelab systems; sixteen hours for entry flight operations; twenty-six hours of proximity operations; sixty-three hours operating the RMS; fourteen hours for ascent flight operations; seventeen hours for orbiter guidance, navigation and control systems; over eleven hours for DPS; thirty-eight hours for orbiter subsystems; and seventy-four hours on crew systems.[53]

Two of the experiments carried on Spacelab 2 originated from the United Kingdom and required the science crew to visit the UK during 1983 for familiarisation purposes. Tony England was the crew representative for the amateur (ham) radio experiment, and made a couple of visits to a group in Wales who were interested in participating during the mission. Further visits to Birmingham University and the Rutherford Appleton Laboratory in Chilton, England, provided personal contact with those who had developed the experiments and the understanding to be able to interface with that experiment if something went wrong. The science crew also got to

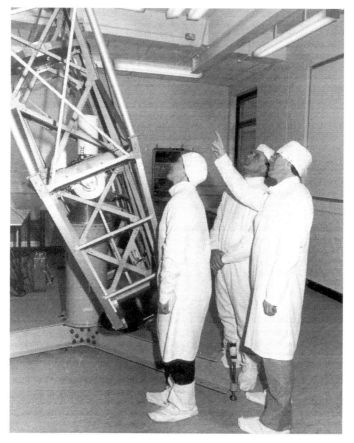

The astronauts of Spacelab 2 inspect the science hardware being developed at the University of Birmingham in England. [Photo credit University of Birmingham.]

know those who would remain on the ground, a key factor in working together in the event of a system failure or problem of some kind during the mission.

Spacelab 2 was the first two-shift operation without a Spacelab laboratory module and this presented its own unique challenges. Because of the close proximity of the STS 51-G crew's training, the 51-F crew were unable to have the correct configuration of the aft flight deck in the fixed-base Shuttle Mission Simulator until after the 51-G mission had flown in June 1985. As a result, it was not possible to conduct full-on training for several complex integrated crew operations until late in the training flow. This led to a more demanding schedule for an already fatigued crew, who recommended that a second simulator should be brought on line to cope with the expected rise in launch rate and the overlapping crew training programmes.

The crew noted that Spacelab training was adequate, but the effort required was much greater than it needed to be, and there were definite weak areas. The simulators only approximated Spacelab systems, with some procedures not available until the week before the flight. However, Henize's experience and association with the Spacelab 2 payload eased the liaison between the PIs and the flight crew.

Spain or Earth orbit?

The problems during launch on 29 July led to the first (and, to date, only) Shuttle Abort-to-Orbit (ATO) situation. The first stage performance (SRB and SSME) was reported to the crew as low, and meant that the SSME had to be throttled up to compensate for the problem. At 3 minutes 30 seconds into the flight, the first signs of a main engine problem were noted and at 5 minutes 45 seconds, the No. 1 engine failed and shut down, triggering the Abort-to-Orbit sequence using the remaining two engines. Challenger limped to orbit, which was gradually raised over the following twenty-four hours to allow the flight crew to accomplish most of the planned mission objectives. The events of the 51-F ATO were detailed in the companion volume *Disasters and Accidents in Manned Spaceflight* [David J. Shayler, Springer-Praxis, 2000, pp. 139–144] and are not reproduced here. However, the three former scientist-astronauts aboard the mission offered their own perspectives of these events.

On the 12 July launch attempt, Tony England was seated on the mid-deck, his main concern during the ascent being the safety of the payload specialists. The two PS were not as familiar with the systems as he was, and as the launch sequence progressed, England mentally reviewed his responsibilities and recalled his emergency training, just in case. When the abort situation arose, the crew was aware that this might not be a simple engine shutdown. Sitting out on the pad, England, like the rest of the crew, was more concerned with performing his assigned tasks than worrying about not reaching space that day. It was not until after they got out of the orbiter that the disappointment set in, but only for a short while. According to England, the crew went to Disneyworld to relax in the days after the pad abort while the engineers recycled the mission to attempt the launch again as soon as possible.

Up on the flight deck, Musgrave's concerns over the method of Shuttle ascent were affirmed by the abort. His intense training for flight deck activities on ascent and entry added to his knowledge (and concerns) about what could go wrong leaving Earth and getting home again. When the ATO call came in, Musgrave briefly thought that the simulation controllers were throwing in an abort situation on a real flight, but their training and knowledge of the systems gave the flight deck crew (Fullerton, Bridges and Musgrave) a good indication of what had happened, even before Houston gave the ATO call. Musgrave was already working on supporting Fullerton and Bridges with calls and procedures, should they need to abort the mission and fly over the Atlantic to an emergency landing in Rota, Spain.

He was so engrossed in the flight manuals on his lap that he was not immediately aware of a call from Karl Henize, who was sitting beside him in the fourth flight deck seat. He asked Musgrave where they were going, finding it hard to believe that after the pad abort he still might not make it into space. Although no one else seemed concerned, he wanted reassurance that he would finally get there. Without thinking, and deep in concentration over transatlantic abort sites, Musgrave replied "Spain," causing both Fullerton and Bridges to quickly glance in amazement at the flight

The Abort-to-Orbit control is clearly displayed in this photo of the flight deck controls of Challenger after the STS 51-F launch suffered the loss of a main engine.

engineer. Musgrave had automatically answered Henize's question without really hearing him, but quickly calmed the situation – and Henize's fears – by telling him to relax. The astronomer would indeed make it to orbit – just. Years later Musgrave did state that it was "frightening" to some degree, as not all the engineering information was available to the crew. They were not fully aware of what the ground were seeing, or what was happening at the back of their vehicle. The reason for the abort (either on the pad or in-flight) was unknown to the crew; they were only aware of the shutdown or failed engine situation and had to follow procedures to overcome the problem.[54]

During the ATO situation, England, down on the mid-deck, was fully aware of what had happened and what was unfolding up on the flight deck. Once placed in a stable orbit, the concern for the crew was that somebody would find something that would force an early termination of the mission. The crew was aware that the controllers and engineers would analyse the data and determine the available options for either a short flight or a full-length mission. The main concern on orbit was over how much fuel would be needed to raise the orbit sufficiently to warrant staying there and what quality of scientific data would be obtained (or lost) as a result. Fortunately, the call came up from the ground that they were indeed safe in orbit and although some work needed to be done, a full mission would be flown. It had been a close call, but three former scientist-astronauts were in orbit on the same mission. Musgrave had survived a second ride of the solids and England had been the only astronaut to have left the astronaut programme without a flight and then come back and fly a mission (in 1997, John Glenn became the first flown astronaut to retire, return and fly, aged 77). Karl Henize, an avid astronomer since he was a boy, had finally made it to space aged fifty-eight, although he often said that the launch abort had added ten years to his real age!

Karl flying high

The primary goal of the mission was the solar observations and every ninety-minute orbit gave the crew about fifty-five minutes in sunlight and thirty-five minutes in shadow. Before sunrise, Henize had to get the IPS ready and in the correct position to find the sun (and other stars as required) to allow the system to lock on and gain an

Spacelab 2 science payload deployed from the payload bay of Challenger during STS 51-F.

optical hold on the solar disc. This took anywhere between five and ten minutes, but having only one target to track (the sun) made life a little easier, particularly as there were other tasks to set up or perform prior to each data run. Intermittently, he operated one of the experiments (No. 9, the British Coronal Helium Abundance Spacelab Experiment – CHASE) and he also had about five minutes on each daylight pass to take the opportunity to look out of the window at the Earth. On night passes, he was busier than during the daylight pass. For thirty-five minutes, he reported to MCC-Houston about how the previous pass had gone and received instructions for the next daylight pass. Because of this, Henize, a well-respected astronomer who had waited years to fly into space, did not have the opportunity to look out of the window at the stellar field in detail during the night passes, nor to look closely at the night side of Earth and its illuminated cities, much to his regret. He did at least manage to record an audio diary of his activities and experiences on orbit.

His diary mentioned a couple of especially interesting sites while flying over China and, "as a closet geographer with an interest in maps," he was particularly interested in this area – still relatively unknown even in the mid-1980s. Henize thought the Tibetan plateau was beautiful, spotted with moderate sized lakes everywhere. This surprised him, as did the variance of colours. He took personal credit for having discovered one large oval basin in the Western province of China that reminded him of the oval shape of great red spot of Jupiter, covered as it was by a reddish desert and clearly visible from orbit. Photographs of Henize looking out of the window from Challenger clearly show the delight on his face at viewing the Earth and stars from this vantage point.

The Instrument Pointing System (IPS) gave the crew a few headaches for the first half of the mission. When asked some six years after flying his only mission, Henize concluded that if a re-flight of Spacelab 2 had been possible, then flying a mission without equipment failures would have been a significant improvement. Other than that, he felt the mission had gone pretty much as designed, so any re-flight would not have required significant changes.

Taking the last chance to fly

When Tony England returned to NASA, he did not really fit into either the new Shuttle mission specialist selections, nor with the older groups. He considered himself to be "always kind of an odd ball," and back at NASA, this feeling continued. He decided to get his flight and then return to work in science somewhere, though he had not intended to stay at NASA nine years to achieve this goal. His assignment to 51-F was to a flight that was complicated, but one which he thoroughly enjoyed. He also enjoyed working with this crew, but not the delays that pushed the flight back to 1985.

Both England and Musgrave were hoping for an EVA on the mission, as they were the contingency EVA crew. But they did not get the chance to go outside. If they had, they would have needed to move a considerable amount of trash bags out of the airlock into the mid-deck to allow them access into open space.

In the first few days, with several failures of the IPS and other systems and hardware, plus the recovery from the Abort-to-Orbit through a difficult programme

The 51-F crew are wearing sunglasses in a humorous crew photo aboard Challenger. Commander Fullerton's head is at centre with England next to him. Around them are, from bottom right clockwise: Bartoe, Musgrave, Bridges, Henize and Acton.

of planned manoeuvres, there was potential for friction between the crew members. But England was impressed that not once did he encounter or hear of any argument about finding space to work in the crowded flight deck (though Henize jokingly said he had to fight through arms and legs in order to get a good view out of the window). According to Tony England, "Karl recognised that this was probably going to be his only flight and he wanted to take as many opportunities as he could to look out of the window. Loren Acton felt the same way." England and Bartoe were as busy on their shift as the others, but had less opportunities to look outside until several days into the mission, when the computers controlling the payload bay overheated and were temporarily shut down. With nothing to do, England grabbed a camera and secured a bag around one of the overhead windows, spending two orbits (about three hours) with his head inside the bag to observe and photograph the Earth. After that, England looked for other opportunities to look outside, just like his colleagues. In hindsight, England felt he would have set aside more time to look out the window, if given the chance to fly the mission again.

STS 51-F was the first opportunity for former scientist-astronauts (now mission specialists and payload specialists – scientists with a part-time astronaut role) to work together on the flight deck rather than the more spacious Spacelab science modules used on Spacelab 1 and Spacelab 3. Crew orchestration of time and roles had to be more precise due to the limited volume, in an area used to both fly the vehicle and

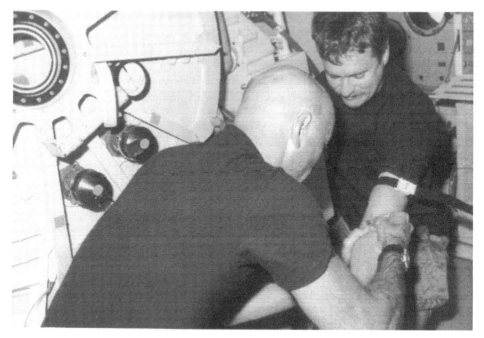
Musgrave draws blood from England during one of the flight's secondary experiments.

operate the experiments. During the assigned shifts, either Fullerton, Bridges or Musgrave occupied the "piloting role" in the front flight deck seats, while the payload and mission specialists occupied the whole of the aft flight deck. Musgrave assisted in manoeuvring the spacecraft as required, allowing the science team to gather the data. On this flight, Musgrave acted more as a pilot-astronaut than a mission specialist, or indeed scientist-astronaut. Close cooperation with pilots Fullerton and Bridges was necessary to fulfil the mission objectives, even before the added complication of recovery from the ATO situation and systems difficulties. Either England or Henize would manage the IPS, cooling and power supplies, allowing the payload specialists to work with their specific experiment. However, if the need arose, both Henize and England would assist the PS in gathering data.

Though each shift was technically twelve hours, there was a hand-over period which included briefings and updates, usually on anomalies. These were reflected in new flight plan updates teleprinted up to the crew each day, further amending the crew activity plans that had been devised during training. This essentially required the outgoing crew to provide a "heads-up" to the incoming shift on any such changes. Working twelve-hour shifts could be tiring, but Henize recalled being in a lot better shape on orbit than after twelve hours working shifts on the ground. Unlike others who found working in space caused extra stress, Henize found the pure joy of working in microgravity relaxing.

During their preparation for the mission, the crew thought it would be a good idea to take some humorous video footage of activities on board the vehicle. Some ideas

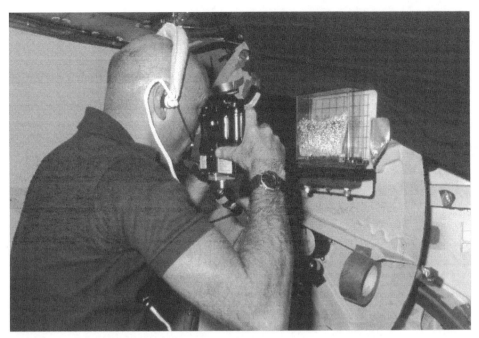

Musgrave photographs a plant growth experiment – one of the secondary experiments on Spacelab 2.

they devised did not work in real time, but others have become vintage footage of "Fun in Space" videos. These staged shorts included Fullerton and Musgrave (both bald) arguing over a hair brush, a scene dubbed "Hair Wars" by the crew; the crew showing the effects of sleeping unrestrained in the mid-deck when the Orbital Manoeuvring Engines were fired; and demonstrating the pages of updates the crew received each day via the teleprinter. Part of an "official experiment" involved taste-testing carbonated drinks from two leading drink suppliers, using specially designed dispensing devices on top of the cans. Henize and England were the "test subjects" and neither was really enthused by the event. "Gas in the stomach is something you really don't want in space," England recalled. Henize simply disliked the taste.

For the landing, England swapped seat positions with Henize to return on the flight deck. From his three years of experience in the simulators at Rockwell just after returning to NASA, he was well aware of each stage of entry and landing and of what Fullerton, Bridges and Musgrave were doing, so he did not have to spend much time worrying about it. Instead, he carried an 18-kg movie camera around to record cabin views during entry and descent. As soon as the crew began to feel the pull of gravity once again, England quickly discovered just how much heavier an 18-kg camera felt after eight days in microgravity. The de-orbit burn was initiated in the dark over South Africa, emerging into "daylight" over Australia. During this phase, the plasma sheath glowed over the flight deck windows. For England, this was a pretty and memorable sight.

Henize tried a Pepsi drink while England samples a Coca-Cola, in an evaluation of carbonated beverages during STS 51-F. Both drinks were deemed unsatisfactory in taste and possible consequences.

For any Shuttle landing, official explanations for the post-flight activities indicating that the crew are "performing tests and shutting down the systems" are not totally accurate. England recalled the time taken by the medical staff helping him walk back and forth across the mid-deck floor to build up tolerance for 1 G, so that when the crew walked down the steps, they did not stumble and fall on national TV. Later missions changed this policy, so that returning crews – especially those on longer missions or returning residents from Mir or ISS – would enter a transfer van directly from the side hatch of the orbiter while on the runway and would be taken to crew quarters instead of walking down a set of steps. England readapted very well, but once back on the ground the strongest emotion he recalled was one of sadness. It was so evident that one of the medics even asked him if there was anything wrong, as he was physically fine. England replied, "After all these years [of waiting to fly in space], it's sad to have it over."[55]

After the crew returned to Houston, they met most of the principal investigators, who talked enthusiastically "for several hours" about the success of a mission that could have been terminated before entering orbit, or shortly thereafter. England also felt that the mission of STS 51-F went too quickly. Towards the end of the flight, the crew were obtaining useful data on some of the instruments, although others never worked at all, and they were eager to extend the mission or re-fly the instruments to allow their full potential to be explored.

The STS 51-F flight crew report noted, "This was a mission of challenge, both in the pre-mission development and during the flight itself. The crew is proud to have been part of a great team that was able to meet the challenge to produce a very significant scientific success." After a long wait, a launch abort and ATO events, to have finally achieved such success from the mission gave the whole crew a great sense of pride and achievement.

ANOTHER TRIP INTO SPACE?

The flight of STS 51-F was the final flight to include "rookie" scientist-astronauts from the 1967 selection. Though they had gone through the long wait and uncertainty over their flights, they had achieved their goal, and while some had already decided to move on to new careers and objectives in life, others still longed for a second or third flight into space. By the end of 1985, Parker and Garriott were both assigned to new missions. For Musgrave, Thornton and (for a time) England, the desire to return to space was still strong enough to keep them in the astronaut programme. Although the line of crews and new astronauts – all awaiting their chance to fly – was much longer and their chances of making another flight were slimmer, they would wait it out for a while again. However, seventy-three seconds of a cold January day in 1986 would drastically and tragically change their fates.

428 The Long Wait

England takes the opportunity to look out of the window during the mission.

REFERENCES

1. AIS telephone interview with Ed Gibson, 3 January 2004.
2. David J. Shayler interview with Karl Henize, Houston, Texas, August 1988, AIS Oral Archives.
3. *NASA News*, JSC 78-03, 16 January 1978.
4. *NASA News*, JSC 78-33, 26 July 1978.
5. Joseph P. Allen, NASA JSC Oral History Project, 18 March 2004, p. 24.
6. *Commercial Space*, March 1987, p. 38 article *Melting Pot* by J. Lowndes.
7. STS-1 mission commentary tape transcript 0181, 04-14-81, 18.16.30 GMT, p. 1.
8. *NASA News*, JSC 82-012, 1 March 1982.
9. AIS telephone Interview with Bill Lenoir, 24 June 2005.

10 *Modifications – STS-4 to STS-5*, Rockwell International News Release update, 1982. Copy on file AIS archives.
11 Reference 5, p. 32.
12 *A Two Million Mile Space Odyssey*, Joseph P. Allen and Thomas O'Toole, Readers Digest **124**, January 1984, pp. 1923. Condensed from Omni magazine, June 1983.
13 *Operational Medicine Report*, STS-5. Crew Health Assessment and Support Activities, NASA JSC Life Sciences Division, Report Number E-989-82-05, December 1982. Copy in AIS archives.
14 *STS-EVA Report No. 1: STS-5*, November 1982, compiled by David J. Shayler and Keith T. Wilson, AIS Publications, January 1984; also STS-5 Air-to-Ground Transcripts, November 1982.
15 Reference 5, p. 22.
16 J. Allen NASA Oral History, 18 March 2004, pp 23-29; also AIS interview with J. Allen, Houston, Texas, August 1988.
17 Memo from John A. Wegener, CG3 Chief, Training Integration Branch; Subject STS-6 EVA, dated 17 December 1982, (CG3 82-TS-73). Copy on file AIS archives.
18 AIS interview with Story Musgrave, JSC Houston, Texas, August 1988.
19 STS-6 Post-flight Crew Conference transcript, 22 April 1983, NASA JSC, Houston, Texas; private discussions with Paul Weitz, May 2003, London, England; also AIS telephone interview with Story Musgrave, 23 June 2005.
20 AIS interview with Jennifer Thornton, San Antonio, Texas, 2 July 2000.
21 AIS telephone interview with Bill Thornton, San Antonio, Texas, 16 February 2006.
22 STS-8 Press Kit, August 1983, NASA Release 83-53, p. 7.
23 STS-8 Flight Crew Report, July 1984, copy on file AIS archives.
24 STS-8 Post-flight Training Report, MOD, Training Division, NASA JSC, 2 December 1983, TD252/A173, copy on file AIS Archives.
25 Reference 23, p. 6.
26 Reference 23, pp. 9–10, Medical DSOs.
27 AIS interview with Bill Thornton, 5 August 1988, NASA JSC, Houston.
28 Owen Garriott email to Colin Burgess, 4 February 2004.
29 *Newsweek* magazine, 25 November 1983, article *A Giant Workshop in the Sky*. No authorship of article given.
30 *Newsweek* magazine, 2 December 1983, article *Not For Astronauts Only*. No authorship of article given.
31 Garriott quote from *Space Flight News* interview, February 1988 edition
32 Garriott quote from JSC Oral History transcript, 6 November 2000
33 From Time magazine 6 December 1983, article *Guinea Pigs in Orbit* by Frederic Golden. Reported by Jerry Hannifin and Geoffery Leavenworth, Houston.
34 Online NASA website, NASA Science News, item, *Ham Radios in Space*. *http://science.nasa.gov*. No authorship given.
35 Website: Ham Sats, *http://www.qsl.net/ve6bpr/page2.htm*. No authorship given.
36 Garriott e-mail to Colin Burgess, 5 February 2004.
37 *Understanding Space: Scientific Eyes in Orbit* by Francis French, *Spaceflight* magazine, October 2003. Based on R. Parker talk given at Reuben H. Fleet Center, San Diego, 1 June 2002.
38 STS-9 Post-flight Mission Report.
39 STS 51-A Flight Crew Report, (CB-85-048) dated 28 March 1985, pp 23-26 EVA training section, copy in AIS Archives.
40 NASA JSC Oral History, Joseph P. Allen, 18 November 2004, Part 4, pp. 1–2.

41 *Exploring Space*, David J. Shayler, Hamlyn Children's Books, 1994, foreword by Joe Allen, p. 2.
42 Joe Allen's EVA comments in *Exploring Space*, previously cited, p. 32.
43 *Walking in Space*, David J Shayler, Springer-Praxis 2004, pp. 264–267.
44 AIS interviews with Joe Allen, August 1988 and August 1989.
45 Briefing by the Astronauts of Shuttle mission 51-A to the Subcommittee on Science, Technology and Space, 99th Congress, 1st Session, 28 January 1985, pp. 20–21.
46 *Entering Space*, Joe Allen and Russell Martin, Orbis London, 1984, pp. 223–238.
47 Reference 44; and Reference 5, pp. 19–20.
48 STS 51-B Technical Crew Debrief, Flight Crew Operations Directorate, June 1985, JSC 20556, copy on file AIS archives.
49 Spacelab 3 brochure, NASA EP203 16M484, Marshall Space Flight Center, 1984, pp. 18–19.
50 NASA distributes Spacelab payload for analysis following Shuttle landing, by Craig Covault, *Aviation Week and Space Technology*, 13 May 1985, pp. 18–21.
51 STS 51-B Flight Crew Report, undated (1985), copy on file AIS Archives.
52 Karl Henize interview with Andy Salmon, 22 July 1991, NASA JSC.
53 Copies of hand-written notes from Karl Henize, dated April-June 1984, AIS archives.
54 AIS interviews with Story Musgrave and Karl Henize, August 1988; AIS telephone conversation with Story Musgrave, 23 June 2005.
55 All recollections from Tony England are sourced from the AIS telephone interview of 14 July 2005.

11

Ending of Eras

By the end of December 1985, eleven of the seventeen scientist-astronauts selected by NASA in 1965 and 1967 had flown in space at least once, and four had made a second flight. Harrison Schmitt was the only one of the seventeen to make it to the Moon, while Kerwin, Garriott and Gibson flew long-duration missions on Skylab. The remainder reached orbit on the Space Shuttle. Indeed, Garriott's second mission was aboard the Shuttle, on the first Spacelab mission in 1983, and in 1985 he was assigned to a third mission, manifested for 1986. Of the 1967 intake, Allen, Musgrave and Thornton had all flown twice and Parker was scheduled to make his second flight in early 1986. England, Henize, and Lenoir had made one flight each. The remaining six had resigned from the programme between 1965 and 1973, when the prospect of making even one flight into space looked very remote. Of those who had stayed the course and made it to space, most now wanted a second or third bite at the cherry, while others, having achieved their dream as an astronaut, had decided it was time to move on to other goals and careers, particularly as their chances of a return to space looked slim. Then, on 28 January 1986, the loss of Challenger and her seven astronauts brought home the dangers and risks of space flight to the Americans. The future of the scientist-astronauts, the Shuttle, the astronaut corps and even of NASA itself looked in doubt.

MOVING ON – LIFE AFTER SPACE FLIGHT

Having waited for years for a precious seat on Skylab or Shuttle missions, the choice for the scientist-astronauts by the mid-1980s came down to staying and holding out for another flight, or taking an opportunity to branch out into a new career while they were still young enough to do so.

Joe Kerwin – Skylab–Shuttle–Space Station

Joe Kerwin had been involved with the Shuttle since completing his Skylab assignments in the mid-1970s and undertook CB technical assignments in the design of the crew station, the controls and medical monitoring equipment. He was also on the selection board for the first group of mission specialists from 1976–78, becoming their first "boss". In addition, he became the lead astronaut for planning in both the operational Shuttle missions following the Orbital Flight Tests, and for the roles and responsibilities of the crew during rendezvous and the deployment and retrieval of satellites using the Canadian-built Remote Manipulator System. According to some sources, Kerwin was an early candidate for assignment on the STS-13 (later re-designated STS 41-C) Solar Max mission, but before the crew could be announced, he was named as NASA's senior science representative in Australia. He returned to JSC in January 1984 and became Director of the Space and Life Science Directorate, a position he held for the next three years.

In his 2000 Oral History, Kerwin remarked: "I can't think of anything spectacular to tell you about those years. Everything went pretty much routinely, waiting for space station. Waiting for a flight on Skylab, at least from my perspective, that was the big thing. We were very happy when [President Ronald W.] Reagan announced the permanent presence in space in 1984, authorising NASA to begin the process of developing a space station. I think we have not progressed in the most stellar fashion in the last sixteen-and-a-half years. We should be there by now."[1] Expressing his disappointment at not seeing a permanent American space station in orbit by 2000, Kerwin stressed that he was not blaming the administration or leadership, but rather the promotion of the station. This was a major new space project, and the agency had tried to broaden cooperation with new partners, expand their objectives and develop a station that was "all things to all people." But they could never quite get control of the schedule and budget. Kerwin believes that NASA might have done it had it not been for the loss of Challenger two years into the space station programme. The combination of trying to recover from Challenger and delivering a space station was too much in too short a time frame. At the time of Kerwin's interview, it was still six months before the first resident crew would be launched to ISS. Despite the involvement of the Russians since 1993, there were still significant hurdles to overcome before ISS could truly be seen as the next step after Mir, or even Skylab.

Kerwin would be involved with ISS, but not from within NASA. In April 1987, he retired from the US Navy with the rank of captain and also resigned from NASA, accepting a position with Lockheed Missile and Space Company (based at their Houston office) as chief scientist for the space station project. He became manager of EVA systems at Lockheed between 1987 and 1990 then manager of Manned Systems Programmes until 1998, when he joined Wyle Laboratories (also in Houston) as Senior Vice President for Life Sciences, Systems and Services. Wyle is a support contractor for the medical effort at JSC, and as Kerwin explained, "My people are the extra hands and feet for the research laboratories, the medical operations and training, the medical equipment development, and building the crew healthcare system and exercise hardware for Space Station. So I'm now (2000) a supporter, a contractor rather than a manager at NASA, but still on the team, still helping the space station to get on with it."

Astronaut Office – circa spring 1984

At the time of Reagan's announcement of an American space station programme (April 1984), an internal CB memo documented the manpower requirements for the Flight Crew Operations Directorate and the assignments of the active astronauts.[2] Of the former scientist-astronauts, Garriott (Space Station) and Lenoir (Mission Development) were assigned as Branch Chiefs, while Allen (51-A), Thornton (51-B), and Musgrave, England and Henize (51-F) were all assigned as "mission specialist crew under training".

In the Mission Development Branch (under Lenoir), Bob Parker held four assignments, for issues concerning Science and Technology (Astro Shuttle missions), and for Payload and Operations (Spacelab, New Pallets, Payload Communications and Data – all as lead astronaut). Garriott was also working on science and technology issues for the (Shuttle-borne) Earth Observation Missions (EOM), while Thornton found himself working on space medicine issues and Musgrave was assigned to the Shuttle Avionics Integration Laboratory (SAIL) in the Systems Development and Test Branch.

Garriott held the position of Program Scientist for Space Station for two years from 1984. In his 2000 Oral History, he explained that his role was to advise on the conduct of science, the type of science that should be carried out on the Space Station, and how such science would be conducted on the various configurations that were being studied. He also commented on the way that the station was placed in orbit. Instead of several dozen Shuttle flights, he argued that a heavy-lift launch vehicle – derived from the Shuttle – could probably have done the job quicker, cheaper and more safely. This would have meant an earlier completion of station construction, so the science research could have begun many years ahead of when it actually did.

After Spacelab 1

On 7 June 1984, the crews were assigned to the first missions under the EOM and Astro series. Owen Garriott was assigned to STS 51-H as MS 2, and Bob Parker to STS 61-E as MS 3. Both had flown on the first Spacelab mission in 1983. STS 51-H was scheduled for launch in November 1985, a re-flight of nine Spacelab 1 experiments in the fields of space plasma physics, solar physics, atmosphere physics, and astronomy and Earth observations. The configuration, termed Earth Observation Mission 1 (EOM-1), was to include the first flight of a short Spacelab habitable module and a pallet of experiment hardware in the payload bay. STS 61-E was manifested for March 1986 and its mission included deployment of the Intelsat VI-I communication satellite and the ASTRO-1 astronomy package, designed to view Halley's Comet among other targets. For Garriott and Parker, this was an opportunity to resume their training for a return to space. For others, it was time to leave NASA and move on.

Lenoir departs – and comes back

In August 1984, it was announced that Bill Lenoir would be leaving NASA that September to take a position with Booz, Allen and Hamilton, an aerospace management and consulting firm based in Bethesda, Maryland. Lenoir had been told he would fly on the joint US/West German Spacelab D1 mission, so he had begun German lessons (along with Bonnie Dunbar) and was looking at early issues for the flight and payload. However, when a partial crew announcement was made on 14 February 1984, Lenoir's name was not among them. It was his own choice not to fly the mission, having decided that a second mission was not where he wanted to take his career. He was concerned about the length of time such a dedicated science mission would require him to be away from home. As this was a joint mission with the West Germans, most of this time would have to be spent in Europe, and would have required the crew to be in West Germany for more than three weeks at a time. With his family in the United States, and with all the work required at NASA installations as well, this did not seem to Lenoir to be the right thing to do. At the same time he realised that, at the age of forty-four, he had to decide between a new career or remaining where he was. While he was debating whether to fly or leave, he was head-hunted for a position with a major consulting firm, which was looking for an astronaut to lead a new consulting group for the space station. "This was one of those one-time opportunities that was perfectly suited for me," Lenoir recalled.[3]

After the loss of Challenger, former Shuttle astronaut Dick Truly returned to NASA from the US Navy in February 1986 to become the Associate Administrator for Spaceflight (essentially to head up the Shuttle programme). His primary job was to get the Shuttle flying again. In January 1989, Jim Fletcher retired as NASA Administrator and was succeeded by Truly. A good friend of Bill Lenoir, Truly eventually persuaded the former scientist-astronaut to return to NASA in July 1989, even though it was not financially beneficial to Lenoir. He would take on Truly's former role as Associate Administrator for Spaceflight, responsible for all US government civilian space launch activities. The role later expanded to take in the two offices for Space Shuttle and Space Station. Lenoir returned on the understanding that Truly would fire him, "If I screwed up or if it was politically needed to save the agency. I wanted a set of objectives – three seemed like a nice number – which I could work towards. I also told him (Truly) that I was willing to wave the flag and work into my savings (Lenoir worked on a negative cash flow) but would not go into debt." For the next three years, Lenoir was in charge of all of NASA's manned missions, successfully overseeing flights from STS-28 in August 1989 through to STS-49 in May 1992 (eighteen missions). In February 1992, Truly was essentially fired by President George Bush after being reportedly at odds with the White House over his suggestions regarding a possible reorganisation of the space agency. For this and other reasons, Lenoir decided to leave NASA in May 1992 and move on.[3]

Lenoir rejoined Booz, Allen and Hamilton in June 1992, spending the next eight years as Vice President of the Applied Systems Division. He retired from the company on 1 April 2000 but continued to work as a consultant for them in a part-time capacity over the next three years, providing continuity to the business area that he led for the company. He fully retired in 2003 but became busier than ever, though he was no longer being paid for his activities![4]

Joe Allen and the ISF

In the middle of June 1985, Joe Allen announced his intention to resign from NASA the following month to assume a position as Executive Vice President for Space Industries Incorporated (SII), a Houston-based firm that was pursuing "ventures in the utilisation and commercial use of space". The founder and president of Space Industries was Dr. Maxime Faget the former Director of Engineering and Development at NASA JSC and a leading engineer, who was instrumental in the design and development of all US manned spacecraft for NASA from Mercury to the Space Shuttle. Caldwell Johnson, a leading engineer and designer of early US spacecraft and chief designer of the US docking system for the Apollo-Soyuz Test Project, was also involved in the company. Allen first became acquainted with Max Faget during his participation in the "Outlook for Space" study in 1974, and quickly came to admire him. It was the study group's recommendation to develop an "outpost in space" that intrigued Allen.

A decade later, in 1984, SII signed a memorandum of understanding with NASA to design and construct the world's first man-tended space platform, and

Former scientist-astronaut Joe Allen in 2002. [Credit Francis French/Reuben H. Fleet Science Center, San Diego.]

was negotiating agreements with NASA to launch the platform in 1989 when Allen joined the company. While still an astronaut, Allen had been working with a small group of engineers on the idea that an outpost in space (not the large space station that NASA had envisaged) would be a cheaper and faster route to achieving regular access to the microgravity environment. These engineers later formed SII and their idea of developing the Industrial Space Facility (ISF) – a man-tended, zero-gravity laboratory and manufacturing plant – intrigued Allen as much as his years as an astronaut had: "I thought it was a long shot at best, [but] the boldness of this project appealed to me no end."[5]

Designed to be launched inside the payload bay of the Shuttle, ISF was essentially a mini space station. It would mostly fly unmanned, but would have regular servicing visits by Shuttle crews every six months or so. The facility was planned to be expandable using Shuttle-compatible docking ports to potentially cluster up to six modules together. It also had the option to fit a life support system, allowing more permanent occupation. Measuring 10.67 m long by 4.42 m in diameter, each module could house up to 11,000 kg of commercial and industrial manufacturing experiments and facilities in a suite of thirty-one payload racks (based on those used on the NASA space station for compatibility). Power came from two solar arrays of 140 m area, which would generate 20 kW of electrical power. A 30.5-m gravity-gradient boom would provide attitude control with minimum disturbance. With up to four docking ports for attaching external facilities, a Shuttle could remain linked to the station for a week to ten days while the experiments and research facilities were upgraded, serviced, exchanged or removed.[6]

In the early stages of Project Space Station, and with the realisation that the Shuttle did, in fact, work as designed, the 1979–85 period saw a boom in private enterprise interest in space exploitation and commercialisation. Indeed, many former NASA employees had moved to the private sector to join both new and established aerospace and enterprise companies trying to get a foothold in what was expected to be a space business boom. Even the space station itself was envisaged as a key step in developing a free enterprise system, both by sub-contracting and as the end user. When Space Industries Inc. signed the memorandum with NASA, the company was widely expected to be a leader in this new use of space, utilising the skills and experiences of its employees and its contacts within NASA to develop hardware that was already compatible with the Shuttle, and less complicated and demanding on resources than the Space Station. In 1985, a Center for Space Policy report indicated that space pharmaceutical manufacturing alone could generate an estimated $20 billion by the year 2000.

However, the loss of Challenger in January 1986 and the cancellation of most of the planned commercial payloads on the Shuttle (as well as a new launch pricing policy) seriously hindered any foreseeable attempts to use the Shuttle as a viable commercial launch system. At the same time, when it was found that ground-based bio-engineering production systems were able to produce similar highly purified biological samples at far less cost, the idea of specialist space-based industrial facilities became a much less attractive proposition than it had been just two years before.

As a result, SII amended the profile of the ISF to make it more accessible to the

In April 2005, Joe Allen (centre) was inducted into the KSC Astronaut Hall of Fame, along with Gordon Fullerton (left) and Bruce McCandless (right).

Shuttle, reducing its operational orbit to 300 km and developing a transfer system to utilise excess water from the Shuttle fuel cells to help reduce logistics supply costs. The plan was to utilise Shuttle flights to deploy two ISF units but, despite requesting a 1990 launch date, NASA re-manifested the facilities from the original 1989 schedule to 1992 at the earliest. Though still behind the project as a complementary system to Shuttle and Space Station, it soon became clear that NASA was only providing minimal launch and service support until its own space station was up and running. Clearly, the space agency did not want SII reaping any rewards from space industrial sectors until they had a facility to support such work themselves, in part justifying the need for the space station.

SII subsequently changed its plans during the late 1980s, marketing the ISF not as a commercial production facility, but as a test bed for station hardware, systems and operations. By this point, however, Space Station Freedom (as it was now known) had encountered serious cost overruns and delays and was under threat of cancellation unless something was done quickly. Congress told NASA to look seriously at the ISF idea to gain early access to space facility operations, but the reduced NASA budget resulted in considerable friction between the space agency and SII for several years. SII argued that NASA should lease the ISF until their own space station was ready (which was now estimated to be in 1996 or 1997). NASA countered by saying that such a lease would strip resources and funds from the development of a station that was already considerably over budget and years overdue. The arguments continued into 1989, when National Research Council and National Academy of Public

Administration reports were released that were less than enthusiastic about supporting ISF. Space Industries were not happy, pointing out that these two panels were comprised mainly of "fundamental scientists", and not "commercial entrepreneurs".

The re-direction of the Shuttle programme in the early 1990s, together with the involvement of the Russian Space Agency, saw the majority of NASA's resources now focused on the International Space Station. This effectively cancelled further development of the SII facility, while the growth of ISS also signalled the death of the highly successful Shuttle-borne Spacelab modules. At the time of the ISF saga, SpaceHab Inc. was proposing its own mid-deck augmentation module, which has since flown successfully on several Shuttle and space station missions. In March 1989, SII Vice President Joe Allen stated that the overriding problem with the ISF/NASA situation was "too many cooks in the commercial space policy stew [in which] the recipe never quite gets finished."

Despite an attempt to market ISF as an additional, cheap, and launch ready facility (in two years from the go-ahead), it quietly slipped from the agenda. The development of Shuttle-Mir, the demise of Spacelab missions, and a revised manifest loaded with ISS assembly flights, all contributed to ISF being postponed indefinitely. SII redirected its efforts towards smaller microgravity experiments on sounding rockets, smaller Shuttle payloads, and offering its unique ground-based facilities and expertise to other aerospace companies.

Joe Allen moved to Washington in 1991 as the Chief Executive Officer for Space Service International. In 1993, when SII took over Calspan-SRL, Allen joined the new company, based in Washington, serving as a Director (1993), then President and Chief Executive Officer (1994–7) and finally as Chairman of the Board (1995–7). Calspan, conceived in 1940, had over fifty years experience in the fields of technology and science, crash and flight data research, system engineering, wind tunnels and transportation science. In 1997, the company merged with Veda International Inc., and the following year became the Veridian Corporation. Allen became Chairman of the Board of the new company between 1997 and 2003. In 2003, Veridian linked up with General Dynamics (GD), operating as part of the Advanced Information Systems division of GD. Joe Allen retired from the company in 2004 but has remained actively involved with the Challenger Science Center Foundation since its inception in 1986.

CB points of contact for Flight Data File – November 1985

On 19 November 1985, a CB memo from Story Musgrave to all current astronauts listed new points of contact for specific issues regarding the Flight Data File (FDF), with effect from STS 61-B. Owen Garriott, now working on the EOM mission that would include a small laboratory module, took responsibility for issues relating to the Spacelab modules with regard to activation and deactivation, in-flight maintenance and operation checklists. Parker, currently in training for the Astro pallet-only mission, took these roles for the Spacelab Pallet issues as well as for Data Processing Systems/software (DPS) and the Command Data Management System (CDMS).

By far the most assignments (fourteen) were given to Musgrave, who was also expecting to be named to his third mission. Some of these were sole assignments, others were shared (see below).

AFTER CHALLENGER

At the start of 1986, only Tony England, Owen Garriott, Karl Henize, Bob Parker and Bill Thornton remained as active astronauts from the seventeen selected in the two scientist-astronaut groups. All were now flight experienced and though they hoped for another flight, only two (Garriott and Parker) were actually in training. Less than a month into the new year Challenger was lost, and though they did not wish to be seen to be deserting the astronaut programme in the immediate aftermath of such a tragedy, thoughts of new goals and careers that may have been in the back of their minds took on more immediate significance after the accident.

Tony England – Losing Sunlab and back to teaching

In October 1985, Tony England had served as Capcom during the STS 51-J classified DoD mission (the maiden flight of Atlantis) and the STS 61-A Spacelab D1 German mission. When he came back to NASA in 1979, he had already decided that once he had achieved his first flight, he would probably return to a teaching position somewhere. But the delays meant it would be six years before England could seriously look at this option.

Anticipating problems in qualifying and using the Instrument Pointing System (IPS) and software on Spacelab 2, NASA had planned follow-on missions for the system to gather more data once it had been qualified on Spacelab 2. These missions became Spacelab pallet missions that were designated Sunlab, but the Challenger accident delayed and finally cancelled these plans. England felt it was not right to leave NASA prior to the Shuttle returning to operational status, but by then any chance of him flying a second mission had gone. It is likely that he would have flown on one of the Sunlab missions in 1987, had the programme continued as planned.

Prior to the loss of Challenger, Sunlab 1 was manifested for launch as STS 71-O in September 1987, a seven-day mission using Columbia. England may well have taken a leading role (payload commander) on that flight. Sunlab 2 would have flown in 1989 and Sunlab 3 probably in 1990/1991.

England moved to the Space Station office for a year, "just to be useful," and also took the position of visiting professor at Rice University for a year. In June 1988, with the Shuttle about to resume flights with STS-26 in September, and in preparation for a new academic year, England announced his resignation from NASA for the second time to assume the position of Professor in Electrical Engineering and Computer Science at the University of Michigan, Ann Arbor, beginning in October 1988. He remains there to this date, teaching and conducting research into environmental remote sensing.[7]

Karl Henize – new mountains to climb

In flying STS 51-F, Karl Henize finally realised his dream of making it into space. He also gained the "honour" of becoming the oldest man to get there (until Vance Brand flew in 1990 aged fifty-nine, Story Musgrave in 1996 aged sixty-one, and then John Glenn in 1998 aged seventy-seven). At fifty-eight years of age, Henize would become (and still remains) the oldest rookie to fly into space. Part of his "award" for this achievement was "The Slayton Trophy", an underarm crutch that was proudly displayed in the Henize household for some years.

Henize was always frustrated that the loss of the Apollo–Saturn hardware deprived America of its heavy lift capability for launching large facilities and space station modules. He was also enthusiastic about re-flying astronomy payloads using the same hardware. "Somehow, we seem to operate so inefficiently. We built these three good telescopes (Spacelab 2) to do good astronomy and once you have spent millions of dollars developing them, you should use them on several missions. But NASA just never realises that. The original plan was to re-fly them, but now we have flown it once, NASA feels that there are other things of a higher priority. But re-flying these experiments is relatively cheap and inexpensive and should bring in lots of good science. I get rather exasperated sometimes the way we (NASA) build good systems, good scientific instruments, then leave them behind and develop new things."[8]

In April 1986, with no prospect of a return to orbit, Henize transferred from the Astronaut Office to the Space Science Branch at JSC as a senior scientist, working on space debris. This was a growing problem whose solution, to Henize, was to simply stop littering up space. There is no way, realistically, to go into orbit and clear everything up, and while natural forces will gradually drag debris back into the atmosphere, this can take anything from a year to several centuries, depending on the apogee. The fear (in 1986) was not related to possible damage to the Space Shuttle or even the Space Station, but that the number of tracked items was continually rising. In the early years, a number of spent rocket bodies with propellant on board had a tendency to explode, altering the nature of the risk from a single body to the potentially greater hazard of several hundreds or thousands of small, fast moving pieces. Both the Americans and the Soviets had contributed to this problem and fifty per cent of the items being tracked in the mid-1980s were fragments from exploded rocket parts or military satellite tests.

NASA led the way in ensuring that rocket stages did not explode so frequently once abandoned in orbit by venting all propellants and making sure that the tanks were not adjacent to each other. This practice was soon followed by other space-faring nations in Europe and Japan. The next stage is simply not to leave spent rocket stages in space, by de-orbiting them – something still not accomplished some twenty years after Henize was working on the problem. According to Henize, "Anyone who flies in space ought to be concerned about limiting the amount of junk that's up there." He was working on a computer tracking system to address the problem and indicated that there were plans to fly a telescope on the Shuttle that would get close up to space debris, to determine exactly how much of the small-diameter items that could not be tracked using ground-based systems was up there.

Karl Henize proudly displays a model of the Spacelab 2 payload at his home in Nassau Bay in Texas during August 1988. [Credit Astro Info Service.]

With regard to astronomy on ISS, Henize pointed out that the Shuttle debris telescope could be adapted to fly on ISS, but that "serious astronomers" were more committed to flying astronomical instruments on unmanned spacecraft. This was mainly because the size of telescope required to keep up with the state-of-the-art in astronomy would be seriously compromised by the contaminations or vibrations of a crewed vehicle. However, the idea of man-tended activities such as the Hubble servicing missions appealed to Henize.

Henize had developed his skills as a satellite tracker in the mid-1950s, identifying likely sites around the world that could be used to track satellites well before any had actually been launched. This certainly helped in his new role. One day, he read an article in *Popular Mechanics* magazine which stated that the USAF was still tracking the glove seen floating out of the Gemini 4 capsule during the EVA by Ed White in June 1965. Henize contacted the editor, pointing out that such a small object would have re-entered years before. The Air Force investigated the claim, which ironically was referred back to Henize himself at JSC, as he was deemed to be the best authority in such matters! A retraction was printed in the magazine some time later.[9]

Henize was an avid mountaineer and had climbed Mount Rainier in Washington State in 1991. In early 1993, he was invited to join a British expedition to climb Mount Everest. In mid-September, Henize travelled to Tibet to join the expedition, which planned to ascend the north face of Earth's tallest mountain. During his second day after reaching the advanced base camp (6,700 m or 22,000 feet), Henize developed signs of extreme high altitude sickness. Despite the valiant efforts of the other members of the team to save him, they could not get him down from the mountain in time. He died of High Altitude Pulmonary Edema (HAPE) at 5,500 m (18,000 feet) at 1:00 a.m. on 5 October, just twelve days short of his sixty-seventh birthday. He was buried nearby above the Changste Glacier. It was three-and-a-half days before news of the tragedy arrived back at the Henize home. On 16 October 1993, a memorial was held in his honour at JSC. Though he never flew in space again after STS 51-F, he rests at perhaps the highest cemetery on Earth, forever close to his beloved space frontier.

Owen Garriott – EOM and SPEDO

Owen Garriott was assigned to the first Earth Observation Mission (EOM-1) and had a good chance of a re-flight on EOM-2 to utilise his experience and ease the training flow (a policy that was followed when the EOM missions became ATLAS in the early 1990s). When he was assigned to the mission the flight was designated STS 51-H, but numerous changes in the launch sequence and delays to several other flights caused the mission to slip into 1986 and be rescheduled as STS 61-K. The launch was set for September 1986, beginning a seven-day flight that would carry a payload merged with that of EOM-2. EOM-3 was manifested for STS 81-F in February 1988 and eight more flights were planned for the series, ending with mission EOM-11.

Interviewed in 2006, Garriott recalled his assignment to the EOM crew and the combination of the EOM-1 and 2 missions, but stated that he never did any mission training. In fact, he remained in his position as Project Scientist for Space Station until he resigned from NASA and recalls completing only "negligible training" for the EOM mission assignment.[10]

With the loss of Challenger, it was clear that Garriott's mission would probably be delayed by about five years. Already aged fifty-six in 1986, the delay meant he would have been about sixty-one by the time EOM was expected to fly in 1991. "I was getting to the point where I was a little doubtful that I would have enough time to really finish with that crew ... [and] there would be no opportunities for another career beyond that. So it was essentially a career decision from my standpoint [to leave NASA]. It was going to be too long before I had the chance to fly again. If I'd had a chance to fly in late 1986 or 87, then I would have stayed for another opportunity. But when that was clearly going to be delayed five years, I just decided that from a personal perspective it'd be better to leave that to somebody else."[11]

The EOM-1/2 mission was finally flown as ATLAS-1 in 1992, the first of what was intended to be a series of ten such missions flown annually over the eleven-year solar cycle. But a number of Shuttle missions were required to support Shuttle-Mir and the first phase of the International Space Station programme, so the series was cut to just three. ATLAS-1 (STS-45) flew in March 1992, followed by ATLAS-2 (STS-56) in April 1993 and ATLAS-3 (STS-66) in November 1994.

In June 1986, Garriott resigned from NASA to become Vice President of Space

Programs at Teledyne Brown Engineering in Huntsville, Alabama. The company was involved with payload development and processing on Spacelab missions for NASA, and Garriott remained in the post until 1993. While at Teledyne Brown, he worked with Senior Systems Engineer Frank Echols, a veteran of thirty years of spacecraft design and instrumentation, on a proposal to extend the orbital life of the Shuttle beyond thirty days by using photovoltaic arrays to provide electrical power for independent flight. The concept was known as the SPEDO (Solar Powered Extended Duration Orbiter) configuration.[12]

SPEDO's photovoltaic array would provide an average of 12 kW of power which, combined with a nine-tank cryogenic Extended Duration Orbiter pallet kit and four double-panel arrays of 6 m × 12 m, could theoretically support an independent orbiter flight of up to eleven weeks (84.9 days). Flown in conjunction with a Spacelab long module and a SpaceHab augmentation module, this configuration would have been capable of supporting a Shuttle docked to ISS during the early construction phase. This man-tended configuration could have provided a platform for early science studies, prior to the deployment of the larger and more complicated ISS solar truss system. The independent flight capability of the EDO series of missions (planned by NASA from 1992 – the first flew as STS-50) offered further extension beyond the maximum duration of seventeen days, using off-the-shelf hardware and systems. The array was of the type tested on STS 41-D in 1984, the Spacelab module was flight-proven, and the EDO kit and SpaceHab modules were already manifested. It was seen as a way to extend the life of the Shuttle and conduct useful orbital research prior to permanent residency aboard ISS and the delivery of its laboratory modules and solar arrays. Garriott and Echols forecast that a SPEDO configuration could be ready for launch in three years from approval.[13]

Like SII's Industrial Space Facility, however, the idea of a solar-powered orbiter never really gained much ground within NASA. According to Garriott, the SPEDO configuration was "too large for NASA to consider having a solar-powered flight, and also they were biased against it because it would compete with ISS, which was also a solar-powered spacecraft. So SPEDO was never really given serious consideration, as much as anything because of its competitive nature with ISS."[14]

In May 1993, Garriott left Teledyne Brown to become co-founder and president of Immutherapeutics Inc., also in Huntsville. The company initiated Federal Drug Agency (FDA) approved human trials for a tumour therapy. Garriott remained there until September 2000, when he became the interim director of the newly-created National Space Science & Technology Center (NSSTC) in Huntsville, and was responsible for managing the centre until July 2001 when a permanent director was found.

Since leaving NASA, Garriott has also devoted his time to a number of charitable causes, including co-founding the Enid (Oklahoma) Arts and Science Foundation in 1992, and has participated in research activities on new microbes. He has experienced other extreme environments, such as very alkaline lakes and deep sea vents, and has participated in undersea dives aboard Russian Mir submersibles to a depth of 2,300 metres, near the Azores in the Atlantic Ocean. He has also completed three trips to Antarctica.

RETURN-TO-FLIGHT AND A RETURN TO SPACE

On 29 September 1988, the Shuttle programme finally returned to space, some twenty months after the loss of Challenger. STS-26 would deploy the third Tracking and Data Relay Satellite (TDRS-C) and the five-person Discovery crew would complete a range of mid-deck and payload bay experiments. The mission lasted just over four days and successfully qualified the changes made to the programme since the tragic short flight of STS 51-L. By this point, only three scientist-astronauts remained in active service; Story Musgrave, Bob Parker and Bill Thornton. Over the next decade, as the Shuttle programme changed and moved towards supporting the construction of the space station, so the prospects of further flights diminished for the surviving members of the pioneering astronaut selections of the pre-Shuttle era. Indeed, in addition to the three former scientist-astronauts, only Vance Brand and Bruce McCandless from the 1966 Group 5 pilot selection remained in training for a future mission. All the other astronauts selected between 1959 and 1969 were either in management roles, had retired or were about to retire from NASA, or had passed away.

Bill Thornton

Back in 1988, Bill Thornton believed that crew assignments were based on need rather than politics and, fortunately, so did those assigned to selecting astronauts for Shuttle missions. Thornton felt that he flew his two space missions exactly when he should have flown and, after the loss of Challenger, was philosophical about the need to "return the Shuttle to routine flight operations and solving all the engineering problems" before flying himself, even if this meant his chances of a third mission would be slimmer. "When we get a bit of breathing space, then yes, once again I could make some real contributions towards long-duration flight ... I have no idea if and when this might occur and I am certainly not getting any younger, but as long as I can perform, it is a possibility."[15]

His Astronaut Office (CB) assignment at this point was on long-duration space flight issues, and over the next few years he worked on exercise and Space Adaptation Syndrome countermeasure equipment for the extended duration Shuttle programme and space station. According to his 1994 official NASA biography, he "continued his work in space medicine while awaiting his next flight opportunity. Thornton has worked on problems relating to extending mission durations in the Space Shuttle, the space station, and in space exploration, and designed the necessary exercise and other hardware to support such missions. He continued his analysis and the publication of results from studies of neurological adaptation, and the study of neuromuscular inhibition following flight, osteoporosis in space and on Earth, and post-flight orthostatics. He has completed designs for exercise and other countermeasure equipment for the Extended Duration Orbiter (EDO), and for Space Station Freedom, including improved treadmills, rowing machines, isotonic exercise devices, and a bicycle. Much of this is currently scheduled for flight."

What was not included in this account was mention of his work in designing an affordable (and more importantly, reliable) waste management system for long-

Return-to-flight and a return to space 445

Bill Thornton poses for the media at a school in the West Midlands, England, with the first British astronaut, Helen Sharman, during a visit to the UK in 1991. This was just a few months after Sharman had spent a week on Mir. [Credit Astro Info Service.]

duration Shuttle flights. The Shuttle system, while much improved over the Apollo "toilet-bag" and Skylab toilet procedures, was still temperamental and prone to breaking down. For a long-duration mission on Shuttle in which five to seven members of a mixed crew could be confined for over two weeks, having a reliable toilet system helps not only with hygiene but also with the well-being of the crew. On a space station, months of orbital life could be either a great experience or a nightmare depending on whether the toilet system functioned properly. Thornton's work – certainly appreciated by the other astronauts – was every bit as important as more "attractive" areas of extended duration space flight.

Thornton also helped to design and develop a new *type* of waste management system.[13] Designed under a joint Detailed Test Objective, the Improved Waste

Collection System (IWCS) was developed as a cooperative effort between ILC Space Systems, SSPSG Systems Division and Whitmore Enterprises Inc. It was designed and developed between 1983 and 1986 and was certified ready for flight in 1986, but was put on hold due to the Challenger accident. In 1989, it was manifested as DTO-0329 and assigned to STS-35 (Astro-1). The unit was flown on that mission in December 1990, installed next to the Shuttle WCS on the mid-deck. It was tested during Flight Days 7, 8 and 9 and performed flawlessly, receiving favourable comments from grateful crew members.

The main difference between the IWCS and the Shuttle commode was that instead of a slinger and vacuum/airflow suction device, it had a compactor device – hand-cranked by the user – that used pistons to move the waste to the end of a chamber. The piston was wiped clean by a highly absorbent wiper pad (added prior to use) which also wiped the chamber clean during operation and absorbed all waste moisture. A series of filters cleaned any odours from the chamber. Despite some minor difficulties, the unit was given a good report by the crew. Although this one flight proved the system, it was not developed further, much to Thornton's frustration, with NASA deciding to proceed with a more expensive commercially developed unit. Though developed at a fraction of the cost, Thornton's system never flew again. NASA once again wasted money and time flying a unit that was more corporate gloss than engineering practicality.

Thornton's frustrations over the increasing bureaucracy of the space programme and space industry while trying to develop cost-effective, workable and, more importantly, useful items for space flight continued. Not having the opportunity to fly a third mission was a great loss both to the astronaut and the programme. He was selected initially to participate in long-duration Apollo Applications Program missions, but never got the opportunity to fly such a mission on Skylab or any long-duration Shuttle mission. He was also never assigned to Spacelab life science or to any of the EDO Shuttle missions. On 27 May 1994, NASA announced Thornton's resignation, effective at the end of the month, but did not reveal what his next position would be. At least he would have more time to devote to writing about his work in the space programme over the previous thirty years. As Thornton himself indicated in the official press release. "Due to my work, I haven't really had the opportunity or the time to do any writing about my technical work, other than a few reports, and none at all about other matters."

In 1995, Thornton became a clinical professor of medicine (cardiology) at the University of Texas Medical Branch in Galveston. After four years on the teaching staff, he developed a computer-based self-teaching system that provided hands-on training in seeing, hearing, and feeling a patient's symptoms. This was subsequently developed for more extensive and broader use. When Thornton began work at UTMB, there was minimal equipment for the students to use, and the extra work involved in developing support material for students took up much of his time. In addition, in the late 1990s the Thorntons moved from their residence in Friendswood near JSC to a new property on the outskirts of San Antonio. This meant commuting about 500 km each week.

On 11 October 2002, Thornton's system was dedicated in the Lucile Matthews

In 2004, Bill Thornton was the proud recipient of the North Carolina 2003 Award for Science. [Credit North Carolina Department of Cultural Resources.]

Wallace Memorial Thornton Self-Teaching Laboratory at UTMB. The Physical Diagnosis Self-Teaching System (PDSTS) integrates three of the traditional elements of cardiac examination – inspection (observing carefully), palpitation (feeling) and auscultation (listening) into a complete interactive teaching/learning experience. Thornton explained: "This lab will be an essential resource at UTMB for the establishment of basic physical diagnostic training consistent with the realities of twenty-first century medical education, in which practicing on patients is replaced by high fidelity simulations."[16] The system was greatly appreciated by faculty members and students alike, and the School of Medicine Class of 2000 recorded a personal message of thanks at the end of their graduation year: "Your excellent and innovative efforts in teaching cardiovascular physical diagnosis will make us and future UTMB students better physicians, no matter what our speciality choices."[17]

Thornton left UTMB in 2003 with a desire to complete interrupted work on his publications in space medicine and to devote time to the development of a new clinical system for the University of North Carolina. He was also awarded the 2003 North Carolina Award for Science for his life's work dedicated to the betterment of medicine – on Earth and in space.

Thornton was always hopeful of another space flight, but several things happened that prevented him from getting that coveted third trip. In 2006, Thornton reflected on an observation that Karol Bobko had made during the 1972 SMEAT ground test for Skylab. Bobko said, "Bill, they may forgive you for telling them they are wrong, but they'll never forgive you for proving it." When Thornton came back after STS 51-B and expressed his strong concerns that the animal cages would not work as designed and that some of his space motion sickness studies were wasted, he had a strong feeling that he had sealed his fate with regard to a third mission.

AN ASTRONOMER FOR ASTRO

Bob Parker had been named to the crew of Astro-1 as MS 1 in June 1984 and began training for the mission from that time. He was named with fellow astronomer Jeff Hoffman (MS 3) from the 1978 group and 1980 astronaut Dave Leestma who would serve as MS 2/FE. Their announcement to the flight allowed them to work with Principal Investigators to begin preliminary discussions over the science objectives of the mission in line with CB guidelines and STS safety rules. The following January, Jon McBride and Dick Richards were named as commander and pilot respectively for the mission and began working with Leestma to develop the flight operational requirements for the mission. They would also coordinate with Parker and Hoffman over the orbital manoeuvring requirements for the science objectives. In October 1985, the payload specialist assignments were announced, with Ron Parise and Sam Durrance in the primary positions and Ken Nordsieck as back-up. As each Astro PS was scheduled to make two flights, Nordsieck was expected to fly on Astro 2 and 3 with either Parise or Durrance.

Forty days from Halley's Comet

On 28 January 1986, the Astro-1 crew was in Building 5 at JSC practising launch and entry profiles in the simulators. It was about forty days before their scheduled launch on STS 61-E and their planned observations of Halley's Comet from above the atmosphere. Taking a break in their training, the crew witnessed the horrifying events unfolding during the launch of Challenger via a small TV and soon realised that they were not going to fly in forty days, nor any time soon.

Parker recalled the immediate post-Challenger activities for the Astro-1 mission and crew during his Oral History interview in 2002. Looking back over sixteen years, he thought it amazing how the launch rate was scheduled in 1986. The Astro crew "had to go to see Halley's Comet" in March to allow for a re-flight in November. This put pressure on the previous mission (STS 51-L) to launch before Astro, which was compounded by the fact that the mission before that (STS 61-C) had slipped through Christmas 1985 and had experienced difficulties both in getting off the ground and in returning. There were also two important launch windows in May to launch the Ulysses and Galileo probes, and it seemed to Parker at the time that making the launch window was the overriding priority.

The crew had expected to launch in March to observe "spring-time stellar targets" before recycling to fly the second mission in November (with the same crew apart from one payload specialist), almost a production line of space flights. But in January, all plans came to a grinding halt and the controllers, trainers and crew found themselves with a possible two-year delay and no confirmed mission. They entered what was termed "generic" training to keep up their proficiency, but without the intensity of preparing for an upcoming launch. It was a much lighter schedule and some of the team, including Parker, went off to do other things.

Temporary duty in Washington

In February 1988, Parker was assigned a temporary duty at the Office of Space Flight at NASA HQ in Washington DC. He was to serve as Director of the Space Flight/Space Station Integration Office, which had been established in 1987 to develop the integration of the fledgling space station programme into the Space Transportation System (Shuttle). He replaced the former director, E. William Land, who had retired in January 1988. "In those days, Space Station was one major code [directorate] and Manned Space Flight was another," Parker recalled, "so I was the Manned Space Flight interface with someone else who was in Space Station. I did that for a year." In early June 1989, he returned to JSC to resume training for STS-35 (then planned for a launch in the spring of 1990).[18]

In the 1986 manifest, Astro-1 was the next mission due to fly after STS 51-L. When the new manifest was issued, the priority was to reconfirm the Shuttle design and verify the changes incorporated into the programme after the loss of the Challenger. In addition, several classified DoD missions had to be launched as soon as possible, two much-delayed Tracking and Data Relay Satellite payloads had to be deployed (as did the Magellan, Galileo and Ulysses space probes and the Hubble Telescope) and the Long-Duration Exposure Facility that had been deployed in 1984 and should have been retrieved in 1985 had to be recovered. This meant that the "science" missions were delayed and Astro-1 gradually slipped into 1990, some eighteen months after the resumption of flights.

On 30 November 1988, NASA announced that the Astro-1 crew and mission – unfortunately no longer able to observe Halley's Comet – would fly as STS-35 under the command of Vance Brand (he had replaced McBride, who had retired from NASA), with pilot Guy Gardner replacing Dick Richards (who had been reassigned to another crew) and MS Mike Lounge replacing Dave Leestma (who was in training for STS-28).

Astro-1 flies – eventually

In the summer of 1990, everything seemed to conspire to prevent the launch of the Astro-1 mission. Originally set for a 16 May launch, a faulty orbiter coolant system

450 Ending of Eras

Robert Parker briefs payload specialists Parise and Durrance during training for STS 61-E Astro-1 in October 1985. They are in the 1-G mock-up of the Shuttle middeck at JSC.

valve caused a delay until 30 May. Then the launch was scrubbed due to a major hydrogen leak in the 43-cm orbiter/ET disconnect umbilical during tanking operations and a minor leak in one of the Mobile Launch Platform's two tail service masts. The crew was not on board during this incident, but they were awakened in the crew quarters at the Cape and promptly told to go back to sleep because they were not going to launch that day.

After the stack had been returned to the VAB to replace the umbilical, the launch was reset for September, only for another hydrogen leak to be revealed during tanking in Columbia's aft compartment. More pumps and seals in the SSME system had to be replaced, and the launch was rescheduled yet again, this time to 18 September. A third hydrogen leak was recorded during the next tanking and the delay forced NASA to move the vehicle to Pad B to make way for STS-38. It was then returned to the VAB again due to tropical storm Klaus.

STS-35 was finally launched on 2 December 1990, nearly five years after it was originally scheduled to fly. It carried the Astro-1 payload of telescopes, mounted on a pallet in the payload bay, which would be operated on a two-shift system. The crew experienced further difficulties with the computer consoles, due to the failure of the visual display units which aimed the telescope mounts and the instruments. Within twenty-four hours, the ground team came up with procedures to guide the telescopes and allow the collection of data from the instruments. On top of all this, there was also a clogged drain which prevented waste water dumps during the mission. Still, the crew

Parker (left) and Durrance conduct a simulation of Astro-1 payload activities during training for STS-35 (formerly STS 61-E).

were able to present a "Classroom in Space", in which the astronomer-astronauts (Parker and Hoffman – who had taught in class before coming to NASA) put together a small field lesson from space demonstrating what astronomy is all about. Parker really enjoyed this, especially the interaction with several children in a school near Goddard and a second near the Marshall Space Flight Center. After a flight of 8 days 23 hours 5 minutes 8 seconds and 144 orbits, STS-35 at least managed to land without incident. The mission had been scheduled for a further twenty-four hours in flight but, in keeping with the pattern of the rest of the mission, impending bad weather at Edwards AFB in California had caused the crew to return a day earlier than planned.

Parker's role on Astro-1

Assigned as MS 3 for the mission and a member of the Red Team (with Gardner and Ron Parise), Parker worked a twelve-hour shift during the mission. During the ascent phase of the flight, he rode in Seat 5 on the mid-deck, exchanging places with Jeff Hoffman for re-entry and returning in Seat 3 on the flight deck. While each of the three

The STS-35 crew during emergency egress training at KSC in Florida during 1990. Left to right: Pilot Guy Gardner, PS Ron Parise, Sam Durrance (hidden from view at front of the M-113 rescue vehicle) Commander Vance Brand, MS Bob Parker, MS Jeff Hoffman and MS Mike Lounge.

major telescopes aboard had its own payload specialist (only two flew each mission), Parker and Hoffman surmised that they might get three flights out of the series by staying with the payloads, utilising their experiences with the science team and the hardware. But with the loss of Challenger, extended delays to the second flight and the eventual cancellation of Astro-3 (in part due to the Shuttle–Mir programme), this would not occur.

The payload specialists were more involved in the science than either Parker or Hoffman and though both were astronomers who "understood the science" they were not involved in its planning due to the division of labour on the mission. Their primary role was to operate the Instrument Pointing System (as with Henize and England on Spacelab 2 five years before) and to aid in the pointing control of the telescopes. When the computer control was lost, Parker's former skills at guiding telescopes and "sitting there for hours guiding a star on a crosshair" proved invaluable. This was something the crew had insisted upon during the development of the Astro package in case there was a computer failure that prevented remote pointing. Direct control of the tele-

STS-35 astronaut Robert Parker uses fire-fighting equipment at the JSC Fire Pit as part of the emergency egress exercises all crews perform during Shuttle mission training.

scopes allowed them to observe intended targets and secure a host of scientific data that would have been unobtainable had the crew been unable to control the telescope from the aft flight deck of the Shuttle.

According to one post-flight release, the Hopkins UV instrument conducted over 100 observations of hot-stars, galactic nuclei and quasars; the UV Imaging Telescope collected over 900 images of supernovae, planetary nebulae, galaxies and clusters of galaxies; the Wisconsin UV Photo-Polar Meter experiment obtained data on over seventy objects including galactic clusters and supernova remnants; and the Broad Band X-Ray Telescope collected data on over seventy-five objects, including active galactic nuclei, quasars and accretion disks. In all, 390 observations of 135 space objects were carried out by the crew, making Astro-1 a "highly productive mission" and providing the impetus to support the flight of Astro-2, which finally occurred in March 1995.[19] According to Parker's 2002 Oral History, the astronomer-astronaut believed, "We observed maybe a third of what we had intended to. Everybody put a good face on it, but it was a far cry from what it was supposed to be."[18] Without the presence of the human crew on board, the achievements would have been far less.

Back to Washington

In January 1991, once his assignments on STS-35 had been completed, Parker left the Astronaut Office and moved back to Washington to become the Director of the

Parker uses a manual pointing controller for the Astro-1 mission's Instrument Pointing System from the aft flight deck of Columbia during the STS-35 mission.

Division of Policy and Plans for the Office of Space Flight at NASA Headquarters. From January 1992 to November 1993, he was Director of the Spacelab and Operations Program and from December 1993 to August 1997, he was Manager of the Space Operations Utilization Program.

In Washington, his concerns became policies and decision-making about future programme directions, rather than near-term activities. Parker felt it was important to conduct a term at Headquarters as well as at the field centres, knowing it was important to understand what Headquarters was all about for any astronaut contemplating moving into a management role, which many would consider after leaving the Astronaut Office. Coming from a science background and an astronaut career (with flying experience), moving on to a management role would require dealing with bureaucrats and politicians, policy makers and accountants. It would entail a completely different set of skills than for preparing to fly a mission.

One of Parker's concerns during his final years in Washington was the construction of the space station. He realised that building the station would seriously reduce the number of scientific Spacelab flights that fell under his responsibility. It was a challenging balancing act, trying to schedule interesting science on Shuttle missions when there were no guarantees of flying Spacelab missions while the station was being constructed. Could even small science experiments be conducted on the construction missions, or on the space station during its early years? Proposals submitted to JSC came back stating that every mission and every kilo launched on every flight was

Bob Parker in 2002. [Credit Francis French, Reuben H. Fleet Science Center.]

needed for the space station, and Parker soon realised he was getting nowhere. In 1997, an opportunity came his way to work at JPL in California, overseeing an office where there was a lot of day-to-day paperwork, from overseeing contracts to the implementation of contracts.

Therefore, from August 1997 to August 2005, Parker served as Director of the NASA Management Office at the Jet Propulsion Laboratory in Pasadena, California. He was responsible for "on-site oversight of the NASA contract with JPL, and leadership in negotiations of NASA contract requirements with JPL and the California Institute of Technology, the organisation that operates JPL. The director also enables management and technical support for NASA field centres and Headquarters offices that have work performed at JPL."[20] In Parker's words, "My particular job, apart from overseeing the overseers, if you will, was basically keeping NASA and JPL happy with each other."

On 31 August 2005, Bob Parker finally retired from NASA after thirty-eight years serving as an astronaut and manager. In his 2002 Oral History, he was asked about his span of time and various duties at NASA, and whether there was anything he would have liked to have pursued a little longer. In reply he said, "No ... at the time of course, but in retrospect, I would say no, although I might have liked to have had a bit more success in continuing the science on the Shuttle during this period [of station construction]."

SIX MISSIONS AND THIRTY YEARS

Bob Parker was the last of the scientist-astronauts to leave NASA, but one member of the group stands out as the most flown and longest serving active astronaut – Story Musgrave. Between 1983 and 1996 he flew on six Shuttle missions, finally retiring from the Astronaut Office in 1997 after thirty years as an active astronaut, when it became clear that he would not get a seventh mission.

After flying on STS-6 in 1983 and STS 51-F (Spacelab 2) in 1985, he completed two DoD missions in 1989 (STS-33) and 1991 (STS-44), participated in the first Hubble Service Mission (STS-61) in 1993 and finally flew on STS-80 in 1996. He became the only astronaut to fly on each orbit-capable vehicle of the Shuttle orbiter fleet, including Challenger. It is a unique record which will not be matched.

Education and mission support

In the Shuttle years, Musgrave continued to provide support to a number of missions in between training for his own. In 1984, he had been CB representative in Mission Control during STS 41-B and PAO support for the ABC TV network for STS 41-C. In 1985, after completing his STS 51-F flight crew assignments, he was assigned to SPAN for the STS 61-A Spacelab D1 mission. From STS 61-B onwards, he became the CB point of contact for Flight Data File issues for ascent/entry systems; ascent pocket checklist; crew personal calculator; contingency de-orbit preparations; orbit operations; entry pocket checklist; in-flight maintenance checklist; IUS deployment; orbit operations checklist and orbit pocket checklist; post-insertion and systems data. His experiences on STS 51-F, and his background in maths and statistics, operational analysis and computer programming, chemistry and medicine, coupled with his previous astronaut/piloting qualifications, made him an ideal choice for these assignments.[21]

In his career at NASA, Musgrave became perhaps the most pilot-orientated scientist-astronaut of the group. As well as flying in space, he managed to accumulate over 18,000 hours flying time in 160 different types of civilian and military aircraft, including over 7,500 hours in jet aircraft. He also held FAA ratings for instructor, instrument instructor, glider instructor and airline transport pilot. As an accomplished parachutist, he logged over 600 free falls, including 100 experimental free-fall descents as part of the study of human aerodynamics. His flying achievements were honoured in 1987 when Northrop Aircraft Division Management Club recog-

STS-33 crew during pre-flight breakfast. Left to right: MS Sonny Carter, MS Kathy Thornton, MS Story Musgrave, Commander Fred Gregory and Pilot John Blaha.

nised him as the first person to fly 6,000 hours in the Northrop-built T-38 jet, which is used by astronauts to travel across the continental United States. The milestone flight occurred on 4 August 1987. Musgrave is the only person to have achieved, or is likely to achieve, that many hours in a T-38. The record was the equivalent of flying non-stop for 250 days, taking off at the stroke of midnight on New Year's Eve and flying until midnight on 7 September.[22] Musgrave also continued his passion for education by earning his sixth degree, this time a Master of Arts in Literature from the University of Houston.

After STS-33 in 1990, Musgrave became Head of the Mission Support Branch of the CB and began a tour as Space Capcom in Mission Control, supporting the missions of STS-38 (DoD), STS-31 (Hubble deployment mission), STS-41 (Ulysses deployment) and STS-35 (Astro-1). After training and flying on STS-44 in 1992, he returned to mission support roles in SPAN for STS-50 (USML-1) and STS-49 (Intelsat re-boost) before returning to astronaut training for STS-61. He returned to the Capcom console in 1994, supporting STS-59 (SRL-1), STS-62 (USMP-2), STS-64 (Lite), STS-65 (IML-2), STS-68 (SRL-2), STS-66 (ATLAS 3), STS-63 (Near Mir), STS-67 (Astro-2), STS-71 (Mir docking 1), STS-70 (TDRS-G), STS-69 (WSF-2), STS-73 (USML-2), STS-72 (SFU retrieval), STS-74 (Mir docking 2) and STS-75 (USMP-3) over the next two years, before resuming training for his sixth and final space mission. One of the longest serving Capcom-experienced astronauts, he was the prime contact between the crew in space and the world below, sharing their victories and setbacks, taking care of the crews' best interests and suffering the good and the bad times himself frequently on the ground.

Military Musgrave

Musgrave was named to his third mission (STS-33) on 30 November 1988. This was a classified military Shuttle mission that was launched on 22 November 1989 and landed on 27 November, after a flight of 5 days 6 minutes 48 seconds and 79 orbits. Six months later, on 24 May, Musgrave was named to the crew of STS-44, another Department of Defense (DoD) Shuttle mission. This flight launched on 24 November 1991 and landed on 1 December, after a flight of 6 days 22 hours 50 minutes 44 seconds and 110 orbits.

Military missions on the Shuttle were, by the nature of the cargo, more classified than the other Shuttle flights. Little information and few photographs would be released on the activities of the crew or the purpose of the mission. In 2005, Musgrave did comment that "after you have done what you can do, you can tell people you deployed something – though [this is] no big deal since amateur astronomers watch you do it." There were two such Shuttle flights in 1985 (STS 51-C and 51-J), a single DoD mission in 1988 (STS-27), two flights in 1989 (STS-28 and STS-33) and another two in 1990 (STS-36 and STS-38). They were all more secretive than the scientific or

The crew of STS-44. At rear (left to right): MS Mario Runco, MS Story Musgrave, Pilot Terence Henricks. At front: Commander Fred Gregory, PS Tom Hennen and MS Jim Voss.

commercially-orientated missions. By 1991, the DoD was pulling away from Shuttle missions and transferring its payloads to expendable launch vehicles. However, three remaining missions (STS-39 and 44 in 1991, and STS-53 in 1992) were linked to military research in technology and procedures, but were more scientifically orientated, receiving a less classified status than those that preceded them.

The primary payloads of STS-33 and STS-44 were military satellites. The STS-33 crew deployed the Mentor (SIGINT) signal intelligence satellite, via an IUS, into a geosynchronous orbit. Two years later, as part of the STS-44 crew, Musgrave supported the deployment of a Defense Support Program early warning satellite called Liberty, also by IUS into geosynchronous orbit.[23] Though reports of crew activity on the STS-33 mission were restricted, with both Story Musgrave and Sonny Carter aboard (both medical physicians), a number of human physiological experiments seemed to be part of the mission's secondary objectives. The pilot of the mission, John Blaha, remarked on a personal tape during the mission that there was not much time for "tourism" or looking out of the window until they were given a three-hour wave-off for landing due to high surface winds. Apart from recording his perspective from orbit, Blaha managed to observe his crewmates as they took the opportunity for extra "free-time in orbit": "Story Musgrave is suspended next to my right shoulder, trying to photograph the clouds. The geologists at JSC will really love Story's pictures." Then Musgrave ran out of film and hurried down to the mid-deck for more, floating up through the open hatchway back into the flight deck. Musgrave asked Blaha if he was going to document the views out of the window as well as talking about them: "Look at all the little ripples in the clouds down there, John; all the little waves and rhythms of thunderstorms and lightning."[24]

On STS-33, Musgrave flew as MS 2/FE, repeating the role he had performed on STS 51-F four years earlier. This additional experience enabled him to repeat those duties for STS-44. His crew task assignments included photo equipment (still, movie and TV cameras), medical issues and Earth observations. As flight engineer, his orbiter-related tasks included DPS, Main Propulsion System (ET/SRB/SSME), OME/RCS, APUs and Hydraulics, EPS, Environmental Control Life Support Systems, communications and instrumentation. He was also assigned as prime crew member on eight DSOs and served as back-up to the DPS payload.[25] "I'm the chief cook and bottle washer on this flight," Musgrave explained in the pre-flight crew press conference at the end of October 1991.

Originally, STS-44 was just a US deployment mission and thus a relatively short flight, but when the mission extension and the medical DSOs were added, Musgrave felt these were a big bonus for flying the mission. He had not really been a doctor within NASA, despite his medical qualifications. He had been able to do some experiments prior to astronaut selection on negative pressure and also worked on the Spacelab simulations, but had flown as an engineer on his previous flights. This time, he got the chance to actually perform some medical experiments in space.[26]

After the deployment of the military satellite early in the flight, the crew settled down to their secondary objectives. These involved the US Army Terra Scout military observation programme of experiments (for which Musgrave assisted US Army payload specialist Tom Hennen as a support crew member) and a range of medical

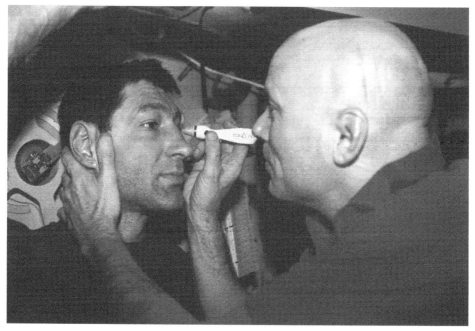
Musgrave conducts one of the medical DSO objectives during STS-44. His test subject is Mario Runco.

experiments aimed at supporting the gradual increase of Shuttle mission durations up to thirteen days, as part of the Extended Duration Orbiter programme. The ten-day mission was unavoidably shortened to just under seven days by the failure of a critical element of the navigation system – Inertial Measurement Unit No. 2. As a result, the crew tried to squeeze as much work as possible into their remaining time on orbit.

One of the more good-natured crews to fly the Shuttle, they soon adopted nicknames for each other. As the veteran commander on his third space flight, Fred Gregory became known as "Dad", while Musgrave, with his experience and new space flight endurance record, became known as "Granddad". Jim Voss picked up the moniker 'Bilge Man' for his work in troubleshooting a humidity separator in the lower mid-deck of Atlantis, while. Mario Runco became known as "Spock" for his affinity to Star Trek. Payload Specialist Tom Hennen used a trash compactor to save space on the flight and was promptly awarded the nickname "Trash Man". Musgrave was also called "Dr. Story" or "Dr. Cryo" for his work on the mid-deck experiments

The DSOs Musgrave was assigned to as primary crew member were:[27]

- DSO 316: Bioreactor Flow and Partial Trajectory in Microgravity: All planned activities were successfully completed and an additional test performed.
- DSO 472: Intraocular Pressure: Early data for fluid shifts were recorded. However, late flight data were not recorded due to the shortened mission.
- DSO 478: In-flight Lower Body Negative Pressure: Though this unit was used

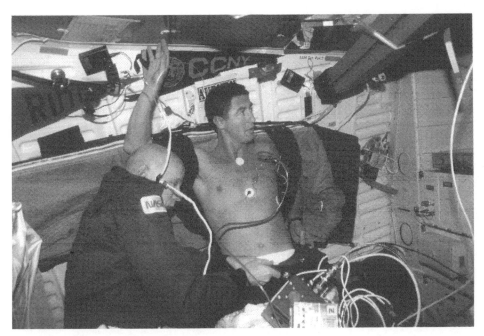
Musgrave works with pilot Tom Henricks on the STS-44 medical DSO programme.

throughout the flight by Musgrave, with Henricks, Runco and Voss assisting, no protocol combinations were completed, again due to the shortened mission.
- DSO 603: Orthostatic Function During Entry, Landing and Egress: Pre-flight data were recorded, and further data were collected during entry, egress and post landing.
- DSO 613: Changes in Endocrine Regulation of Orthostatic Tolerance Following Space Flight: Pre-flight and post-flight data were collected.

In addition he was responsible for:

- DSO 901 Documentary Television;
- DSO 902 Documentary Motion Picture Photography; and
- DSO 903 Documentary Still Photography.

These were all accomplished, with large numbers of images taken during the flight.

Despite the shortening of the mission, Musgrave still surpassed the cumulative duration record for time spent in space on the Shuttle, accumulating over 596 hours on four flights. However, Musgrave, like his colleagues, was disappointed to have come home early: "For all of us, it was an incredible disappointment. It really hurt that we didn't get everything done, but that's the way it is."

As with all missions, the post-flight crew report indicated to engineers, mission managers and training staff exactly what worked and what didn't. Not all of the crew's

462 Ending of Eras

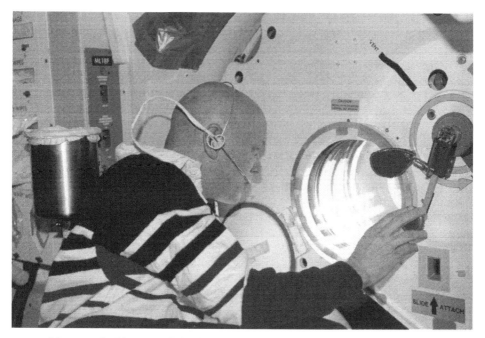

Musgrave looking out of the side hatch window on the mid-deck of Atlantis.

recommendations get accepted or taken up. As Musgrave recalled, "Life is a compromise. You try, you suggest some things which you know are the ideal and what would be best, but some of them may be too expensive or too hard to apply. If you are a long-term player in the programme, which of course I was, you get to know what would get accepted and what would not. A lot of astronauts were there only five years, so they did not know what the outcome of their recommendations was. I knew what came out of the Skylab programme in terms of science operations in space and the whole integration and payload processing process."[26]

Servicing Hubble

Musgrave had been assigned to Shuttle EVA development issues in 1972 and had worked on Hubble servicing by Space Shuttle crews for about seventeen years. So it was obvious that, when the Hubble Space Telescope experienced the failure in its mirror focusing, the first servicing mission would involve the installation of corrective optics and other tasks, Musgrave would work on developing the EVA programme for the flight. On 16 March 1992, three months after returning from STS-44, Story Musgrave was assigned to his fifth space flight as payload commander for STS-61, the first Hubble Service Mission. Tom Akers, Jeff Hoffman and Kathy Thornton were named as the other three EVA mission specialists for the mission in August 1992. Like Musgrave, they all had previous EVA experience. The commander (Dick Covey), pilot (Ken Bowersox) and MS 2/FE (ESA astronaut Claude Nicollier) were named

What the well-dressed payload commander should wear. Musgrave waits his turn as fellow crew members finish donning their launch and entry suits prior to their departure for Launch Pad 39B for the lift-off of Shuttle Endeavour on mission STS-61.

later. Due to the complexity of the mission, NASA assigned a back-up EVA astronaut with previous EVA experience (Greg Harbaugh) to the flight to support the crew in their preparations and to replace an injured crewmember should the need arise. Fortunately this option was not required, but Harbaugh's work and input were of considerable assistance to the prime crew, helping them to master all their tasks on time.

As payload commander (PC), Musgrave had overall responsibility for the planning, integration and on-orbit coordination of all payload/Shuttle activities on the assigned mission, while the flight commander retained overall responsibility for mission success and safety. In working as a PC, Musgrave also continued to identify training issues and operating constraints prior to and during the training process. The role of PC was seen by NASA (in the early 1990s) as a foundation for the development of the Space Station mission commander concept. Musgrave, the former aviation electrician and aircraft crew chief turned mathematician, doctor and astronaut, was once asked why he gave up being a surgeon to become an astronaut. He quickly replied with a wide smile, "Why, so I could operate on the Hubble of course!"

Back in the pool

Musgrave was also designated MS 4 and assumed the EV 2 and IV 2 role for the mission's EVA programme. His primary crew responsibilities also included the galley on the mid-deck, 35 mm photographic equipment, medical issues, on-orbit stowage, and FDF issues concerning EVA. His back-up crew responsibilities included the Hubble Space Telescope, payload communications, 70 mm photographic equipment, in-flight maintenance, dealing with the Public Affairs Office and flight plan issues in the FDF.

The crew trained hard for their mission to ensure the eyes of Hubble could finally peer clearly deep into the universe. Building upon past missions and their own previous flight experiences, the crew strove to mould themselves into an effective team for the rendezvous, RMS operations, EVAs and the deployment. The EVA training for Musgrave and his colleagues was far more extensive and varied than for any previous mission, with all four crew members and the back-up trained on all EVA tasks. Multiple training facilities were used to master these spacewalking skills, including:

Weightless Environment Training Facility (WETF) at JSC, Neutral Buoyancy Simulator (NBS) at Marshall Space Flight Center (MSFC), High Fidelity Mechanical Simulator (HFMS) at the Goddard Space Flight Center (GSFC), Precision Air Bearing Floor (PABF) using modified Apollo-era fixtures at JSC, 193 hours of Joint Integrated Simulations (JIS) between JSC/GSFC/MSFC, Manned Thermal Vacuum Tests (MTVT) at JSC, including three round-the-clock operations of the last two EVA days and deployment timeline (59 hours) (with Claude Nicollier as RMS operator), and using the Manipulator Development Facility (MDF) at JSC to further assist in developing astronaut and RMS coordination. The MDF consisted of a high-fidelity RMS simulator in a mock-up orbiter payload bay and the use of a full-scale HST balloon mock-up, with a mock-up astronaut figure on the end of the RMS Virtual Reality Simulator at JSC

For STS-61, there were 105 EVA-related training tasks. As a crew, the five EVA astronauts logged 350 hours in total in the water tanks at Houston and Marshall, setting a new crew record.

On 10 March, the EVA crew's training took them across the Atlantic to England to visit the British Aerospace Space Systems Ltd facilities at Filton in Bristol and familiarise themselves with the new solar arrays being fabricated there. The four EVA astronauts were accompanied by ESA astronaut Claude Nicollier, primary RMS operator for the mission. Covey, Bowersox and Harbaugh remained in Houston during the training trip.[28]

During simulated EVA activities on 28 May 1993, Musgrave was slightly injured

Musgrave in the White Room of Pad 39B preparing to enter Endeavour for his fifth ride into space and the first mission to service the Hubble Space Telescope.

in Thermal Vacuum Chamber B at JSC. In wanting to maximise his training time during one session in the chamber, he suffered a mild case of frostbite in his right-hand fingers through his EVA gloves, as the chamber was cooled to minus 130 degrees. During the equipment test, he had been practising with the tools and other instruments he would use during the EVAs on the mission. He communicated that his hand felt cold during the eight-hour simulation, but elected to complete the chamber run. After leaving the chamber and taking off his suit, Musgrave reported numbness and discolouration of the fingers. Under examination from NASA physicians, he began appropriate therapy to improve the condition as quickly as possible. Fortunately this had no lasting impact on his mission training or his assignment to STS-61.

The crew was due to visit Marshall Space Flight Center at the end of June to work in the Neutral Buoyancy Simulator (water tank) developing and evaluating EVA timelines. The injury was expected to restrict Musgrave's suited activities, but as he was far ahead of the other crew members in training due to his long involvement with the Hubble mission, this was not expected to be a problem. To prevent a repeat of the incident in space, an extra layer of insulation was added to every EVA glove.[29]

Flight Day (FD) 1 of the mission called for ascent into orbit and corrective burns to prepare for rendezvous with the HST. FD 2 would include checking out the RMS and the EVA equipment and further refinements to the flight trajectory. FD 3 would see the rendezvous, grapple and berthing of the telescope. FD 4 was scheduled as

EVA 1 Day, with Musgrave and Hoffman changing the rate sensor units in the telescope's gyros. FD 5 would see Thornton and Akers exchange the twin solar arrays, supported by Musgrave and Hoffman from within the orbiter. On FD 6, Musgrave and Hoffman were again to go outside, to change the WFPC-II and two magnetic sensing systems. During FD 7 Akers and Thornton would install the corrective optics, and on FD 8 Musgrave and Hoffman would install solar array drive electronics. The telescope would be released on FD 9 with the crew taking a half-day off in turn on FD 10 as they prepared for entry and landing on FD 12.[30]

Just prior to departing to the Cape for the launch of STS-61, Musgrave reflected on the complexity of the mission and his apprehensions over the flight plan. "I'm not overconfident. I'm running scared. This thing is frightening to me. I'm looking for every kind of thing that might get out and bite us." This approach was one of a veteran astronaut well used to the twists and turns of space flight, the unforgiving environment an astronaut lives and works in and of the technology they have to deal with every day on their mission. According to Musgrave, running scared keeps you thinking about how to get the job done. He had also commented on why he wanted to fly in space again: "I'm here because I love space, I love being in space, and I like the space business. You can't separate out space and the space business. I've been an astronaut for twenty-six years, and I've had roughly twenty-six days in space. I've got only one day in space for every 365 days down here. It's not a very good ratio, but I still love space so much that I would continue to do it even if the ratios were worse."[31] Musgrave also revealed that, "The night before launch, I go down and lie by the ocean and look at the stars. Sometimes I see satellites overhead and think 'Tomorrow, you're going to be one of those. See that streak? That's you tomorrow'."[32]

Improving Musgrave's ratio

STS-61 was launched on 2 December 1993. The crew aboard orbiter Endeavour successfully grappled the HST into the payload bay as planned during FD 3 and the following day began their extensive EVA programme with Musgrave and Hoffman completing the first of the mission's five EVAs.

> *EVA 1–4 Dec:* Musgrave and Hoffman complete a 7 hour 54 minute EVA to successfully change the rate sensing units, the electronic control unit and eight fuse plugs that protect the telescope's electrical circuits, restoring the telescope to a set of six fully-functioning gyros.
>
> *EVA 3–6 Dec:* Musgrave and Hoffman conduct the third EVA of the mission in 6 hours 47 minutes, completing a flawless installation of the WFPCII after removing the original instrument from Hubble and stowing it in the payload bay of Endeavour for the return to Earth. They also installed two magnetometers.
>
> *EVA 5–8 Dec:* Musgrave and Hoffman complete a 7 hour 21 minute EVA to install replacements for the solar array drive electronics that control the pointing of the solar arrays, which in turn provide power for the telescope. Shortly after

Story Musgrave and Jeff Hoffman during the first of five STS-61 EVAs. Musgrave is in the foreground whilst Hoffman works at the space telescope on the end of the RMS.

installation, Musgrave gently pushed on each array to help unfold the primary deployment mechanism, which had become stuck. They then installed the Goddard High-Resolution Spectrograph Redundancy Kit and placed two Mylar covers over the original magnetometers to contain any contamination or debris that might come off the instrument and protect it from UV degradation.

At the end of their third and final EVA, and the final excursion of the mission, Musgrave perched on top of the RMS and paid tribute to the team that had assisted in the preparation of the EVAs: "You're in our hearts ... you're in our heads. What we've done and what we're going to do is simply a reflection on what you have given us."

468 Ending of Eras

Hoffman is reflected in the helmet visor of Musgrave as he photographs the veteran astronaut during one of their EVAs.

The telescope was successfully released into orbit by Nicollier using the RMS on 10 December, ending a highly successful STS-61 mission. Endeavour landed on 13 December at the Shuttle Landing Facility at KSC, after a flight of 10 days 19 hours 58 minutes 37 seconds and 163 orbits.

The last flight

Musgrave was named to the crew of STS-80 on 17 January 1996. The sixteen-day science mission was to deploy and retrieve a pair of science satellites – the ORFEUS-SPAS astronomy satellite and the Wake Shield Facility (making its third flight to grow thin film semiconductor material in the near-perfect vacuum in orbit). There were also two planned EVAs, designed to refine techniques that would be employed on the ISS during its construction.

Musgrave was assigned as MS 3 and rode on the flight deck (Seat 3) during ascent. He was supposed to swap with Tammy Jernigan for the ride home and sit on the mid-deck Seat 5, though this did not happen. For the first time Musgrave was not part of the EVA crew, but used his experiences as IV crew member for the planned EVA. His primary crew responsibilities were for the Wake Shield Facility and for Earth observations.

The launch of STS-80 occurred on 19 November 1996 and Columbia's crew successfully deployed and retrieved both satellites during their mission. Musgrave

Back in the "deeper pool". STS-61 astronaut Story Musgrave, with helmet visor raised, holds onto one of the handrails on Hubble during one of the mission's five EVAs, more than ten years after he participated in the first Shuttle EVA from STS-6.

supported preparations for the EVA, giving a video downlink explaining the EVA mobility unit components and operations, before the two scheduled EVAs were cancelled due to a jammed hatch door latch. Despite the cancelled EVA, the crew's disappointment was alleviated somewhat by being given an extra day in orbit to conduct science and observe the Earth. Bad weather also played its part in making STS-80 the longest space flight of the Shuttle to date.

Musgrave's feelings about not being required to perform an EVA on this mission were more disbelief than disappointment. But, having worked as back-up for Skylab and as a Capcom for over twenty-five missions, he had learned to live with such successes and failures. His disbelief was due in part to his past involvement in

470 **Ending of Eras**

The crew of STS-80. At rear, left to right: Commander Ken Cockerel, MS Tamara Jernigan, Pilot Kent Rominger. At front: MS Tom Jones and Story Musgrave, veteran of six Shuttle missions and the only astronaut to have flown into orbit on all five orbiters.

designing the mechanisms that opened the EVA hatch. Once the airlock was repressurised and opened into the mid-deck, Musgrave went inside to see if he could determine what was wrong. He knew the design of the mechanism could not fail on its own and as he touched the handle it was obvious that it was jammed against something in the gears, which were not accessible to the crew. Post-flight inspections revealed that a screw had indeed jammed in the gear mechanism. It was replaced and the hatch opened perfectly.

Habitability on his last flight was better than before, as there were fewer crew members (five), no shift system, no extensive EVAs planned and few mid-deck experiments to crowd up the area. It was a lot more comfortable than on his other missions, especially compared to STS-61.

For Musgrave, the opportunity to remain in space was always greeted with sheer pleasure. He was never happy with delays and extensions from the programme's point of view, because they were expensive and took the programme a little off track in terms of getting the vehicle back on the ground. "But I love every time I get delayed in coming home – even getting one extra revolution (orbit). I am in the suit and not that comfortable, and we wave off and I love it because I get another hour and a half in space. When we wave off a day, I get another twenty-four hours up here. For the sake of the programme I don't like to have that happen, but on the other hand, when an act of God tells me I have to do it, then I just love it."[26]

During descent, having no major responsibilities for entry and landing and riding on the flight deck instead of the mid-deck as scheduled, Musgrave again took the opportunity to stand up, becoming the first person to witness the plasma and the fire at entry interface through the overhead windows all the way down to Mach 10. Holding onto the shoulder of Tom Jones in the centre seat (MS 2), he was able to observe and take extensive video footage of the phenomena all the way down. "You can't believe you don't evaporate in a flash with that amount of plasma. I call it fire. Technically, it's plasma, but to the naked eye, it's fire."[26] The plasma shock wave and high Mach aerodynamics were captured on film as Columbia cut through the upper atmosphere. The experiment was not an official one, but one which Musgrave wanted to record from an engineering perspective and, as a scientist-astronaut, from a humanities angle. There was no operational requirement for the event to be filmed, but Musgrave went ahead and did it anyway.

Tom Jones stated in his 2006 book, *Sky Walking: An Astronaut's Memoir*, that the flight rules called for Musgrave to be downstairs and strapped into his mid-deck seat well before they reached Mach 20 on descent, but as he had no official duties to perform during re-entry he decided to stay on the flight deck to observe and film the spectacular light show."

"We were at 1 G now, yet Story stood braced against Taco's [Ken Cockrell's] seat in front and my seat in the middle rear," Jones wrote. "His weight with suit and harness was probably something like 250 pounds and increasing steadily with our deceleration. I shook my head at his sheer stamina. While we focused on the cockpit tasks, Story braced the camera against the overhead window frame to capture the full view of the plasma display outside. I kept expecting him to call it quits, thump me on the shoulder, and clamber downstairs to his seat (or just collapse to the floor). But he made no move to leave. This was his sixth re-entry, his last, and he looked determined to get the most out of it."[33]

Musgrave would remain standing, braced against the seats, throughout the entire landing procedure. Wearing his cumbersome orange LES, this would have been a mighty effort under normal circumstances, but after more than two weeks of weightlessness it was bordering on superhuman. As Jones told the authors, "Our max g-level was about 1.7 during the Mach 12–20 time frame, and again during the turn around, the HAC lining up for final approach. After eighteen days in free fall, it felt more like 4 or 5 G to me."[34]

Columbia came home on 7 December 1996 after 17 days 15 hours 53 minutes 18 seconds and 279 orbits, with the STS-80 crew having set a programme record that

Musgrave works with a pair of computers dedicated to the Wake Shield Facility operations on the aft flight deck of Columbia during STS-80.

will probably not be surpassed by any other Shuttle mission. Musgrave extended his Shuttle duration record and, at sixty-two, became the oldest man to fly into space (until 1998 when former senator John Glenn, at the age of seventy-seven, returned to orbit as a crew member of STS-95 some thirty-six years after his pioneering Mercury orbital mission in 1962).

"You can't fly anymore!"

For the majority of astronauts, flying in space is a memorable experience. They are happy and proud to make one, two, three or even four flights and relieved to get back safely each time. A few will complete five missions and fewer still complete six or more because that usually means spending many years at NASA, missing opportunities and promotions outside the space programme. John Young was selected in 1962, flew six missions and only retired as an active astronaut in 2004, aged seventy-four. Though he was removed from the active flight roster in 1987, there was always a slim chance he would get his coveted seventh mission. He had in fact been scheduled to make his record seventh mission in September 1986 on the Hubble deployment mission, but this was delayed by the loss of Challenger, and he was later replaced as mission commander by Loren Shriver when the crew was reassigned to STS-31.

For Story Musgrave, who in 1997 was "only sixty-two", flying into space was not

Story Musgrave photographed at Princeton University, New Jersey, in June 2004. [Credit: Hart Sastrowardoyo.]

just a job – it was a dream, a passion in the very being of the man. He was more than ready to go again with ISS on the horizon, and with all his experience it was reasonable to assume that it would not be long before Story was back in orbit, perhaps as a commander of a long-duration space station crew, bringing the story of the scientist-astronauts full circle. But it was not to be. A condition of flying STS-80 was that it would be his last flight and he was told that he would not get another flight after his sixth. Having been told that STS-80 was his last mission, he was forced to accept that he had to finally hang up his space boots as he could not fly anymore. He decided to focus his efforts on educating the world about space, "to communicate the heart and soul of what human space flight is about." Ironically, a year later, John Glenn had his flight on the Shuttle, and this raised some interesting comments from observers and the media about why Musgrave had been forced to leave the year before when he was fifteen years younger than the veteran Mercury astronaut. Musgrave clearly had the mission experience that could have been utilised on a Space Station assembly mission or on a long flight, something he was enthusiastic about.

Musgrave had volunteered for, and worked hard to try to secure, a long-duration flight on the Mir space station. He was nominated to back up Norman Thagard on the first American long-duration flight to the station. At the time he was preparing for

STS-61 and would have left for Russia only a couple of months after returning from the Hubble mission. Musgrave would have served as Thagard's back-up and then would have launched on STS-71, taking over from Thagard and then coming home on STS-74 after a four- to six-month flight, "But nothing happened and I kept pressing them (NASA). Even in the middle of getting ready for Hubble, I said I needed a language instructor, I need language tapes to play in my car, I needed language video. I said 'If you commit to me I will commit to you and I will be ready', but NASA would never commit to me."[26] Thagard was eventually backed up on the Mir mission by Bonnie Dunbar who, though eligible for long-duration flight to Mir on STS-71/74, did not do so, partly because of several delays that prevented her extensive training with the Russian Mir resident crew with whom she would have served.[35] Clearly Musgrave's disappointment and frustration in not securing a long-duration mission to Mir was in no way appeased by flying STS-80. In taking that flight, he had reluctantly agreed not to fly anymore and so lost the opportunity of a long-duration ISS mission. Such a mission would have required a two- or three-year commitment to the Russian training programme. The lack of NASA commitment to his Mir mission revealed that there was no guarantee that he would be put forward for resident crew training for ISS, and even if he was he would have to pass the Russian exams before confirming his flight seat. STS-80 therefore offered a better chance to return one last time to the place he loved so much – space.

On 2 September 1997, thirty years after arriving at Houston to begin astronaut training, Story Musgrave reluctantly retired from NASA, ending an era of scientists flying as career astronauts in NASA that encompassed the American manned space programme from Gemini to Space Shuttle and almost to ISS. Many were sad to see Musgrave leave, taking with him the "spirit and the passion" that had driven so many astronauts across the decades with NASA, from the pioneering years to placing men on the Moon, through getting the Shuttle flying to supporting the space station. At a time when NASA needed personalities like Musgrave to generate inspiration among the public, such pioneers were no longer under consideration for the new-look NASA in a new century. The "Right Stuff" era of test pilots had long gone. So, too, it seemed, had the era of scientist-astronauts.

ALL GOOD THINGS COME TO AN END

Musgrave's retirement from the astronaut programme and NASA signalled the end of the era of NASA scientist-astronauts. For over thirty years they had worked to be recognised as "astronauts", not only for their academic and work experience, but for their devotion and dedication to supporting the NASA programme. They became jet pilots, competed with the pilot-astronauts, completed the astronaut training programme, supported other missions, developed new equipment, hardware, procedures and experiments, and let science take a back seat to operations and engineering. Most of all, they waited and hoped. But they also proved to the sceptics that scientists could be both astronauts and pilots, engineers and specialists in mission operations, and managers and administrators. They also turned the role of scientist-astronaut into the

Story Musgrave's biography *The Way of Water* was released in 2004. He is shown here in New York with his biographer, Anne Lenehan. [Credit: Karl Tate/*collectspace.com*.]

mission specialist and forged the way for the assignments of payload commander and ISS science officer. But were these really the new generation of scientist-astronauts?

REFERENCES

1. Joseph P. Kerwin, JSC Oral History Project, 12 May 2000, p. 51.
2. *FCOD Manpower Requirements*, CB, 27 April 1984, JSC History Archive, Cliff Charlesworth Collection, copy on file AIS Archives.
3. AIS telephone interview with Bill Lenoir, 24 June 2005.
4. Private email from Bill Lenoir to Dave Shayler, 20 April 2006.
5. *An Ex-Astronaut's Spaceship Comes In*, Paulette Thomas, The Wall Street Journal, 12 February 1988.
6. ISF Press information supplied by Joe Allen, 1988, held in AIS archives.
7. AIS telephone interview with Tony England 14 July 2005.
8. Interview with Karl Henize by Andy Salmon, JSC, 22 July 1991.
9. Information supplied by Vance K. Henize, *Astronaut Karl G. Henize: A personal history* – main page from the website of Vance K. Henize, *http://spacsun.rice.edu/~henize/karlhenize.html*, last accessed 5 June 2002.

10 Correspondence between Owen Garriott and D. Shayler, 13 April 2006.
11 Owen K. Garriott JSC Oral History, 6 November 2000, pp. 72–73.
12 *Long-duration space flight: Faster, Cheaper, Better* by Owen K. Garriott and Frank L. Echols, Teledyne Brown Engineering. Original paper 10 July 1992, reprinted in *Aviation Week and Space Technology Forum*, 26 October 1992, pp. 69–71.
13 *Stretching the Shuttle*, David J. Shayler, Orbiter 83, pp. 47–67, AIS Publications, March 1995.
14 AIS telephone interview with Owen Garriott, 27 June 2005.
15 Bill Thornton interview with Dave Shayler, NASA JSC, Houston, Texas, 5 August 1988.
16 UTMB News Release, 11 October 2002.
17 UTMB News Release, 6 August 2001.
18 Parker Oral History interview transcript, 23 October 2002, pp. 40–41; NASA News 88-006, JSC 1 March 1988; 89-033, JSC 7 June 1989.
19 STS-35 Mission Highlights, NASA educational publication, MH-002/12-90 1991.
20 NASA News 5 Aug 1997.
21 CB Memo – Astronaut Office point of contact for Flight Data File, from Story Musgrave, 19 November 1985, in the Cliff Charlesworth collection, JSC History Files, University of Clear Lake, Copy on file AIS archives.
22 *Space News Round Up*, NASA JSC, issue dated 20 November 1987.
23 *Who's Who in Space*, 3rd Edition, 1999, by Michael Cassutt, published by MacMillan, p. 205.
24 *Space and People*, transcript of notes recorded by John Blaha aboard STS-33, 1989, copy on file AIS Archives.
25 STS-44 Crew Task Assignments, PAO Contact Book for STS-44, NASA JSC, PAO files – copy on file AIS archives.
26 AIS telephone interview with Story Musgrave, 23 June 2005.
27 STS-44 Mission Report, January 1992 (NSTS-08273), NASA LBJ Space Center, PAO Contact files; copy on file AIS Archives.
28 British Aerospace Space Systems Ltd Press Release 10 March 1993.
29 NASA News, 1 June 1993 93-038.
30 STS-61 Press Kit, December 1993; HST First Servicing Mission Training and Operations Workshop 30 August–1 September 1993; copies on file AIS Archives.
31 *Adventure in Space* by Elaine Stott, Hyperion Books, 1995, pp. 13–18.
32 The Houston Post, 12 December 1993, *He's ready for a close encounter*, by Dan Feldstein pp. A-1 and A-16.
33 Tom Jones, *Sky Walking: An Astronaut's Memoir* (Smithsonian Books/Collins), New York, 2006, pp. 199–200. Extract used with permission of author Jones, given 22 April 2006.
34 Tom Jones email correspondence with Colin Burgess, 22 April 2006.
35 *Women in Space*, David Shayler and Ian Moule, Springer/Praxis 2005, pp. 320–322.

12

Science Officers on ISS

"America has always been greatest when we dared to be great. We can reach for greatness again. We can follow our dreams to distant stars, living and working in space for peaceful, economic and scientific gain. Tonight I am directing NASA to develop a permanently-manned space station, and to do it within a decade." – US President Ronald Reagan's State of the Union Address to Congress, 25 January 1984.

The desire for a permanently-manned space station had existed for decades. Indeed, the idea had featured in the concepts of pioneering space theorists Konstantin Tsiolkovsky and Hermann Oberth at the beginning of the twentieth century. Both Sergey Korolyov and Wernher von Braun had always placed a space station at the focal point of sustained manned space exploration, and so too had the military, whose plans had included bases on the Moon. The race for space was diverted into a race to the Moon in the 1960s before the age of space stations began in the early 1970s with the launch of Soviet Salyut stations and the American Skylab. While the Soviets continued with their quest for space via an expanding space station programme, the Americans switched their attention to the Space Shuttle reusable multi-launch system. The Shuttle was originally intended as part of a large infrastructure that included space stations and extended exploration of the Moon, but was the only part of this grand scheme to actually receive any funding. Thus, with nowhere to fly to, Shuttle was instead marketed as a versatile launching system capable of a multitude of tasks, including its potential use as a science platform for a variety of research fields. It was not until 1998 that President Reagan's space station plan finally saw its first hardware reach orbit.[1]

The concept of the "permanently manned space station" had constantly evolved during the two decades before Reagan directed NASA to actually begin building one. Budget limitations were always a major factor in the design and in selling the idea to Congress and the American public. The heady days of seemingly limitless funding that Apollo had enjoyed prior to the first manned lunar landing were long gone. Even after

Reagan committed NASA to the station, its development over the next decade saw the concept, called Freedom, grow in size, complexity and cost. A change of US administration came at a time when NASA was divided about the direction of manned space flight in the 1990s and beyond. The desire to push the boundaries of exploration further to the Moon and Mars was still floundering, but there was a hope that, with international cooperation from ESA, Canada and Japan, a truly international space station, rather than just an American one, could be developed. When the Soviet Union collapsed in 1991 and Russia began to look for external cooperation to support its struggling finances, the idea of joining the space station programme seemed ideal. Russian experience, hardware and facilities added to the mix would enable construction of the station to begin in 1998. In fact the first element launched was Russian, and it would be mainly Russian components that formed the nucleus of the early station until the Shuttle could deliver the American-developed hardware to allow a permanent crew to reside on the station. At least that was the plan.

BUILDING A DREAM

In 1998, ISS prime contractor the Boeing Company, issued a Press Information Book that overviewed the plans for the station at the start of construction.[2] A significant amount of this information had changed as the station itself changed, but the underlining objectives of the programme remained:

- Finding solutions to crucial problems in medicine, ecology and other areas of science.
- Laying the foundation for developing space-based commerce and enterprise.
- Creating greater worldwide demand for space-related education at all levels by cultivating the excitement, wonder and discovery that the ISS symbolises.
- Fostering world peace though high-profile, long-term international cooperation in space.

To achieve all of this, the ISS would require over 100 elements to be constructed during forty-five missions, using American, Russian, European and Japanese launch systems and spacecraft and elements from all sixteen partners in the programme. According to the press information, nearly 500 tons of structure, equipment and supplies needed to be placed into orbit. More space walks would be required in five years of assembly than the combined total of spacewalks since Leonov's pioneering effort in the early 1960s.

The research facilities would include:

- US Laboratory – where US studies into human research, fluids and combustion, biotechnology, and materials science would be conducted.
- ESA Columbus Orbital Facility – where European astronauts would perform research into microgravity studies, fluid physics, biology and physiology.
- Japanese Experiment Module with Centrifuge Facility – where Japanese astro-

nauts would conduct experiments in the pressurised module and unpressurised exposed facility, featuring a dedicated RMS and experiment logistics module.
- Two Russian research modules based on earlier Mir-type modules – where Russian cosmonauts would perform research into Earth, space, and human life sciences, fundamental biology and microgravity.

The ISS programme was to be divided into several phases:

- *Phase 1 (1994–1998)* included Shuttle dockings with Mir and over twenty-seven months of continuous residence by seven US astronauts on the Russian station. Several Russian cosmonauts would also fly on the Shuttle. This was intended to provide invaluable knowledge and experience in both international cooperation and coordination in space operations. Many lessons learned from this phase had direct application to subsequent phases.
- *Phase 2 (1998–2000)* The initial assembly and habitation of ISS up to and including the arrival of the first three-person resident crew and independent operations without a docked Shuttle.
- *Phase 3 (2000–circa 2004)* The assembly completion phase, with the delivery and attachment of all the additional laboratories, solar arrays and other facilities, and supplies and logistics to support a seven-person crew on permanent basis.
- *Phase 4 (circa 2004–2024)* Progressing from construction to utilisation, the ISS would be completed and fully operational, with rotating resident crews and numerous visiting crews conducting a growing programme of science, research and observations. Estimates of the operational life of the early space station varied from a few months to several years. The Russian Mir station, planned for a five-year operational phase, actually operated for fifteen years and it was hoped that ISS would exceed that record.

When fully assembled, ISS would have a mass exceeding 475 tonnes and would house a permanent crew of six or seven in 1,300 cubic metres of pressurised modules (about the equivalent of two Boeing 747 jumbo jets). There were to be six laboratories, two habitation modules and two logistics modules and the whole structure would measure 108.5 m across and 88.3 m long. The research expected to be completed aboard this huge space complex would focus on microgravity, life sciences, space sciences, Earth sciences, engineering research and technologies and the development of commercial products, as well as educational outreach objectives.

From imagination to reality

On 20 November 1998, the first element of ISS was launched from the Baikonur cosmodrome in Kazakhstan. The Functional Cargo Bloc (FGB) or control module was named Zarya (meaning "Dawn"). This module was intended to provide electrical power, computer commands and attitude control to the early space station, to keep it in its operational orbit and to host the docking of the Zvezda (meaning "Star") Service Module which would allow permanent habitation of the station. The module would then serve as a fuel depot and storage facility as the station grew in size and would offer

The US laboratory module (Destiny) for ISS is shown under construction in the fall of 1997.

additional docking ports for visiting Russian spacecraft. On 4 December 1998, the American Shuttle Endeavour lifted off from KSC in Florida on the STS-88 mission, to deliver the first American element to the station, which was called Unity. Over the next four years, the combination of Russian and US assembly and supply missions gradually expanded the station towards its assembly complete target.[3]

Table 6. ISS main element launches.

Date	Element
1998.20.11	Zarya (Russian) control module
1998.04.12	Unity (US) Node 2/Pressurised Mating Adapter 1 and 2
2000.12.07	Zvezda (Russian) service module
2000.11.10	Integrated Truss Structure Z1; PMA-3
2000.30.11	ITS P6 2001.07.02 Destiny (US) laboratory module
2001.12.07	Quest (US) airlock
2001.15.09	Pirs (Russian, meaning "Pier") docking compartment
2002.08.04	Central truss segment (S0)
2002.07.10	ITS S1
2002.24.11	ITS P1

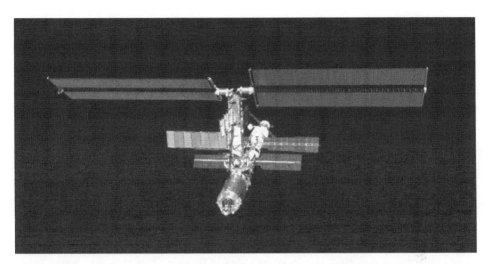

Destiny laboratory is shown attached to the ISS in February 2001. With the arrival of the lab, an expanded programme of science could be delivered and operated by the permanent crew of three astronauts and cosmonauts.

Unfortunately the loss of Columbia on 1 February 2003 grounded the Shuttle fleet and seriously affected the construction and expansion of the space station. The July 2005 Return-to-Flight mission of STS-114 (a logistics flight) was a success, but further tile damage to the Shuttle delayed the next mission well into 2006 and pushed any further assembly flights towards the end of the year. The loss of Columbia also affected the supply of logistics to the station, and as a result the two-person resident crews of ISS-7 through ISS-12 were more like caretakers than fully-operational science crews.[4]

Table 7. ISS resident crews (2000–2006).

Expedition	Crew	Launched	Landed	Duration (D:H:M:S)
One	Shepherd–Gidzenko–Krikalev	2000.31.10	2001.21.03	140:23:38:55
Two	Usachev–Voss–Helms	2001.08.03	2001.22.08	167:06:40:49
Three	Culbertson–Dezhurov–Tyurin	2001.10.08	2001.17.12	128:20:44:56
Four	Onufrienko–Walz–Bursch	2001.05.12	2002.19.06	195:19:38:12
Five	Korzun–Treshchev–Whitson	2002.05.06	2002.07.12	184:22:14:23
Six	Bowersox–Pettit–Budarin	2002.23.11	2003.03.05	161:01:14:38
Seven	Malenchenko–Lu	2003.26.04	2003.27.10	184:22:46:09
Eight	Foale–Kaleri	2003.18.10	2004.30.04	194:18:33:43
Nine	Padalka–Fincke	2004.19.04	2004.24.10	187:21:16:09
Ten	Sharipov–Chiao	2004.14.10	2005.24.04	192:19:00:59
Eleven	Krikalev–Phillips	2005.14.04	2005.10.10	179:23:00:00
Twelve	McArthur–Tokarev	2005.03.10	2006.08.04	189:19:53:00
Thirteen	Vinogradov–Williams	2006.29.04	2006.24.09	178 days (planned)

Science on ISS

With the arrival of the US Destiny Lab in 2001, science capabilities aboard ISS increased significantly. Over the next two years, Shuttle re-supply missions delivered, exchanged and returned an array of experiment racks, research facilities and experiments to and from the station. Some experiments in bioastronautics, Earth sciences, fundamental biology, physical science, space product development, and habitability remained on the station for several months, being operated by successive crews and providing extensive sources of data for the principal investigators (PIs).[5]

ISS SCIENCE OFFICER

In 2002, in recognition of the expected expansion of scientific activities on the space station, NASA identified a new "duty assignment" for one NASA astronaut on each ISS main expedition crew, the Science Officer. Initially, they would focus solely on US research, but over time NASA planned to discuss the role with other partners, possibly expanding the concept as the station itself was enlarged and incorporated more scientific research facilities from other member countries. The NASA ISS science officer would work with the US research committee to "understand and meet the requirements and objectivities of each ISS experiment." This was intended to achieve maximum scientific research return and to be the point of contact between the serving ISS resident crew and NASA-sponsored PIs, as well as the developers of payloads, those who integrate these payloads into the mission, and the training teams who prepare the crews. The first person to receive this assignment was ISS-5 flight engineer Dr. Peggy Whitson, who was named as NASA's pioneering ISS science officer in September 2002.[6]

Science officer – a job description

On 2 January 2003, NASA issued the first version of the Description of Duties document for the agency's science officer representative on ISS resident crews.[7] It was stated that only one member of each resident crew would be designated the NASA ISS science officer, even when there was more than one NASA flight crew member in that crew. The primary duty of the assigned crew member would be to "oversee the onboard activities required for NASA investigations on the ISS."

In order to qualify for the role, the designated ISS science officer (SO) should have acquired the "education and credible professional experience to understand and take responsibility for the NASA investigations on board the ISS." The SO also had to be capable of understanding both the technical issues and the scientific basis of each experiment assigned to their mission. An assigned ISS science officer would have to have good communication skills and was expected to interact with the PIs about both the objectives of their experiments and their current status. On top of all this, the SO would have to communicate the merits of the scientific investigations to both the

A close-up of Destiny taken on the STS-100 mission in May 2001 during the second residency which greatly expanded the science programme on the station.

general public and the media and would still have to perform their other assigned duties and roles as part of the resident crew.

The role, developed specifically for resident crews of at least three people, would be divided into three areas of responsibility; pre-launch, in-flight and post-flight.

Pre-launch responsibilities: Upon assignment to an increment crew, a NASA astronaut would be designated the agency's ISS science officer. The SO would assist the increment commander with the development of crew task assignments directly related to scientific investigations for that crew. Both crew members would review the crew payload training schedule and ensure that the different crew members assigned to execute the scientific programme were adequately prepared and trained. Expected to develop a "professional relationship" with the PIs, the SO would arrange meetings with them and spend as much working and training time in their laboratories as the training template would allow.

In-flight responsibilities: During the mission, the NASA SO would assume responsibility as the point of contact for all issues relating to the NASA investigations and experiment hardware on that specific residency mission. Each SO would be the primary operator for many of the NASA experiments, depending upon the final training programme, the allocations of tasks to the assigned crew member and the availability of time on orbit during the mission. To support this assignment, the designated SO would develop a working knowledge of the whole programme of investigations and experiments assigned to their particular expedition. Where possible, the scope of science objectives of previous expeditions and follow-on crews would be used to provide an effective and smooth flow of science research on the station.

The SO would also ensure, to the best of their ability under actual flight conditions and the constraints of flight operations (which always affect even the most well planned and trained-for programme), that the ISS laboratory facilities were operated both smoothly and efficiently to achieve the best possible results for all experiments. This would be done in cooperation with the expedition commander (Russian or American) and the lead and shift flight directors to ensure that crew members and consumable resources were managed to support this effort.

At times, space flight requires reactions and decisions to be almost second nature, and this includes determining priorities among experiments and investigations as the science process develops in real time. Where direct communication with the lead increment scientist on the ground is not possible, the NASA ISS SO would assume responsibility for onboard decisions relating to the investigations. These decisions would be made in collaboration with the expedition commander, the flight director and the lead increment scientists as appropriate. The SO would also be responsible for in-flight repairs of any failed experiments, including assigning a crew member with the appropriate skills to effect such repairs.

Upon the completion of each experiment, the NASA SO would be responsible for ensuring that all the related data (disks, tapes, notes and images) and all samples were preserved correctly and packed for a safe return to Earth, according to documented procedures. The NASA SO would also serve as the main crew spokesperson for media interviews regarding the accomplishment of NASA experiments aboard the station during the given expedition.

Post-flight responsibilities: The role of NASA science officer would continue well after the return to Earth. The astronaut would participate in formal debriefings with the principal investigators and work with the PIs on their post-flight analysis of experimental observations. Other support duties would include conveying any lessons learned (or not learned) to the ISS Program Office and to managers and administrators at the NASA field centres and Headquarters, as well as participating in NASA-sponsored public affairs and outreach activities. As an integral member of the scientific team, the NASA ISS SO would be encouraged to discuss his or her participation in the analysis and to report on experiment findings and their implications. This could include joint authorship of journal publications, participation in technical seminars at research institutes and the presentation of findings at professional conferences.

NASA's first ISS science officer

When NASA announced that Peggy Whitson, at the time already resident on the station, would be the first designated ISS science officer, NASA Administrator Sean O'Keefe stated that she was the obvious choice for the honour. She had earned a doctorate in biochemistry and had received numerous awards and fellowships. Prior to her selection as an astronaut, Whitson had worked as a NASA researcher and had helped to develop experiments for the Space Shuttle. As the project scientist for the

The first NASA science officer, Expedition 5 Flight Engineer Peggy Whitson, working at the Microgravity Science Glove Box in the Destiny Laboratory in 2002.

Shuttle–Mir programme, she also served as the co-chair of the US–Russian Mission Science Working Group. After a career that included the pursuit of scientific knowledge, developing experiments for space flight, working with international partners and training as an astronaut, she was the logical choice for the role of science officer on ISS.

After hearing she was to be named the first ISS science officer, Whitson commented: "It's like being the leader of a team. I was responsible for making sure all the science got done, and that follow-up work was finished. It was my job to make sure science had a high priority while in space. I obviously don't mind the new title, in spite of the fact that my many supportive friends have sent an incredible amount of Star Trek/Mr. Spock-related emails!"

During her mission (184 days between 5 June and 7 December 2002), Whitson conducted science investigations in bioastronautics research (eleven experiments), physical sciences (nine experiments), space product development (five experiments), and space flight development (four experiments). The work she conducted aboard the station included activating and verifying the Microgravity Science Glove Box. Her SO role had to fit in with her duties as a flight engineer and her assignments on the station's robotic arm system, as well as qualifying and training for EVA using Russian EVA hardware and systems. During her pre-flight interview, Whitson revealed that she considered herself to be both a scientist and a builder and explorer, and that becoming an astronaut had focused all of her previous skills and experiences and brought her many new challenges and demands. During the flight, she remained busy with the "mundane" chores of housekeeping and maintenance activities, punctuated with good measures of "fun," such as the robotics and science work and a 4 hour 25 minute EVA. When she was preparing to leave the station at the end of her residency, Whitson admitted to mixed feelings; eager to see family and friends after such a long time away, but also realising that she was not going to live aboard the ISS anymore.

Saturday Morning Science – ISS Science Officer Two

The second designated NASA ISS science officer was Don Pettit, flight engineer on the sixth expedition crew (161 days between 23 November 2002 and 3 May 2003). His background before selection as an astronaut included earning a PhD as a chemical engineer and thirteen years working at Los Alamos National Laboratory. Having long been interested in the space programme, Pettit first tried to join the astronaut corps in 1983. He had been interviewed four times and his persistence (and additional career experiences) finally paid off with his selection in 1996.[8]

Pettit was in training as back-up crew member (for Don Thomas) for ISS-6 and in line to fly to the station as a member of the prime crew for ISS-13, but when Thomas was grounded due to medical disqualification three months prior to launch, Pettit was reassigned to fly the mission. Fortunately, the ISS residency training programme meant he had performed a significant amount of training with the prime crew over the previous eighteen months, making the transition smoother. His main challenge now was to come to terms with flying in space two years before he expected to.

During Expedition Six, Pettit would oversee a package of experiments in bio-astronautics research (ten experiments), physical sciences (six experiments), space product development (two experiments) and space flight development (three experiments). The crew was launched on the Shuttle Endeavour (STS-113 in November 2002) and were scheduled to return aboard STS-114 in the spring of 2003, but during their residency, the orbiter Columbia and her crew of seven were lost in February 2003, grounding the Shuttle for over two years. This meant that logistics would have to be supplied to the station in smaller amounts using the Russian Progress tankers, and the resident crews were reduced to two persons from ISS-7 onwards. The ISS-6 crew returned to Earth aboard Soyuz TMA-1 in May 2003.

A reduced role – ISS science officers 2003–5

One factor in the assignment of a science officer on the reduced two-person "caretaker" crews was the need to perform maintenance work and housekeeping chores at a time when re-supply levels were restricted by the limits of the Soyuz and Progress vehicles. Further expansion of the station (and its science programme) was also delayed by the grounding of the Shuttle fleet. With only one American astronaut

The second US science officer, ISS-6 Flight Engineer Don Pettit, pictured in Destiny.

flying with one Russian cosmonaut from ISS-7 through to ISS-13, the NASA astronaut had to combine the role of FE or ISS commander with that of science officer, although the science was restricted to baseline science due to other responsibilities. The assigned science officers aboard ISS since 2003 have been:

Expedition	**Science officer**	**Other role**
Seven	Ed Lu	Flight engineer
Eight	Mike Foale	ISS commander
Nine	Michael Fincke	Flight engineer
Ten	Leroy Chiao	Flight engineer
Eleven	John Phillips	Flight engineer
Twelve	Bill McArthur	ISS commander
Thirteen	Jeffrey Williams	Flight engineer

Is the "science officer" really a science officer?

There still seems to be a division in the CB between pilots and "scientists" although the term mission specialist has helped narrow the divide somewhat. But in today's programme, are there really any scientists in the astronaut corps? There are relatively few MSs who have served up to ten years between graduation or medical school and their selection as astronauts. Many of them are not what could be called "pure" scientists but worked in areas other than research. Dave Wolf has spent a considerable amount of time working on a cell bioreactor both before and following his selection as an astronaut, while Mike Gerhardt has logged many hours participating in decompression research. Then there is Franklin Chang-Diaz, who has worked on his VASIMIR propulsion system for years.

When the science officer role was instigated in 2002, the plan was to have one per residency crew of three crew members, but the following year Shuttle Columbia was lost and the resident crew of three was consequently reduced to a caretaker crew of just two. As a result, "science" was severely limited. It was hoped that programmes like the "Saturday Morning Science"-type of experiments conducted by Garriott on Skylab and Pettit on ISS could be continued, but this has not happened to date due to a lack of time available to the two-person crew. Perhaps the term "science officer" should be used only when an astronaut is able to perform more than baseline science on board the ISS – nominally when the crew of three is reintroduced.

ISS is an important international facility for good science to be conducted in a unique environment and to tell the world about it. Principal investigators are the best people to inform the media and public about their work and perhaps flying more PIs to conduct their own research after construction of the ISS is completed may be an option. However, this then raises more concerns for requisite training and preparation, in the same way as flying ill-prepared payload specialists or space flight participants.

Simple research and demonstrations were completed by Pettit on ISS-6, and despite being always busy there was still time to fit in the smaller demonstrations. This was due to them being well thought out, pre-planned and capable of evolving during the flight. A direct link between the PI and the astronauts to improve communications would be an advantage, as would a "basic chemistry set" for "Saturday Morning Science" demonstrations, as time allows. It has to be remembered that field science (which ISS research is) is not as well-defined as in a pristine laboratory-controlled environment.

There was interest among the astronauts to pursue more science, but always at the forefront of planning is the need to maintain and control the station, support (eventual) ISS construction and visiting missions, and receive the Progress re-supply missions.

So is the science officer really a science officer? At the moment, probably not – at least not until the Shuttle delivers more research facilities and crew numbers are increased to three. The caretaker crew includes a caretaker science officer, who may be able to perform some limited research, or send emails to tell the world about life and activities on orbit, but full-time "science" gathering has to wait for a larger commitment and, probably, the completion of ISS construction. It is simply a matter of priorities and resources. On Earth, a laboratory is usually built and tested prior to operational use, while scientists trying to complete pristine experiments do not normally have to share floor space with workers in muddy boots. It's the same on ISS: a reduced crew, limited power, logistics management and incomplete construction do not help in supporting a crucial science programme. Hence there are relatively few media reports about science being carried out on the station.

With Skylab, there was already a solar telescope package in place, as well as an Earth resources package and a host of other experiments. The difficulty involved in recovering from the loss of the micrometeoroid shield and solar wing meant science had to wait until later, although the first resident crew worked hard to get as much done as they could in the time they had. The second crew charged ahead and once the third crew had settled in, they provided the best results of the programme – although their station was not under construction, and no longer needed to be repaired. There was a limited programme of maintenance but there were no re-supply facilities, save for what the new crew could bring with them. By comparison, on Salyut and Mir the cosmonauts were constantly rushing around trying to keep the stations flying. Serious failures affecting Salyut 7 in 1984 and Mir in 1997 took precious research time away from the flight plan while these setbacks were overcome. Mir spent fifteen years in orbit and was obviously deteriorating rapidly towards the end of its orbital life. ISS on the other hand is only seven years old but already the paint is wearing off some of the EVA handrails. Will it survive for another decade? Scientists planning research on the station certainly hope so.

Future roles?

The Shuttle performed its long-awaited Return-to-Flight mission (STS-114) in the summer of 2005. Hopes were high for a full resumption of flight operations to complete the construction of the station before the Shuttle's retirement in 2010. However, in-flight incidents indicated that the problem of loose foam falling from the external tank had not been completely resolved and the Shuttle was further grounded until at least mid-2006. Soyuz TMA will remain the sole method of transporting crews to the station until the Shuttle is fully returned to flight-operational status. Until that time, NASA astronauts will continue to be termed science officers, but with the science programme restricted, just how much science is being completed, and is the designation of science officer simply another name for a scientist-astronaut?

ARE SCIENCE OFFICERS TODAY'S SCIENTIST-ASTRONAUTS?

This question was put to some of those selected as NASA scientist-astronauts in the 1960s. They were also asked whether they felt that the role of mission specialist or even payload commander matched what a scientist-astronaut was expected to do during AAP or Skylab.

For Bill Thornton, the answer was simple: "In a word, no!" Tony England, on the other hand, thought it was much the same kind of role: "I had argued really strongly while I was at NASA for the guest investigator or the payload specialist and there was a movement away from that again. My understanding is that it's not really survived into the space station era and I think that is a real mistake. Flying in space is not that hard and as long as you have somebody around who can keep you out of trouble, I think it's really important to use people who spend most of their time thinking about the problem you are trying to solve and the science you are trying to do, rather than careers as an astronaut. That practical opinion did not make me very popular in the CB, but I still feel that way and I think it's a missed opportunity."[9]

In the 1970s, the designation of scientist-astronaut changed to senior scientist-astronaut with the arrival of the new mission specialist candidates. When asked if the roles were the same or different, former astronaut and NASA assistant administrator Bill Lenoir replied, "In the early years, the scientist-astronaut's *intended* and *foreseen* mission was different than the mission specialist's. I think the scientist-astronaut's *actual* mission was exactly the same as the mission specialist's. I think that what we learned was that the role is not as important as we thought. When there are only four of you in orbit, there are a lot of other things to do besides the science – preparing your bed, making your food and so on. I think we just learned that it's a much broader role than a pure science focus, which is why in the latter Shuttle missions many of the mission specialists were chosen for their operations skills rather than their technical expertise."[10]

Expanding on this, Lenoir was asked if the re-naming of the scientist-astronauts to mission specialists for Shuttle changed the way that non-pilot astronauts were looked upon in NASA and the CB, since they no longer need to fly jets and more science and engineering was an integral part of many early Shuttle missions. Lenoir observed: "Yes and no. I think in the first order it didn't really change much. I think when we got that first group in 1978 that contained non-pilots, the military non-pilots were nevertheless aviators. They were navigators or weapons systems officers, and in the respective military services, the pilots and those support people had mutual respect for each other. The scientists who came in with no aviation experience whatsoever were seen as different animals. They were going to get training in the back seat of the airplane, but initially the pilots couldn't trust them until they got to know them, because they had no military background. Having never been in the back seat of an aeroplane, or never held a 'stick', nobody really knew how they were going to behave."

It is clear that the potential for scientists as astronauts was lost along with the cancellation of the AAP/Skylab programme. The three manned Skylab missions produced some of the most outstanding and significant developments in space flight, while the results from the main scientific research programmes of solar science, Earth resources, biomedics, astronomy and technical developments came not only from the scientific ability of scientist-astronauts Kerwin, Garriott and Gibson, but also from their skills at adapting to the role of astronaut. This included learning to fly jets, foregoing their primary science interests, becoming multi-skilled engineers and, particularly, influencing their pilot colleagues to take part in the science data gathering as a crew working together. Skylab was not just about the scientist-astronauts, but also the efforts of the commanders, pilots, back-up and support crews, and hundreds in the team on the ground. Often left in the shadow of Apollo, the full benefits of Skylab are becoming apparent in today's ISS programme. However, many who work on ISS do not realise just how significant the pioneering work aboard Skylab was to the work they are doing in orbit today.

By the Shuttle era, most of the scientist-astronauts had retired, or were battling to get some good work done on their missions. The value of Garriott and Parker on Spacelab 1, Thornton on STS-8 and Spacelab 3, Parker on Astro-1 and Musgrave, England and Henize on Spacelab 2 was reflected in each mission's success. But just as important were their many years of unsung work on the ground in countless hours of meetings, simulations, committees and memos, trying to change the mould not only of NASA but of the larger space community.

So are NASA's science officers on ISS today's science-astronauts? Probably not, one feels. The original idea was to have selected specialists in fields of science applicable to space flight exploration – astronomy, the space sciences, geology, medicine, physics, electrical engineering, nuclear physics and so on. At that time, the Apollo programme was going to expand to include a range of additional flights to the Moon and around the Earth after the national goal of the first landing had been achieved. Most of the effort in the 1960s was aimed towards that goal and though the follow-on plans for AAP, Space Station, Moon bases, Shuttle and manned Mars flights looked good on paper in the excitement of trying to get to the Moon, the hardware, promotion, and public and political support was not as well organised and significant as it was at the height of the Cold War.

What replaced these projected visions was a tighter, budget-orientated programme, featuring reusable equipment, returns from the investment in space, commercialisation, utilising less equipment to gain more from each flight and a multi-task role for the astronauts. With the realisation that Shuttle would not be as cheap and frequent as first intended, and that the space station would not be ready "in a decade," NASA changed from a commercial promotion agency to an international cooperation agency. For the astronauts, past careers and academic qualifications were not considered purely to be able to fly a doctor on a medical mission or an astronomer on a observing mission, but to demonstrate that the individual had the ability to learn, because once they entered the astronaut programme, they went right back to school and became a team player learning a multitude of roles and assignments.

From 1978, the new NASA "science astronauts" became known as mission specialists, a broader and more operational role than purely scientific research. In the early 1990s, the role of payload commander was created to allow leading MSs to take a more active role in integrating payload and mission objectives into the planning of crew activities. This would evolve again for the longer duration Space Station programme once the science hardware was up and running.

In the first selection of candidates for the Space Shuttle, eleven of the twenty mission specialists held PhDs, and three others held a doctorate in medicine. Since then, there have been a further eleven selections. In the most recent (Group 19 – May 2004), three former teachers were selected as "educator-astronauts", joining former payload specialist teacher Barbara Morgan (STS 51-L), who was selected for MS training by NASA in June 1998.

At the time of their selection to the various groups, the MS candidates had achieved a total of thirty-eight PhDs and fifteen medical doctorates:

Group	Year	Selected	MS	PhDs	MDs
9	1980	19	11	2	2
10	1984	17	10	1	2
11	1985	13	7	3	0
12	1987	15	8	2	1
13	1990	23	16	8	2
14	1992	19	15	7	4

[*Note:* Dan Barry held both a PhD and an MD, and Richard Linnehan held a Doctor of Veterinary Medicine. Not included in the figures is pilot candidate Scott "Doc" Horowitz, who had also acquired a PhD.]

15	1994	19	9	5	0
16	1996	35	25	8	4

[*Note:* Lee Morin holds both an MD and a PhD.]

17	1998	25	17	7	1
18	2000	17	10	5	1
19	2004	11	9	2	1

[*Note:* None of the educator MS candidates held a PhD.]

The scientist-astronauts used their experience from the pioneering days of NASA to try to influence changes in the Shuttle's Spacelab design that would benefit crews in orbit, although not necessarily the scientist on the ground, as they would have done on AAP. In some cases, this was not a popular approach. Trying to influence scientists on the ground to design equipment and work procedures that could be successful off the Earth was not as easy or as straightforward as it had been in the early days when NASA was younger, more focused and, publicly and outwardly, an "exciting" agency. For the new mission specialists, their training and roles on Shuttle flights required adapting to the new NASA way and not (with a few exceptions) continuing their role in research science or engineering.

As the number of "science" experiments on the Shuttle increased from the late 1980s, replacing the military, commercial and planetary payloads removed after the loss of Challenger in 1986, six mission specialist astronauts from the 1978 and 1980 selections formed an Astronaut Science Support Group on 29 February 1988. It was designed to provide direct interaction with prospective experimenters on both Shuttle and future space station missions.[11] Focusing on increasing the scientific and engineering flexibility of experiments in space without violating the strict Shuttle operating guidelines, the group believed their experience in the design, development and operation of experiments gained from past Shuttle missions would improve the return of data and simplify the repair of equipment in space, thus maximising the scientific return from each experiment or investigation. As CB points of contact between the scientific "customers" NASA, and the Astronaut Office, they specialised in the areas of:

- plasma and space physics (Franklin Chang-Diaz, also the group leader);
- biological material processing (Mary Cleave);
- material processing (Bonnie Dunbar);
- astrophysics and remote sensing (Jeff Hoffman);
- EVA, satellite servicing and space construction (Jerry Ross); and
- life sciences (Rhea Seddon).

During the 1990s, the potential for multiple Shuttle flights supporting the same or similar experiment payloads (such as the Space Radar Laboratory, ATLAS Earth observation missions, Astro astrophysical payloads and Spacelab long module and pallet missions) was curtailed and finally cancelled by the involvement of the Shuttle in the Soviet Mir programme, and the diversion of limited resources to construct the International Space Station.

As the ISS programme increased, the science role of mission specialists decreased, to that of manager of logistics, operator of the RMS, member of an EVA team and centre-seat flight engineer. On ISS construction and logistics missions, the science payload on the Shuttle decreased significantly due to mass limits, time constraints and the workload of the crew in other areas. In the era of ISS orbital operations, during Phase 1 (Shuttle-Mir) between 1995 and 1998, there were ten missions related to rendezvous and docking operations with Mir and sixteen "scientific" Shuttle missions. During the first period of ISS construction between 1998 and 2003, there were sixteen assembly missions and just five scientific missions. With the Shuttle grounded from 2003 to 2005 and delayed again until 2006, and with all further flights devoted to ISS construction (apart from a probable Hubble servicing mission), the opportunity to fly mission specialists for science operations has diminished to almost zero in a decade.

The background of payload specialists is a completely different story. Originally intended to provide non-astronaut scientists (including foreign citizens) with the opportunity to perform experiments on the Shuttle in Spacelab, or to accompany national payloads into space, their role and participation was always at odds with different departments within NASA and with political and commercial circles. The flights of some of the payload specialists were seen more as a public affairs exercise than pure scientific need, although some participants have undoubtedly contributed to some great science. Some payload specialists have managed to fly more than once and three have paid the ultimate price in taking on the challenge. But it is interesting to note that no payload specialist has made it to the ISS, thanks primarily to changes in NASA policy, restrictions in the scientific capabilities of the station (as a consequence of delayed launches), and the loss of both Challenger and Columbia.

On the other hand, the Soviets have launched "paying customers" on Soyuz taxi missions and flown European astronauts on Soyuz exchange missions. None, however, has yet performed an extended mission (although Thomas Reiter is scheduled to join the ISS-13 resident crew in 2006). Part of the problem was the old agreements signed between the partners, in which American and Russian crew members would be joined by European and Japanese crew members *when their hardware was attached to ISS*. Since this is still to be achieved, NASA will only launch international researchers to the station as long as it is safe and prudent to do so, providing at least one American remains on board along with one Russian at all times. With the current two-person crews, there has simply been no room for any others.

Memories from orbit

Memories of vivid events in one's life can remain with you forever and this has to be true for a flight into space. After being selected to fly at a time of pioneering the exploration of space, struggling to be accepted as competent astronauts, waiting long years to fly and proving their worth on more than one occasion, what remains for the scientist-astronauts as their overriding memory of their participation or contribution in man's greatest adventure?

For Bill Thornton, it was the work he did on SMEAT and Skylab that was most rewarding in the early years, along with the achievement of making it to orbit. "Just getting there with my lab on STS-8 was the highest point ... just making it. It was not that I did not want to look out the window – there were some memorable scenes during the brief time I had – but being there and being able to make the measurements, in particular with the tools I developed myself, were for me another high point." He said he had learned more in one orbit than in all the years of research on the ground combined. Asked if there was a high point for him on 51-B, Thornton replied, "Yep. I got to look out the window a lot more on 51-B."

His greatest disappointment during his years in the CB was the inability to work with the Life Sciences Group. "It was a constant confrontation, and I would be the first to step up and say that it was understandable and predictable. I now know that they believed firmly in what they were doing, as I believed in what I was doing, and they believed I had no business doing it. This is still true today. So my greatest disappointment was not being able to work with them in a meaningful way. It was a constant confrontation, for which I will accept more than my share of the blame."

Bill Lenoir's most memorable impression on his first and only flight was his first look out of the window at the Earth. "I could say, hey, the guys who drew the maps did it right!" Asked about his personal contributions to the space programme, he answered the question in two ways: "As an administrator, working at Headquarters, it was getting the Space Shuttle under cost control ... [and] under schedule control so we flew when we said we would. As an astronaut, it was developing the flight engineer role, developing the deployment procedures for satellites and putting together the whole aspect of how we performed EVAs from the Shuttle."

For England, it was the Apollo era that gave him the greatest satisfaction in his first term at NASA, and his only space flight during his second term. "It's fun to tell people that I was the crewman in the *Apollo 13* movie who went away and worked on the lithium-hydroxide canisters. It makes a nice story. But I think the science we did on Apollo 16 was probably better because I was involved in it. Not that somebody else couldn't have done as well, but I know that even the choice of the landing site and some of the experiments – the way we approached the experiments – I had an influence in, so I am very happy with that involvement. For Spacelab 2, I felt I made a contribution and, given the situation, we got as much out of the mission as we could have, so I am happy with those experiences." For England, as with others, a most particular joy was the view of the Earth from orbit. "I had spent a couple of seasons in Antarctica, and we were in high-inclination orbit so we could see over Antarctica. The southern aurora was active at the time, so you could see the green curtain and the waves in the curtain, which you can't see from the surface of the Earth. While we were up there, the annual meteor showers were going on, so while I watched the southern aurora, I could see shooting stars below. That was a scene that was otherworldly." As for going to NASA and coming back a second time, he felt, "That was not the wisest thing to do for a science career, but I do not regret it at all. The benefit for me was the adventure and it was well worth the cost."

Owen Garriott commented that the most rewarding thing for him was the long-duration flight of fifty-nine days on Skylab, demonstrating "that you can indeed live and work and conduct very good science in space. Though we take that for granted now, in 1973 that was not an accepted fact, so the science that was done and the demonstrations that were made of the ability to live and work in space were the contributions that were the most satisfying to me."[12]

Story Musgrave, the final scientist-astronaut to leave the CB, who flew six missions on the Shuttle over a period of thirteen years during a thirty-year career at NASA, very much lived the dream: "The spirit of it, over thirty years, was my calling, and I did pretty much the best I could every minute. I had the passion for it and I did make a difference. Some people were sad to see me go because they knew that some of the spirit and the passion would go."[13]

And indeed, some of it has. The NASA these men joined in the mid-1960s was an agency with the spirit and passion to reach for the Moon and beyond, to believe in the impossible and make it happen. The NASA of today is, naturally, very different, operating in a different era. The commitment is certainly there, but the spirit and the passion has been knocked a few times. One can only hope that with the Vision of Exploration announced by President George W. Bush, that same passion will once again make NASA reach for the impossible and make dreams a reality. In a way, they have done that with ISS. In the 1970s, the idea of a large space station was a far-off dream. In the 1980s, it became a Presidential objective, but in the 1990s, it almost became an embarrassment. In the first decades of the new century, it is ISS that will carry mankind forward towards new dreams and goals. The role of a scientist on that station, or indeed as part of any exploration crew, will be an enduring legacy of the NASA Scientist-Astronauts, Classes of 1965 and 1967.

REFERENCES

1 *The Space Station*, Ed Theodore, R. Simpson, IEEE Press, 1985; *The Space Station Decision*, Howard E. McCurdy, John Hopkins University Press, 1990; *Creating the International Space Station*, David Harland and John Catchpole, Springer-Praxis, 2002; *The Continuing Story of the International Space Station*, Peter Bond, Springer-Praxis, 2002.
2 *Boeing Press Information Book, International Space Station*, 1998 with amendments through September 2000.

3 *The International Space Station: From Imagination to Reality*, Ed. Rex Hall, British Interplanetary Society (BIS), 2002
4 *The ISS, From Imagination to Reality*, Vol. 2, Ed. Rex Hall, BIS 2005.
5 *Research in Orbit* by Andy Salmon, from The International Space Station, BIS 2002, previously cited, pp. 119–143.
6 *NASA Administrator Names Whitson First NASA ISS Science Officer*, NASA News Release #H02-175, 16 September 2002.
7 NASA ISS Science Officer Description of Duties, Version 1, 2 January 2003.
8 Scientist Profile: Don Pettit, Astronaut, interview by David S.F. Portree, Earth & Sky Radio Show, March 2004.
9 AIS telephone interviews with Tony England, 14 July 2005, and Bill Thornton, 17 February 2006.
10 AIS telephone interview with Bill Lenoir, 24 June 2005.
11 NASA News (JSC) 88-004, 29 February 1988.
12 AIS telephone interview with Owen Garriott, 27 June 2005.
13 AIS telephone interview with Story Musgrave, 23 June 2005.

Appendix 1

Chronology of the NASA Scientist-Astronaut Programme

1957 Oct 4	The Soviet Union opens the space age with the launch of Sputnik 1
1958 Oct	NASA is created as an American civilian space agency to manage and develop US civilian man in space programmes
1959 Apr	NASA selects America's first astronauts – seven test pilots chosen for the one-man Project Mercury programme
1960 Mar	The Soviet Union selects its first team of cosmonauts – twenty military pilots, including Yuri Gagarin, primarily for the one-man Vostok programme
1961 Apr 12	Cosmonaut Yuri Gagarin becomes the first person to fly in space. A single orbit aboard Vostok is completed in 108 minutes
1961 May 5	Astronaut Alan Shepard becomes the first American in space – but not into orbit. He completes a 15-minute sub-orbital mission aboard Mercury 3 (Freedom 7)
1961 Aug 6	Cosmonaut Gherman Titov spends twenty-four hours in space aboard Vostok 2
1962 Feb 20	Astronaut John Glenn becomes the first American to orbit Earth – three times – in Mercury 6 (Friendship 7)
1962 Sep 17	NASA selects a second group of nine pilot astronauts
1962 Dec	Dr. Homer Newell, NASA's Director of the Office of Space Sciences, expresses the hope that scientists would be among NASA's next (third) group of astronauts
1963 Jan 6	A National Academy of Sciences-sponsored Space Science Summer Study report recommends that "trained scientist-observers be assigned important roles in future US space missions"
1963 Jun 16	Valentina Tereshkova, a former sports parachutist and not a career pilot, becomes the first woman in space, giving evidence that the Soviets were gathering biomedical data on non-pilot subjects and may be preparing to launch a crew with scientists, not pilots

1963 Jun 18	NASA announces the recruitment for a third group of astronauts, but still with a major focus on piloting skills rather than academic qualifications
1963 Sep	Robert A. Voas, Human Factors Assistant at MSC, confers with Dr. Homer Newell about the best qualifications for specialists to enter the astronaut programme in order to fulfil Apollo and AAP missions. Geology, geophysics, medicine and physiology were identified
1963 Oct	NASA selects fourteen new pilot astronauts
1964 Apr	The Space Science Board of the National Academy of Sciences defines appropriate scientific qualifications for selecting NASA's first group of scientist-astronauts
1964 Oct	The Soviets fly Voskhod 1 with three crew members, including a scientist (Feoktistov) and a doctor (Yegorov), commanded by an experienced Air Force pilot (Komarov)
1964 Oct	NASA announces that recruitment would begin for astronauts with a scientific background
1965 Jun 28	NASA names six new astronauts in its fourth selection since 1959. They are Garriott, Gibson, Graveline, Kerwin, Michel and Schmitt; all scientists
1965 Jul 29	Garriott, Gibson, Graveline and Schmitt commence a 53-week USAF jet pilot training course at Williams AFB, Arizona (Kerwin and Michel were already qualified). With the exception of Graveline, who resigned in August 1965, they graduate in July 1966
1966	Group 4 astronauts complete academic and survival training with Group 5 astronauts
1966	During the summer, Kerwin, Ed Givens of the 5th Astronaut Group and USAF Captain Joe Gagliano complete a 163-hour manned test of the Block I CSM in Altitude Chamber A at MSC in Houston
1966 Sep 26	NASA announces it will begin selection of a second group of scientist astronauts
1967 Aug 4	Eleven new astronauts are named by NASA in the sixth selection since 1959. The second group of scientist astronauts are: Allen, Chapman, England, Henize, Holmquest, Lenoir, Llewellyn, Musgrave, O'Leary, Parker and Thornton
1968 Mar	Group 6 astronauts commence a USAF pilot training programme at five air bases across the United States. After graduating the course in 1969, they would complete their survival and environmental training
1968 Apr 23	O'Leary resigns from NASA after only eight months due to his distaste for pilot training and separation from his science research projects. He takes up a position as Assistant Professor of Astronomy, University of Texas. O'Leary's experiences and frustrations were expressed in his autobiography, *The Making of an Ex-Astronaut* (1970)
1968 Jun 16–24	Astronaut Kerwin (commander), Vance Brand (CMP) and Joe Engle

	(LMP) complete a 177-hour simulated mission in a Block II CSM in Altitude Chamber A at MSC, Houston
1968 Sep 6	Llewellyn resigns from NASA after only one year due to his inability to progress sufficiently in the required USAF jet pilot training programme. He becomes Professor of Chemistry at the University of South Florida, Tallahassee, Florida
1968 Aug	Schmitt (with Group 5 astronaut Don Lind) performs a full ALSEP deployment demonstration on a simulated lunar surface wearing an Apollo EMU
1969 Jun 10	Thornton's graduation from jet pilot training is delayed due to an eye problem that would ground him for over 8 weeks until he adjusts to wearing the corrective glasses employed to resolve the problem.
1969 Jul	Garriott serves as shift (Maroon Team) Capcom for Apollo 11. Schmitt serves as an incidental Capcom during the mission
1969 Aug 6	Chapman is named to the support crew of Apollo 14. As mission scientist, he was Capcom during the flight in January–February 1971
1969 Aug 18	Michel resigns from NASA after four years due to "too few opportunities to fly on a mission"
1970 Mar 26	Schmitt is named back-up LMP Apollo 15; Allen, Henize and Parker are named as support crew members, with Allen also serving as mission scientist. They worked to support the flight during July–August 1971, including serving as incidental Capcom
1971 Mar 3	Chapman and England are assigned as support crewmen to Apollo 16. England is also named mission scientist for the April 1972 mission
1971 Aug 13	Schmitt is named LMP Apollo 17, the last planned mission to the Moon. He was reassigned from the cancelled Apollo 18 mission, replacing Joe Engle, to allow a geologist to reach the Moon. Parker is named to the support crew and as mission scientist
1972 Jan 16	NASA names Kerwin, Garriott and Gibson as science pilots on the three Skylab missions. Musgrave is named back-up science pilot for the first manned flight and Lenoir to the second and third. Henize, Thornton and Parker are assigned as support crew members (and Capcom) for all three manned missions
1972 Jul 14	Chapman resigns from NASA after almost five years to accept a position with AVCO Research Laboratory, Massachusetts, as a principal research scientist
1972 Jul 26	Thornton participates as a Skylab Medical Experiment Altitude Test crewmember. He completes a 56-day ground simulation (in an altitude chamber) of a Skylab mission to obtain baseline medical and operational data for the three manned missions
1972 Aug 14	England resigns from NASA after five years due to lack of flight opportunities, to take up a position with the US Geological Survey in Denver, Colorado
1972 Dec 6	Schmitt is launched as LMP Apollo 17, the sixth manned lunar

	landing mission. He landed on the Moon on 12 December and made three surface EVAs, and a stand-up EVA during the return flight. Allen served as incidental Capcom for Apollo 17
1973 Feb 19	Parker is designated Skylab Program Scientist for all three manned missions in addition to his support crew/Capcom roles
1973 Jul 28	Garriott is launched aboard Skylab 3 and spends 59 days in space, completing three EVAs
1973 Sep	Holmquest resigns from NASA after six years to take the post of Assistant Dean for Academic Affairs in the College of Medicine at Texas A&M University, Navasota, Texas. He had been on extended leave of absence since May 1971
1973 Dec 7	In a reorganisation at JSC, Garriott is assigned as Deputy Director, Science and Applications at JSC Houston. Schmitt is named Chief of Scientist Astronauts
1973	Allen takes a two-year leave of absence from NASA to serve on the President's Council in International Economic Policy
1974 Feb 6	Henize survives a landing mishap of his T-38, at Bergstrom AFB in Texas, when the undercarriage collapses. Though the aircraft suffers some damage, Henize is not injured
1974 May	Schmitt is named NASA Assistant Administrator for Energy Programs at HQ, Washington DC
1974 Aug 15	Garriott is named Acting Director, Science and Applications, JSC Houston. Parker becomes Chief of the Scientist Astronaut Office, replacing Schmitt
1974 Sep 30	Henize participates in the fourth Learjet Spacelab simulation, an airborne science programme as part of the Spacelab development programme. The tests ended on 4 October
1974 Oct 1	Musgrave begins a one-week ground-based Spacelab Medical Development test (SMD-I), conducting a series of twelve biomedical experiments as part of the Spacelab development programme
1975 Jun	Parker participates in the ASSESS-I space science simulation flights aboard the NASA Convair 990 aircraft, part of the Spacelab development programme
1975 Aug 1	Allen begins a three-year term as NASA's Assistant Administrator for Legislative Affairs at NASA HQ in Washington DC.
1975 Aug 30	Schmitt resigns from NASA after ten years to run for the US Senate. On 2 Nov 1976, he was elected as New Mexico Republican Candidate for the Senate. He was defeated in the November 1982 election
1976 Jan 26	Musgrave participates in the second Spacelab Medical Development test (SMD-II), a ground-based test as part of the Spacelab development programme. Tests end 1 February
1977 May 17	Thornton participates in a seven-day ground-based Spacelab Medical Development test (SMD-III), a life sciences programme which ended 24 May, as part of the Spacelab development programme
1977 May	Henize participates in the ASSESS-II space science simulation flights

	aboard the NASA Convair 990 aircraft, part of the Spacelab development programme
1977	Kerwin, Gibson and Parker serve as members of the Group 8 astronaut selection panel, the first Shuttle era selection
1978 Jan 16	NASA names 35 new astronauts to train for the Space Shuttle programme. In addition to fifteen pilot candidates, there are twenty mission specialist candidates (a new designation), including the first six women selected by NASA for astronaut training
1978 Mar	Gibson returns to NASA CB as a "senior scientist-astronaut", to prepare for flights aboard the Space Shuttle as a mission specialist. He is also assigned to the STS-1 support crew and would serve as a Capcom during the launch phase
1978 Jul 28	Allen returns from NASA HQ to JSC as a "senior scientist astronaut", to train as a mission specialist for future Shuttle flights
1978 Aug 3	Garriott and Parker are named as mission specialists to Spacelab 1
1979 Jun 3	England returns to NASA CB as a "senior scientist-astronaut", to prepare for future Shuttle flights as a mission specialist. He is assigned technical duties for the Spacelab 2 payload
1979 Oct	Holmquest writes a letter to George Abbey requesting to return to CB astronaut duties for a possible assignment as a mission specialist on a Shuttle flight. He acknowledged that mistakes were made on his first tour (1967–1973). He did not return to CB
1980 Oct 31	Gibson resigns from NASA a second time (after two years), when the prospect of an early Shuttle flight was not forthcoming. He had hoped for an MS assignment on STS-5 or 6
1980	STS CB planning documents indicate that astronauts Henize and Thornton could be flying as PS rather than MS, though both eventually flew as mission specialists
1981 Apr 14	Allen serves on the support crew of STS-1, acting as entry and landing Capcom on the "Silver" (Entry) flight control team for the first Shuttle mission
1982 Mar 1	Allen and Lenoir are named as MS for STS-5, a satellite deployment mission; Musgrave is named MS to STS-6, a TDRS deployment mission
1982 Apr	Kerwin is reassigned as NASA Representative to Australia; he had been preparing for a possible assignment as MS on a Shuttle mission (STS-13/Solar Max)
1982 Nov 11	Allen and Lenoir are launched aboard Columbia on the first "operational" Shuttle flight STS-5, the first mission to deploy a commercial comsat. A planned EVA was cancelled due to several technical difficulties. Landed 16 November after 122 hours 14 minutes
1982 Dec 21	Thornton is added to the STS-8 crew, with the specific purpose of conducting on-orbit studies of space adaptation syndrome (space sickness)
1982 Feb 22	England and Henize are named as MS to the Spacelab 2 mission

	(which became STS 51-F). Henize had been working on the payload since 1977. Thornton is named to the Spacelab 3 dedicated microgravity mission (which became STS 51-B)
1983 Apr 4	Musgrave, serving as an MS, is launched aboard Challenger on her maiden flight (STS-6) to deploy the first TDRS. Musgrave also participated in the first Shuttle-based EVA (4 hours 17 minutes) in the payload bay, to evaluate Shuttle EVA hardware and procedures. Mission landed on 9 April after 120 hours 24 minutes
1983 Aug 30	Thornton launched as MS aboard Challenger (STS-8) to study the crew's adaptation to space flight
1983 Sep 21	Allen is named MS to Rick Hauck's Shuttle crew. Originally assigned to STS-16, schedule shifts result in changes to the mission profile before the mission was designated STS 51-A
1983 Nov 17	Musgrave is named MS on the Spacelab 2 mission (which became STS 51-F)
1983 Nov 28	Garriott and Parker fly as MS aboard STS-9 (Columbia, carrying Spacelab 1), landing on 8 December after 247 hours 47 minutes. This was Garriott's second space flight
1983 Dec	Kerwin is named NASA Director of Life Sciences at JSC
1984 Jun 7	Garriott is named MS for the STS 51-H Earth Observation Mission (EOM) 1. Parker is named MS for the Astro-1 mission (originally STS 61-E, but which flew as STS- 35 following several delays in the manifest)
1984 Sep 31	Lenoir resigns from NASA after almost seventeen years to take up a senior position with the management consultancy company Booz, Allen and Hamilton Inc. in Alexandria, Virginia. He had been selected to fly on the first German Spacelab mission and had started language training, but did not wish to commit to extended training visits to Europe
1984 Nov 8	Allen flies his second mission as MS on STS 51-A, a satellite deployment and retrieval mission. He completed two EVAs totalling 11 hours 42 minutes, supporting the recovery of two stranded satellites, including a 2 hr 22 min untethered flight on an MMU
1985 Apr 29	Thornton flies his second mission, as MS aboard STS 51-B (Spacelab 3). Mission ended on 6 May after 168 hours 9 minutes
1985 Jul 1	Allen resigns from NASA after almost eighteen years service, to join Space Industries in Houston as an Executive Vice President
1985 Jul 29	England, Henize and Musgrave fly as MS aboard Challenger, with the Spacelab-2 payload (STS 51-F). An Abort-to-Orbit situation was overcome, allowing the mission to be completed successfully after 190 hours 45 minutes, landing on 6 August. It was Musgrave's second spaceflight
1986 Jan 28	Challenger is lost 73 seconds after launch, killing seven astronauts aboard, including teacher Christa McAuliffe. The Shuttle manifest is

	suspended pending a full review and evaluation of the future of the programme
1986 Apr	Henize transfers from the CB to the JSC Space Sciences Branch as a senior scientist, working on space debris issues
1986 Jun 11	Garriott resigns from NASA after twenty-one years, to become Vice President, Space Programs, Teledyne Brown Engineering, Huntsville, Alabama
1987 Mar 31	Kerwin resigns from NASA after more than twenty-two years to join Lockheed
1988 Mar 1	Parker becomes Director of Spaceflight, Space Station Integration Office
1988 Aug	England resigns from NASA a second time (after ten years) to become Professor of the Electrical Engineering and Computer Science Department at the University of Michigan. He was a candidate for the Sunlab series of Spacelab missions to re-fly Spacelab 2 experiments which were cancelled in the wake of the 1986 Challenger accident
1988 Sep 29	STS-26 completes a successful mission, returning the Shuttle to flight following the Challenger tragedy
1988 Nov 30	Parker is named as MS to STS-35, the new designation of STS 61-E (Astro 1). Musgrave is named MS to STS-33, a DoD mission
1989 Jul	Lenoir returns to NASA as Associate Administrator for Space Station Freedom and is later promoted to Associate Administrator for Space Flight
1989 Nov 22	Musgrave flies his third mission as MS aboard STS-33 (Discovery), a classified DoD mission. Duration was 120 hours 6 minutes, ending on 27 November
1990 May 24	Musgrave is named MS to STS-44, a dedicated DoD mission
1990 Dec 2	Parker flies as MS STS-35 (Astro 1). Duration was 215 hours 6 minutes, ending on 10 December
1991 Jan	Parker is reassigned as Director of Division Policy and Plans, Office of Spaceflight, at NASA HQ, Washington DC.
1991 Nov 24	Musgrave flies his fourth mission as MS STS-44 (DoD). Duration was 166 hours 52 minutes, landing on 1 December
1992 Jan	Parker becomes Director of the Spacelab and Operations Planning Program at NASA HQ in Washington DC until November 1993
1992 Feb 21	Musgrave is named Payload Commander MS for STS-61 – the first Hubble Space Telescope service mission
1992 May	Lenoir resigns from NASA a second time after almost three years as Associate Administrator for Space Flight
1993 Dec	Parker serves as Manager of the Space Operations Utilization Program at NASA HQ in Washington DC until August 1997
1993 Dec 2	Musgrave flies his fifth mission as MS/PC STS-61 (HST-SSM-1), during which he completes three EVAs in a mission lasting 259 hours 59 minutes, landing on 13 December

1994 May 31	Thornton resigns from NASA after twenty-seven years to take up a teaching post
1996 Jan 17	Musgrave is named as MS to STS-80
1996 Nov 19	Musgrave flies his sixth mission as MS on STS-80, a space sciences mission that lasted 423 hours 54 minutes, landing on 6 December
1997 Aug	Parker serves as Director of the NASA Management Office at the Jet Propulsion Office in Pasadena, California for the next eight years
1997 Sep 2	Musgrave reluctantly resigns from NASA after thirty years of active service as an astronaut "to pursue other interests." The only man to fly all five orbiters, he is the last of the seventeen scientist-astronauts selected in 1965 and 1967 to retire from active flight status, ending an era
1998 Nov	First element of ISS is launched
2000 Oct	First resident crew to ISS is launched
2002	ISS resident Peggy Whitson is nominated first Space Station Science Officer while in orbit
2003 1 Feb	Columbia is lost, with all seven astronauts, following a two-week science research mission designed to support ISS science research fields
2005 Jun 28	The fortieth anniversary of the selection of the first group of NASA scientist-astronauts
2005 Jul 26	The STS-114 Return-to-Flight mission is launched to ISS. Although the mission is a success, problems with ET foam debris delay subsequent missions into 2006.
2005 Aug 31	Parker retires from NASA after 38 years serving as an astronaut and manager. He became the final member of the two scientist-astronaut groups to retire from NASA ending a 40-year programme

Appendix 2

Scientist-Astronaut Careers and Experience

506 Appendix 2

Name	Born	Selected	Group	Flights	First	Second	Third	Fourth	Fifth	Sixth	Hours	EVAs	Astronaut status	Note
Allen J.	1937	1967	6	2	1982	1984					314	2	Former 1985	
Chapman	1935	1967	6	0	—						0	0	Former 1972	
England	1942	1967	6	1	1985						190	0	Former 1972/1988	[1]
Garriott	1930	1965	4	2	1973	1983					1651	3	Former 1986	
Gibson E.	1936	1965	4	1	1973						2017	3	Former 1974/1980	[2]
Graveline	1931	1965	4	0	—						0	0	Former 1965	
Henize	1926	1967	6	1	1985						190	0	Former 1986; deceased 1993	[3]
Holmquest	1939	1967	6	0	—						0	0	Former 1973	
Kerwin	1932	1965	4	1	1973						673	1	Former 1984; mgr 1984–1988	[4]
Lenoir	1939	1967	6	1	1982						122	0	Former 1984; mgr 1989–1992	[5]
Llewellyn	1933	1967	6	0	—						0	0	Former 1968	
Michel	1934	1965	4	0	—						0	0	Former 1969	
Musgrave	1935	1967	6	6	1983	1985	1989	1991	1993	1996	1281	4	Former 1997	
O'Leary	1940	1967	6	0	—						0	0	Former 1968	
Parker	1936	1967	6	2	1983	1990					463	0	Former 1991; mgr 1991–2005	[6]
Schmitt	1935	1965	4	1	1972						302	4	Former 1975	[7]
Thornton W.	1929	1967	6	2	1983	1985					313	0	Former 1994	

Notes:
[1] England resigned from the Astronaut Office in 1972 only to return in 1978, retiring a second time ten years later.
[2] Gibson resigned from the CB in 1974 but returned in 1978 for two years, leaving a second time in 1980.
[3] Henize left the CB in 1986 but remained at NASA's JSC Space Sciences Branch until his death while ascending Mt. Everest in 1993.
[4] Kerwin was NASA Representative in Australia for two years from 1982; upon his return, he became Director of Space and Life Sciences until leaving NASA in 1988.
[5] Five years after resigning from astronaut status, Lenoir returned to NASA as an Associate Administrator, initially for Space Station but later for Spaceflight.
[6] Parker relinquished his astronaut status after his second space flight, taking a management role at NASA HQ and, from 1997, at JPL California, until he retired in August 2005.
[7] After leaving NASA, Schmitt ran for Republican nomination for New Mexico in the US Senate. He was elected in 1976 but subsequently defeated in 1982

Appendix 3

Spaceflight Records and EVA Experience

SPACEFLIGHT RECORDS

Name	Group	Flights	Missions	dd:hh:mm:ss	Career total	Notes
Gibson	4/1965	1	Skylab 4	84:01:16:00	84:01:16:00	[1]
Garriott	4/1965	2	Skylab 3	59:11:09:04		[2]
			STS-9	10:07:48:17	69:18:57:21	
Musgrave	6/1967	6	STS-6	05:00:24:36		[3]
			STS 51-F (19)	07:22:46:22		
			STS-33	05:00:07:50		
			STS-44	06:22:52:28		
			STS-60	08:07:10:13		
			STS-80	17:15:54:26	50:21:15:55	
Kerwin	4/1965	1	STS-2	28:00:49:49	28:00:49:49	[4]
Parker	6/1967	2	STS-9	10:07:48:17		[5]
			STS-35	08:23:06:05	19:06:54:22	
Allen J.	6/1967	2	STS-5	05:02:15:29		
			STS 51-A (14)	07:23:45:59	13:02:01:28	
Thornton	6/1967	2	STS-8	06:01:09:33		[6]
			STS 51-B (17)	07:00:09:53	13:01:19:26	
Schmitt	4/1965	1	Apollo 17	12:13:51:59	12:13:51:59	[7]
England	6/1967	1	STS 51-F (19)	07:22:46:22	07:22:46:22	[8]
Henize	6/1967	1	STS 51-F (19)	07:22:46:22	07:22:46:22	[8]
Lenoir	6/1967	1	STS-5	05:02:15:29	05:02:15:29	

Notes:
1. Skylab 4, the third manned mission, set a world endurance record held until 1978 (Salyut 6 1st resident crew – 96 days). Gibson jointly held the US record for one flight/career total until 1995 (Norman Thagard on Mir – 115 days)
2. Skylab 3, the second manned mission, set a world endurance record held until Skylab 4. STS-9 carried the Spacelab 1 long-module configuration
3. Musgrave became the only scientist-astronaut to fly more than two missions (he flew six) and the only person to have flown on each of the orbital class Shuttle vehicles (Columbia, Challenger, Discovery, Atlantis and Endeavour). STS 51-F flew the Spacelab 2 pallet-only configuration. STS-80 set the record for the longest Shuttle flight
4. Skylab 2, the first manned mission, set a world endurance record until surpassed by Skylab 3
5. As well as flying on Spacelab 1 (STS-9), Parker flew on the Astro 1 mission (STS-35)
6. STS 51-B also carried the Spacelab 3 long-module configuration
7. During Apollo 17, Schmitt became the only scientist-astronaut to walk on the Moon (12th man). He also stayed for 75 hours 00 minutes on the lunar surface and 72 hours 44 minutes in lunar orbit during the mission
8. STS 51-F carried the Spacelab 2 pallet-only configuration

EVA EXPERIENCE

Name	Group	EVAs	Missions	Date	Location	hh:mm	Career total
Musgrave	6/1967	4	STS-6	1983 Apr 7	OV099/Earth orbit	4:17	
			STS-60	1993 Dec 4	OV105/Earth orbit	7:54	
				1993 Dec 6	OV105/Earth orbit	6:47	
				1993 Dec 8	OV105/Earth orbit	7:21	26:19
Schmitt	4/1965	4	Apollo 17	1972 Dec 11	LM/Lunar surface (12th man)	7:12	
				1972 Dec 12	LM/Lunar surface	7:37	
				1972 Dec 13	LM/Lunar surface	7:15	
				1972 Dec 17	CM/deep space stand up	1:06	23:10
Gibson	4/1965	3	Skylab 4	1973 Nov 22	OWS/Earth orbit	6:33	
				1973 Dec 29	OWS/Earth orbit	3:38	
				1974 Feb 3	OWS/Earth orbit	5:19	15:30
Garriott	4/1965	3	Skylab 3	1973 Aug 6	OWS/Earth orbit	6:31	
				1973 Aug 24	OWS/Earth orbit	4:30	
				1973 Sep 22	OWS/Earth orbit	2:45	13:46
Allen	6/1967	2	STS-5	1982 Nov	OV102/Earth orbit	Cancelled	
			STS 51-A	1984 Nov 12	OV103/Earth orbit (MMU)	6:00	
				1984 Nov 14	OV103/Earth orbit	5:42	11:42
Kerwin	4/1965	1	Skylab 2	1973 Jun 7	OWS/Earth orbit	3:25	03:25

Appendix 4

Profiles of the Seventeen

Allen IV, Joseph ("Joe") Percival: (Mission Specialist STS 5, STS 51-A)
Born: 1937 Jun 27 in Crawfordsville, Indiana, USA.
Qualifications: (1959) BA in maths/physics; (1961) MS in physics; (1965) PhD in physics from Yale University.
NASA career: (1967 Aug 4) selected as a NASA scientist-astronaut, Group 6; (1967–1969) astronaut academic, simulator, survival and jet pilot training programme. Worked on early Skylab development issues; (1971) support crew member and Capcom, Apollo 15; (1972) Capcom Apollo 17; (from 1973) Shuttle development issues. Worked on early studies for Spacelab simulations. Early candidate for assignment to Spacelab 1; (1974–1975) participated in NASA's Outlook for Space study; (1975–1978) Assistant Administrator for Legislative Affairs, NASA HQ, Washington; (1978) returned to JSC as a senior scientist-astronaut; (1980–1981) Technical Assistant to Director of Flight Crew Operations, JSC; (1981) support crew and Capcom, STS-1; (1982 Nov 11–16) flew as MS STS-5 (122 hours), the first operational Shuttle mission. Planned demonstration EVA cancelled; (1983) CB support for STS-8, and 41-B; (1984 Nov 8–14) flew as MS STS 51-A (191 hours) satellite deployment and retrieval mission. Completed two EVAs totalling 11 hours 42 minutes, including a 2 hr 22 min flight on an MMU (Unit 3); (1985 Jul 1) resigned from NASA to join Space Industries Inc.
Allen logged over 313 hours in space on two missions.

Chapman, Philip ("Phil") Kenyon
Born: 1937 Mar 5 in Melbourne, Australia (became US citizen in 1967 May).
Qualifications: (1956) BS in physics from Sydney University, Australia; (1964) MS in aeronautics and astronautics, MIT; (1967) PhD in physics from MIT.

NASA career: (1967 Aug 4) selected as a NASA scientist-astronaut, Group 6; (1967–1969) astronaut academic, simulator, survival and jet pilot training programme; (1971) support crew member, Capcom and mission scientist, Apollo 14; (1972) support crew, Apollo 16. Also worked on early Skylab development issues; (1972 Jul 14) resigned from NASA to join AVCO Research Laboratory.

Chapman never flew into space.

England, Anthony (Tony) Wayne: (MS STS 51-F/Spacelab 2)
Born: 1942 May 15 in Indianapolis, Indiana, USA.
Qualifications: (1965) BS in geology from MIT; (1965) MS in geology from MIT; (1970); PhD in geophysics from MIT.
NASA career: (1967 Aug 4) selected as a NASA scientist-astronaut, Group 6. Aged 25 he was, and remains as of 2005, the youngest person ever selected by NASA for astronaut training. Worked on early Skylab development issues; (1967–1969) astronaut academic, simulator, survival and jet pilot training programme; (1970) support crew, Apollo 13; (1972) support crew, Capcom and mission scientist, Apollo 16; (1972, Aug 14) resigns from NASA; (1979 Jun 3) returns to NASA CB as a senior scientist-astronaut. Shuttle development roles; (1985 Jul 29–Aug 6) flew as MS STS 51-F/Spacelab 2, a solar physics, astronomy and Earth sciences mission (190 hours). CB support for STS 51-J and 61-A; (1988 Oct) resigned from NASA when he was not assigned to a second mission. Became Professor of Electrical Engineering and Computer Science Department, University of Michigan.

England logged over 190 hours in space on one mission.

Garriott, Owen Kay: (Science Pilot Skylab 3; MS STS-9/Spacelab 1)
Born: 1930 Nov 22 in Enid, Oklahoma, USA.
Qualifications: (1953) BS in electrical engineering from the University of Oklahoma; (1957) MS in electrical engineering, Stanford University; (1960) PhD in electrical engineering, Stanford University.
NASA career: (1965 Jun 28) selected as a NASA scientist-astronaut, Group 4; (1965–1967) astronaut academic, simulator, survival and jet pilot training programme; (from 1967) development issues for Apollo Applications Program (AAP, later Skylab); (1967) served on the Group 6 Astronaut Selection Board; (1968) Chief AAP; (1969) Capcom Apollo 11; (1973 Jul 28–Sep 25) flew as Science Pilot Skylab 3 (1427 hours), the second manned mission, setting new endurance record of 59 days 11 hours 9 minutes 4 seconds. Completed three EVAs totalling 13 hours 44 minutes; (1974–1976) various management roles at JSC, including Deputy Director, Science and Applications, Chief of Scientist-Astronauts, and Acting Director, Science and Applications; (1976) took a one-year sabbatical at Stanford University, returning as a senior scientist-astronaut; (1983 Nov 28–Dec 8) flew as MS STS-9/Spacelab 1 (247 hours), a multi-discipline scientific research flight and the first flight of the dedicated research module. Subsequently served as Space Station Freedom Program Scientist at JSC and trained for two years for two Spacelab Earth Observation Missions (EOM-1

in 1985 and EOM-2 in 1986). Scheduling problems and the loss of Challenger resulted in their cancellation; (1986 Jun) resigned from NASA to become Vice President of Space Programs, Teledyne Brown Engineering in Huntsville, Alabama.

Garriott logged over 1674 hours in space on two missions.

Gibson, Edward ("Ed") George: (Science Pilot Skylab 4)

Born: 1936 Nov 8 in Buffalo, New York, USA.

Qualifications: (1959) BS in engineering from University of Rochester, NY; (1960) MS in engineering from the California Institute of Technology; (1964) PhD in engineering from CalTech.

NASA career: (1965 Jun 28) selected as a NASA scientist-astronaut, Group 4; (1965–1967) astronaut academic, simulator, survival and jet pilot training programme; (from 1966) became involved in the AAP (Skylab) programme; (1969) support crew and Capcom, Apollo 12; (1973 Nov 16–1974 Feb 9) flew as Science Pilot Skylab 4 (2017 hours), the third and final manned mission, which set a world endurance record of 84 days 1 hour 15 minutes. Completed three EVAs totalling almost 16 hours; (1974 Nov) resigned from NASA to join Aerospace Corporation, based in Los Angeles, researching in solar physics; (1978) returned to CB at JSC as a senior scientist-astronaut and is assigned support and Capcom role for STS-1; (1980 Oct) resigned from NASA a second time when he was not assigned to an early Shuttle flight. He joined TRW Inc., in Redondo Beach, California.

Gibson logged over 2017 hours in space on one mission.

Graveline, Duane Edgar

Born: 1931 Mar 2 in Newport, Vermont, USA.

Qualifications: (1952) BS from University of Vermont; (1955) MD from University of Vermont; (1958) MS in public health from John Hopkins University.

NASA career: (1965 Jun 28) selected as a NASA scientist-astronaut, Group 4; (Aug) resigned from NASA "for personal reasons," taking a position with the State of Vermont Department of Health.

Graveline never flew into space.

Henize, Karl Gordon: (MS STS 51-F/Spacelab 2)

Born: 1926 Oct 17 in Cincinnati, Ohio, USA.

Died: 1993 Oct 5 of respiratory failure during an ascent of Mount Everest.

Qualifications: (1947) BA in mathematics from the University of Virginia; (1948) MA in astronomy from the University of Virginia; (1954) PhD in astronomy from the University of Michigan; (1966) principle investigator of Experiment S-033 (UV Astronomical Camera) flown on Gemini 10, 11 and 12.

NASA career: (1967 Aug 4); selected as a NASA scientist-astronaut, Group 6; (1967–1969) astronaut academic, simulator, survival and jet pilot training programme; (1971) support crew and Capcom, Apollo 15; (1973–1974) support crewman and

Capcom for the three manned Skylab missions; (1973) "crew member", Learjet Spacelab simulation mission 3; (1974) "crew member", Learjet Spacelab simulation mission 4; (1977) "crew member", ASSESS-II airborne Spacelab simulation tests; (from 1977) worked on Spacelab 2 payload; (1985 Jul 29–Aug 6) flew as MS STS 51-F/Spacleab 2 (190 hours), a solar physics, astronomy and earth sciences mission; (1986 Apr) left CB to become a senior scientist in the JSC Space Science Branch, working on space debris issues.

Henize logged 190 hours in space on one mission.

Holmquest, Donald Lee

Born: 1939 Apr 7 in Dallas, Texas, USA.

Qualifications: (1962) BS in electrical engineering from the Southern Methodist University; (1967) MD from Baylor College of Medicine and PhD in physiology, also from Baylor College of Medicine.

NASA career: (1967 Aug 4) selected as a NASA scientist-astronaut, Group 6; (1967–1969) astronaut academic, simulator, survival and jet pilot training programme; (1969–1971) completed support work on early Skylab development issues (1971 May–1973 Sep) took a leave of absence and completed training in nuclear medicine when it became clear that he would not be assigned to a Skylab mission; (1973 Sep) resigned from NASA to become Director of Nuclear Medicine at the Navasota Medical Center, Texas.

Holmquest never flew into space.

Kerwin, Joseph ("Joe") Peter: (Science Pilot, Skylab 2)

Born: 1932 Feb 19, in Oak Park, Illinois, USA.

Qualifications: (1953) BA in philosophy from the College of Holy Cross, Worcester, Massachusetts; (1957) MD from Northwestern University Medical School, Chicago, Illinois; (1958) designated US Naval Flight Surgeon; (1962) earned US naval aviator wings.

NASA career: (1965 Jun 28) selected as a NASA scientist-astronaut, Group 4; (1965–1967) astronaut academic, simulator, and survival training programme; (from 1966) provided support in the CB AAP (Skylab) branch office; (1967–1968) served on the committee for operational readiness of the Lunar Receiving Laboratory (LRL) at JSC; (1968–1969) CB representative for the development of the Apollo post-flight biological isolation garments (BIG); (1968 Jun) served as "commander" for an eight-day (177 hours) simulated spaceflight in an Apollo CM (2TV-1), held in Altitude Chamber A, Space Environment Simulator Laboratory; (1970) Capcom, Apollo 13; (1973 May 25–Jun 22) flew as science pilot, Skylab 2 (672 hours), the first manned mission, setting a new endurance record of 28 days 0 hours 49 minutes 49 seconds. Completed one 3 hr 30 min EVA; (1973 Dec 7–1977) served as CB Chief, Life Sciences Branch; (1974–1975) participated in the Outlook for Space 1980–2000 study group; (from 1974) assigned to support roles in Shuttle development, including RMS and EVA issues; (1977) member of the Astronaut Selection Board for Group 8, the first

Shuttle era selection; (1981–1982) Chief of On-orbit branch of CB. He was a leading contender for assignment to the STS-13 Solar Max repair mission as MS; (1982 Apr–1984 Jan) NASA Senior Scientific Representative in Australia; (1984 Jan–1987 Apr) Director of Life Sciences at JSC; (1987 Apr) retired from NASA to join the Lockheed Space and Missile Company as Chief Scientist for the space station at their Houston office.

Kerwin logged over 672 hours in space on one mission.

Lenoir, William ("Bill") Benjamin: (MS, STS-5)

Born: 1939 Mar 14 in Miami, Florida, USA.

Qualifications: (1961) BS in electrical engineering from MIT; (1962) MS in electrical engineering from MIT; (1965) PhD in electrical engineering from MIT.

NASA career: (1967 Aug 4) selected as a NASA scientist-astronaut, Group 6; (1967–1969) astronaut academic, simulator, survival and jet pilot training programme; (from 1969) assigned to AAP/Skylab development issues; (1973) back-up science pilot for Skylab 3 and 4, the second and third manned missions. Would have been science pilot for a twenty-day fourth closeout visit to Skylab in 1974, had it not been cancelled due to budget cuts and the extension of the third manned mission from 59 to 84 days; (from 1974) assigned Shuttle development issues; (1974 Sep–1976 Jul) assigned to the NASA Satellite Power Team, working on the adaptation of space power systems on Earth; (1982 Nov 11–16) flew as MS STS-5 (122 hours), the first "operational" Shuttle flight. Planned demonstration EVA cancelled. CB support role STS-6; (1984 Sep 1) resigned from NASA when he was not assigned to a second space flight, to work with the management consultant firm Booze, Allen and Hamilton Inc., in Alexandria, Virginia. One of his projects was as consultant for Space Station Freedom; (1989 Jul–1992 May) NASA Associate Administrator for Space Station Freedom programme, promoted later in 1989 to Associate Administrator for Spaceflight; (1992 Jun) returned to Booz, Allen and Hamilton as Manager, Applied Systems Division.

Lenoir logged 122 hours in space on one mission.

Llewellyn, John Anthony ("Tony")

Born: 1933 Apr 22 in Cardiff, Wales (became US citizen 1966 Feb).

Qualifications: (1955) BS in chemistry from University College, Cardiff; (1958) PhD in chemistry from University College, Cardiff.

NASA career: (1967 Aug 4) selected as a NASA scientist-astronaut, Group 6; (1967–1968) astronaut academic, simulator and jet pilot training programme; (1968 Aug) resigned from NASA due to his inability to progress through the USAF jet pilot training programme. Took a position as Professor of Chemistry, University of Southern Florida, Tallahassee, Florida.

Llewellyn never flew into space.

Michel, Frank Curtis

Born: 1934 Jun 5 in LaCrosse, Wisconsin, USA.

Qualifications: (1955) BS in physics from CalTech; (1957) graduated USAF jet pilot training; (1962) PhD in physics from CalTech.

NASA career: (1965 Jun 28) selected as a NASA scientist-astronaut, Group 4; (1965–1967) astronaut academic, simulator, and survival training programme; (1966–1969) worked on a number of CB technical assignments, including Apollo lunar surface experiments and Apollo Applications Program; (from 1968 Apr) spending eighty per cent of his time teaching at Rice University, Houston for one year; (1969 Aug 18) resigned from NASA due to too few opportunities to fly into space. Took a faculty position in the Space Physics and Astronomy Department of Rice University.

Michel never flew into space.

Musgrave, Franklin Story: (MS STS-6; MS STS 51-F/Spacelab 2; MS STS-33; MS STS-44; MS STS-61; MS STS-80)

Born: 1935 Aug 19 in Boston, Massachusetts, USA.

Qualifications: (1958) BS in maths and statistics from Syracuse University; (1959) MA in operational analysis and computer programming from University of California at Los Angeles; (1960) BA in chemistry from Marietta College; (1964) MD from Columbia University; (1966) MS in physiology and biophysics from University of Kentucky; (1987) MA in literature from University of Houston.

NASA career: (1967 Aug 4) selected as a NASA scientist-astronaut, Group 6; (1967–1969) astronaut academic, simulator, survival and jet pilot training programme; (from 1969) worked on AAP/Skylab issues; (1973) back-up science pilot, Skylab 2, the first manned mission; (1973–1974) incidental Capcom, Skylab 3 and Skylab 4; (from 1974) worked on Shuttle development, specifically in EVA and Spacelab issues; (1974 Oct) participated in Spacelab Medical Demonstration 1 ground test, part of the Spacelab development programme; (1976 Jan) participated in the second Spacelab ground simulation, SMS-II; (1977) participated in tests of the Shuttle EMU suit; (1979–1982) worked in SAIL, testing Shuttle computer software; (1983 Apr 4–9) flew as MS, STS-6 (120 hours), a TDRS deployment mission. Performed the first Shuttle-based demonstration EVA, of 4 hours 17 minutes; (1985 Jul 29–Aug 6) flew as MS, STS 51-F/Spacelab 2, a solar physics, astronomy and Earth sciences mission (190 hours); (1989 Nov 22–27) flew as MS, STS-33, a classified DoD mission. Served as Head of the Mission Support Branch of the CB; (1990–1991) CB support for STS-36, 31, 41, 38 and 35; (1991 Nov 24–Dec 1) flew as MS, STS-44 (166 hours), another classified DoD mission. CB support for STS-50, 49, 46, and 47; (1993 Dec 2–12) flew as MS, STS-61 (259 hours), the first Hubble Space Telescope Service Mission. Completed three EVAs totalling over 22 hours. CB support for STS-62, 64, 65, 68, 66, 63, 67, 71, 73, 72, 74 and 75; (1996 Nov 19–Dec 7) flew as MS, STS-80, a space sciences research flight; (1997 Sep) informed that he would not be assigned to a seventh mission due to his age (he was 62), he left NASA after thirty years' service to become a consultant and public speaker. He was the only person to fly into orbit on all five Shuttle orbiters.

Musgrave logged over 1280 hours in space, and in excess of 26 hours on EVA in six missions.

O'Leary, Brian Todd

Born: 1940 Jan 27 in Boston, Massachusetts, USA.

Qualifications: (1961) BA in physics from Williams College; (1964) MA in astronomy from Georgetown University; (1967) PhD in astronomy from the University of California at Berkeley.

NASA career: (1967 Aug 4) selected as a NASA scientist-astronaut, Group 6; (1967–1968) astronaut academic, simulator, and jet pilot training programme; (1968 Apr 23) resigned from astronaut programme due to his inability to progress though the required USAF jet pilot training programme. He resumed his academic career, taking up a variety of teaching posts.

O'Leary never flew into space.

Parker, Robert ("Bob") Alan Ridley: (MS STS-9/Spacelab 1; MS STS-35/Astro 1)

Born: 1936 Dec 14 in New York City, USA.

Qualifications: (1958) BA in astronomy and physics from Amherst College; (1962) PhD in astronomy from CalTech.

NASA career: (1967 Aug 4) selected as a NASA scientist-astronaut, Group 6; (1967–1969) astronaut academic, simulator, survival and jet pilot training programme; (1971) support crew member and Capcom, Apollo 15; (1972) mission scientist, support crew member and incidental Capcom, Apollo 17; (1973–1974) Skylab program scientist and member of support crew for all three manned missions; (from 1974) began working on Shuttle development issues; (1974 Aug 15–1978) Chief of Scientist-Astronauts, CB; (1975) crew member for ASSESS-I Spacelab airborne simulation programme; (1977) member of the Astronaut Selection Board for Group 8, the first Shuttle era selection. CB support role for STS 41-C; (1983 Nov 28–Dec 8) flew as MS, STS-9/Spacelab 1 (247 hours), a multi-discipline scientific research flight and the first flight of the dedicated research module; (1988–1989) Director of the Office of Spaceflight Space Station Integration Office, NASA HQ, Washington DC; (1990 Dec 2–10) flew as MS, STS-35/Astro 1 (215 hours), an astronomical research flight; (1991 Jan–1997 Jun) returned to NASA HQ as Director of Division Policy and Plans, Office of Spaceflight, subsequently becoming Deputy Associate Administrator for Operations and Director for Spacelab Operations; (from 1997 Jun) Director of Program Requirements, NASA JPL; (Aug 2005) retires from NASA after a 40-year career.

Parker logged over 462 hours in space on two missions.

Schmitt Jr., Harrison ("Jack") Hagen: (LMP Apollo 17)

Born: 1935 Jul 3 in Santa Rita, New Mexico.

Qualifications: (1957) BS in science from CalTech; (1964) PhD in geology from Harvard University.

NASA career: (1965 Jun 28) selected as a NASA scientist-astronaut, Group 4; (1965–1967) astronaut academic, simulator, survival and jet pilot training programme; (from 1966) supported preparations for the Apollo lunar landings with lunar navigation and geology training; (1968 Aug) performed ALSEP deployment simulation at MSC, Houston, wearing full Apollo EMU; (1969 Jul) incidental Capcom Apollo 11; (1969–1970) took part in analysis of the first lunar samples returned by the Apollo 11 and 12 missions; (1971) back-up LMP, Apollo 15, serving as incidental Capcom during the mission; (1971 Aug) reassigned from the cancelled Apollo 18 to LMP, Apollo 17; (1972 Dec 6–19) flew as LMP, Apollo 17 (301 hours), the sixth and final Apollo lunar landing mission. The 12th man to walk on the Moon, he completed three surface EVAs totalling over 22 hours. He remained on the surface for a total of 75 hours, and in lunar orbit for over 72 hours, and also completed a 1-hr 6-min stand-up EVA from the CM on the way back from the Moon; (1974 Feb) Chief of Scientist-Astronauts, CB; (1974 Jan 29–May 12) Special Assistant to the NASA Administrator to coordinate NASA's energy research; (1972 May 13–1975 Aug 30) NASA Assistant Administrator for Energy Programs, HQ, Washington DC; (1975 Aug 30) resigned from NASA to enter politics, as Republican senator from New Mexico, serving in the US senate from 1977–1983.

Schmitt logged over 301 hours in space on one mission and became the only scientist-astronaut to reach the Moon.

Thornton, William "Bill" Edgar: (MS STS-8; MS STS 51-B/Spacelab 3)

Born: 1929 Apr 14 in Faison, North Carolina, USA.

Qualifications: (1952) BS in physics from the University of North Carolina; (1963) MD from the University of North Carolina; (1964–1967) worked on medical issues for the USAF Manned Orbiting Laboratory Programme.

NASA career: (1967 Aug 4) selected as a NASA scientist-astronaut, Group 6; (1967–1969) astronaut academic, simulator, survival and jet pilot training programme; (from 1969) working on technical issues for AAP/Skylab missions; (1972 Jul 26–Sep 20) science pilot on 56-day SMEAT ground simulation; (1973–1974) support crew member and Capcom for all three manned Skylab missions. PI on two Skylab experiments: M 074 Specimen Mass Experiment and M 172 Body Mass Measurement. Developed a "prototype space treadmill" for the crews to condition their lower limbs, which became known as "Thornton's Revenge"; (from 1974) worked on Shuttle development issues related to life sciences; (1977 May) "crew member" of third Spacelab ground simulation SMD-III. Applied as PS for Spacelab 1. CB support role for STS-4 and STS-5. Developed SAS experiments for crews of STS 4–7; (1982 Dec) added to the crew of STS-8 to study space adaptation syndrome; (1983 Aug 30–Sep 5) flew as MS, STS-8 (145 hours); (1985 Apr 29–May 6) flew as MS, STS 51-B/Spacelab 3 (168 hours), a mixed discipline research mission. Continued his work on space adaptation for long duration and space station missions; (1994 May 31) retired from NASA to continue medical interests and teaching at Department of Medicine, University of Texas-Galveston.

Thornton logged over 313 hours in space on two missions.

Appendix 5

Where Are They Now?

In the transition to the twenty-first century, many of the seventeen men selected in NASA's two scientist-astronaut groups in 1965 and 1967 are still actively pursuing many other goals, while their ongoing work in science and medicine is both prodigious and worthwhile.

Allen IV, Joseph P., completed his second space mission in 1984, and released his book, *Entering Space: An Astronaut's Odyssey* (co-authored with Russell Martin) the same year. The following year, he announced his retirement from NASA. Allen then began his educational and business career, serving as a guest research assistant at Brookhaven National Laboratory and as a member of the physics faculty at the University of Washington in Seattle. In 1994, he became president and CEO of Space Industries, where he worked alongside such innovative people as Max Faget to design and develop modest but powerful spacecraft, including the bus-sized International Space Facility. The following year, he was elected chairman of Calspan SRL, which merged with Veda International in 1997 to form Veridian, a provider of advanced information-based technology systems, with Allen remaining as chairman of the new corporation. Veridian became part of the Advanced Information Systems Division of General Dynamics Corporation in 2003. Allen retired from the company in 2004. He is also the Chairman of the Challenger Center for Space Science Education and a director of the United Space Alliance in Houston. He resides in Washington DC with his wife Bonnie, and they have two grown-up children.

Chapman, Philip K., took a position with AVCO Everett Research Laboratory as a senior research associate following his resignation from NASA in July 1972. Five years later, he left AVCO and joined the Arthur D. Little Company in Massachusetts, where he spent the next four years working on the Solar Powered Satellite (SPS). He was also elected President of the L5 Society Space Advocacy Group in 1977. In the early 1980s, he worked as a freelance consultant to industry and government, and he married Maria Tseng in 1984. More recently, he was involved in the Rotary Rocket Company, working to develop a single-stage manned launch vehicle called the Roton. However, the company failed, and Chapman is currently devoting himself to full-time novel writing.

England, Anthony W., retired from NASA on 31 August 1988, after a nine-year appointment as Senior Scientist-Astronaut. He then joined the faculty of the University of Michigan in Ann Arbor, where he is currently a professor in both the Department of Electrical Engineering and Computer Science (1988–present) and the Department of Atmospheric, Oceanic and Space Sciences (1994–present), and is also the Director of the Center for Spatial Analysis. In 1995, he was also appointed the university's Associate Dean of the Rackham School of Graduate Studies. England is a member of the National Research Council's Space Studies Board and chair of their Committee on Research and Analysis. He still enjoys cross-country skiing, sailing and flying, with his wife Kathi, who is also a pilot.

Garriott, Owen K., left NASA in June 1986 and became a consultant for a number of aerospace companies. He also served as a member of several NASA and National Research Council committees. In January 1988, he became Vice President of Space Programs at Teledyne Brown Engineering, a position he held until his resignation in 1993. He also remarried, and his new wife Eve brought three grown-up children into their family. More recently, he has accepted the post of Adjunct Professor in the Biological Sciences Department at the University of Alabama in Huntsville, and works on the collection of extremophilic microbes from various corners of the globe. He devotes considerable effort to various charitable activities, including the Astronaut Scholarship Foundation (Chairman) and is co-founder of the Enid (Oklahoma) Arts and Sciences Foundation. He and Eve still live in Huntsville, and travel extensively.

Gibson, Edward G., resigned from NASA in December 1974 to conduct research on Skylab solar physics data as a senior staff scientist with the Aerospace Corporation in Los Angeles. Beginning in March 1976, he served for one year as a consultant to the German company ERNO Raumfahrttechnik GmbH, to help integrate NASA operational concerns into the Spacelab design. The following March, he returned to the Astronaut Office, where he was involved in the selection (and led the training) of a new astronaut class. Post-NASA, he took up management roles with TRW and Booz, Allen & Hamilton, in energy development, space station development, and planning for the Space Exploration Initiative. In 1990, he formed his own consultancy firm, the Gibson International Corporation. Currently, he is a Senior Vice President with the Science Applications International Corporation (SAIC) in Sioux Falls, South Dakota, and resides part-time in Scottsdale, Arizona. He is the author of several fiction and non-fiction books.

Graveline, Duane E., returned to practising medicine, as a family doctor in Burlington, Vermont, following his enforced return to civilian life. During this time, he also served as a flight surgeon for the Vermont Army National Guard. Prior to the first Space

Shuttle flight in 1981, he made a temporary return to NASA as Director of Medical Operations, subsequently assisting on the first four Shuttle missions. Following his retirement from medical practice at the age of sixty, Graveline became a prolific author of medical and science-fiction thrillers, with nine published novels to his credit. In late 2003, following the publication of his latest non-fiction book, *Lipitor, Thief of Memories: Statin Drugs and the Misguided War on Cholesterol*, he began working with Kennedy Space Center medical personnel to establish liaison biomedical research projects at a hospital located near Merritt Island in Florida. The outcome of this endeavour was a KSC funding package that was submitted to NASA Headquarters. It contained four proposals, uniquely his, for which he will be principal investigator. Pleased to be involved once again in space research, he and his second wife Suzanne (née Gamache) now happily reside in Merritt Island.

Henize, Karl G., retired from the astronaut corps in 1986, and took on a role as a senior scientist with NASA's Orbital Debris Group, involved in exploring the hazards posed by space debris. Late in 1993, realising a long-held dream, he joined a British expedition attempting to climb the north face of Mount Everest. At 22,000 feet, two days after the team set out, he began suffering worsening symptoms of extreme high altitude sickness at an advance camp. Despite a valiant rescue attempt, he died of high-altitude pulmonary oedema at the 18,000 feet level. He is buried nearby, above the Changste Glacier. Henize died on 5 October 1993, twelve days short of his 67th birthday, and is survived by his wife, Caroline, children Kurt, Marcia, Skye and Vance, and four grandchildren.

Holmquest, Donald L., took a leave of absence from NASA in May 1971, in order to take up a position as Assistant Professor of Radiology and Physiology with the Baylor College of Medicine, taking additional training in nuclear medicine. He held this position until 1973, concurrently serving as the Director of Nuclear Medicine in the Nuclear Medicine Laboratory at the Ben Taub General Hospital in Houston. He officially resigned from NASA in September 1973 to become Director of Nuclear Medicine at Eisenhower Medical Center in Palm Desert, California. In 1974, he was appointed Associate Dean for Academic Affairs in the new College of Medicine at Texas A&M University. Two years later, he became Director of Nuclear Medicine at the Navasota Medical Center, also in Texas, and the following year took on the same position at the Medical Arts Hospital in Houston, while at the same time studying for a law degree. He then took on a whole new career as an associate, and then senior partner, at the law firm of Wood, Lucksinger & Epstein. Following the firm's dissolution, he became a principal in his own firm of Holmquest & Associates, where he practised general health law with an emphasis on the complex regulatory and organisational interface between physicians and health care institutions. He is currently the Chief Technical Officer for eMedicalResearch Inc., a Houston-based developer of software applications for the medical industry.

Kerwin, Joseph P., served in various NASA management positions after Skylab, including NASA Representative in Australia (1982–83) and Director of Space and

Life Sciences at the Johnson Space Center (1984–87). He retired from the Navy, left NASA, and joined Lockheed in 1987, where he managed the Extravehicular Systems Project, providing hardware for Space Station Freedom, from 1988 to 1990. With two other Lockheed employees, he invented the Simplified Aid For EVA Rescue (SAFER), recently tested for use by space walking astronauts on the International Space Station (ISS). He then served on the Assured Crew Return Vehicle team and served as Study Manager on the Human Transportation Study, a NASA review of future space transportation architectures. In 1994–95, he led the Houston liaison group for Lockheed Martin's FGB contract, the procurement of the Russian "space tug" which became the first element of the ISS.

He joined Systems Research Laboratories (SRL) in June 1996, serving as Program Manager of the SRL team which bid for the Medical Support and Integration Contract at the Johnson Space Center. The incumbent, KRUG Life Sciences, was selected, and they recruited him to replace its retiring President, T. Wayne Holt. He joined KRUG on 1 April 1997. The following March, KRUG Life Sciences became the Life Sciences Special Business Unit of Wyle Laboratories of El Segundo, California. In 2003, Wyle was awarded the ten-year, billion-dollar bioastronautics contract by NASA to manage its future medical work in support of human space flight. After managing that program, Kerwin resigned from Wyle in July 2004. He still serves on the Board of Directors of the National Space Biomedical Research Institute (NSBRI) as an industry representative. He and his wife have three daughters and five grandsons.

Lenoir, William B., took on responsibility for the direction and management of mission development within the Astronaut Office following his November 1982 flight on the STS-5 mission. He resigned from NASA in September 1984 and took on a position with the management and technology consulting firm of Booz, Allen & Hamilton, Inc., in Arlington, Virginia, in their Space Systems Division. Lenoir returned to NASA in June 1989 as the Associate Administrator for Space Flight, responsible for the development, procurement and operation of the Space Shuttle and Space Station; management of all US government civil launch capabilities; US Spacelab operation; and planning for future space flight, transportation and system engineering programs. He resigned from NASA a second time in April 1992 in order to take on the position of Vice President and member of the board at Booz, Allen & Hamilton (this time in Bethesda, Maryland). He served as Vice President of the Applied Systems Division until retiring on 1 April 2000 but continued to work for the company on a part-time basis until 2003.

Llewellyn, J. Anthony, withdrew from the scientist-astronaut programme, having completed the initial academic training, on 23 August 1968, for what NASA says were "personal reasons." However, the NASA news release from that date cited Llewellyn's "trouble learning to fly jets," and his inability to qualify as a jet pilot. He returned to Florida State University, and was an associate professor until 1972 (in the latter two years, he was also Dean of the School of Engineering Science). Following this, he moved to Tampa and joined the College of Engineering faculty at the University of South Florida (USF), where he was a professor in the Department of Energy Conversion and Mechanical Design. Since 1981, he has been a professor in the university's departments of Chemistry and Mechanical Engineering, and is currently USF's director of Academic Computing Technologies.

Underwater exploration always held a fascination for Llewellyn, and he was involved in several underwater projects in the 1970s, including the role of coordinator for the Racon Corporation's "Scientist in the Sea" project in 1973. Three years later, he was an aquanaut with the National Oceanographic and Atmospheric Administration (NOAA), and he currently serves as a scientific consultant on the marine environment and on energy issues.

Michel, F. Curtis, resigned from NASA in August 1969, soon after the first manned lunar landing, and returned to teaching and research. He accepted a full-time position as Associate Professor of Space Science at Houston's Rice University, and in 1970 was appointed Professor of Space Physics and Astronomy, and of Physics. The following year, he took a twelve-month sabbatical at the Institute for Advanced Studies at Princeton University in New Jersey, and in 1974 was elected chairman of Rice University's Space Physics and Astronomy Department. In 1979, Dr. Michel took on advanced studies as a Guggenheim Fellow at the University of Paris Polytechnic School in France, and then spent two years at the Max-Planck Institute of the University of Heidelberg in Germany, having also been granted an Alexander von Humboldt Foundation research award. He returned to Rice University in 1984 and is currently the Andrew Hays Buchanan Professor of Astrophysics in the Departments of Space Physics and Astronomy, and of Physics (Emeritus), a position he has held for more than thirty years. He works largely on relativistic astrophysics, with particular emphasis on the functioning of radio pulsars. In 1994, he revisited the Max-Planck Institute in Germany with the Humboldt Foundation, studying Extraterrestrial Physics. More recently at Rice University, he directed and taught a challenging Natural Science foundation course directed at non-science undergraduates. Michel is the author of numerous papers and publications on space physics. On the home front, he and his second wife Bonnie are now proud grandparents to Greg, Brent and Dexter.

Musgrave, F. Story, spent a total of just under 1,282 hours in space on six space flights. In August 1997, at the age of 62, Musgrave reluctantly left NASA to pursue other interests after he was told he wouldn't be flying any more missions. He moved to central Florida and became a consultant to Walt Disney Imagineering, working on ideas for new resorts, parks and pavilions, films, television shows and other forms of media. He also works with California-based Applied Minds Inc., again in a technical and creative capacity. In recent years, he has travelled abroad showing and commentating on images of our planet from space. He also enjoyed a cameo appearance as a Capcom in the Brian DePalma film, *Mission to Mars*, for which he was a consultant. Having re-married in 2006, and now the father of seven children, Musgrave remains a staunch advocate and visionary for the continuing exploration of space. His recreational interests include flying, photography, scuba diving, parachuting, gardening and running. His biography, *Story: The Way of Water*, by Anne Lenehan, was published in 2004.

O'Leary, Brian T., became a research associate at Cornell University in New York State, following his resignation from NASA in April 1968. He held the position of Assistant Professor of Astronomy for two years, during which time he was also a visiting associate at the California Institute of Technology. In the interim, he had also penned a controversial book called *The Making of an Ex-Astronaut* which expressed his dissatisfaction with the treatment of the scientist-astronauts in the Astronaut Office. In 1971, he took on duties as a teacher, researcher and lecturer at California State University and taught technology assessment and energy policy at the University of California's Berkeley School of Law. For three years from 1972, he was Assistant Professor of Astronomy and Science Policy Development at Hampshire College in Massachusetts. In 1973, he took on additional duties as an experimenter with the Mariner Venus–Mercury Television Science team, and in 1975 served as a special consultant on nuclear energy matters to the House Interior Committee in Washington DC. Over the next five years, he was a member of the research faculty in the physics department at Princeton University in New Jersey. For two years from 1980, he became a freelance writer, consultant and lecturer, before becoming senior scientist at Science Applications International Corporation in Redondo Beach, California. In 1987, he left that position and served for a year as chairman of the board of directors of the Institute for Security and Cooperation in Outer Space.

O'Leary has been a special consultant to the US House of Representatives Subcommittee on Energy and the Environment, and was senior advisor to presidential candidate Morris Udall. He has also assisted presidential candidates George McGovern, Walter Mondale and Jesse Jackson on economic and environmental issues and policies. In 1987, he once again took up full-time study, writing and lecturing, and became widely recognised for his many books and articles on the frontiers of science, space, energy and culture. He moved from Phoenix, Arizona to Hawaii in 1995, but now resides in Ecuador.

Parker, Robert A.R., resigned from active astronaut status on 30 November 1992. Subsequent to his flying career, he took up the position of Director of the Spacelab Operations Program, and then held several senior management positions in NASA's Office of Space Flight in Washington DC, eventually becoming Director of Space Operations and Utilization. In 1997, he was appointed Director of NASA's Management Office at the Jet Propulsion Laboratory in Pasadena, California, a position in which he not only provided leadership, but was responsible for negotiations of the NASA contract requirements with JPL and the California Institute of Technology (CalTech), the organisation that operates JPL. Parker resigned his position at JPL on 31 August 2005 and has now retired from NASA. He is married to the former Judy Woodruff of San Marino, California. They have five children and seven grandchildren.

Schmitt, Harrison H., became NASA's Chief of Scientist-Astronauts in February 1974, and also managed their Energy Programs office. He resigned from NASA in August 1975 and successfully sought election as a US senator in his home state of New Mexico, subsequently serving a six-year term from November 1976. In his last two years in the Senate, among other appointments, he held the position of Chairman of the Subcommittee on Science, Technology and Space. Since 1982, he has worked as an aerospace consultant, company director, and lecturer on matters related to space, science, technology and public policy. In 1985, he married Teresa Fitzgibbon.

Schmitt was appointed Adjutant Professor of Engineering at the University of Wisconsin in 1994, a position he still holds. Today, he continues to speak on business, public and governmental initiatives, particularly in the fields of space, risk, geology, energy, technology and policy issues of the future. He is also a freelance writer, contributing non-fiction articles on space and the American Southwest to numerous books and magazines. He is a member of the Independent Strategic Assessment Group for the US Air Force Phillips Laboratory. His corporate board memberships include Orbital Sciences Corporation and the Draper Laboratory. Schmitt is also a founder and the chairman of Interlune-Intermars Initiative, Inc., advancing the private sector's acquisition of lunar resources and He-3 fusion power. In 2006, his book *Return to the Moon* was published by Copernicus Books, an imprint of Springer Science in association with Praxis Publishing Ltd. The book reviews the exploration, enterprise, and energy in the human settlement of space with emphasis of the colonisation of the Moon.

Thornton, William E., took medical retirement from NASA on 31 May 1994 having flown twice as a mission specialist aboard Space Shuttle Challenger. The following year, he became a clinical professor of medicine in cardiology at the University of Texas Medical Branch (UTMB) in Galveston. After four years of teaching at UTMB, he developed a computer-based, self-teaching system that could provide hands-on training for seeing, hearing and feeling patient signs. It is now in extensive and expanding use. In 2003, Dr. Thornton left UTMB in order to complete some interrupted work and a publication in space medicine. He also wanted to pursue the development of a new clinical system with the University of North Carolina. Today, he remains active in the world of space medicine and associated research. He is also involved in preservation efforts concerning his birthplace of Faison in North Carolina, including his family's home and woodlands, and the development of a youth library named after his parents, Will and Rosa Thornton. In 2003, among his numerous awards, publications and more than fifty patents, Bill Thornton proudly received the North Carolina Award for Science. He and Jennifer, who have two grown-up sons and seven grandchildren, live in Fair Oaks Ranch, Texas.

Bibliography

Both authors have followed the lives and careers of the seventeen NASA scientist-astronauts for many years and have assembled an extensive resource archive. This personal reference source has been extensively used in the compilation of this book. In addition, a number of personal interviews, direct communications, reports and books have been consulted and are listed here for reference and further research.

Scientist-astronauts oral interviews:

Both authors have been in regular informal communication with many of the scientist-astronauts for several years. The more formal recorded interviews and unrecorded discussions related to this project are listed below:

Astronaut	*Date(s)*	*Interviewer*	*Location*
Allen	1988 Aug	D. Shayler	Webster, Texas
	1989 Aug	D. Shayler	League City, Texas
England	2003 Mar 4	C. Burgess	Telephone interview
	2005 Jul 14	D. Shayler	Telephone interview
Garriott	2002 Oct 12	D. Shayler	Houston, Texas
	2005 Jun 27	D. Shayler	Telephone interview
Gibson	2004 Jan 3	D. Shayler	Telephone interview
Henize	1988 Aug	D. Shayler	Nassau Bay, Texas
	1989 Aug 15	D. Shayler	Nassau Bay, Texas
	1991 Jul 22	A. Salmon	NASA JSC, Houston, Texas
Lenoir	2005 Jun 24	D. Shayler	Telephone interview
Michel	1992 Jun 26	D. Shayler	Rice University, Houston, Texas
Musgrave	1988 Aug	D. Shayler	NASA JSC Houston, Texas
	2005 Jun 23	D. Shayler	Telephone interview

528 Bibliography

Thornton W.	1988 Aug	D. Shayler	NASA JSC, Houston, Texas
	1998 Dec 16	D. Shayler	Friendswood, Texas
	2000 Jun 30	D. Shayler	San Antonio, Texas
	2000 Jul 1	D. Shayler	San Antonio, Texas
	2006 Feb 16	D. Shayler	Telephone interview
[Thornton J.	*2000 Jul 2*	*D. Shayler*	*San Antonio, Texas]*

Unrecorded discussions:

Allen	1994 Jun	D. Shayler	UK book tour
Garriott	2004 Sep 20	C. Burgess	Huntsville, Alabama
Gibson	2001 May	D. Shayler	Autographica, Northampton UK
	2003 May	D. Shayler	Autographica, Northampton UK
Graveline	2005 May 20	C. Burgess	Merritt Island, Florida
Thornton W.	1991	D. Shayler	UK speaking tour

NASA Johnson Space Center Oral History Project:

Joe Allen, 23 January 2003, 16 March 2004, 18 March 2004 and 18 November 2004
Owen Garriott, 6 November 2000
Ed Gibson, 1 April 2000
Joseph Kerwin, 12 May 2000
Bob Parker, 23 October 2002
Jack Schmitt, 14 July 1999 and 16 March 2000

BIOGRAPHICAL RECORDS

The authors have collected and referred to a selection of official NASA biographical data releases on all seventeen scientist-astronauts from 1965 to date. This material has been supplemented by access to a number of personal records, media reports and NASA archives on each member of the two selections and their flown missions. It has been further enhanced by extended personal correspondence with most of the subjects, spread out over several years.

REPORTS

SMEAT Final Report October 1973, NASA TMX-58115
Skylab 1/2 Technical Crew Debrief, 30 June 1973, JSC-08053
Skylab 1/3 Technical Crew Debrief, 4 October 1973, JSC-08053
Skylab 1/4 Technical Crew Debrief, 22 February 1974, JSC-08809
MSFC Skylab Mission Report-Saturn Workshop, Skylab Program Office, NASA Technical Memorandum NASA TM X-64814 October 1974

BOOKS BY NASA SCIENTIST-ASTRONAUTS

1969	*Introduction to Ionospheric Physics*, Owen Garriott (with Henry Rishbeth), New York Academic Press
1971	*The Making of an Ex-Astronaut*, Brian O'Leary, The Scientific Book Club (UK Edition)
1973	*The Quiet Sun*, Edward G. Gibson, NASA SP-303
1981	*The Fertile Stars*, Brian O'Leary, Everest House
1982	*The New Solar System* (Eds. Brian O'Leary, Andrew Chaikin, Kelly J. Beatty), Sky Publishing Corporation
1983	*Project Space Station: Plans for a Permanent Manned Space Center*, Brian O'Leary, Stackpole Books
	Space Ship Titanic (Fiction), Brian O'Leary (with Richard Duprey), Dodd, Mead and Company
1986	*Entering Space: An Astronaut's Odyssey*, Joseph P. Allen with Russell Martin, Revised and Enlarged Edition, Stewart, Tabor & Chang
1987	*Mars 1999: Exclusive Preview of the US-Soviet Mission*, Brian O'Leary, Stackpole Books
1989	*Exploring Inner and Outer Space: A Scientist's Perspective on Personnel and Planetary Transformation*, Brian O'Leary, North Atlantic Books
	Reach (Fiction), Edward Gibson, Doubleday
1991	*Theory of Neutron Star Magnetospheres*, F. Curtis Michel, University of Chicago Press
1992	*In the Wrong Hands* (Fiction), Edward Gibson, Spectra
	The Second Coming of Science: An Intimate Report on the New Science, Brian O'Leary, North Atlantic Books
1994	*The Greatest Adventure*, Edward Gibson (Ed.), C. Pierson Publishing
	Miracle in the Void: Free Energy, UFOs and Other Scientific Revelations, Brian O'Leary, Kamapua' Press
2003	*Reinheriting the Earth: Awakening to Sustainable Solutions and Greater Truths*, Brian O'Leary, Truth Seeker Publications
2004	*Lipitor, Thief of Memory: Statin Drugs and the Misguided War on Cholesterol*, Duane Graveline, Infinity Publications
	Return to the Moon: Exploration, Enterprise, and Energy in the Human Settlement of Space, Harrison H. Schmitt, Copernicus Books

In addition, former scientist-astronaut Duane Graveline has written a series of self-published works of fiction:

1995	*Icarus Destiny 1996 Caducean Triangle*
1997	*The Ark*
1998	*Brothers of the Perseids*
1999	*Mindchange*
2000	*Our Father*
2001	*Twinkleseed*

2002 *Spacedoc*
2002 *Pathway*
2003 *Cradle*
2004 *The Phosphene Chronicles*

REFERENCE SOURCES

1977 *Skylab, Our First Space Station*, (Ed.) Leland F. Belew, NASA SP-400
A House in Space, Henry S.F. Cooper Jr., Angus & Robertson
The All-American Boys, Walter Cunningham, Macmillan Publishing

1983 *Living and Working in Space: A History of Skylab*, William David Compton and Charles D. Benson, NASA SP-4208

1984 STS EVA Report No. 1 STS-5 November 1982, David J. Shayler and Keith T. Wilson, AIS Publications
STS EVA Report No. 2 STS-6 April 1983, David J. Shayler and Keith T. Wilson, AIS Publications

1985 *The Real Stuff: A History of NASA's Astronaut Recruitment Program*, Joseph D. Atkinson, Jr & Jay M. Shafritz, Praeger Books
NASA Astronaut Biographical Data Record Book, Group 4, David J. Shayler, AIS Publications

1986 *America's Astronauts and Their Indestructible Spirit*, Dr. Fred Kelly, Aero/Tab Books

1987 *NASA Astronaut Biographical Data Record Book*, Group 6, David J. Shayler, AIS Publications
Challenger Shuttle: Aviation Fact File, David J. Shayler, Salamander Books
Science in Orbit: The Shuttle and Spacelab Experience 1981–1986, Valerie Neal, Tracy McMahan and Dave Dooling, NASA Marshall Space Flight Center Publication

1989 *Where No Man Has Gone Before: A History of Apollo Lunar Exploration Missions*, William David Compton, NASA SP-4214

1992 *The Shuttlenauts* (Volume 1 mission data; Volume 2 flight crew assignments) David J. Shayler, AIS Publications
Men and Women of Space, Douglas B. Hawthorne, Univelt Inc.

1993 *To a Rocky Moon: A Geologist's History of Lunar Exploration*, Don E. Wilhelms, University of Arizona Press

1994 *Deke! US Manned Space from Mercury to the Shuttle*, Donald K. Slayton with Michael Cassutt, Forge Books
A Man on the Moon: The Voyagers of the Apollo Astronauts, Andrew Chaikin, Michael Joseph

1995 *Adventure in Space: The Flight to Fix the Hubble (Juvenile)*, Elaine Scott and Margaret Miller, Hyperion Books

1996 *The Shuttlenauts* (Volume 3 – biographies) David J. Shayler, AIS Publications

1997	*Walking to Olympus: An EVA Chronology*, David S.F. Portree and Robert C. Trevino, NASA Monographs in Aerospace History Series #7
1999	*Who's Who in Space: The International Space Station Edition*, Michael Cassutt, Macmillan.
	Exploring the Moon: The Apollo Expeditions, David M. Harland, Springer-Praxis
	The Last Man on the Moon, Eugene Cernan and Don Davis, St Martin's Press
2000	*Disasters and Accidents in Manned Spaceflight*, David J. Shayler, Springer-Praxis
2001	*Skylab: America's Space Station*, David J. Shayler, Springer-Praxis
	Taking Science to the Moon: Lunar Experiment and the Apollo Program, Donald A. Beattie, New Series in NASA History, John Hopkins University Press
2002	*Apollo: The Lost and Forgotten Missions*, David J. Shayler, Springer-Praxis
	Apollo 17: The NASA Mission Reports Volume 1, (Ed.) Robert Godwin, Apogee Books
2004	*The Story of the Space Shuttle*, David M. Harland, Springer-Praxis
	Walking In Space, David J. Shayler, Springer-Praxis
	Story: The Way of Water, Anne E. Lenehan, The Communications Agency
2006	*Sky Walking: An Astronaut's Memoir*, Tom Jones, Collins/Smithsonian Books

Index

Italicized entries refer to figures or figure captions

Abbey, George 336, 365, 501
Abuzyarov, Zyyadin 46
Acton, Loren 135, *415*, 415, *417*, *423*
Advanced Orbiting Solar Observatory 101
Akers, Tom 462, 466
Albrook AFB (Panama) 109, 186
Aldrin, Edwin (Buzz) 35, 129, 195, 206, 223, 278
Alekseyev, Vladimir 44
Alexander, W. Carter 321, *321*, *327*
Algranti, Joe 56
Allen, Bonnie Jo 124–126, 519
Allen, David 123, *124*
Allen, Harriett 122
Allen, Joseph P. 123
Allen, Joseph P., Jr. 123
Allen, Joseph P. III 122
Allen, Joseph P. IV xviii, xix, *31*, *120–121*, 122–126, *123*, 161, 173, *174*, *180*, 184, *187*, *189*, 190, 207, 210, 211, 212, 245, 252, 256, 280, 286, 287, 294–296, 333–355, *340*, *341*, *343*, *344*, *345*, *347*, *348*, *350*, *351*, *354*, *355*, 368, 396, 398, 399–405, *406*, *408*, 431–438, *435*, *437*, 498–502, 511, 519

America's Astronauts and Their Spacecraft (book) xviii
Ames Research Center, California 294, 297–298, 300–301, 316, 321–322, 326, 410
Anders, William 35, 100, 201, 204, 250
Apollo missions:
  Apollo 1 21, 202–203, 251, 260
  Apollo 2 21, 250
  Apollo 5 177
  Apollo 7 204, 260
  Apollo 8 21, 195, 260, 347
  Apollo 9 21, 260, 347
  Apollo 10 21, 260
  Apollo 11 191, 195, 204, 206, 207, 210, 214, 225, 228, 246, 337, 499, 512, 518
  Apollo 12 21, 204, 207–210, 255, 337, 518
  Apollo 13 xviii, 21, 206, 207, 208, 210, 212, 214, 255, 283, 494, 512, 514
  Apollo 14 xx, 21, 191, 193, 194, 204, 210, 212, 255, 499, 512
  Apollo 15 21, 129, 194, 195, 207–216, 246, 252, 256, 337, 354, 511, 513, 517–518
  Apollo 16 xx, 21, 207–216, 246, 265, 499, 512
  Apollo 17 xviii, 21, 194, 207, 210, 212, 214–241, *220*, *222*, 255, 264, 333, 499, 500, 511, 517–518
  Apollo 18 195, 212–214, 499, 518

534 Index

Apollo missions (*cont.*)
   Apollo 19 195, 212–214
   Apollo 20 195, 206, 212–214
Apollo-Soyuz Test Program (ASTP) 333
Apollo Telescope Mount (ATM) 18, 100, 101, 189, 245, 252, 254, 257, 264, 265, 269, 270
Armstrong, Neil 35, 191, 195, 199, 206, 223
Association of Space Explorers xix
Artsebarsky, Anatoly 47
Atkinson, David J. 53, 54, 88
Atkov, Oleg 46
Avis, Frederick 152

Bagian, Jim 339
Barrett, Alan H. 145
Barry, Dan 492
Bartoe, John-David 135, *415*, 415, *417*, *423*, 423
Bassett, Charles 35
Bean, Alan 35, 70, 100, 101, 172, 250, 251, 257, *270*, 270, *271*, 271, *272*, 273, 319
Beattie, Donald 118–121, 165
Bechis, Kenneth 45
Beckman, John 296
Belov, Nikoley 3
Belyayev, Pavel 83
Bentwaters AFB (UK) 72
Beregovoi, Georgy 42
Berger, Dorothy 133
Berger, Leonard 133
Bergstrom AFB, Texas 500
Beriya, Sergey 43
Berry, Charles A. 7, 8, 34, 38, 56, 120, 166
Birky, Carl William 88
Birky, C. William 53–54, 88, 165
Birky, Pauline 88
Blaha, John 141, *457*, 459
Bless, Bob 161
Bloom, Otto 72
Bloom, Samuel Morris 72
Bluford, Guion 329, 366, 370, 371, *372*, 373, *374*
Bobko, Karol (Bo) 253, 261, *262*, 340, 341, 356, *357*, *360*, 361, 448
Boesen, Dennis 45
Boone, Daniel 136

Borman, Frank 35, 195, 196, 252
Bowersox, Ken 462, 464
Brand, Vance 202, 203, 208, 212, 215, 257, 276, 278, 340, 340, 345, 349, 355, 440, 444, 449, *452*, 425
Brandenstein, Dan 366, 370, *372*, *374*
Brezhnev, Leonid *269*
Bridges, Roy 135, *415*, 415, 416, 419, *423*, 424, 425
Brinkley, David 70
Brookhaven National Laboratory 125
Brooks AFB, Texas 38, 53, 66, 82, 83, 119, 157, 164
Brown, Allan H. 34
Bryan, John Neely 139
Buchan, John 126
Burdeyev, Mikhail 45
Burroughs, Edgar Rice 126
Bush, George 434
Bush, George W. 237, 495
Bykovsky, Valery 51

California Institute of Technology (CalTech) 53, 65, 72, 73, 78, 89
Cambridge University (UK) 61
Carleton E. Tucker Award 145
Carlson, Loren 34
Carpenter, Scott *10*, 20, 35, 100, 250
Carpentier, Bill 56
Carr, Jerry xix, 207, 257, *275*, 300
Carter, Sonny *457*, 459
Cassidy, Butch 77
Cernan, Eugene 35, 190, 212–241, 217, *218*, *235*, *238*
Chaffee, Roger 35, 130
Chandler, Doris 292
Chang-Diaz, Franklin 488, 493
Chapman, Colin R. 126
Chapman, Peter H. 129
Chapman, Philip xx, *120*, *121*, 126–130, *127*, 172, 173, *174*, *176*, 177, *178*, *179*, *181*, 182, *183*, 184, 185, *187*, 188, *189*, 190, 191, *192*, 192–197, 207, 210, 252, 255, 256, 280, 333, 498, 499, 511, 512, 519
Chapman, Phyllis 126
Chappell, Charles (Rick) 395
Charles, Prince of Wales 407

Chiao, Leroy 487
Clark, Jack 56
Clark, John F. 34
Clarke, Arthur C. 126
Clark, Robert 308, *315*
Cleave, Mary 493
Clinton, Bill 143
Cockrell, Ken *470*, 471
Collins, Michael 35
Conrad, Charles (Pete) 35
Contra, Louis 64
Cooper, Gordon *10*, 20, 35, 251
Cooper, Kenneth 164
Copernicus, Nicolas 66
Covey, Dick 462, 464
Cowing, Patricia 321
Craft, Harry 377
Crippen, Bob 253, 258, 260, 261, *262*, 336, 361, 365
Culbertson, Philip E. 296
Cunningham, Walter 35, 85, 99, 245, 246, 247, 251, 252, 257
Cushing, Arlene 81

Darling, Bonnie Jo 124, 125
DeBakey, Michael 141, 142
Dessler, Alexander 73, 74
Dewey, Thomas 154
Dick, Kenneth 296
Douglas, Bill 192
Dryden, Hugh 6, 27
Duke, Charles M. 212
Duke, Michael B. 53, 54, 79, 89
Dunbar, Bonnie 434, 474, 493
DuPraw, Ernest 53, 54, 89, 165
Durrance, Sam 448, 450, *451*, *452*

Echols, Frank 443
Edwards AFB, California 9, 338, 339, 343, 361, 373, 394, 451
Eglin AFB, Florida 163
Eisele, Donn 35
Eisenhower, Dwight D. xliii, 2, 4, 5, 6, 11
Ellingson, Harold V. 55
Ellington AFB, Texas 115
Ellsworth AFB, South Dakota 133
Eltgroth, Peter 165

Enders, Lawrence J. 53
England, Alice 131
England, Betty 131, 132
England, Ethan 131
England, Herman 131, 132
England, Kathi 134, 520
England, Michael 131
England, Tony xviii, *120*, *121*, 121, 130–134, *132*, 173, 194, 207, *208*, 210, 212, 252, 255, 256, 280, 286, 287, 333, 334, 336, 342, 396, 414, *415*, 415, *417*, 417–422, *423*, 423, *424*, 424, 425, *426*, 427, *428*, 431, 433, 439, 489, 494, 498–503, 512, 520
Engle, Joe 100, 104, 202, 203, 215, 217, 240, 248, 250, 330, 498, 499
Enos (chimpanzee) 82
European Space Agency (ESA) 376, 377
Evans, A.G. 148
Evans, Ron *104*, 215, 217, 218, 225, 233–241, *234*, *238*
Explorer 1 (satellite) 129

Fabian, John 98, 329, 363
Faget, Maxime 34, 56, 190, 435, 519
Fairchild AFB, Washington 186
Fatkullin, Mars 42, 43
Feltrop, Fr. Victor 68
Feoktistov, Konstantin 35, 41, 498
Ferris, Frederick L., Jr. 34
Feynman, Richard 73, 167
Fincke, Michael 487
Fisher, Anna 329, *337*, *338*, 396, *399*, 399, 400, 401, *408*
Fitzgibbon, Teresa 525
Fletcher, James 194, 284, 285, 434
Foale, Mike 487
Foran, Fr. "Fuzzy" 68
Foss, Ted *104*
Fowler, William (Willy) 73, 75
Frank, Joseph 64
Frazier, Donald 166
Freeman, Theodore (Ted) 35
Fulbright, William 124
Fullerton, Gordon 221, 223, 253, 340, *415*, 415, 419, *423*, 424, 425, *437*
Fursov, Sergey 47

Gagarin, Yuri xlv, 15, *15*, 16, 25, 44, 51, 61, 129, 199, 336, 497
Gagliano, Joe 202, 498
Galilei, Galileo 194, 195
Gamache, Suzanne 86
Gants, Kenneth 7
Gardner, Dale 366, 370, 371, *372*, 373, *374*, 374, 396, *399*, 399–401, *402*, *404*, 405, *406*, 406, 407, *408*, 449, 451, *452*
Garriott, Donna Jean 60
Garriott, Eve 520, 522
Garriott, Helen Mary 61
Garriott, Linda 61
Garriott, Mary Catherine 59
Garriott, Owen xv, xviii, 53–62, *57*, *59*, *69*, 93–100, 102, 108, 114, 115, 172, 207, 240–245, 250, 251, 256–258, 264, 270–274, *270*, 270, *271*, 271, *272*, *273*, *274*, 279, 280, 283, 286, 287, 296, 329, 333, 334, 342, 375, *376*, 376, 377, *378*, *379*, 379, 380, *381*, 383, *388*, 389, *390*, 390, *391*, 392–395, 401, 427, 431, 433, 438–443, 488–491, 495, 498–503, 512, 513, 520
Garriott, Owen (Snr.) 59
Garriott, Randall Owen 61
Garriott, Richard 61
Garriott, Robert 61
Gatenby, Pamela 129
Gemini missions:
  Gemini IV 441
  Gemini IX 190
  Gemini X 21
  Gemini XI 21, 201
  Gemini XII 21
Gerhardt, Mike 488
Gibson, Alexander Calder 62
Gibson, Calder 62
Gibson, Calder Alexander 62
Gibson, Edward xviii, 53–55, *57*, 57, 62–66, *69*, 93, 94, 97, 100, 102, 108, 114, 115, 172, 207, 208, 210, 240, 244, 245, 250, 251, 256–258, 264, 272, 273, 274, *275*, *276*, 280, 283, 286, 287, 329, 333, 334, *335*, 336, 431, 490, 498, 499, 501, 513, 520
Gibson, Everett 166
Gibson, Geraldine 62
Gibson, Helen 62

Gibson, Jannet Lynn 64
Gibson, John Edward 64
Gibson, Julie Ann 64
Gilruth, Robert 3, 11, 30, 32, 33, 85, 204, 255
Givens, Edward G. 101, 108, 202, 250, 251, 498
Glenn, John xix, *10*, 20, 26, 35, 82, 191, 421, 440, 472, 473, 497
Glennan, Keith T. 2, 6
Goldwater, Barry 390
Good, Shirley Anne (Lee) 69
Gorbatko, Viktor 41
Gordon, Richard (Dick) 35, 93, 212–215
Gorton, Slade 405
Gould, Roger *357*
Graveline, Carole Jane 84, 86
Graveline, Duane xx, 53, 54, 57, *57*, 69, 80–88, *84*, 247, 498, 513, 519
Graveline, Edgar 80
Graveline, "Mamere" 80
Graveline, Norman 81
Graveline, "Papere" 80
Graveline, Suzanne 521
Gray, Paul 145
Grechko, Georgy 47
Gregory, Fred 407–409, *457*, *458*, 460
Greene, Jay 401
Griggs, David 415
Grindeland, Richard 321
Griner, Carolyn 292, *293*
Grissom, Virgil (Gus) *10*, 13, 31, 26, 35, 130
Grushin, P.D. 43
Guili, R. Thomas 166
Gulyayev, Rudolf 42, 43

Habinger, Eugene 94
Hagan, Ethel 76, 77
Halley, Edmund 62
Halley's Comet 448, 449
Harper, D. 296
Hartsfield, Hank 207, 253, 258, 260, 340
Hauck, Bob 124
Hauck, Rick 363, 396, 398, *399*, 399, 400, *408*, 502
Haughney, Louis C. 297
Hawley, Steven 329

Heinlein, Robert 68
Hemingway, Ernest 68
Henize, Caroline Rose 138, 521
Henize, Claire 136
Henize, Fred R. 135–137
Henize, Karl xviii, xix, xx, xxi, 21, 94, 120, *121*, 135–139, *136*, 173, *186*, 190, 207, 209, 211, 245, 252, 256, 260, 261, 280, 286, 287, 294, 295, 303–307, 333, 334, 335, 342, 396, 414, *415*, 415, 416, *417*, 419–422, *423*, 423, 424, 425, *426*, 431, 432, 439, 440, 491, 498–503, 513, 514, 521
Henize, Kurt 138, 521
Henize, Mabel 135–137
Henize, Marcia 138, 521
Henize, Skye 138, 521
Henize, Vance 521
Henize, Wilson 136, 137
Hennen, Tom 458, 460
Henricks, Terence 458, *460*, 460
Herzberg, Gerhard 149
Hess, Wilmot (Bill) 120, 245, 246
Hibbs, Albert 167
High Frontier, The (book) 168
Hoagland, Dick 194
Hodge, John 204
Hoffman, Jeff 329, 448, 451, *452*, 452, 462, 466, *467*, *468*, 493
Holmquest, Charlotte Ann 141
Holmquest, Donald *120*, 139–142, *140*, 173, 181, *189*, 207, 252, 256, 333, 335, 498, 500, 501, 514, 521
Holmquest, Lillie Mae 139, 140
Holmquest, Sidney 139
Holt, Al 273
Hone, Vivien 161
Horowitz, Scott 492
How Green Was My Valley (book) 147
Hubble Space Telescope 462, 464–466, 472, 474, 493, 503, 516
Hussein, King of Jordan 390

Ilyin, Yevgeniy 45
Ingold, Keith 149
Interplanetary Flight (book) 126
Introduction to Outer Space (paper) xlii
Irwin, James 212
Ivanyan, Gurgen 46

Jackson, Jesse 524
Jacobs, Ed *33*
James, L.B. 254
Jeffries, Jeff *33*
Jernigan, Clarence 56
Jernigan, Tammy 468, *470*
Jet Propulsion Laboratory (JPL), California 145, 455, 504, 524
Johns Hopkins University 82
Johnson, Caldwell 435
Johnson, Lyndon B. xliv, xlv, 5, 6, *16*, 16
Johnston, Mary Helen 292, *293*, 407
Jones, Thomas *470*, 471

Kaminsky, Beverly 72, 73
Kapryan, Walter 218
Katys, Georgy 41, 42, 44
Katys, Petr 41
KC-135 (aircraft) 107
Keffer, H. 34
Keldysh, Mstislav 42, 43
Keleher, Fr. William 68
Keller, Don 53, 54, 89
Keller, Samuel 395
Kelly AFB, Texas 82
Kelly, Bob *33*
Kelly, Fred xx, *33*, 83, 85, 86
Kennedy, John F. xv, xliv, xlv, 14, 15, *16*, 16, 17, 26, 129, 190, 214, 249, 283
Kennedy, Robert F. xliv
Kerwin, Edward 67
Kerwin, Joanne 70
Kerwin, Joseph 33, 53, 54, 56, *57*, 57, 66, 67, 67, *69*, 69, 93–102, 106, 108, *114*, 115, 199, *200*, 200–209, 209, 240, 244, 248–258, 264–267, *267*, 268–270, 278, 280, 283, 286, 287, 300, 303, 329, 333, *337*, 338, 342, 395, 431, 432, 490, 498–503, 514, 515, 521, 522
Kerwin, Kristina 70
Kerwin, Marie 67, 68
Kerwin, Shirley Ann (Lee) 521
Killian, James xliii, 5
King, Elbert 167
Kerwin, Sharon 70
Kerwin, Shirley Ann (Lee) 70, 71
Kisilyov, Aleksandr 45
Kohl, Helmut 391, 392

Kohoutek Comet 278
Kolomitsev, Ordinard 42, 43
Komarov, Vladimir 35, 130, 498
Kordelewski, Krzysztof 193
Korolyov, Sergey 2, 3, 12, *15*, 25, 40, 51, 52, 477
Kraft, Christopher 56, 213, 301, 303, 334, 335
Kranz, Eugene 299, 300
Kreutz, Kathi 133, 134

Lackland AFB, Texas 72, 164
Laredo AFB, Texas 72
Laika (dog) 1, 3, 82
Land, E. William 449
Langley Research Center 8
Latysheva, Irina 46, 47
Lauritsen, Thomas 73
Lazarev, Vasiliy 45
Lebedev, Valentin 47
Lee, William A. 30
Leestma, David 448, 449
Lenehan, Anne *475*, 524
Lenoir, Barbara 143
Lenoir, Benjamin B. 143
Lenoir, Elizabeth May 145
Lenoir, Iona 143
Lenoir, Samantha Elen 145
Lenoir, Samuel S. 143
Lenoir, Gen. William 143
Lenoir, William B. *31*, *120*, 121, 142–146, *144*, 173, 188, 207, 252, 257, 276, 278, 280, 286, 287, 299, 302, 304, 329, 333, *340*, 340–356, *341*, *343*, *344*, *345*, *348*, *350*, *353*, *356*, 368, 431–434, 490, 494, 498–503, 515, 522
Lenoir, William Ballard 143
Lenoir, William Benjamin, Jr. 145
Leonov, Aleksey 83, 130
Lewis, C.S. 68
Lichtenberg, Byron *376*, 376, 377–388, *391*, 391
Lind, Don 100, 108, 114, 205, 206, 244, 245, 250, 252, 257, 261, 276, 278, 280, 407, 408, *409*, *414*, 499
Linnehan, Richard 492
Llewellyn, Ceri Eluned 150
Llewellyn, David 148

Llewellyn, Gareth Roger 149
Llewellyn, J. Anthony xx, *120*, 121, 146–150, *147*, *174*, 177, 185, 252, 498, 499, 515, 522, 523
Llewellyn, John 147, 148
Llewellyn, Morella 147, 148
Llewellyn, Richard 147
Llewellyn, Roger 148
Llewellyn, Sian Pamela 149
Llewellyn, Valerie Mya 149
Lord, Douglas 293
Los Alamos Scientific Laboratory 5
Louden, James A. 285
Lounge, Mike 449, *452*
Lousma, Jack 100, 172, 208, 257, *270*, 270, *271*, 271, *272*, 340
Lovelace Clinic, New Mexico 38
Lovell, James 35, 70
Low, George 31, 204
Lu, Ed 487
Lucid, Shannon 329

Making of an Ex-Astronaut, The (book) 161, 498, 524
Man in Space (book) 7
Mann, Charles 149
Marana AFB, Arizona 72
Mark, Hans 335
Massachusetts Institute of Technology (MIT) 5, 129, 133, 144, 145, 146, 252
Mattingly, Ken 201, *209*, 300, 301–303, 340, 368
Melua, Arkediy 46
Menzies, Robert G. 126
Menzies, Robert T. 307
Merbold, Ulf *376*, 376–380, *383*, 387, 389, *391*, 391
Mercury missions:
    Mercury-Redstone 3 (MR-3) 497
    Mercury-Atlas 6 (MA-6) 497
    Mercury-Atlas 9 (MA-9) 30
McArthur, Bill 487
McAuliffe, S. Christa 502
McBrien, Vincent 68
McBride, Jon 448, 449
McCandless, Bruce 100, 108, 114, 172, 244, 245, 250, 257, *437*, 444
McDivitt, James 35

McGovern, George 524
McNair, Ron 329
Michel, Beverly 72, 75, 247
Michel, Bonnie 523
Michel, F. Curtis 53, 54, *57*, 57, *69*, *71*, 71–75, 93–95, 99–103, *105*, 106, 108, *110*, *111*, *114*, 115, 173, 189, 190, 199, 200, 243–251, 255, 278, 333, 498, 499, 516, 523
Michel, Frank J. 71, 72
Michel, Jeffrey B. 75
Michel, Viola Olivia 71
Mighty Joe Young (movie) 401
Milton, Daniel 53, 54, 75, 79, 89
Mishin, Vasily 52
Mission to Mars (movie) 523
Mitchell, Edgar *112*, 212
Mondale, Walter 524
Moore, Jesse 416
Morin, Lee 492
Morukov, Boris 46
Morgan, Barbara 491
Morgan, Clifford 34
Morrison, Dennis 310
Münch, Guido 160
Musgrave, Marguerite 150, 151, 152
Musgrave, Percy, Jr. 150, 151, 152
Musgrave, Percy, III 152
Musgrave, Story xix, *120*, 135, 150–154, *151*, 173, *174*, 174, 181, 182, 183, 187, *188*, *189*, 207, 252, 256, 257, 259, 266, 268, 280, 286, 287, *289*, 294, 306, *308*, 310–316, *311*, *312*, *313*, 333, 340–342, 356–364, *357*, *359*, *360*, 363, *365*, 396, *415*, 415, 419–422, *423*, *424*, 424, *425*, 425, 427–440, *441*, 441–456, *457*, 457, *458*, 458, 459, *460*, 460, 461, *462*, 462, *463*, 463, *465*, 465, 466, *467*, 467, 468, *469*, 469, *470*, 470, 471, *472*, 472, *473*, 473, 474, *475*, 491–504, 516, 517, 523, 524
Musgrave, Tom 152

National Academy of Sciences (NAS) xv, 5, 26, 27, 29, 32, 34, 35, 37, 38, 52, 83, 117–119, 142, 150, 157, 160, 168, 497
National Advisory Committee for Astronautics (NACA) xliii, 2

National Aeronautics and Space Act, 1958 xliii, 2, 5, 6
National Air and Space Museum, Washington, DC xix, 101, 197, 279
National Oceanic and Atmospheric Administration (NOAA) 146
National Research Council (NRC) 117, 118, 119
Naugle, John 292
Nelson, George (Pinky) 329
Newell, Homer E. 27, 35, 255, 497, 498
Newmeyer, James 143, 144
Newton, Isaac 405
Nicollier, Claude 307, 462, 464, 468
Nikolayev, Andrian 45
Nikolayev, Sergey 51
Nixon, Richard M. xliv, 195, 213
Nolan, Bernard 303
Nordsieck, Ken 448
North, Warren J. 34, 35
North American Aviation (NAA) 202
Norwegian Geological Survey 78

Oberth, Hermann 477
O'Keefe, John A. 167
O'Keefe, Sean 484
O'Leary, Brian T. *120*, *121*, 154–157, *155*, 161, 173, *174*, 176, 177, *180*, 181–185, 252, 255, 285, 498, 517, 523
O'Leary, Fred 154, 155, 156
O'Leary, Fred, Jr. 154
O'Leary, Joyce 154
O'Leary, Judith 154
O'Leary, Mabel 154
O'Neill, Gerald 167, 168
On Top of the World (book) 194
Overmyer, Bob 234, 253, *340*, 340–352, 356, 407, 408, *409*, 410, 411

Paine, Thomas 213, 214
Parise, Ron 448, *450*, 451, *452*
Parker, Alice 158
Parker, Allan 158, 159
Parker, Brian 159
Parker, Joan 159
Parker, John 159
Parker, Judy 159, 160, 524

Parker, Kimberly 59
Parker, Peter 158, 159
Parker, Robert *120*, 157–161, *158*, 173, *174*, 181, *184*, 207, 210–212, 252, 256, 258, 280, 286, 287, 294, 296, 298–303, 306, 329, 333, 334, 342, 375, 376, 376, 377–380, 387, 388, 389, 391, 392, 393, 427, 431, 433, 438, 439, 444, 448, 449, 450, 451, *451*, 452, *452*, 453, *453*, 454, *454*, 455, *455*, 456, 491, 498–504, 517, 524
Parker, Samuel 159
Parker, Sarah 159
Parker, Vernie 159
Pensacola NAS, Florida 186
Perrin AFB, Texas 186
Peterson, Don 11, 253, 340, 341, 356, *357*, 357, *360*, 361
Pettit, Don 486, *487*, 488
Phillips, John 487
Pimentel, Geroge 168
Pitzer, Kenneth 74
Pogue, Bill 100, 250, 257, 272, 274, *275*, 336
Polyakov, Valeriy 45
Pool, Sam L. 320
Popovich, Pavel 51
Porvatkin, Nikolay 45
Post, Wiley 165
President's Science Advisory Committee (PSAC) xliii, 5
Prinz, Diane 417
Purpura, Dominick 153

Quiet Sun, The (book) 273

Randolph AFB, Texas 182, 183
Ranger 8 (probe) 79
Rea, Donald 156
Reagan, Ronald 391, 392, 432, 433, 477, 478
Real Stuff, The (book) 39
Ream, Bud 56
Reese AFB, Texas 185
Reiter, Thomas 493
Reno, Janet 143
Rensik, Judy 329

Return to the Moon (book) 525
Reuben H. Fleet Science Center 239, 393, 435, 455
Rice University, Texas 103, 115, 172, 190, 245, 247, 430, 523
Richard, Louie *178*
Richards, Dick 448, 449
Ride, Sally 329, 363, 365
Right Stuff, The (movie) 119
Riley, Jack 247
Roberts, Bob 208
Rockwell, Norman 150
Rominger, Kent *470*
Roosa, Stuart 193
Ross, Jerry 363, 493
Rowell, Lyman 81
Runco, Mario 458, *460*, 460, 461

Sagan, Carl 174
Salyut missions:
    Salyut 2 44
    Salyut 3 44
    Salyut 4 44
    Salyut 5 44
    Salyut 7 44, 489
Sawin, Charles *308*, 315
Schirra, Walter (Wally) *10*, 20, 21, 35, 120, 204
Schmitt, Alexandra (Sandy) 77
Schmitt, Armena 77
Schmitt, Ethel 76–78
Schmitt, Harrison Ashley 76–78
Schmitt, Harrison H. (Jack) xviii, 53, 54, *57*, 57, *69*, 76, *77*, 77–80, 94, 97, 100, *112*, *113*, 114, 115, 172, 173, 194, 199, 205–209, 212–241, *217*, *218*, *219*, 221, *228*, *229*, *231*, *235*, *238*, 238, 239, 239, 243, 244, 248, 250, 251, 256, 283, 286, 333, 342, 431, 498–500, 517, 518, 525
Schmitt, Paula 77
Schweickart, Russell (Rusty) 35, 99, 129, *200*, 257
Scott, David 35, 129, 194, 195, 209, 210, *211*, 212
Scott, Sheila 194
Scully-Power, Paul 363–366
Seddon, Margaret (Rhea) 329, 493
See, Elliot 35

Sells, Saul B. 7, 8
Sembach Air Base (Germany) 72
Senkevich, Yuri 45
Seven Little Australians (book) 126
Sevier, Jack 29
Sharman, Helen *445*
Shaw, Brewster *376*, 378, 380, 392
Shea, Joseph 30, 31, 56
Shepard, Alan xlv, *10*, 19, 26, 34, 35, 56, 61, 93, 99, 100, 120, 121, 129, 146, 158, 164, 191, 194, 200, 201, 212, 240, 244–246, 250, 257, 274, 368, 497
Shoemaker, Eugene *28*, 28, 29, 31, 34, 78, 79, 118, 205
Shriver, Loren 472
Shuttle missions:
  STS-1 337, 338, 363, 365, 501, 511, 513
  STS-2 339, 344, 365, 368, 370, 380
  STS-3 339, 340, 341, 344
  STS-4 306, 339–341, 344, 365, 366, 368, 371, 518
  STS-5 124, 339–357, 368, 396, 399, 403, 407, 501, 502, 511, 515, 518
  STS-6 339–343, 356–362, 366, 368, 370, 397, 456, 501, 502, 515, 516
  STS-7 339, 342, 363–369, 372, 396, 407, 518
  STS-8 339, 342, 359, 360, 363–372, 375, 396, 399, 408, 413, 491, 494, 501, 511, 518
  STS-9/Spacelab 1 359, 360, 368, 375, 377, 512, 517
  STS 41-B 456, 397, 398, 401
  STS 13/41-C 342, 349, 357, 398, 432, 456, 501, 515
  STS 41-F (cancelled) 398
  STS 41-H (cancelled) 397
  STS 51-E (cancelled) 398, 416
  STS 51-H (cancelled) 433, 442, 502
  STS 61-E (cancelled) 433, 448, 502, 503
  STS 61-K (cancelled) 442
  STS 81-F (cancelled) 442
  STS 41-D 398, 443
  STS 41-G 365, 366, 397
  STS 51-B 407, 410, 413, 433, 448, 465–468, 494, 502, 518
  STS 51-G 418
  STS 51-F/Spacelab 2 xix, 135, 414–433, 440, 442, 456, 459, 502, 503, 512, 514, 516
  STS 51-J 439, 458, 512
  STS 61-B 438, 456
  STS 61-C 448
  STS 51-L 444, 449, 491
  STS-10 360
  STS-12 359
  STS-26 xix, 439, 444, 503
  STS-28 449, 434
  STS-33 456, 457, 458, 516
  STS-36 516
  STS-31 457, 472, 516
  STS-41 457, 516
  STS-38 450, 457, 458, 516
  STS-35 449–457, 502, 516, 517
  STS-45 442
  STS-49 434, 457, 516
  STS-50 443, 457, 516
  STS-46 516
  STS-56 442
  STS-61 456–474, 503, 516
  STS-62 457, 516
  STS-59 457
  STS-65 457, 516
  STS-64 457, 516
  STS-68 457, 516
  STS-66 457, 442, 516
  STS-63 457, 516
  STS-67 457, 516
  STS-71 457, 516
  STS-70 457
  STS-69 457
  STS-73 457, 516
  STS-74 457, 516
  STS-72 457, 516
  STS-75 457, 516
  STS-80 456, 468–474, 504, 516
  STS-95 472
  STS-106 46
  STS-113 486
  STS-114 481, 486, 489
Simon, George 417
Skuridin, Gennady 42
Sky Walking: An Astronaut's Memoir (book) 471
Slayton, Donald (Deke) *10*, 19, 20, 27, 34, 35, 39, 56, 85, 93, 100, 117, 118, 120, 122, 134, 138, 158, 165, 184, 185, 191,

Slayton, Donald (Deke) (*cont.*)
    192, 193, 194, 196, 201, 209, 212, 214, 215, 240, 246, 247, 253, 255, 256, 257, *272*, 279, 280, 285, 342
Sloan, Alfred 143
Smith, Charlie 219
Smith, Chester 219
Sorokin, Aleksey 45
Soviet Academy of Sciences 39–44, 46
Soyuz missions:
    Soyuz 1 130
    Soyuz 6 22
    Soyuz 9 22
    Soyuz 13 22
    Soyuz 16 22
    Soyuz 19 22
    Soyuz 22 22
    Soyuz TMA-1 486
Space Shuttles:
    Challenger xix, 73, 357, 360, 362, 373, 397, 401, 402, 414, 419, 422, 431, 434, 436, 472, 492, 493, 502, 503, 525
    Columbia xviii, 338–364, 365, 375, 384, 393, 394, 395, 401, 450, 468, 471, 481, 486, 488, 493, 501, 504
    Discovery 400, 401
    Endeavour 466, 468
Space Task Group (STG) 8–11
Special Committee on Life Sciences 9
Speed, Robert 34
Sputnik spacecraft:
    Sputnik (I) xliii, 1, 3, 4, 5, 61, 69, 78, 80, 82, 129, 138, 155
    Sputnik II xliii, 1, 61, 82
    Sputnik III 3, 61
Stafford, Tom 35
Stalin, Josef 41
Stead AFB, Nevada 111
Stepanov, Yuri 47
Stevenson, Bob 356, 363–366
Story, Joseph 150
Story: The Way of Water (book) 524
Story, William Wetmore 150
Sullivan, Kathryn 329
Swahlen, Percy Hypes 123
Swigert, Jack 108

Taft, William H. 136

Taking Science to the Moon (book) 165
Taylor, Anna Stroud 163
Tereshkova, Valentina 51, 497
Thagard, Norman 329, 339, 363–368, 372, 407, 408, *409*, 410, 413, 473, 474
Thatcher, Margaret 407
Thomas, Don 486
Thornton, James F. 164
Thornton, Jennifer 164, 366, 370, 371, 525
Thornton, Kathy 457, 462, 466
Thornton, Rosa Lee 162, 163, 525
Thornton, William xviii, xix, *120*, *121*, 161–165, *162*, 167, 173, 174, 177, 183, 186, 192, 207, 252, 256, 260, 261, 262, 262–266, 280, 294, 316–320, *320*, *321*, 323, *324*, 324, *325*, 325, 326, 327, 327, 333, 339, 342, 347, 348, 353, 355, 363, 365–375, *367*, *370*, *372*, *374*, 396, 408, *409*, 409–413, *414*, 414, 427, 431, 432, 439, 444, *445*, 445, 446, *447*, 447, 448, 489–504, 518, 525
Thornton, William E. (Snr.) 162, 163, 525
Thornton, William S. 164
Tifft, William 53, 54, 89, 90
Tikhonravov, Mikhail 2, 3
Titov, Gherman 26, 51, 497
To a Rocky Moon (book) 79
Tollerton, Carole Jane 81
Travis AFB, California 72
Trinh, Eugene 407
Truly, Cody 370
Truly Richard (Dick) 253, 258, 286, 287, 365, 366, *370*, 370, *371*, 373, *372*, *374*, 434
Truman, Harry S. 154
Truitt, Bessie 60
Tseng, Maria 520
Tsiolkovsky, Konstantin 2, 477
Turner, Ethel 126

United States Geological Survey (USGS) 28, 29, 34, 53, 75, 78–80, 89, 499

Van Allen, James 26, 27
Van Allen Radiation Belts 129
Van Den Berg, Lodewijk 407, 408, *409*
van Hoften, James 329

Van Kirk, Patricia 153
Van Vogt, A.E. 68
Vaughan, Simon xix
Verne, Jules 126
Vis, Bert xix
Voas, Robert 30–32, 498
Volk, Julie Ann 64
Volynov, Boris 41
von Braun, Wernher 1, 16, *18*, 23, 155, 477
von Buedeingen, Richard 168
Voskhod missions:
  Voskhod (1) 20, 35, 41, 45, 51, 52, 497
  Voskhod 2 83, 130, 497
  Voskhod 3 41
  Voskhod 4 41, 42
Voss, James 458, 460, 461
Vostok missions:
  Vostok (1) 3, 497
  Vostok 2 3, 51, 347
  Vostok 3 3, 51
  Vostok 4 3, 51
  Vostok 5 51
  Vostok 6 51

Walker, Dave 396, *399*, 399, 400, *408*
Walker, Evan 27
Walker, Mary Helen 61
Walter Reed Army Hospital 86
Walter Reed Medical Center 82, 119
Wang, Taylor 407, 408, *409*, 413, *414*
Waters, Aaron 34
Waters, Joe 146
Weaver, Lee 294, 295, 304
Webb, James 29
Weitz, Paul *110*, 172, 207, 248, 250, 257, 264, 267, 329, 340, 341, 356, *357*, 357, 361

Wells, H.G. 126
Wells, Nancy 124
Wells, Nick 296
Wheeler, Iris 80, 81
Whitaker, Ann 292, 293
White, Edward H., II 35, 130, 441
Whitson, Peggy xvii, 482, 484, *485*, 485, 486, 504
Wilhelms, Don 79
Williams AFB, Arizona 80, 85, 94, 96, 98, 181, 498
Williams, Bill *321*, 321, *327*
Williams, Clifton (C.C.) 35, 251
Williams, Jeffrey 487
Wilmore, Peter *417*
Wolf, David 488
Wolfe, Tom xliv
Woltz, Joanne 143
Woodruff, Robert 53, 54, 90
Woomera, South Australia 127
Worden, Al *112*, 209, 212, 213
Wright, Frank Lloyd 68
Wright-Patterson AFB, Ohio 38, 82
Wyatt, Philip 53, 54, 90

Yann, William 143
Yardley, John 301, 303
Yates, Dornford 126
Yegorov, Boris 20, 35, 45, 498
Yershov, Valentin 42–44, 46
Young, John W. 35, 142, 212, 250, 300, 336, 338, *376*, 377, 379, 380, 393, 394, 472

Zenit (satellite) 3

Made in the USA
Lexington, KY
17 March 2011